Making 20th Century Science

Other Books by Stephen G. Brush

Boltzmann's Lectures on Gas Theory (translator)
Kinetic Theory, 3 volumes (editor)
History in the Teaching of Physics: Proceedings of the International Working Seminar on the Role of the History of Physics in Physics Education (coeditor)
Resources for the History of Physics (editor)
Introduction to Concepts and Theories in Physical Science (coauthor, second edition)
The Kind of Motion We Call Heat: A History of the Kinetic Theory of Gases in the 19th Century (2 volumes)
The Temperature of History: Phases of Science and Culture in the Nineteenth Century
Maxwell on Saturn's Rings (coeditor)
Statistical Physics and the Atomic Theory of Matter, from Boyle and Newton to Landau and Onsager
Maxwell on Molecules and Gases (coeditor)
The History of Modern Physics: An International Bibliography (coauthor)
The History of Geophysics and Meteorology: An Annotated Bibliography (coauthor)
The History of Modern Science: A Guide to the Second Scientific Revolution, 1800–1950
History of Physics: Selected Reprints (editor)
The Origin of the Solar System: Soviet Research, 1925–1991 (coeditor)
Maxwell on Heat and Statistical Mechanics (coeditor)
A History of Modern Planetary Physics (3 volumes)
Physics, the Human Adventure: From Copernicus to Einstein and Beyond (coauthor)
Choosing Selection: The Revival of Natural Selection in Anglo-American Evolutionary Biology, 1930–1950

Making 20th Century Science

How Theories Became Knowledge

Stephen G. Brush with Ariel Segal

Oxford University Press is a department of the University of Oxford.
It furthers the University's objective of excellence in research, scholarship,
and education by publishing worldwide.

Oxford New York
Auckland Cape Town Dar es Salaam Hong Kong Karachi
Kuala Lumpur Madrid Melbourne Mexico City Nairobi
New Delhi Shanghai Taipei Toronto

With offices in
Argentina Austria Brazil Chile Czech Republic France Greece
Guatemala Hungary Italy Japan Poland Portugal Singapore
South Korea Switzerland Thailand Turkey Ukraine Vietnam

Published in the United States of America by
Oxford University Press
198 Madison Avenue, New York, NY 10016

Library of Congress Cataloging-in-Publication Data
Brush, Stephen G., author.
Making 20th century science : how theories became knowledge / Stephen G. Brush.
pages cm
Includes bibliographical references and index.
ISBN 978-0-19-997815-1 (alk. paper)
1. Science—Methodology—History—20th century. 2. Science—History—20th century. 3. Science--Methodology—History—19th century. 4. Science—History—19th century. I. Title. II. Title: Making twentieth century science.
Q174.8.B78 2015
509'.04—dc23
2014014220

9 8 7 6 5 4 3 2 1
Printed in the United States of America
on acid-free paper

To my granddaughter

Rebecca Nicole Roberts

CONTENTS

LIST OF ILLUSTRATIONS

PREFACE

Historians have chronicled the observations, experiments, and theories of scientists from antiquity to the present. This book could not have been written without surveying their publications. But only a few historians have presented evidence to answer the question: why were these theories accepted, at least for a while, as valid knowledge? Was it because the theories successfully *explained* the observations and experiments, or because they successfully *predicted* the results of observations and experiments not yet done?

This question seems to have been left for philosophers to answer. Yet it calls for historical research plus, in some cases, interviews with scientists. Philosophers sometimes seem more interested in discussing whether scientists *should* accept theories because of predictions or explanations, rather than what they actually *do*. So I have to persuade philosophers to consider historical evidence, and to convince historians that they should answer—in their reception studies—questions of interest to philosophers.

Of course the first thing I needed for my project was access to a good library. I was able to use the Library of Congress and the Princeton University libraries for short periods of time. The Niels Bohr Library at the American Institute of Physics in College Park, Maryland has a unique collection of textbooks, which happens to be just what I needed to study the reception of physics theories in the early twentieth century; the University of Maryland library, also in College Park, owns Max Born's personal library. The University of Pennsylvania Library and the Chemical Heritage Foundation, both in Philadelphia, have excellent collections of older chemistry books.

If you already know what book you need to look at, because someone else has cited it, you may have to rely on interlibrary loan. I have to thank the librarians at three institutions for efficiently obtaining books from other libraries for me: McKeldin Library at the University of Maryland, the Institute for Advanced Study at Princeton, and the Brandywine Hundred Library in Wilmington, Delaware.

What about archives of unpublished letters and manuscripts? In general I have not used these sources, for two reasons: First, to search the archives of the hundreds or thousands of scientists who *might* have recorded their opinions of one of the theories included in my project would be impractical. Therefore I have included only a few such documents, mainly those of Einstein and his correspondents that have been published in the *Collected Papers of Albert Einstein*. Second, published comments on a theory are likely to have more influence on the scientific community and (through textbooks) on the next generation.

During the three decades I have worked on this project, I have enjoyed valuable assistance from many historians, philosophers, and scientists: Peter Achinstein (Chapter 3), Stephen Adler (Chapter 9), Gar Allen (Chapters 1, 13, 14), Ralph Alpher (Chapter 12), Gustav Arrhenius (Chapter 1), Francisco Ayala (Chapter 14), John Beatty (Chapter 14), Richard Bellon (Chapter 2), Vincent Brannigan (Chapter 1), Dieter Brill (Chapter 11), L. M. Brown (Chapters 3 and 9), Louis Brown (Chapter 8), David Cassidy (Chapter 3), Matt Chew (Chapter 14), John Connerney (Chapter 1), David L. Cooper (Chapter 10), David P. Craig (Chapter 10), James Crow (Chapter 14), Lindley Darden (Chapters 3, 13, and 14), Bibhas De (Chapter 1), Alex Dessler (Chapter 1), Igor Dmitriev (Chapter 5), Tim Eastman (Chapter 1), C.W.F. Everitt (Chapter 11), John Gaffey (Chapters 1 and 15), Joseph Garratt (Chapter 10), George Garratty (Chapter 14), Owen Gingerich (Chapter 12), Thomas Gold (Chapter 12), George Gorin (Chapter 5), O. Wally Greenberg (Chapter 9), Ivan Gutman (Chapter 10), J. L. Heilbron (Chapter 7), Sandra Herbert (Chapter 14), Robert Herman (Chapter 12), Robert B. Hermann (Chapter 10), Norris Hetherington (Chapter 12), Richard Highton (Chapter 14), Roald Hoffman (Chapter 10), Gerald Holton (Chapter 11), Ruth Kastner (Chapter 13), Margaret Kivelson (Chapter 1), Alexei Kozhevnikov (Chapter 7), Helge Kragh (Chapters 2 and 3), Larry Laudan (Chapters 1, 2, and 3), Aleksey Levin (Chapters 9 and12), Richard Lewontin (Chapter 14), Jane Maienschein (Chapter 13), David Matthews (Chapter 1), Deborah Mayo (Chapter 3), Robert McColley (Chapter 1), Edward McKinnon (Chapter 3), Arthur I. Miller (Chapters 3 and 11), Peter Morris (Chapter 10), Gonzalo Munevar (Chapters 3 and 7), Ludmilla Nekoval-Chikhaovi (Chapter 5), Norman F. Ness (Chapter 1), Sally Newcomb (Chapters 2, 5, and 12), Mary Jo Nye (Chapter 5), David O'Brochta (Chapter 14), Denis Papadopoulos (Chapter 1), D. J. Pasto (Chapter 10), Lewis Pyenson (Chapter 11), Anya Plutynski (Chapter 14), Duncan Porter (Chapter 2), Helmut Rechenberg (Chapter 3), Alan Rocke (Chapter 5, 6, and 10), William K. Rose (Chapter 12), Theodore Rosenberg (Chapter 1), David Rudge (Chapter 14), Christopher T. Russell (Chapter 1), Halley Sanchez (Chapter 11), Mendel Sachs (Chapter 11), Carl Sagan (Chapter 1), Eric Scerri (Chapters 3 and 5), Wilfried Schroeder (Chapter 1), S. S. Schweber (Chapter 9), Ezra Shahn (Chapter 14), Sason Shaik (Chapter 10), Dudley Shapere (Chapter 9), Stanley Shawhan (Chapter 1), V. Betty Smocovitis (Chapter 14), George A. Snow (Chapter 9), Michael Sokal (Chapters 1, 5, and 11), Carol Sokolski (Chapter 14), Katherine Sopka (Chapter 3), David Stern (Chapter 1), Roger Stuewer (Chapter 7), Frank Sulloway (Chapter 11), Frederik Suppe (Chapter 3), Roger Thomas (Chapter 14), Virginia Trimble (Chapters 1 and 12), Ron Westrum (Chapter 1), Polly Winsor (Chapter 14), John Worrall (Chapter 3), and Nick Zimmerman (Chapter 14).

In a class by himself is my excellent assistant Ariel Segal, who tracked down many missing facts and references essential to this book.

Funding for my research was provided by the Institute for Physical Science and Technology and the General Research Board at the University of Maryland, the Institute for Advanced Study at Princeton with the aid of the Andrew Mellon Foundation, the National Science Foundation, the National Endowment for the Humanities, and a fellowship from the John Simon Guggenheim Foundation.

For permission to reprint substantial portions of articles previously published in journals and books, I thank the American Association for the Advancement of Science, publisher of *Science* (Chapter 11); the American Philosophical Society, publisher of *Choosing Selection* in its *Transactions* (Chapter 14); Elsevier Science Ltd., publisher of *Studies in History and Philosophy of Science* (Chapters 6 and 10); the Geological Society (London), publisher of *The Age of the Earth from 4004 BC to AD 2002* (Chapter 12); Kluwer Academic Publishers, publisher of *Journal of the History of Biology* (Chapter 13); the University of Chicago Press, publisher of *Isis* (Chapter 5); the Philosophy of Science Association (Chapter 3); the MIT Press, publisher of *Perspectives on Science* (Chapter 12); the Regents of the University of California (Chapter 7); and Springer Science and Business Media, publisher of *Physics in Perspective* (Chapter 11).

Part One discusses general issues such as the role of prediction and explanation in science, the concept of "reception" as used by historians of science, and the debate about whether science is "socially constructed." Part Two applies these concepts in the history of atomic and molecular chemistry and physics, especially the role of quantum mechanics. Part Three covers relativity theory and cosmology. Part Four discusses a selected sequence of theories in biology including chromosome theories of heredity and the revival of Darwin's theory of evolution by natural selection. Part Five summarizes the results and compares the success of prediction and explanation. Notes for the chapters are followed by a selected bibliography and an index.

Making 20th Century Science

PART ONE

The Reception and Evaluation of Theories in the Sciences

1

Who Needs the Scientific Method?

> Most Americans have never met a scientist, and despite having been 'taught science' at school, most have no real idea of how a scientific consensus is reached. . . . Every adult should have a base of scientific understanding about how the world works. But understanding the process through which scientific knowledge develops is equally critical.
>
> —Bruce Alberts, *former president of the U. S. National Academy of Sciences (2010)*

> Among the work to be done is to achieve some understanding of what is actually involved in rational acceptance and proof in science, of what, in Boyle's words, deserves 'a wise man's acquiescence.' . . . This job involves exploring the diverse range of contexts, historical and contemporary, in which inquiry is carried out.
>
> —Arthur Fine, *professor of philosophy of science at Northwestern University and later the University of Washington (1996)*

How do theories become scientific knowledge? I try here to answer that question by using several examples from the history of modern science. I have selected examples from chemistry, physics, astronomy, and biology. All of them are well known and most have been studied carefully by historians, so we know something about how the theories were developed and tested. The one thing often missing is the final stage of the process: the adoption of the idea by the relevant scientific community. We know that it *was* accepted, but often we don't know much about *why* it was accepted.

The process of acceptance (or rejection) of new theories—their reception, as historians call it—has been widely discussed but not well understood. Most reception studies focus on one of a few famous cases—the theories of Darwin, Freud, and Einstein—and tell us more about the response of the public or by the scientific community as a whole than about the *reasons* why *experts* accepted those theories. When authors do suggest reasons they often just give their own opinions, rather than citing publications or letters that would provide reliable evidence for the views of experts at the time the new theory was published.

I will be primarily concerned with two major reasons for accepting a new idea: (1) it leads to a successful *prediction* of empirical facts not yet known, or (2) it provides a successful *explanation* of facts already known but not understood. I will also touch on two other possible reasons: (3) the idea is so beautiful (usually in the sense of mathematical elegance) that it "has to be true," and (4) its acceptance is socially constructed (for example, to advance the interests of the scientists themselves).

The first reason corresponds to the so-called scientific method, sometimes called the hypothetico-deductive method, which is often found in science textbooks and in the publications of philosophers of science. One might expect that if a young scientist follows the method and makes a successful prediction, he or she would gain some favorable recognition, and perhaps a career boost.

That's not necessarily true. The counterexample described in the following section led me to undertake the project described in this book.

1.1 THE RINGS OF URANUS

One of the most striking features of the solar system is the set of rings around the planet Saturn. For three centuries after the discovery of the most prominent of these rings by the Dutch scientist Christiaan Huygens in 1655, no one was able to establish the existence of rings surrounding any other planet, despite numerous telescopic observations that revealed satellites accompanying Mars, Jupiter, Uranus, and Neptune. In the 1970s, respected theorists explained why other planets don't have rings.

In 1972 Bibhas De, a graduate student from India working with the Swedish scientist Hannes Alfvén at the University of California in La Jolla, did some calculations based on Alfvén's theory of the origin of the solar system. Alfvén, who had won the Nobel Prize in 1970 for his research in space physics, proposed that the planets were formed from clouds of ionized gas that fell toward the sun and were stopped at different distances before they condensed. De deduced from this theory that the planet Uranus has rings. This was essentially a *prediction* that the rings existed and should eventually be discovered. De submitted a short paper on this prediction to *Icarus*, a journal of solar system astronomy. The paper was rejected.

Five years later, rings around Uranus were discovered. De sent the same paper back to *Icarus*, asking that it be published so he could get the credit for a successful prediction. It was again rejected. Although he did succeed in getting his paper published in another journal, he rarely gets any recognition for this achievement.

It might seem that Bibhas De was following the scientific method as described in textbooks: propose a hypothesis, use it to derive a testable prediction, make the test by a suitable observation or experiment, and then—depending on the result—reject, revise, or (tentatively) accept the hypothesis. If the prediction is confirmed, other scientists should be willing to give serious consideration to the hypothesis. Isn't that the way science is supposed to work? Did De lack the command of English needed to explain his theory? Did he, perhaps unintentionally, offend the editor or referees by denigrating accepted theories and their prominent authors? Was he discriminated against as an Asian unfamiliar with or disrespectful toward American customs? Even well-established scientists have had their papers capriciously rejected.

Or is there something wrong with the assumption itself: that scientists normally follow the scientific method and give credit to others who follow it successfully?

Although several scientists and philosophers have denied that there is a unique scientific method, the remarkable case of Bibhas De's prediction of Uranian rings inspired me to look closely at a number of famous theories: what predictions did they make, and which of those were confirmed? Did other scientists accept the theories *because* these predictions were confirmed? It appeared that while most of the theories did make successful predictions, the empirical tests often did not come until *after* the idea had been accepted by leaders in the field. So I generalized the question: *why* were these theories accepted?

Those who believe that theories are accepted because their predictions are confirmed often also believe that this is more likely to be true in the "hard" sciences like physics and chemistry; the knowledge gained in those sciences is therefore more valid and reliable than that obtained in "soft" sciences like biology, psychology, and sociology. Thus Newton's laws help us to send rockets to the moon and planets, and Einstein's relativity theory led to the atomic bomb, while theories of revolution have a weak track record.

From the history of science we learn that the knowledge accepted by a scientific discipline is not eternal truth. All physicists understood, in the middle of the nineteenth century, that space was filled with an invisible ether that transmitted light as transverse waves. By the middle of the twentieth century they no longer believed this, but we still want to understand how they came to accept the ether theory, as well as how they came to reject it.

To clarify what is meant by "theories became knowledge": an idea is a qualitative statement, such as "light is transmitted by waves in an invisible ether" or "organic evolution is driven primarily by natural selection." Before an idea can become scientific knowledge it must be incorporated into a theory. I use the term "theory" as scientists understand it—a proposition, often quantitatively expressed, based on one or more ideas put forth as a basis for reasoning or argument, which may or may not be regarded as valid. The theory *becomes* knowledge if and when it is regarded as valid by the practitioners of the discipline. The thrust of this book is to illustrate just how, and by what criteria, practitioners come to regard a theory as valid.

How do we find out what criteria are used? From the statements, mostly published, by scientists themselves. That's not as easy as it sounds, since research papers often don't explain why a theory is considered valid. Private communications (letters and e-mails) are better—but not easy to find, except for the most famous scientists whose papers have been preserved and are available to historians. Monographs and review articles are good sources. Introductory textbooks (biology, physics, etc.) generally present only a simplified and perhaps inaccurate version of knowledge. They are nevertheless useful. Historians tend to denigrate textbooks because they are often so out-of-date that when they are published, the advanced knowledge for which they claim to provide an approximation has already been rejected or significantly changed by the scientists. Yet they can indicate what was accepted by the practitioners in the recent past.

The word "became" in the subtitle of this book should also be noted. I do not claim that the results from my study of a few episodes in a short period of time (between around 1870 and around 1970) can be generalized to conclusions about the inherent nature of scientists and their behavior in the present and future. One reason why these episodes may not be typical is that many involve the validity of radical new theories; prediction-testing may be more useful for the refinement of theories whose basic principles have already been accepted. On the other hand, some of the cases were selected precisely because it was generally believed that they did in fact involve predictions whose confirmation led to the acceptance of radical new theories.

1.2 MAXWELL AND POPPER

Maxwell

The scientific method defined in the previous section, also known as the hypothetico-deductive method, was widely discussed by scientists and philosophers in the nineteenth century. Physicist James Clerk Maxwell invoked it in 1860 when he showed that his kinetic theory of gases has a serious defect: thermal properties calculated from the theory are clearly in disagreement with experimental results. At first he concluded that the result "overturns the whole hypothesis, however satisfactory the other results may be." Moreover, the theory predicted the absurd result that the viscosity of a gas is the same at all densities, and increases rather than decreases as the temperature increases, contrary to what was then known about the viscosity of fluids.

Did Maxwell actually follow the scientific method? He did not abandon his refuted theory; instead, he continued to work on it. He showed by his own experiments that his absurd theoretical results about viscosity were actually correct, and that what was then known was wrong. But better experiments on specific heats sharpened the disagreement with his theory, and Maxwell refused to accept the ad hoc assumptions that others proposed in order to force the theory to accommodate the experimental data for diatomic molecules. Instead, Maxwell concluded that we simply do not know enough about intramolecular structure and motions to be able to solve the problem; for the time being we must recognize that we are in a "state of thoroughly conscious ignorance which is the prelude to every real advance in knowledge."

Maxwell was not only a great scientist; he was also very sensitive to philosophical and methodological issues. Thus he realized that while the scientific method may sometimes be a useful tool, it should not prevent us from relying on our own judgment about which scientific hypotheses to accept. The kinetic theory, for Maxwell, was the best way to describe gases, but it might fail to account for those properties of gases that depend on the details of their behavior at the atomic level. Sometimes the limits of validity of a theory can be determined only by trial and error.

Popper

> Popper reportedly would enter his lecture hall on the first day and announce to the students, "I am a professor of scientific method, and I have a problem. There is no such thing as scientific method!"
>
> —PHILOSOPHER THOMAS NICKLES *(2006)*

In the twentieth century, the most-frequently mentioned version of the scientific method was the falsificationism of the philosopher Karl Popper. (That doesn't mean it was actually followed by all scientists.) In a 1934 book and many later publications, Popper argued that science advances most rapidly when a scientist proposes a new idea, and uses it to predict something that would not be expected according to the current views of other scientists. The scientist then tries to falsify (disprove) it by experiment. If it survives several experimental tests, it may be tentatively accepted.

Popper stipulated that predicting another occurrence of a cyclic phenomenon doesn't count. Thus, if Halley's Comet has appeared several times at 75-year intervals, one doesn't get credit for predicting it will return again in 75 years. More generally, the predicted event or experimental result should be one that is *not* expected on the basis of currently accepted theories.

According to Popper and his followers, confirming a new fact predicted by a theory is better evidence than explaining an already-known fact. The reason is neatly summarized in the following 2010 statement about global warming by the Nobel Prize-winning economist Paul Krugman:

> While it's relatively easy to cook up an analysis that matches known data, it is much harder to create a model that accurately forecasts the future. So the fact that climate modelers more than 20 years ago successfully predicted the subsequent global warming gives them enormous credibility.

But that *prediction* was not enough to convince a self-described climate-change skeptic, physicist Richard A. Muller. What did persuade him was his *explanation* based on his own analysis (with his daughter Elizabeth) of land temperatures over the past 250 years, using "sophisticated statistical methods developed by . . . scientist Robert Rohde." He concluded in 2011 "that global warming was real and that the prior estimates of the rate of warming were correct." Then in 2012 he went further and asserted: "Humans are almost entirely the cause." Data for this period cannot be ascribed to solar variability, but do match the known greenhouse gas increase. The observed warming is, he claims, accurately *explained* when factors like volcanic activity (which tends to produce temporary cooling) are taken into account.

Contrary to what many philosophers and scientists seemed to believe, Popper asserted that while an experimental test could *refute* a hypothesis by falsifying its predictions, it could not *confirm* a hypothesis by verifying its predictions. This is because, he pointed out, you don't know whether some other hypothesis could have led to the same predictions.

The statement that a hypothesis is not necessarily true just because it leads to a correct prediction was not original with Popper; in fact it goes back to Aristotle. Logicians have given a rather strange name to the mistaken belief that a correct prediction does confirm the hypothesis: the fallacy of affirming the consequent. Popper avoided the fallacy, but many of his followers did not.

How could two different hypotheses lead to the same correct prediction? A well-known example from the history of early astronomy is the prediction of planetary positions: both the geocentric (Ptolemaic) and heliocentric (Copernican) hypotheses give approximately the same predictions for the observed motions to Mars, Jupiter, and so on, and both are approximately in agreement with observations. Astronomers accepted the heliocentric system for other reasons (Section 2.2).

What if the prediction is falsified by experiment? Does that mean that the hypothesis is wrong? Physicist, chemist, and historian Pierre Duhem argued in 1906 that it is impossible to refute an hypothesis by a single experiment, because one is always testing that hypothesis in combination with one or more assumptions needed to predict the result in a particular situation. If the prediction is falsified, all you know is that the hypothesis *and/or* at least one of the auxiliary assumptions is wrong. As an example, Duhem cited the long-running battle between the wave and particle theories of light, apparently settled in 1850 when physicists Armand Fizeau and Jean Foucault disproved the prediction from Newton's particle theory that the speed of light should be greater in glass than in air. But Newton's prediction assumed not only that light is composed of particles, but also that those particles have special properties that govern their refraction at an interface. Another particle theory might still be correct, and indeed a decade later Duhem was vindicated by the success of Einstein's light-quantum hypothesis (Chapter 7).

Despite these objections, Popper argued that the most important method for evaluating a scientific theory is the fate of its predictions. These predictions should not only contradict what the previously accepted theory predicts or implies. He wrote:

> We have no reason to regard the new theory as better than the old theory—to believe that it is nearer to the truth—until we have derived from the new theory *new predictions* which were unobtainable from the old theory (the phases of Venus, the perturbations, the mass-energy equation) and until we have found that these new predictions were successful.

Popper argued that a theory or hypothesis that cannot be refuted by any test—one that cannot be falsified—is neither true nor false, but it is *unscientific*. In effect, he used the property of *falsifiability* as a definition of "scientific." While I disagree with much that Popper wrote, I do endorse one of his statements:

> If anyone should think of scientific method as a way which leads to success in science, he will be disappointed. There is no royal road to success.

If only he had applied this insight to his own method.

1.3 WHAT IS A PREDICTION? A MERCURIAL DEFINITION

Now we encounter a confusing linguistic problem. Physicists and some other scientists use the word "prediction" to mean not just the forecast of a future event or previously unknown phenomenon; it may also refer to the deduction from theory of a known fact. This seems quite reasonable—after all, if a fact F supports a theory T, how can it make a difference, logically speaking, whether we were aware of F before T was proposed? Either F supports T or it doesn't.

But philosophers do want to make a distinction between those two situations. They use the term "novel prediction" when the prediction refers to a fact not yet known. Physicists, if they recognize the distinction at all, call this "prediction in advance." Philosophers use "explanation" or (what seems a rather derogatory term) "accommodation" for non-novel predictions. The question, which mainly concerns philosophers rather than physicists, is: does a successful novel prediction count more, other things being equal, than a successful explanation? Do you get extra credit for novelty?

Consider the early empirical tests of Einstein's general theory of relativity. In the 1920s and 1930s, physicists usually mentioned three predictions derived from the theory: the bending of a light ray by the sun's gravitational field, the gradual shifting of Mercury's orbit (advance of the perihelion), and the shift of spectral lines in a strong gravitational field. The Mercury orbit-shift had been known for several decades but not satisfactorily explained by Newtonian gravitational theory; it was still counted as a prediction because it could be deduced from relativity theory. As we will discuss in Chapter 11, the success of the Mercury orbit prediction was not considered less important evidence for relativity just because it was not novel. Indeed, for some physicists it was *more* convincing.

Many opponents of relativity asserted that their own theories should be preferred to Einstein's if they could *explain* (deduce) light bending, even though they had not predicted it in advance. They were taking advantage of the fallaciousness of affirming the consequent, even though they didn't use those words. Supporters of relativity had to show that the alternate theories predicted other consequences that were not observed. Even today it is still possible that a new theory could match all the confirmed predictions of general relativity and show itself to be superior in other ways (e.g., consistency with quantum mechanics)—just as general relativity matches all the confirmed predictions of Newtonian mechanics and is also superior in other ways.

The major advantage of the novelty of a prediction is publicity: probably most physicists ignored general relativity until news of the eclipse test hit the headlines. Those who considered themselves experts on theoretical physics then had to learn something about the theory and give it serious consideration. At that point one had to consider the evidence furnished by the phenomenon of light bending on its own merits, aside from novelty, and weigh it against the great disadvantage of having to give up well-established ideas about space, time, and gravity. Some opponents of relativity asserted that their own theory, which did not abandon those ideas, should be preferred to Einstein's as long as they could explain (deduce) light bending from their own theories, even though they had not predicted it in advance.

Here is a more explicit example of the ambiguous meaning of the word "prediction": physicist Graham Farmelo, in his biography of physicist P.A.M. Dirac, writes that Dirac used his version of quantum mechanics, tweaked to be consistent with relativity, to describe the scattering of a photon by a single electron. This theory

> enabled Dirac to make the first prediction of quantum mechanics: using a graph, he compared observations of electron scattering with his "new quantum theory" and showed that it was in better agreement than the classical theory.

This was clearly not a novel prediction, since the experimental data were available before he published his theory.

An example from biology: Sehoya Cotner and Randy Moore published a book Arguing for Evolution, in which the introduction was called "Evolution as a Predictive Science" and Chapters 2 through 9 each started with a section of "Predictions." Out of a total of 43 predictions listed, I counted only two that could be considered novel predictions (see end of Section 2.4):

1. Darwin's statement "In Madagascar there must be moths with proboscises capable of extension to a length of between ten and eleven inches!" This is because a Madagascan orchid [*Angraecum sesquipedale*], which he examined in 1862, had "a whip-like green nectary of astonishing length," so an insect would have to go to the bottom of a foot-long spur to get nectar and pollinate the orchid. Such an insect was not found until 1903.
2. Darwin's statement in *The Descent of Man* (1871) that humans originated in Africa. This was confirmed by Raymond Dart in 1924, and supported by other evidence found later.

The other 41 predictions were postdictions or explanations of previously known facts; Cotner and Moore usually introduce them with the phrase "If evolution has occurred, we predict that" and then give evidence that the prediction is correct.

Darwin also made what I would call an *implied* prediction: he explained why beetles inhabiting the Madeira Islands in the Atlantic Ocean are wingless; other scientists tested this explanation by reproducing in a laboratory experiment the causal factor postulated by Darwin (Section 14.7).

The result of the confusion about the meaning of the term "prediction" has been to persuade many philosophers and writers of popular works on science that the true scientific method requires proposing and testing *novel* predictions—citing as evidence for that assertion the statements of physicists about the importance of predictions but not realizing that the physicists themselves (and perhaps also biologists) were usually including deductive explanations as predictions.

A few scientists have become famous in part because one of their predictions was confirmed contrary to their own expectations. In two cases, they wanted to disprove a theory they disliked by showing that one of its consequences was clearly wrong. The mathematical physicist Siméon Denis Poisson, an advocate of the particle theory of light, proved that the wave theory developed by physicist Augustin Fresnel predicted a bright spot at the center of the shadow cast by a solid circular disk when struck at a 90° angle by a beam of light. Since such a bright spot had never been observed, he thought, he argued that the nonexistence of the spot refuted the wave theory. Instead, Fresnel, with the help of physicist and astronomer François Arago, did the experiment and found that the spot does exist. It is now called the Poisson bright spot.

In the other case, Albert Einstein, Boris Podolsky , and Nathan Rosen proved that quantum mechanics predicted a phenomenon now called "entanglement." Two particles, which interact and then remain completely separated (e.g., they fly off to different galaxies) still influence each other. Thus a measurement on one can determine a property of the other. Contrary to their expectation that this prediction would be refuted and thereby expose the absurdity or at least the incompleteness of quantum mechanics, the "EPR" prediction revealed one of the theory's most sensational and potentially useful applications.

1.4 HIERARCHY AND DEMARCATION

Popper's definition of scientific method has sometimes been used to determine where each science stands on a hierarchical scale ranging from hard to soft. Many scientists (especially physicists) assume that physics is the hardest science because its theories are rigorously tested before being accepted. The other physical sciences—chemistry, astronomy, geology, meteorology, oceanography—try to follow the scientific method but are supposedly softer because of the more complex nature of their subject matter and the difficulty of performing definitive experimental tests of their predictions under controlled laboratory conditions in a reasonable amount of time. Sociology is often considered the softest science, because these difficulties are aggravated when one tries to study groups of interacting human beings. Other fields may be completely off the scale—not scientific at all because their theories are not and perhaps cannot be empirically tested; there is no experiment that could *falsify* them (prove them wrong).

In a 1953 lecture Popper recalled how, as a student in Vienna after World War I, he was very impressed by the confirmation of Einstein's general relativity theory by eclipse observations—not that the theory was necessarily true (for its postulates were still very hard to believe) but that it could be submitted to a clear-cut test, and that the theory (Popper assumed) would have been rejected if it failed the test. By contrast, three other revolutionary theories of the day—Marx's theory of history, Freud's psychoanalysis, and Alfred Adler's individual psychology—were so flexible that they could explain everything: no conceivable test could refute them.

Popper worked with Adler for a short period in 1919, investigating psychologically disturbed children, and reported to him a case "which to me did not seem particularly Adlerian, but which he found no difficulty in analysing in terms of his theory of inferiority feelings." Yet Popper thought that any case could be explained equally well by Freudian or Adlerian theory; hence there was no way to choose between them or to test either of them.

From this experience, Popper developed his theory that a scientific theory must be capable of disproof by some specified experiment whose result is not already known. A theory that is not falsifiable—that is so flexible it can explain anything—may or may not be true, but it is not scientific. He concluded that Marxism and psychoanalysis were not falsifiable and therefore were pseudosciences, not sciences.

Bibhas De's conclusion (based on Alfvén's theory) that the planet Uranus has rings really was a novel prediction, so according to Popper's doctrine its later confirmation should have made scientists take it seriously and subject it to further tests, not just ignore it. Conversely, Einstein's explanation of the shift of Mercury's orbit, which was explained but not predicted in advance by his theory, should not have counted as much as the confirmation of his novel prediction of light bending.

Intrigued by these two apparent violations of the scientific method, I decided to find out whether there were any other episodes in which novelty *did* count—where scientists preferred one theory over another just because, other things being equal, it made a successful novel prediction of a fact that the other theory could explain only after the fact was known. Because of my own background I started with physics and chemistry, recalling several famous successful predictions.

For example, surely a theory (especially if proposed by a respected scientist) that revealed the existence of a new particle ought to get lots of credit. So I looked at the following:

1. Einstein's light quantum hypothesis (the photon)
2. Dirac's relativistic quantum theory of the electron (the positron)
3. Yukawa's theory of nuclear forces (the yukon, later called π meson)
4. Gell-Mann's SU(3) symmetry-group theory (the omega-minus, $\Omega-$)

My study of the physics literature did not turn up any positive evidence that physicists regarded these theories more favorably because the particles were discovered *after* the predictions were published rather than *before*. In fact, physicist Ernest Rutherford expressed exactly the opposite view in the second case. At the 1933 Solvay Congress, just after physicist Carl Anderson's discovery of the positron, he said:

> It seems to a certain degree regrettable that we had a theory of the positive electron before the beginning of experiments. . . . I would be more pleased if the theory had appeared after the establishment of the experimental facts.

I will come back to this curious remark later.

In all four cases the discovery of the predicted particle (whether or not *because* of the prediction) had a major impact on theoretical and experimental research. It certainly called attention to the theory that led to the prediction. But that did not prevent the theories from being rejected or substantially modified soon afterward. In particular, Dirac's theory was replaced by another theory (quantum electrodynamics) which, as far as I can tell, did not even attempt to predict (deduce) the existence of the positron, but simply took it as a basic postulate.

1.5 WHAT'S WRONG WITH QUANTUM MECHANICS?

Conversely, there are examples of theories that were accepted because they give satisfactory explanations of known facts, before any of their novel predictions had been confirmed. The quick adoption of quantum mechanics is an example of this unscientific behavior (Chapter 8). Werner Heisenberg's matrix mechanics was published in 1925, Erwin Schrödinger's wave mechanics appeared in 1926; their practical equivalence was quickly demonstrated, and the unified theory was widely adopted by leading atomic physicists by 1928. The reason was that the new theory allowed physicists to solve many of the problems that had baffled the practitioners of the old quantum theory, such as the ionization potential of helium and the effect of magnetic fields on spectra (the anomalous Zeeman effect). Quantum mechanics did make many novel predictions that were later confirmed, but there was not enough time to design, perform, and report experimental tests of these predictions before the theory was accepted.

Most experts on atomic physics were willing to accept this revolutionary theory without demanding a successful novel prediction. There was, however, one significant exception, which will be discussed in Chapter 9.

1.6 WAS CHEMISTRY MORE SCIENTIFIC THAN PHYSICS (1865–1980)? MENDELEEV'S PERIODIC LAW

Having failed to find any cases in which physicists gave extra credit for novelty in evaluating new theories, I turned to chemistry. Following the same approach—looking for a well-known example of a successful prediction—I studied Dmitri Mendeleev's periodic law of the elements. In 1871 he predicted the chemical and physical properties of several elements that should be discovered to fill empty spaces in his table. In 1875 one of them was discovered by P. E. LeCoq de Boisbaudran and named gallium. He initially reported its density as 4.7, but when Mendeleev pointed out that it should have a density about 6.0 (by interpolation from surrounding elements in his table), Le Coq improved his experimental technique and found it to be 5.935. During the next decade two more elements were discovered—scandium and germanium—that fit quite well into two other gaps in Mendeleev's periodic table. Mendeleev was also able to satisfy another of Popper's criteria by predicting facts contrary to what was generally believed at the time: he stated, correctly, that the atomic weight of beryllium should be 9 rather than 14, and that of uranium should be 240 rather than 120 or 60.

Did chemists accept Mendeleev's periodic law because of its successful novel predictions of the properties of newly discovered elements or because it conveniently organized the properties of well-known elements? In order to answer this question I had to look at chemistry textbooks, since little relevant evidence could be found in research journals. (Also, the periodic law was especially useful in teaching the properties of the elements.) I found that textbook writers in the period 1870–1890, at the end of which Mendeleev's law was almost universally accepted, gave several reasons for introducing it and for preferring it to its rivals proposed by Meyer and others. Novel predictions did count in its favor, though not as much as the law's usefulness in systematizing inorganic chemistry. Nevertheless, this was the first case I studied in which a novel prediction was worth more than a deduction of a known fact.

1.7 SCIENTIFIC CHEMISTS: BENZENE AND MOLECULAR ORBITALS

During the same period, August Kekulé was helping to organize organic chemistry with the theory of the quadrivalent carbon atom and his theory of the hexagonal structure of the benzene molecule C_6H_6. These theories initially seemed contradictory, since if one imagines that the C atoms are located at the vertices of a hexagon and each C is bonded to the Cs on each side and to one H, then the Cs each have one valence left over. Kekulé first suggested that the extra valences are used to form double bonds between every other pair of adjacent carbons. This was not satisfactory, and in 1872 he proposed his oscillation hypothesis. As it was later interpreted, the hypothesis stated that the fourth valence of each C oscillates between its two neighbors, synchronously with all the other fourth valences, so that the structure as a whole switches rapidly between the two structures.

Kekulé's oscillation hypothesis assumes that all the Cs and all the CC bonds are equivalent. In particular, it predicts

1. there is only one form ("isomer") of a compound in which one of the Hs or 5 of the Hs are replaced by another atom such as chlorine (C_6H_5Cl or C_6HCl_5)

Moreover, it predicts (if we label the H atoms $H^1 \ldots H^6$)

2. a compound formed by replacing H^1 and H^2 with Cs, for example, cannot be distinguished experimentally from one formed by replacing H^1 and H^6.

Both predictions were confirmed; there are no other experimentally distinguishable isomers of the substitution products of benzene than the ones predicted by Kekulé's theory.

These successes were not sufficient to persuade all chemists to accept the oscillation hypothesis. Several other models could explain the observed number of isomers of substitution products of benzene. Nevertheless, as long as those models did not explain any other phenomena *better*, Kekulé's hypothesis seemed to have an advantage because it had predicted them in advance.

But I think that the scientific character of chemistry (i.e., its use of the scientific method) is illustrated not so much by the reaction to success as by the reaction to failure. Adolf von Baeyer, a student of Kekulé, was able to eliminate the three-dimensional prism structure, one of the competitors to the two-dimensional hexagonal models, by falsifying one of its predictions.

During the first part of the twentieth century many other benzene models were proposed, but none gained wide acceptance because none could be proved empirically superior to the others. Significantly, in view of the comparison with physics I am making here, none gained an advantage because it was based on a better *theory*.

Of course the advent of quantum mechanics brought a dramatic change to the benzene problem and to chemistry as a whole. During the 1930s and 1940s, Linus Pauling's valence-bond approximation was widely accepted—partly because Pauling himself was very successful in presenting it in a way that chemists could use to explain and predict chemical phenomena, and partly because, as Pauling often pointed out, it was quite similar to empirical rules with which they were already familiar. The rival molecular orbital theory, developed by Robert Mulliken and a few others, was seen by chemists as a physicists' theory—too mathematical and not very useful.

Pauling's valence bond description of benzene as a resonance between different structures was widely viewed as being compatible with Kekulé's oscillation hypothesis. Moreover, a new experimental test by Amos Cole and A. A. Levine in the early 1930s, using the ozonization of a benzene derivative, ortho-xylene, seemed to demonstrate the temporary existence of the two Kekulé structures—long enough to produce two different products of the reaction, in a way that could be explained by valence bond theory.

In the 1950s and 1960s, the molecular orbital theory displaced the valence bond approach as the best explanation of benzene and other organic molecules. There were several reasons for this shift; only two will be mentioned here:

1. Charles Coulson became an effective advocate for molecular orbitals because, like Pauling, he could communicate well with chemists and teach them how to use quantum mechanics in a qualitative intuitive way.
2. The theory of aromaticity (benzene-like chemical behavior) was successfully developed on the basis of the molecular orbital theory. This goes back to the work of Erich Hückel, who predicted in 1931 that molecules with $4n + 2$ mobile electrons, where n is an integer, will have aromatic character.

In particular, Hückel's rule predicts that C_4H_4, cyclobutadiene, is *not* aromatic as one might have expected from its similarity to C_6H_6, benzene. A similar but more detailed prediction was derived in the early 1960s from molecular orbital calculations, which also showed that it should be rectangular rather than square. This was a true novel prediction, because C_4H_4 is so unstable that it was not successfully synthesized until 1965. Moreover, early evidence from the infrared spectrum was interpreted to mean that it is square. Thus molecular orbital theorists were in the classic Popperian situation of making a risky contraprediction: predicting a fact contrary to the empirical evidence known at the time, with sufficient confidence in the validity of their theory to hope that further experiments would vindicate them (recall Mendeleev's prediction of the density of gallium). And that is what happened.

Just as in the case of Mendeleev's periodic law, chemists did not adopt molecular orbital theory *primarily* because of its successful predictions; the utility of the theory—in organizing and explaining previously known facts—was a more important factor. Nevertheless, chemists did agree that predicting a new fact counts more in favor of a theory than explaining a known fact of similar importance.

Given this historical background, it is no surprise to find that chemistry textbooks in the late twentieth century recommend the predictive scientific method. Science educators Mansoor Niaz and Arelys Maza analyzed 75 general chemistry textbooks published in the United States, more than half of them in the period 1965–1990. They rated each book on nine criteria, including the extent to which it presented the view that there is no universal step-by-step scientific method. This, along with eight other criteria, represented a consensus on the nature of science among prominent science educators.

Contrary to this consensus, Niaz and Maza found that the overwhelming majority of textbooks still followed the doctrine promulgated by the US National Society for the Study of Education in 1947. The scientific method demands that one follow six steps of making observations, defining the problem, constructing hypotheses, experimenting, compiling results, and drawing conclusions. Niaz and Maza lament that "This oversimplified view of what constitutes the scientific endeavor has proven to be resistant to change and is used in almost all parts of the world." They ascribe this conclusion to a 2004 paper by M. Windschitl.

Niaz and Maza found that only four of the 75 textbooks gave a satisfactory presentation. of their no-universal-method view; 27 mentioned it, and the rest ignored or rejected it.

I don't know if anyone has yet done a similar analysis of textbooks in other sciences. At least in chemistry, the results of Niaz and Maza are consistent with my finding that chemistry is more predictivist than physics and biology.

1.8 THE UNSCIENTIFIC (BUT VERY SUCCESSFUL) METHOD OF DIRAC AND EINSTEIN: CAN WE TRUST EXPERIMENTS TO TEST THEORIES?

> If the observer knows in advance what to expect . . . his judgment of the facts before his eyes will be warped by this knowledge, no matter how faithfully he may try to clear his mind of all prejudice. The preconceived opinion unconsciously, whether he will or not, influences the very report of his senses, and to secure trustworthy observations, it had been recognized everywhere, and for many years, he must keep himself in ignorance of what he might- expect to see.
>
> —*Astronomer* H. N. Russell *(1916)*

What is the difference between chemistry and physics? It is widely believed that physicists put more faith in theories, while chemists put more faith in experiments. Chemistry journals in the nineteenth century reputedly would not accept theoretical papers with no supporting experimental data. Physics journals in the late twentieth century sometimes rejected experimental papers if there was no theory to explain the results—and, as we have seen from the case of the rings of Uranus, a paper based on established empirical facts may still be rejected if it deduces those facts from a theory disliked by the referees.

How do physicists know which theory is right if they don't rely primarily on experiments? Here are three versions of an esthetic criterion, proposed by physicists whose predictions were ultimately quite successful.

As Paul Dirac famously asserted,

> It is more important to have beauty in one's equations than to have them fit experiment . . . It seems that if one is working from the point of view of getting beauty in one's equations, and if one has really a good insight, one is on a sure line of progress. If there is not complete agreement between the results of one's work and experiment, one should not allow oneself to be too discouraged, because the discrepancy may well be due to minor features than are not properly taken into account and that will get cleared up with further developments of the theory.

What is a beautiful equation? According to astrophysicist S. Chandrasekhar, scientists seek theories that display "a proper conformity of the parts to one another and to the whole" while still showing "some strangeness in their proportion."

Albert Einstein proclaimed in 1930 that the significance of his theory of relativity does not depend on "the prediction of some tiny observable effects, but rather the simplicity of its foundation and its consistency." In a 1933 lecture he elaborated on his method:

> Nature is the realization of the simplest conceivable mathematical ideas. I am convinced that we can discover, by means of purely mathematical constructions, those concepts and those lawful connections between them which furnish the key to the understanding of natural phenomena. Experience may suggest the appropriate mathematical concepts, but they most certainly cannot be deduced from it. Experience remains, of course, the sole criterion of physical utility of a mathematical construction. But the creative principle resides in mathematics. In a certain sense, therefore, I hold it true that pure thought can grasp reality, as the ancients dreamed.

These statements may sound arrogant today, since we know that coherent and beautiful physical theories have sometimes been overthrown by *experiments*. But what replaces them? The new theory may have a different kind of beauty, one that does not appeal to the older generation but which others learn to appreciate.

Perhaps now we can understand Rutherford's strange remark at the 1933 Solvay Congress about the discovery of the positron (quoted above in Section 1.4): experiments are not always a reliable test of theories, because the experimenter may be so attracted by the beauty of the theory that he or she will be unconsciously biased toward confirming it. Maybe this is why astronomer Arthur S. Eddington, who was already on record as a strong supporter of relativity before 1919, reported results in agreement with Einstein's prediction, although closer scrutiny of his data suggests that he discarded the results that did not agree with the theory. Physicist Emil Rupp, around 1930, published confirmations of predictions by Einstein and others that turned out to be invalid (Section 7.10).

More recently there is physicist Victor Ninov's false discovery of Uuo (ununoctium, element 118), following a prediction by Robert Smolanczuk that one could make Uuo by bombarding lead with krypton (82 + 36 = 118). Ninov, who had once remarked that "Robert must talk to God," confirmed the prediction. Doesn't God always tell us the truth?

Some mistaken discoveries were probably not deliberate fraud but, as Joachim L. Dagg suggests, attempts by scientists to "cut corners in reaching a conclusion that they genuinely believe to be true." Sometimes they turn out to be right, and the dubious method used to make the discovery is forgotten (chemist John Dalton, biologist Gregor Mendel); other times they are wrong (psychologist Cyril Burt), and the scientist is disgraced and eventually forgotten.

In another notorious case, it appears that physicist Jan Hendrik Schön did commit deliberate fraud in his papers on the electrical properties of unusual materials. According to historian David Kaiser, reviewing Eugenie Reich's book about Schön, *Plastic Fantastic*, he "sought to make his fakes fit in rather than stand out, massaging his data to match established predictions."

In 2011, Harvard psychologist Marc Hauser resigned his position after an internal investigation found him guilty of scientific misconduct in reporting his research on monkeys, including experiments on how well they understand human gestures. The following year, the US Department of Health and Human Services Office of Research Integrity concluded that he "falsified results in a way that supported his theoretical predictions" in a paper published in the journal *Cognition*.

According to journalism professor Charles Seife, a widespread and dangerous kind of fraud occurs in the testing of new drugs. A pharmaceutical company develops a drug and predicts that it will cure a particular disease. This prediction is tested by scientists who are either employees of the company, or who are paid to conduct the tests. In either case the scientist knows that his future income may depend on how often he reports that the prediction has been confirmed.

Perhaps we may conclude from these examples that there is sometimes a motivation for an experimenter to confirm a prediction, made either by himself or by a more eminent scientist.

Finally, I want to mention physicist J. J. Thomson's 1930 story illustrating the dangers of putting too much emphasis on testing a theory rather than exploring "the subject with which the theory deals," as he recommends. At Cambridge University, where Thomson was a professor at that time, the residential colleges employed women to make the beds in the student rooms. When he told one of these bed-makers that, at Oxford University, they had male "scouts" to do that job, she exclaimed that then "the staircases must be very dirty. In the true scientific spirit she determined to test her theory, and went to Oxford when next there was an excursion" [reduced-fare round trip on the railway]. She confirmed her theory by observing all the staircases but didn't notice the buildings, courts, walks and river. She just regarded them as obstacles in the way of getting at staircases on which she had concentrated. She had proved her theory but she had missed Oxford."

1.9 WHY WAS BIBHAS DE'S PAPER REJECTED BY *ICARUS*?

When in 1977 Bibhas De resubmitted his 1972 paper, he argued that it should be published because his prediction—that the planet Uranus has rings—had now been confirmed. When told that his prediction had not been derived from accepted physical ideas, he protested that it was derived from Alfvén's critical velocity effect, which had been experimentally established.

Carl Sagan, editor of *Icarus*, then justified his rejection of De's paper on the grounds that having sent it to a large number of referees, he could not find one who advocated its publication. He wrote to De:

> The essential problem is, as you know, the feeling of all the referees that we are engaged in a fallacy sometimes called the enumeration of favorable circumstances—that is, that erroneous theories, if there are enough of them and if they make a sufficiently large number of predictions, must on occasion make a subsequently validated prediction.

Here Sagan ignored the fact that Alfvén's theory had not, at that time, made any incorrect predictions and had made several correct ones.

Sagan and the referees seem to have been influenced by the then-recent controversy about psychiatrist Immanuel Velikovsky's theory of solar system history. That theory was based on assumptions that contradicted accepted physical facts and principles, but a couple of his many predictions were qualitatively confirmed.

When I showed a draft of my discussion of the De episode and of Alfvén's theories to prominent planetary scientist Christopher T. Russell, he replied:

> Asking a scientist for a prediction is like going to the temple and asking the Sibyl for a prediction. You get a guess, perhaps education. However, this is not anything a scientist can use in the development of understanding. There must be discovery based on physical laws or well understood empiricism with a firm logical or mathematical derivation to a new theorem. You do not do geometry by predicting. If I were the first to predict that the bisector of a vertex of a triangle bisected the opposite side of the triangle, it would be scientifically useless. However, if I was the first to *prove* it, that would be important. Most important would be the proof itself, not the prover.

While not addressing De's case specifically, this statement should be considered by anyone who advocates a scientific method based on predictions.

1.10 THE PLURALITY OF SCIENTIFIC METHODS

> It is a common view that the heydays of theories of scientific methods are truly over and that current conceptions of science leave little, or no, room for a role for methodology. . . . However, methodology is a live and active field of investigation.
>
> —*philosophers* ROBERT NOLA *and* HOWARD SANKEY *(2000).*

I will return to a more detailed discussion of predictions and falsifiability in Chapter 3. Enough has been said to suggest that the insistence on a single scientific method for all sciences is misguided. Popper's method may often be irrelevant for judging major advances in physics, but is followed to some extent in chemistry. Both physics and chemistry have been successful sciences during the last two centuries.

According to philosopher Massimo Pigliucci,

> It is rather ironic that many science textbooks have essentially adopted Popper's view of science as an enterprise dealing in falsificationism . . . Popperian falsificationism has long been superseded in philosophy of science on the grounds that science just doesn't work that way.

So we should put aside our preconceptions about what method *should* be followed and ask instead what methods have actually been followed. As Dennis Flanagan, long-time editor of *Scientific American*, proclaimed, "science is what scientists do, not what nonscientists think they do or ought to be doing."

In particular, what reasons did scientists give for accepting or rejecting specific theories? How did ideas and discoveries move from the research frontier to become part of the established core of knowledge? That is the question I address in this book, without expecting to find a single definitive answer but with the hope of provoking others to at least take the question seriously.

To begin with, a major reason is that the theory gives a satisfactory explanation (preferably a logical *deduction*) of the empirical facts. It seems clear that most space physicists did not think Alfvén's theory provided a satisfactory explanation of the phenomena, regardless of how many successful predictions it made. Conversely, quantum mechanics was a logically coherent theory that allowed remarkably accurate deductions of known empirical facts, and it was accepted before its new predictions had been tested. Comparison of these two examples suggests that we can't expect to find an objective definition of "satisfactory explanation." Alfvén's critics seemed to want a mechanistic-atomistic explanation, while quantum physicists seemed to be willing to abandon what an earlier generation called "mechanism." If Alfvén had derived his formulas from quantum mechanics he might have satisfied his critics but betrayed his own convictions about the proper way to do science. It is precisely at this point that personal, esthetic, psychological, social, and nationalistic factors play a role in science, undermining any attempt to impose a universal scientific method.

Evidence that such nonobjective factors play a role in science is easily found by any observer who looks closely enough. The decision to ascribe a dominant role to one or another factor probably tells us more about the observer than about science itself. During the late twentieth century the doctrine of social construction of scientific knowledge was popular among practitioners of science studies—mostly sociologists, but also a few historians of science. According to this doctrine, a theory or concept is accepted primarily because it has social, political, and financial benefits for a group of scientists. We will examine the rise and fall of social constructionism in Chapter 4.

Finally, students of the rhetoric of science claim that the success of a theory depends on the verbal skills of its advocates. In particular, the neo-Darwinian evolutionary synthesis (Chapter 14) may have profited from the effective rhetoric of

Theodosius Dobzhansky and other biologists. Popper's falsifiability probably owes some of its popularity to its effectiveness as a rhetorical weapon; one can sometimes use rhetoric to reap the rewards of falsifiability without actually allowing one's theories to be falsified.

NOTES FOR CHAPTER 1

001 *"Most Americans have never met a scientist . . . understanding the process through which scientific knowledge develops is equally critical"*: Alberts, "Policy-Making Needs Science," Science, 330 (2010): 128.

001 *"Among the work to be done is to achieve some understanding"* Arthur Fine, "Science Made up: Constructivist Sociology of Scientific Knowledge," in The Disunity of Science: Boundaries, Contexts, and Power, edited by Peter Galison and David J. Stump (Stanford, CA: Stanford University Press, 1996), pp. 231–254, 482–484, on p. 254.

SECTION 1.1

004 For further details of the story sketched in this section see Brush, "Prediction and Theory "Evaluation: Alfvén on Space Plasma Phenomena," *Eos: Transactions of the American Geophysical Union* 71 (1990): 19–33, and "Alfvén's Programme in Solar System Physics," *IEEE Transactions on Plasma Science* 20 (1992): 577–589.

004 *respected theorists explained* D. J. Stevenson, "The Outer Planets and Their Satellites," in The Origin of the Solar System, edited by S. F. Dermott (New York: Wiley, 1978), pp. 395–431; A.G.W. Cameron, "Cosmogonical Considerations Regarding Uranus," Icarus 24 (1975), 280–284.

004 *Alfvén . . . proposed that the planets were formed from clouds of ionized gas* H. Alfvén, *On the Origin of the Solar System* (Oxford: Clarendon Press, 1954). For details and other references see S. G. Brush, *Fruitful Encounters: The Origin of the Solar System and of the Moon from Chamberlin to* Apollo (New York: Cambridge University Press, 1996), Chapter 3.4.

004 *rings around Uranus were discovered* J. L. Elliot, E. Dunham, and D. Mink, "The Rings of Uranus," *Nature* 267 (1977): 328; J. L. Elliot, E. Dunham, and R. L. Mills, "Discovering the Rings of Uranus," *Sky and Telescope* 53 (1977): 412; R. L. Millis, L. H. Wasserman, and P. V. Birch, "Detection of the Rings around Uranus," *Nature* 267 (1977): 330; J. Eliot and R. Kerr, *Rings: Discoveries from Galileo to Voyager* (Cambridge, MA: MIT Press, 1984), 1–20, 58–72.

004 *[De] did succeed in getting his paper published in another journal* B. De, "A 1972 Prediction of Uranian Rings Based on the Alfvén Critical Velocity Effect," *Moon & Planets* 18 (1978): 339–342.

005 *scientists and philosophers have denied that there is a unique scientific method* George-Louis LeClerc, Comte de Buffon wrote in the "Premier Discours" in his *Histoire Naturelle*, vol. 1 (Paris: Imprimerie Royale, 1749), "it is impossible to establish one general system, one perfect method, not only for the whole of natural history but even for one of its branches; for, in order to make a system, an arrangement, in a word, a general method, it is necessary that it includes everything" (as translated by Thierry Hoquet in his article "History without Time: Buffon's Natural History as a Nonmathematical Physique," *Isis* 101 (2010): 30–61, on p. 48. See also P. W. Bridgman, "The Prospect for Intelligence," *Yale Review* 34 (1945) 444–461, on 450; J. B. Conant, *Science and Common Sense* (New Haven: Yale University Press,

1951), p. 45; Haym Kruglak, "The Delusion of the Scientific Method," *American Journal of Physics* 17 (1949): 23–29; Alan Gross, *The Rhetoric of Science* (Cambridge, MA: Harvard University Press, 1990), pp. 40–47, 210.

Science educators, concerned with teaching the nature of science (NOS), have also recently recognized that there are several legitimate ways to do science, not just Popper's. See W. F. McComas, M. Cliugh, and H. Almazroa, "The Nature of Science in Science Education: An Introduction," *Science & Education* 7 (1998): 511–532; N. G. Lederman, F. Abd-el-Khalick, R. L. Bell, and R. S. Schwartz, "Views of Nature of Science Questionnaire: Toward Valid and Meaningful Assessment of Learners' Conceptions of Nature of Science," *Journal of Research in Science Teaching* 39 (2002): 497–521. McComas has prepared an extensive list of historical examples that may be useful in teaching each of these ways: "Seeking Historical Examples to Illustrate Key Aspects of the Nature of Science," *Science & Education* 17 (2008): 249–263. He notes that most of these examples come from physics, astronomy, and chemistry, and suggests that "one of the reasons why some members of the public reject evolution is that it cannot be demonstrated in the laboratory in the same fashion as can many notions in the physical science" (p. 255). I address this problem in Part Four.

005 *I use the term "theory" as scientists understand it* In the text I will follow the usage of scientists themselves; sometimes "theory" means an established doctrine (relativity in physics, evolution in biology), sometimes just a well-developed but untested hypothesis (string theory in physics, Gaia in biology).

005 *textbooks . . . can indicate what was accepted* "The cutting edge of the history of science in recent years has rediscovered books and texts after a long excursion into the social history of science," David Philip Miller, "The Encyclopaedic Life," *Metascience* 11 (2002): 154–159. See also S. G. Brush, "How Theories Became Knowledge: Why Science Textbooks Should be Saved," in *Who Wants Yesterday's Papers: Essays on the Research Value of Printed Materials in the Digital Age*, edited by Yvonne Carignan et al., 45–57 (Lanham, MD: Scarecrow Press, 2005), and works cited in notes 3 and 4 of that essay. J. R. Bertomeu-Sánchez, A. Garcia-Belmar, A. Lundgren, and M. Patiniotis, "Introduction: Scientific and Technological Textbooks in the European Periphery," *Science & Education* 15 (2006): 657–665 and articles by Bernadette Bensaude-Vincent and Kathryn M. Olesko in the same volume. Bertomeu-Sánchez Garcia-Belmar and Bensaude-Vincent, "Looking for an Order of Things: Textbooks and Chemical Classifications in Nineteenth-Century France," *Ambix* 49 (2002): 227–250. John Hedley Brooke, "Introduction: The Study of Chemical Textbooks," in *Communicating Chemistry: Textbooks and their Audiences*, 1789–1939, edited by Anders Lundgren and Bernadette Bensaude-Vincent, 1–18 (Canton, MA: Science History Publications, 2000), and other articles in that book. David Kaiser, editor, *Pedagogy and the Practice of Science: Historical and Contemporary Perspectives* (Cambridge, MA: MIT Press, 2005). J. Z. Young, *The Life of Vertebrates* (Oxford: Oxford University Press, 1950), preface to first edition.

For lists of textbooks see Jane Clapp, compiler, *College Textbooks: A Classified Listing of 16,000 textbooks used in 60 Colleges* (New York: Scarecrow Press, 1960).

Section 1.2

006 *scientists and philosophers in the nineteenth century* Peter Achinstein, "Inference to the Best Explanation: or, Who Won the Mill-Whewell Debate?" *Studies in History and Philosophy of Science* 23 (1992): 349–364. Ralph M. Blake, Curt J. Ducasse,

and Edward H. Madden, *Theories of Scientific Method: The Renaissance through the Nineteenth Century* (Seattle: University of Washington Press, 1960). Larry Laudan, "Theories of Scientific Method from Plato to Mach: A Bibliographic Review," *History of Science* 7 (1969): 1–63. John Losee, *Theories on the Scrap-Heap: Scientists and Philosophers on the Falsification, Rejection, and Replacement of Theories* (Pittsburgh: University of Pittsburgh Press, 2005).

006 *kinetic theory of gases* See S. G. Brush, *The Kind of Motion We Call Heat: A History of the Kinetic Theory of Gases in the 19th Century* (Amsterdam: North-Holland, 1976); *Statistical Physics and the Atomic Theory of Matter, from Boyle and Newton to Landau and Onsager* (Princeton, NJ: Princeton University Press, 1983); *The Kinetic Theory of Gases: An Anthology of Classic Papers with Historical Commentary* (London: Imperial College Press, 2003).

006 *thermal properties [of gases]* Specific heat (amount of heat needed to raise the temperature by one degree) at constant pressure and at constant volume. The ratio of these specific heats is related to the speed of sound in the gas.

006 *"Overturns the whole hypothesis"* J. C. Maxwell, "On the Results of Bernoulli's Theory of Gases as Applied to Their Internal Friction, Their Diffusion, and Their Conductivity for Heat," *Report of the 30th Meeting of the British Association for the Advancement of Science*, Oxford, June and July 1860, Notes and Abstracts, pp. 15–16; reprinted in *Maxwell on Molecules and Gases*, edited by E. Garber, S. G. Brush, and C.W.F. Everitt (Cambridge, MA: MIT Press, 1986), pp. 320–321.

006 *"state of thoroughly conscious ignorance"* Maxwell, "Review of *A Treatise on the Kinetic Theory of Gases* by H. W. Watson," *Nature* 17 (1877): 242–246, reprinted in *Maxwell on Heat and Statistical Mechanics*, edited by E. Garber, S. G. Brush, and C.W.F. Everitt (Bethlehem, PA: Lehigh University Press, 1995), pp. 156–167, quoted from pp. 164–165.

006 *[Maxwell] continued to work on [the kinetic theory]* Maxwell, "Illustrations of the Dynamical Theory of Gases," *Philosophical Magazine*, [series 4] 19 (1860): 19–32; 20 (1860): 21–37; "On the Dynamical Theory of Gases," *Philosophical Transactions of the Royal Society of London* 157 (1867): 49–88; reprinted with related articles and documents in Garber, *Maxwell on Molecules*, pp, 277–550.

006 *"Popper . . . would . . . announce . . . There is no such thing!"* Thomas Nickles, "Heuristic Appraisal: Context of Discovery or Justification?" In *Revisiting Discovery and Justification: Historical and philosophical Perspectives on the Context Distinction*), edited by Jutta Schickore and Friedrich Steinle (Dordrecht: Springer, 2006; *Archimedes*, Vol. 14), pp. 159–182, on p. 177, note 1.

006 *falsificationism of the philosopher Karl Popper* Popper, *Logik der Forschung: Erkenntnistheorie der modernen Naturwissenschaft* (Vienna: J. Springer, 1935, pub. 1934); English translation, *The Logic of Scientific Discovery* (London: Hutchinson, 1959). Popper, *Conjectures and Refutations* (New York: Basic Books, 1962).

007 *statement about global warming* Paul Krugman, "Green Economics: How We Can Afford to Tackle Climate Change," *New York Times Magazine* (April 11, 2010), pp. 34ff, on p. 38. Krugman does not mention Popper.

007 *that* prediction *was not enough to convince . . . Muller* Richard A. Muller, "The Conversion of a Climate-Change Skeptic," *New York Times*, July 30, 2012, p. A17.

007 *goes back to Aristotle* See *Aristotle on Fallacies or The Sophistici Elenchi* with translation and commentary by Edward Post (1886). Aristotle does not use that phrase, which may have been first introduced by J. N. Keynes in 1884. See also C. L. Hamblin, *Fallacies* (London: Methuen, 1970), pp. 35–36.

007 *astronomers accepted the heliocentric system* See below, Section 2.2.

008 *Duhem argued . . . that it is impossible La Théorie Physique, son Objet et sa Structure* (Paris, 1906; 3rd ed. 1933); English translation, *The Aim and Structure of Physical Theory* (Princeton, NJ: Princeton University Press, 1954). The American philosopher Willard Van Orman Quine developed Duhem's view further, concluding that any hypothesis can be considered correct no matter what results are observed in experiments, simply by making appropriate changes in the auxiliary assumptions. This is now called the Duhem-Quine thesis, but unlike Duhem's original insight, Quine's version is only a theorem of abstract logic. In scientific practice one cannot rescue a refuted hypothesis indefinitely many times by inventing suitable auxiliary assumptions or ad hoc hypotheses (hypotheses for which there is no evidence other than their success in saving a refuted theory).

008 *"We have no reason to regard the new theory as better"* K. R. Popper, *Conjectures and Refutations* (New York: Basic Books, 1962), p. 246. Unfortunately for the credibility of Popper's views on science, he has relied in his first example on a notorious myth. It is often stated that Copernicus predicted that in a heliocentric universe the planet Venus would have phases like the moon, although they had not yet been observed. The confirmation of this prediction by Galileo thus removed an objection to the Copernican theory. An alternative version of the myth is that even though Copernicus himself did not make this prediction his theory did. Neither statement is correct, because each unjustifiably assumes that Venus is opaque and shines only by reflecting sunlight, an assumption whose validity was not generally accepted before Galileo's telescope observations. See the detailed discussion by Neil Thomason, "1543—The Year that Copernicus Didn't Predict the Phases of Venus," in *1543 and All That*, edited by G. Freeland and A. Corones (Dordrecht: Kluwer, 2000), pp. 291–331.

008 *"There is no royal road to success"* K. Popper, *Objective Knowledge* (Oxford: Clarendon Press, 1972), p. 265.

Section 1.3

008 *prediction . . . may also refer to the* deduction *from theory* for an example of the confusion caused by this usage see Thomason, "1543—The Year that Copernicus Didn't Predict the Phases of Venus."

In defense of the usage, Henry Margenau wrote: "The word *prediction*, as used in science, does not mean 'forecast' in a temporal sense. *Pre*—implies 'prior to completed knowledge'; it does not contrast with *post*, as does *ante*. The counterpart to *prefix* is not *postfix* but *suffix*. It is therefore unnecessary to coin a new word, *postdiction*, to denote what we should call prediction of the past. The use of this word, though it has been suggested, would seem a bit *preposterous*." Margenau, *The Nature of Physical Theory* (New York: McGraw-Hill, 1950), p. 105.

009 *The Mercury orbit-shift . . . [was] not satisfactorily explained by Newtonian gravitational theory* One could simply change the exponent 2 in Newton's $1/r^2$ force law and thereby explain it, but then the orbits of other planets would no longer agree with the theory. See N. T. Roseveare, *Mercury's Perihelion from LeVerrier to Einstein* (Oxford: Clarendon Press, 1982).

009 *"enabled Dirac to make the first prediction"* Farmelo, *The Strangest Man: The Hidden Life of Paul Dirac, Mystic of the Atom* (New York: Basic Books, 2009), pp. 95–96. See notes for Section 8.9 under "Some other alleged novel predictions."

009 *Sehoya Cotner and Randy Moore published a book Arguing for Evolution: An Encyclopedia for Understanding Science* (Santa Barbara, CA: Greenwood/ABC-Clio, 2011).

010 *"a whip-like green nectary of astonishing length"* Darwin, *On the Various Contrivances by which British and Foreign Orchids are Fertilised by Insects, and on the Good Effects of Intercrossing* (London: John Murray, 1862), pp, 197, 198. Cotner and Moore, *Arguing*, p. xxiii. They do not give any information about the 1903 confirmation, except that the moth with the long proboscis is named *Xanthopan morgani praedicta*.

010 *Darwin's statement . . . that humans originated in Africa . . . confirmed by Dart* See end of Section 2.4.

010 *The other 41 predictions were postdictions* One other prediction, "that evolution is driven by natural selection" (Cotner and Moore, *Arguing*, p. xxiv) is ambiguous: there is ample evidence that natural selection is a major cause of evolution, but they do not attempt to show that other factors (random genetic drift, etc.), which I discuss in Chapter 14, do not make any contribution to evolution.

010 *the spot does exist* According to E. T. Whittaker, the bright spot had been observed in the early part of the eighteenth century by J. N. Delisle (Whittaker, *A History of Theories of Aether and Electricity. I. The Classical Theories* [London: Nelson, 1951], p. 108). According to Max Born and E. Wolf, it had been observed by Maraldi (Born and Wolf, *Principles of Optics*, 3rd ed. [New York: Pergamon Press, 1965], p. 375). Several books on optics and its history have claimed that this episode played an important role in the decision of French and other physicists to accept the wave theory, but John Worrall has refuted this claim: see his paper "Fresnel, Poisson and the White Spot: The Role of Successful Predictions in the Acceptance of Scientific Theories, "in *The Uses of Experiment*, edited by David Gooding, Trevor Pinch, and Simon Schaffer (Cambridge: Cambridge University Press, 1989), pp. 135–157.

010 *proved that quantum mechanics predicts a phenomenon now called "entanglement"* Einstein et al., "Can Quantum-Mechanical Description of Reality be considered complete?" *Physical Review* 48, no. 2 (1935): 696–702.

SECTION 1.4

010 *Popper worked with Adler. Conjectures and Refutations* (New York: Basic Books, 1962; Routledge paperback reprint, 2002), p. 46. In his *Unended Quest: An Intellectual Autobiography* (La Salle, IL: Open Court, 1974), pp. 36–37, Popper stated that it was Marxism rather than psychoanalysis that provided the most impressive example of a nonfalsifiable pseudoscience. See also Michel ter Hark, "Between Autobiography and Reality: Popper's inductive Years," Studies in History and Philosophy of Science 33A (2002): 79–103, where a different view of Popper's relation to Adler is presented.

012 *did not turn up any positive evidence* Brush, "Prediction and Theory Evaluation: Subatomic Particles," *Rivista di Storia della Scienza* [II] 1, no. 2 (1993): 47–152. The first two examples are discussed in Chapters 7 and 8. A minor exception is Arthur Holly Compton, who asserted that his own work on the Compton effect was better evidence for the photon than Einstein's treatment of the photoelectric effect because he had predicted in advance a new phenomenon, the recoil electron, needed to conserve energy and momentum when a photon was scattered by an electron. Einstein, he stated, had invented the light quantum "primarily to explain the photoelectric effect: so "the fact that it does so very well is no great evidence in its favor." Compton, "Light Waves or Light Bullets," *Scientific American* 133 (October 1925): 246–247.

But this argument, which underestimated the value of Einstein's work, did not persuade other physicists (see below, Chapter 7).

012 *Rutherford expressed exactly the opposite view* Discussion remarks, quoted by D. V. Skobeltzyn in *Early History of Cosmic Ray Studies*, edited by Y. Sekido and H. Elliot (Dordrecht: Reidel, 1985), p. 50.

Section 1.6

013 *Mendeleev was also able to satisfy* The new atomic weight for uranium was accepted fairly quickly, but Mendeleev's value for beryllium was resisted for several years by Swedish chemists, who argued that it should be 13.8. The latter value would require beryllium to be trivalent (rather than bivalent, as Mendeleev claimed), and would have made it difficult to fit it into the periodic table. The Swedish chemists finally confirmed Mendeleev's value in 1884.

013 *a novel prediction was worth more* It appears that chemists, unlike physicists, used the word "prediction" to mean "novel prediction," although I have not established that this is always true.

Section 1.7

013 For further details and references see Chapter 9.

014 *there are no other experimentally distinguishable isomers* That was the situation in the nineteenth century; evidence for the transient existence of two ortho isomers was found in the 1930s.

014 *molecular orbital theory displaced the valence bond approach* See below, Chapter 9. These remarks about acceptance of the molecular orbital theory are intended to apply only to the period ending around 1980. An improved version of the valence bond theory was later revived, but that fact does not necessarily affect the *historical* analysis of reasons why the molecular orbital theory was accepted in the 1960s and 1970s. All statements about the acceptance of *any* scientific theory must be regarded as provisional.

015 *Mansoor Niaz and Arelys Maza analyzed 75 general chemistry textbooks* Niaz and Maza, *Nature of Science in General Chemistry Textbooks* (Dordrecht: Springer, 2010).

015 *a consensus on the nature of science among prominent science educators* Niaz and Maza, *Nature*, p. 4. They cite F. Abd-El-Khalick, M. Waters, and A. Le, "Representations of Nature of Science in High School Chemistry Textbooks over the past four Decades," *Journal of Research in Science Teaching* 45 (2008): 835–855; N. G. Lederman, F. Abd-El-Khalick, R. L. Bell, and R. Schwartz, "Views of Nature of Science Questionnaire: Toward Valid and Meaningful Assessment of Learners' Conceptions of Nature of Science," *Journal of Research in Science Teaching* 39 (2002): 497–521; W. F. McComas, "Seeking Historical Examples to Illustrate Key Aspects of Nature of Science," *Science and Education* 17 (2008): 249–263; M. U. Smith and L. C. Schermann, "Defining versus Describing the Nature of Science: A Pragmatic Analysis for Classroom Teachers and Science Educations," *Science Education* 83 (1999): 493–509.

015 *Niaz and Maza lament that "This oversimplified view"* Niaz and Maza, *Nature*, p. 5.

015 *They ascribe this conclusion to a 2004 paper by M. Windschitl* Windschitl, "Folk Theories of 'Inquiry': How Preservice Teachers Reproduce the Discourse and Practices of an Atheoretical Scientific Method," *Journal of Research in Science Teaching* 41 (2004): 481–512.

015 *Niaz and Maza found that only four . . . gave a satisfactory presentation* N. J. Tro, *Chemistry: A Molecular Approach* (Upper Saddle River, NJ: Prentice Hall [Pearson

Education], 2008); R. H. Petrucci, W. S. Harwood, and F. G. Hering, *General Chemistry: Principles and Modern Applications* (New York: Macmillan, 2003); M. D. Joesten, D. O. Johnston, J. T. Netterville, and J. L. Wood, *World of Chemistry* (Philadelphia: Saunders, 1991); R. Boikess and E. Edelson, *Chemical Principles* (3rd ed., New York: Harper & Row, 1985).

Section 1.8

015 *"If the observer knows in advance"* H. N. Russell, "Percival Lowell and his Work," *Outlook* 114 (1916): 781–783, as quoted by N. Hetherington, *Science and Objectivity: Episodes in the History of Astronomy* (Ames: Iowa State University Press, 1988), p. 61.

016 *Chemistry journals . . . would not accept* The most notorious example is the refusal of the Chemical Society of London to publish a periodic table proposed in 1866 by John Newlands. After Mendeleev published his table in 1869, Newlands claimed priority for his own earlier table, with limited success. According to one biographer, his claim to having predicted the existence of germanium before Mendeleev is valid. E. L. Scott, "Newlands, John Alexander Reina," *Dictionary of Scientific Biography* 10 (1974): 37–39.

016 *Physics journals in the later twentieth century sometimes rejected* "Every new particle had to be presented as a previously predicted entity in some not yet tested theory." Thus Cavendish Laboratory physicists argued that the neutron was a confirmation of a prediction by Rutherford. J. Navarro, "New Entities, Old Paradigms: Elementary Particles in the 1930s," *Llull* 27 (2004): 435–464, on p. 442. Amikam Aharoni, "Agreement between Theory and Experiment," *Physics Today* 48, no. 6 (June 1995): 33–37; see also the case of Sheldon Schultz, mentioned by Joel Achenbach, "'Left-Handed' Material said to reverse energy," *Washington Post* (March 22, 2000): A13); *Physical Review Letters* refused to publish a paper until the authors found that their effect had been previously predicted by a theorist. For other views on the differences between chemistry and physics see Dudley Herschbach, "Chemistry: Blithe Sibling of Physics," *Physics Today* 50, no. 4 (April 1997): 11, 13; R. Hoffmann, "Theory in Chemistry," *Chemical & Engineering News*, July 29, 1974: 32–34. For further comments on the relation between theory and experiment in physics see Harry J. Lipkin, "Who Ordered Theorists?" *Physics Today* 53, no. 7 (July 2000): 15, 74, and comments by L. Wolfenstein, T. Wilson, M. Nauenberg, and L. Brown, with reply by Lipkin, in the January 2001 issue; comments by N. F. Ramsey in the September 2001 issue; by M. Riordan in the August 2003 issue; and by Lipkin in the August 2003 issue.

016 *"It is more important to have beauty"* P.A.M. Dirac, "The Evolution of the Physicist's Picture of Nature," *Scientific American* 208, no. 5 (1963): 45–53. See Michael Dickson, "Beauty Doth of Itself Persuade: Dirac on Quantization, Mathematical Beauty, and Theoretical Understanding," in *Discourse on a New Method: Reinvigorating the Marriage of History and Philosophy of Science*, edited by Mary Domski and Michael Dickson (Chicago: Open Court, 2010), pp. 405–421. The title alludes to a famous quotation from Shakespeare, *The Rape of Lucrece*, lines 29–30: "Beauty itself doth of itself persuade/The eyes of men without an orator." See also Christopher Shea, "Is Scientific Truth Always Beautiful?" *Chronicle of Higher Education* 59, no. 21, (Feb. 1, 2013), pp. B14–B15.

016 *According to astrophysicist S. Chandrasekhar* "Beauty and the Quest for Beauty in Science," *Physics Today* 63, no. 12 (December 2010): 57–62, reprinted from the July

1979 issue. Some of his phrases are credited to Werner Heisenberg and Francis Bacon.

016 *does not depend on "the prediction of some tiny effects"* Quoted in A. Pais, *"Subtle is the Lord": The Science and the Life of Albert Einstein* (Oxford: Oxford University Press, 1982), p. 273, from Einstein, "Raum, Äther und Feld in der Physik," *Forum Philosophicum* 1 (1930): 173–180; English translation in same journal, "Space, Ether, and Field in Physics," 180–184. See also *Beyond Geometry: Classic Papers from Riemann to Einstein*, edited by Peter Pesic (Mineola, NY: Dover, 2007), p. 176. For a different version not using the word "prediction" see Gerald Holton, *Thematic Origins, of Scientific Thought*, rev. ed. (Cambridge, MA: Harvard University Press, 1988), p. 254, from Carl Seelig, *Albert Einstein* (Zurich: Europa Verlag, 1954), p. 195.

016 *"Nature is the realization"* Einstein, "The Method of Theoretical Physics," quoted in Holton, *Thematic Origins*, p. 252, from *Mein Weltbild* (1934); see Holton's note 41 on p. 272 for other versions of this statement.

017 *Maybe this is why astronomer Arthur S. Eddington* John Earman and Clark Glymour, "Relativity and Eclipses: The British Eclipse Expeditions of 1919 and their Predecessors," *Historical Studies in the Physical Sciences* 11, part 1 (1980): 49–85 and other works cited in Chapter 11.

017 *Emil Rupp . . . published confirmations . . . that turned out to be invalid* A. P. French, "The strange Case of Emil Rupp," *Physics in Perspective* 1 (1999): 3–21; Jeroen van Dongen, "Emil Rupp, Albert Einstein, and the Canal Ray Experiments on Wave-Particle Duality: Scientific Fraud and Theoretical Bias," *Historical Studies in the Physical Sciences* 37, Supplement (2007): 73–119; "The Interpretation of the Einstein-Rupp Experiments and their Influence on the History of Quantum Mechanics," ibid. 121–131.

017 *Victor Ninov's false discovery* Bertram Schwarzschild, "Lawrence Berkeley Lab Concludes that Evidence of Element 118 was a Fabrication," *Physics Today* 55, no. 9 (September 2002), 15–17. G. Johnson, "At Lawrence Berkeley, Physicists Say Colleague Took Them for a Ride," *New York Times/Science Times*, October 15, 2002, pp. 1, 4. The quotation "Robert must talk to God" is from *The Chronicle of Higher Education*, August 16, 2002. See also T. S. Kuhn, "The Function of Measurement in Modern Physical Science," *Isis* 52 (1961): 161–193 (on measurements of the specific heats of gases by Delaroche and Berard, "confirming" Laplace's theory); Norriss Hetherington, *Science and Objectivity* (Iowa State University Press, 1988) (on gravitational redshift).

017 *"cut corners in reaching a conclusion"* Joachim L. Dagg, "Forgery: Prediction's Vile Twin," *Science* 302 (2003): 783–784, on p. 784. For other examples see Holton, *The Scientific Imagination: Case Studies* (Cambridge: Cambridge University Press, 1978), Chapter 2 on Millikan's oil-drop experiment.

017 *Sometimes they turn out to be right . . . other times they are wrong* These three scientists succumbed to the belief that dimensionless ratios of empirical quantities should be integers (2, 3, 4 respectively). See J. R. Partington, "The Origins of the Atomic Theory," *Annals of Science* 4 (1939): 245–282 (on Dalton); Leonard K. Nash, "The Origins of Dalton's Chemical Atomic Theory," *Isis* 47 (1956): 101–116. R. A. Fisher, "Has Mendel's Work been Rediscovered?" *Annals of Science* 1 (1936): 115–137. L. S. Hearnshaw, *Cyril Burt, Psychologist* (Ithaca, NY: Cornell University Press, 1979).

017 *According to historian David Kaiser* Kaiser, "Physics and Pixie Dust" (review of E. S. Reich, *Plastic Fantastic: How the Biggest Fraud in Physics Shook the Scientific World), American Scientist* 97 (2009): 496–498, on p. 497.

017 *"falsified results in a way that supported his theoretical predictions"* Siri Carpenter, "Government Sanctions Harvard Psychologist," *Science* 337 (2012): 1283.

017 *fraud occurs in the testing of new drugs* Charles Seife, "Is Drug Research Trustworthy?" *Scientific American*, 307, no. 6 (December 2012): 56–63.

017 *J. J. Thomson's 1930 story* Thomson, *Tendencies of Recent Investigations in the Field of Physics* (London: BBC, 1930).

Section 1.9

018 *Alfvén's critical velocity effect* Lars Danielsson, "Experiments on the Interaction between a Plasma and a Neutral Gas," *Physics of Fluids* 13 (1970): 2288–2294.

018 *"The essential problem is"* Carl Sagan's letter to De, July 12, 1977; copy supplied by De and quoted with permission of author.

018 *Alfvén's theory . . . had made several correct [predictions]* In addition to the critical velocity effect, Alfvén predicted the existence of magneto-hydrodynamic waves [now called "Alfvén waves"] in 1942 (confirmed by several experiments *circa* 1960]; field-aligned currents, first predicted by Kristian Birkeland in 1913 but generally ignored until Alfvén revived the idea in 1939; confirmed in an experiment by A. J. Zmuda et al. In 1966, as interpreted by Alex Dessler and W. David Cumming in 1967; electrostatic double layers, a geophysical application of a phenomenon discovered by Irving Langmuir and Harold Mott-Smith in the 1920s, and predicted to be important in space plasma by Alfvén in 1958, confirmed by R. D. Albert and P. J. Lindstrom using data from a rocket probe of an aurora in 1966. For details and references see Brush, "Prediction and Theory Evaluation: Alfvén . . . " and "Alfvén's Programme."

018 *controversy about psychiatrist Immanuel Velikovsky's theory* Michael D. Gordin, "Separating the Pseudo from Science," *Chronicle of Higher Education* September 21, 2012, pp. B10–B12; Gordin, *The Pseudoscience Wars: Immanuel Velikovsky and the Birth of the modern Fringe* (Chicago: University of Chicago Press, 2012.

018 *"Asking a scientist for a prediction"* C. T. Russell, Personal communication (1989), quoted by permission.

Section 1.10

018 *"It is a common view that the heydays of theories of scientific methods are truly over"* Robert Nola and Howard Sanka, "Introduction," to *After Popper, Kuhn and Feyerabend: Recent Issues in Theories of Scientific Method* (Dordrecht: Kluwer, 2000), edited by Nola and Sankey, pp. xi–xix, on pp. xii–xiii.

019 *According to philosopher Massimo Pigliucci* Pigliucci, *Nonsense on Stilts: How to Tell Science from Bunk* (Chicago: University of Chicago Press, 2010), p. 228.

019 *"Science is what scientists do"* Marc Santora, "Dennis Flanagan, 85, Editor of *Scientific American* for 37 Years," *New York Times*, January 17, 2005, p. A18. Flanagan liked the Dutch translation better: "Wetenschap is wat wetenschappers doen." Here is a more explicit version by P. W. Bridgman: "the people who talk most about scientific method are the people who do least about it. Scientific method is what working scientists do, not what other people or even they themselves may say about it . . . Scientific method is something talked about by people standing on the outside and wondering how the scientist manages to do it." *Reflections of a Physicist*, 2nd ed. (New York: Philosophical Library, 1955), page 81.

I would supplement this apparently circular definition by stipulating that an enterprise like physics or chemistry establishes itself historically by a record of important achievements, using whatever methods actually work. That's what makes it a "science." It is generally not possible for another discipline to become a "science" simply by copying the methods of the established sciences; the requisite achievements may require the development of completely different methods.

019 *did not think Alfvén's* theory *provided a satisfactory explanation* See S. G. Brush, "Alfvén's Programme in Solar System Physics," *IEEE Transactions on Plasma Science 20* (1992): 577–589.

019 *rhetoric of Theodosius Dobzhansky* Leah Ceccarelli, *Shaping Science with Rhetoric: The Cases of Dobzhansky, Schrödinger, and Wilson* (Chicago: University of Chicago Press, 2001).

019 *probably owes some of its popularity to its effectiveness as a rhetorical weapon* David L. Hull, *Science as a Process: An Evolutionary Account of the Social and Conceptual Development of Science* (Chicago: University of Chicago Press, 1988).

2

Reception Studies by Historians of Science

2.1 WHAT IS RECEPTION?

An idea or theory becomes knowledge when it is accepted and used by the relevant scientific community, and is taught to the next generation of scientists. Historians of science use the term "reception" to identify this process. My goal here is more specific: I want to find out *why* scientists and teachers received a theory. What reasons do they give for accepting a particular theory? Is it because, as the scientific method demands, predictions based on the theory were confirmed? Or were other reasons more important?

Reception studies published by historians of science sometimes answer such questions. These studies try to go beyond nebulous discussions of the influence or impact of scientific theories to examine the positive or negative reactions of specific scientists and intellectuals, usually those in a particular country during a specified time period. Using the electronic database maintained by the History of Science Society, one can easily retrieve more than 100 reception studies.

Although these studies are very useful, they don't provide all the information I need. As historian Susan Cannon noted in 1976, there are no established standards for reception studies. Historians do not usually focus on the reasons *why* scientists accept a theory, and tend to ignore philosophical distinctions like that between novel and non-novel predictions. They may include instead the views of intellectuals who are not competent to judge the technical merits of a theory but are more articulate in expressing their opinions about it. Moreover, most reception studies deal with one of only three theories: Darwinian evolution, Freudian psychoanalysis, and Einsteinian relativity. Together, those three account for more than half of all published reception studies. Quantum theory, whose impact on science is substantially greater than that of relativity, comes in a distant fourth. For many of the theories discussed elsewhere in this book, there are no published reception studies at all.

Published works provide a useful first approximation to the views of the community. Letters and interviews may give a second approximation, although it is usually not practical to search all of these systematically. But the statements about reception found in many works on the history of science are only *zero*-level approximations: claims are made (e.g., "general relativity was accepted because of the observation of gravitational light bending," or "Maxwell's electromagnetic theory was accepted because of Hertz's experiments on electromagnetic waves") on the basis of statements of only one scientist, or even none at all. One cannot assume that just because an experiment is now seen as confirming a theory, it persuaded scientists to accept that theory when it was performed.

To illustrate the lack of attention given to the reasons for accepting theories, consider a compendium of studies by 11 historians on the reception of relativity in nine countries. In several of these studies one finds general statements to the effect that scientists accepted Einstein's theories because of the negative results of the Michelson-Morley experiment and the positive results of the 1919 eclipse expedition. Since the latter results—observations of the bending of light by the sun—were announced as a confirmation of Einstein's 1915 general theory of relativity, we seem to have here an illustration of the importance of novel predictions in science. But although it is reasonable to suppose that these general statements about the impact of the eclipse observations reflect careful study of the literature, it is disappointing to find that they are rarely backed up by specific documentation. Out of a total of 138 scientists, mathematicians, and engineers named as supporters of relativity (usually not distinguishing between the special and the general theory), only one—the Italian engineer Ferdinando Lori—is actually quoted as giving light bending as a reason to accept relativity, and that was in a newspaper article that also listed the Michelson-Morley experiment and the advance of Mercury's perihelion as evidence. In fact, for about 80% of those mentioned as prorelativists, the historians give no reason at all for their support.

I have found only one serious attempt to construct a general theory of the reception of scientific theories, by historians Thomas Glick and M. G. Henderson. This theory will be discussed at the end of this chapter.

In the next three sections I discuss studies of the reception of three important theories—the Copernican heliocentric system, Newton's universal gravity, and Darwinian evolution—in order to show the nature of the work done by historians and assess its usefulness for my own project.

2.2 THE COPERNICAN HELIOCENTRIC SYSTEM

The reception of the Copernican System has been studied by a number of scholars, going back at least as far as the nineteenth-century philosopher-scientist William Whewell. Since then there have been a number of other books and articles, including those by historians Dorothy Stimson and Francis Johnson. Here are some generalizations that have emerged about the reasons why the theory was accepted.

In the decades after the publication of Copernicus's *De Revolutionibus* (1543), there were two very good reasons why the theory should *not* be accepted. First, it assumed the rotation of the Earth around its own axis every 24 hours, instead of the rotation of the heavens around the Earth. Since it was well known to astronomers that the circumference of the Earth is more than 24,000 miles, this means that a person standing on the ground at the equator must be moving at more than 1,000 miles per hour; at higher latitudes the speed would be several hundred mph. Yet we don't notice the many phenomena that should be produced by this motion: an incredibly strong wind knocking us down, birds and clouds left behind in the sky, or objects dropped from high towers falling to the West rather than straight down.

Copernicus himself answered that objection by pointing out a common experience of travelers: When a ship is floating calmly along, those on the ship suppose that they are stationary while the surrounding water is moving. The experience is even more impressive for airplane travelers.

Second, if the Earth moved around the sun every 365 days, instead of the sun moving around the Earth, then the apparent position of two stars, one behind the other, would change over a period of six months and then change back again after another six months (stellar parallax effect). This effect was not observed by anyone (until the nineteenth century). One might expect that astronomers would not accept the Copernican system unless these objections could be satisfactorily answered.

A third objection, not so obvious to the modern reader, was that the geocentric (geostatic) system was firmly embedded in the medieval synthesis of Aristotelian philosophy and Christian theology. Abandoning geocentric astronomy would undermine the religious worldview still dominant in sixteenth-century Europe, although Copernicus himself did not seem to see any conflict between his heliocentric system and his allegiance to both Aristotle and Christianity.

According to Thomas Kuhn, the Copernican system won by infiltration: astronomers in the late sixteenth century found that it provided a superior method for calculating and predicting *planetary* positions, so they simply set aside the absurd consequences of the hypothetical motions of the Earth and hoped that someone would solve that problem later. Thus heliocentrism was first established as a mathematical theory; only later were its physical assumptions and astronomical features adopted. But the Copernican system retained much of the complexity of the Ptolemaic system, using epicycles and eccentrics but not equants. The notorious preface by theologian and astronomer Andreas Osiander, disclaiming the reality of the Earth's motion, may have served to justify this mathematical view of the theory and arguably could have protected it from religious criticism.

Next, observations by astronomer Tycho Brahe of the supernova of 1572 and various comets from 1577 through 1596 showed that these phenomena must be taking place in the celestial realm, thereby undermining the Aristotelian doctrine that the heavens are pure and immutable. According to historian Robert Westman, astronomer Michael Mästlin's calculations showing that the 1577 comet must go around the sun rather than the Earth helped convert Johannes Kepler to Copernicanism.

Kepler then showed that replacing circular orbits by ellipses made the heliocentric system both simpler and in better agreement with observations. Kuhn argued that this success "would almost certainly have converted all astronomers to Copernicanism."

Galileo Galilei's telescopic observations of the moon and Venus, and his discovery of satellites of Jupiter, provided suggestive though not definitive evidence for the heliocentric system. His proclamation of the principle of inertia (later known as Newton's first law of motion) implied that the Earth's motion would not produce the alarming effects previously expected.

The argument that Galileo himself thought to be a conclusive proof of the Earth's motion was the phenomenon of ocean tides. Many (though not all) historians and scientists consider this proof to be fallacious. Yet unlike the falling stone argument, it did occasionally appear as an argument for the Earth's motion in the seventeenth century.

The explanations by Kuhn and others of reasons for accepting the Copernican system are plausible but not as well supported by primary sources as one might wish. Kuhn gives hardly any primary sources at all for his mathematical infiltration scenario, but states that he relies on Stimson and Johnson. Yet Johnson had concluded (contrary to Kuhn) that, at least in England, most of the eminent English astronomers were "open advocates" of the physical truth of the Copernican theory, "especially so far as the rotation of the earth was concerned." Westman, writing a few years after Kuhn, presented an account of the influential sixteenth-century Melanchthon circle of astronomers at the University of Wittenberg, which does largely agree with Kuhn's interpretation. Historian John L. Russell quotes a prominent mathematician and navigation theorist, Henry Gellibrand, who wrote in 1635 that "the Greatest Masters of Astronomie, which this age hath afforded," accept heliocentrism despite the fact that it is absurd, just so they can save the appearances of the motions of heavenly bodies. Some of the other historians content themselves with telling us who supported heliocentrism but don't tell us what their reasons were, even when they state that those authors did give a list of arguments. Others state the reasons but don't say who asserted them.

Few historians have systematically examined textbooks to find evidence and reasons for the changing cosmological views of astronomers after Copernicus. The importance of the textbook literature is noted by historian James M. Lattis in connection with astronomer and mathematician Christoph Clavius, whose book was first published in 1570 and last revised in 1611:

> Generations of university students learned what they knew of astronomy from Clavius's book. When we read Clavius's text, we gain an understanding of what the average educated person of those days knew and believed about the universe inhabited by humankind. We cannot learn this from the works of the innovators precisely because they were innovators. The average student of the early seventeenth century did not read Galileo, and certainly not Copernicus or Kepler. The dusty, neglected textbooks of yesterday, not the celebrated controversial works, give us a window into the minds of our real ancestors.

If we examined a large number of astronomy textbooks published after 1543, we might find some acceptance of the heliocentric system in the early seventeenth century, with or without the Keplerian improvements. After the notorious Trial of Galileo in 1633, we would expect to find very few favorable references in books published in predominantly Catholic countries, unless the author, like Descartes, stipulated that the Earth's motion was not real but merely a useful hypothesis for doing calculations and "saving the phenomena." By the early nineteenth century it seemed that the Vatican had stopped actively enforcing its ban on heliocentrism, but without explicitly admitting that Copernicus and Galileo were right. Concurrently, we would probably find increasing support for heliocentrism in Protestant countries.

In the late twentieth century the Catholic Church started talking about retrying Galileo and possibly declaring that the Earth really does move. These announcements mainly served to remind the public that the *official* position of the Church was still that Copernicus and Galileo were heretics and that good Catholics should still believe that the Earth is the center of the universe.

Finally, in October 1992, Pope John Paul II announced that a special commission, after deliberating for 13 years, had found Galileo not guilty, and admitted that his original conviction had been a tragic mistake. The Pope recognized that the Galileo case had become "the symbol of the church's supposed [sic] rejection of scientific progress."

From the viewpoint of modern science, the reasons why we now accept the premise that the Earth and planets move around the sun are rather different from the reasons that were persuasive in the seventeenth century. Even the mathematical argument that this premise accurately describes the observed motions of the moon, planets, sun, and stars is considered valid only if we replace the Copernican system by the significantly different theories of Kepler and Newton. So clearly the view that the Copernican system was accepted because it was correct is inadequate. In addition to the reasons given above, we must therefore add rhetorical and psychological reasons.

The academic discipline that studies the art of persuasion is rhetoric, a topic going back to Aristotle. Historian and rhetorician Jean Dietz Moss, in her comprehensive study of the Copernican revolution, notes that the text of *De Revolutionibus* shows sympathy rather than condescension toward its intended audience "and makes it easier for the reader to accept what follows." It "conveys a spirit of congenial, even fraternal reform" in contrast to the "vituperation" of other controversial writings of that time. She suggests that "the manner in which Copernicus presented his argument may help to explain why the Church was slow in condemning the hypothesis, and why it did not criticize him directly." In view of later events, it seems clear that the tolerance of Copernican views displayed by church officials in the sixteenth century allowed those views to be freely circulated and (at least in part) accepted by astronomers.

Moss also argues that Copernicus's conciliatory approach is consistent with the advice of psychologist Carl Rogers: to change another's opinion, first "show sympathetically that the opposing view is understood." Then show how the reader or listener would benefit if he accepted your view.

Historian and psychologist Frank Sulloway, in his analysis of the psychological factors in the reception of revolutionary theories, lists heliocentrism as the first of several radical ideological revolutions. He finds that birth order is an important factor; later-born persons were more than five times as likely to support the Copernican theory in the period 1543–1609 as firstborns. After 1609, with the accumulation of new scientific evidence for the theory, birth order was no longer a significant factor in its acceptance. Galileo was a particularly eminent firstborn.

Finally we need to dispose of a *possible* reason for accepting heliocentrism that has been so widely discussed in other cases that an entire chapter of this book is devoted to it. Did anyone accept heliocentricity because it made one or more confirmed novel predictions?

In a paper published in the proceedings of a 1973 conference on The Copernican Achievement, philosophers Imre Lakatos and Elie Zahar asked: "Why did Copernicus's Research Program supersede Ptolemy's?" Contrary to the implication of this question, the authors stipulate that they are actually concerned with only the normative question of whether it was *rational* to accept Copernicus's program. The reason why scientists actually did adopt it is dismissed as merely a psychological problem.

Lakatos and Zahar argue that Copernicus made only two novel predictions: (1) the phases of Venus (the times when we see only part of the planet would be different, depending on whether it orbits the Earth or the sun); and (2) the (qualitative) existence of stellar parallax (shift in the apparent positions of nearby stars due to the Earth's motion around the sun). The first was confirmed (according to them) in

1616, the second in 1838. Thus by using Popper's criterion (see Chapter 1) heliocentrism was not corroborated until 1616, "*when it was almost immediately abandoned for the new dynamics-oriented physics*." As noted in Section 1.2, Copernicus did *not* predict the phases of Venus, so the Lakatos-Zahar assertion is based on a notorious historical myth.

To avoid the absurd conclusion that it was not rational to accept heliocentrism until 1616, Lakatos abandoned his own previous criterion that judged a research program by its confirmed novel predictions. Instead he adopted Zahar's new definition of "novel," which allows a fact to be considered novel with respect to a particular theory if it was not used in constructing that theory. Lakatos and Zahar claim that "several important facts concerning planetary motions," such as the existence of (apparently) retrograde motion, follow from heliocentrism and lend more support to that theory than to the Ptolemaic system in which "they were dealt with only in an ad hoc manner, by parameter adjustment."

Philosopher Stephen Toulmin, in his comment on this paper, remarks that it represents a significant step away from Popperian doctrine (Section 1.2) toward the view that in real science predictions are less important than explanation. But other philosophers seem to have missed this point, as we will see in the next chapter.

To conclude this section: the Copernican system was accepted by astronomers in the seventeenth century in part for the same reason that Copernicus himself gave for proposing it: the Ptolemaic system was "neither sufficiently absolute nor sufficiently pleasing to the mind." By "absolute" he meant conforming to the fundamental principle attributed to Plato: one must explain the motion of celestial objects by *uniform circular motion* or a combination of several such motions. Ptolemy violated the principle with his equant, which required a planet to move "with uniform velocity neither on its deferent nor about the center of its epicycle." Copernicus could dispense with the equant. Astronomers had to expand their aesthetic senses a bit to accept the ellipse as pleasing to the mind, and they could then dispense with epicycles. The heliocentric system then gave a much simpler and more accurate explanation of planetary motion.

But one prediction was not yet confirmed: stellar parallax was first reliably observed only in the nineteenth century. The delay did not seem to prevent the acceptance of heliocentrism.

2.3 NEWTON'S UNIVERSAL GRAVITY

According to most accounts of the Copernican revolution it was Galileo, Kepler, and Newton who finally persuaded scientists that the Earth moves. Kepler showed that the motions of planets in a heliocentric system could be simply and accurately described by assuming they move in elliptical orbits with speeds depending on their distance from the sun (located at one focus of the ellipse); this was a successful mathematical description but he explained it by invoking the physical idea that the planets are pushed around by the magnetic force of a rotating sun. Galileo made telescopic observations that undermined the Aristotelian cosmology and suggested but did not prove the Earth's motion. Perhaps more importantly for the progress of physics, he invented a principle expressly designed to counter the traditional arguments *against* that motion. Notably, he proposed that a cannonball dropped from a tower on a moving Earth, like one dropped from the mast of a moving ship, would retain the horizontal motion it had before it was dropped, in addition to its vertical motion, and therefore would not get left behind. This "no cannonball left behind" law of inertia became Newton's first law of motion. The revolution that replaced Aristotelian mechanics by Newtonian mechanics was the direct result of the mathematical calculations based on Copernican astronomy.

Something similar was involved in the theory of gravitational motion proposed by Isaac Newton. Just as the Earth's motion was incompatible with Aristotelian philosophy and eventually led to the demise of the latter, the reality of gravitational force as "action at a distance" was inconsistent with Cartesian philosophy and eventually helped to vanquish that philosophy.

When Newton was developing his theory in the period between 1666 and 1687, like most European scientists he accepted the axiom of René Descartes that action at a distance is impossible. The idea that the sun could reach out through empty space to exert an attractive force on the planets (or the Earth could attract the moon) was simply absurd, a remnant of unscientific mysticism. So when Newton derived the Keplerian motions of the planets and the moon, he assumed they move as if pulled by an inverse-square gravitational force, but explicitly stated that he could not accept action at a distance. He was simply describing and predicting empirical facts, but abstained from offering any speculative hypothesis about the nature of gravity itself. His famous motto "hypotheses non fingo" does *not* mean "I don't make hypotheses"—indeed, he made many—but rather "I don't *fake* hypotheses," meaning "I don't propose a hypothesis unless I have good reason to think it might be true." This may have been an indirect criticism of Descartes, who proposed incompatible hypotheses for the nature of light.

Historian Alexandre Koyré gave a good summary of Newton's view of gravity in the context of the general transformation of European science in the seventeenth century:

> The essential feature of the scientific revolution was the mathematization of nature. . . . Newton . . . never admitted attraction as a "physical" force. Time and again he said, and repeated, that it was only a "mathematical force."

Historians continue to debate what Newton really meant in his statements about gravity. According to John Henry, Newton did accept gravity as action at a distance but stipulated that it was not "essential to matter in the same way that extension is." Instead it must be due to divine intervention, and thus provides another proof of the existence of God.

Newton's two major rivals on the Continent, physicist, mathematician, and astronomer Christiaan Huygens and mathematician and philosopher Gottfried Wilhelm Leibniz, recognized his success in deducing Kepler's laws but rejected the concept of gravity as a property of matter, acting at a distance. It was incomprehensible, a reversion to the notorious occult qualities of an earlier unscientific age. Instead, they proposed to explain gravity as the result of impacts of particles in a Cartesian vortex.

But the next generation was seduced away from its Cartesian abhorrence of action at a distance by the translation of Newton's theory into the beautiful mathematical language of Leibniz's calculus—so much more elegant and powerful than Newton's fluxion version of calculus (at least for anyone except Newton himself).

This translation was begun by mathematician Pierre Varignon, who "reduced the questions of physics to the art of solving equations."

Historian Eric J. Aiton, in his comprehensive monograph on the history of the Cartesian vortex theory, asserts that the turning point in the battle in France between that theory and Newtonian gravity was the publication in 1732 of mathematician and astronomer Pierre Louis Moreau de Maupertuis's *Discours sur les différentes Figures des Astres*. In the words of d'Alembert, the author was the first Frenchman who had the courage openly to declare himself a Newtonian: "Maupertuis' Newtonianism sprang from the failure of the Cartesian theory to explain phenomena, especially Kepler's laws; the Newtonian system . . . explained all the phenomena wonderfully." Aiton cites mathematician Johann Bernoulli as a Cartesian who was partly persuaded by Maupertuis's *Discours* to look more favorably on Newton's theory. But Johann Bernoulli remained a Cartesian, unlike his son, mathematician and physicist Daniel Bernoulli, who seems to have grown up as a Newtonian (this is perhaps an instance of Planck's principle, discussed in Section 2.4).

Maupertuis argued that gravitational attraction, having proved itself indispensable by its success in explaining and predicting phenomena, should simply be accepted as an empirical fact; the inconvenient question of its cause should be banished from science to the realm of metaphysics. Moreover, he argued that Cartesian impulsion—pushing by collisions—was no more intelligible than attraction. This argument, Aiton writes, allowed the Cartesians "to abandon the vortices [of Descartes] with a clear conscience and accept universal gravitation as a physical axiom" after its superior predictive power had been demonstrated.

Mathematician, astronomer, physicist, and geodesist Rudjer Bošković (Roger Boscovich) turned the Cartesian objection upside down by arguing that contact action is incomprehensible and nonexistent; not only gravity, but all other properties of matter, should be explained by postulating attractive and repulsive forces. The mathematics of Newton and Leibniz had overthrown the physics of Descartes.

Historian Geoffrey Sutton has traced the sequence of early empirical tests of Newton's gravity theory and their impact on the acceptance of this theory, especially in France. The essays in Volume 2B of The General History of Astronomy provide more technical details on these tests.

Sutton asserts that although the first important evidence beyond the derivation of Kepler's laws, the measurements of the shape of the Earth by Maupertuis and his colleagues, did not come until 1737, the adoption of the Newtonian system was well begun" before that. Newton had predicted that the earth must be approximately an oblate spheroid (onion-shaped, with the polar axis shorter than the equatorial axis), while astronomer and geodesist Jacques Cassini claimed to have shown from his own measurements that it is a prolate spheroid (lemon-shaped, with the polar axis longer than equatorial). The results of the French expeditions are often presented as a dramatic refutation of Cartesian theory as well as a confirmation of Newton, but this is somewhat unfair to Descartes; Huygens had predicted oblateness from Cartesian theory, and it was only a special version of Cartesian physics, developed by John Bernoulli, that predicted prolateness, so the geodetic measurements could not really be considered a crucial experiment to decide between Newton and Descartes. Sutton asserts that "the technical achievement of the calculation of the shape of the earth . . . could only have won over those already firmly leaning in Newton's direction."

Another application of Newtonian theory around this time also seemed to have little effect in converting Cartesians to Newtonians. Sutton notes that "in 1740, Daniel Bernoulli was able to reconcile Newton's theory of the tides with observations," but at that time "the theories set forth in the *Principia* still could not produce predictions about the motion of the planets or their moons in a manner demonstrably superior to Cassini's—or even to Kepler's." Physicist Emilie du Châtelet, in her French translation of Newton's *Principia*, pointed out this evidence that "attraction rather than the pressure of the moon on the terrestrial vortex," as claimed by Descartes, caused the tides, but Sutton implies that this did not help to persuade scientists to accept Newtonian gravity. Other historical accounts of the reception of gravitational theory ignore or minimize the significance of this achievement of Newton; writers on the subject seem more interested in the validity of Galileo's claim that the tidal phenomena prove the Earth's motion.

Aiton, surveying the battle between the Cartesian and Newtonian theories of the solar system, suggests that Newton's advantage over Descartes was like that of Copernicus over Ptolemy: his theory explains many phenomena by a single consistent set of postulates, while his opponent must resort to a motley collection of ad hoc hypotheses. This view was expressed by geodesist and physicist Pierre Bouguer and may have been shared by other Newtonians. But the Newtonians risked losing this advantage when they modified the inverse square law, as did Newton himself, to explain terrestrial, short-range phenomena.

In the mid-eighteenth century, experts on theoretical astronomy were concerned about the failure of Newton's theory to give an accurate value for the motion of the lunar apsides (the two points in the moon's orbit where it is either closest to or farthest from the Earth). This was one of the first attempts to solve a three-body problem (Earth, moon, sun) in celestial mechanics, and it showed how difficult such problems can be. Before 1749 the best efforts of Newton, mathematician and physicist Alexis-Claude Clairaut, mathematician, astronomer, and physicist Leonhard Euler, and physicist and mathematician Jean le Rond d'Alembert could explain only half of the observed result. This apparent failure led to suggestions that Newton's law of gravity should be revised, either by adding an inverse fourth-power term or by resurrecting Cartesian vortices or even magnetic effects. But then Clairaut discovered and corrected an error in his calculations of the moon's motion. (He, Euler, and d'Alembert had made the same mistake.) Now the theory was considered to be in excellent agreement with observation; Clairaut's 1754 tables for predicting the position of the moon "were the first . . . that depended on theoretical developments beyond Kepler's laws . . . At last . . . everyone with any technical competence at all came to accept the correctness of Newton's law," according to Sutton.

The calculation of the motion of the lunar apsides was not a prediction but a theoretical deduction of a known phenomenon. Any suspicion that the theory had been modified in some ad hoc way to give the correct answer was presumably eliminated by the fact that three of the best mathematicians in Europe had independently obtained the same result; more importantly, their calculations were published and available for scrutiny by any skeptic. Of course the *application* of their theoretical calculations to yield tables giving future positions of the moon was a prediction, which was confirmed many times by the safe arrival of ships at their intended destination. Yet as far as I know, no one has suggested that these arrivals made Newton's law of gravity more valid that it was before,

Clairaut's other major contribution to the success of Newton's gravity theory was his prediction of the date of the return of Halley's comet. Astronomer and geophysicist Edmond Halley predicted that the comet of 1682 would appear sometime in the later 1750s, since a similar comet had been previously seen at approximately 75- or 76-year intervals. But as Karl Popper would have asserted, that by itself is not a significant prediction. A more difficult problem is: how would the gravitational force of bodies other than the sun affect the comet's orbit?

Halley suggested that the action of Jupiter on the comet in 1682 would delay its next return, but he was not able to make a quantitative calculation. Such a calculation, like the determination of the motion of the lunar apsides, was a rather difficult and rigorous test of the universality of Newton's gravity, since it depended on the simultaneous interaction of at least three bodies (Saturn as well as Jupiter and the sun might influence the orbit); it would also test the assumption that the inverse-square law is valid far beyond the most distant parts of the solar system known at that time.

Historian Simon Schaffer has described in detail what he calls the "making of the comet," by which he means establishing the premise that not only is the comet expected to appear in the late 1750s the *same* as the one that appeared in 1752, but that the successful prediction of its date of arrival would be a triumph for Newtonian celestial mechanics and for Halley's cometography rather than for the competing research program of the Cassini family of astronomers. In view of the French neglect of or opposition to Halley's cometography in the 1730s and 1740s, Schaffer argues that the consensus by 1759 that Halley should get part if not all the credit "was by no means a natural consequence of the appearance of the comet in January 1759. On the contrary, it had required painstaking and strenuous propagandizing and argument by astronomers such as [Joseph-Nicolas] Delisle [and J.J.L. de Lalande] during the previous half century." But looking ahead to the issue of social construction of scientific knowledge, to be discussed in Chapter 4, it seems impossible to deny—nor does Schaffer do so—that the success of Clairaut's prediction would provide substantial evidence for the validity of Newton's theory of gravity, quite apart from any propaganda.

Clairaut, with considerable help from Lalande and Mme. Nicole-Reine Étable de la Brière Lepaute in doing the actual calculations, was able to predict in November 1758 that the comet would reach perihelion (the point in its orbit nearest the sun) in mid-April 1759. It actually did so about a month earlier. (Compare this error with the uncertainty of three or four years based simply on the intervals observed in the past.) Lalande wrote that the enormous labor had to be done quickly because "it was important that the result was given before the arrival of the comet," in order that no one should doubt the agreement between the observation and the calculation which served as the foundation of the prediction. This is one of the earliest recorded statements of predictivism: the explicit assertion that a *novel* prediction—one announced in advance of the empirical observation—is better evidence than an explanation deduced from the same theory of a fact already known (see Section 1.3 on the terminology used for the distinction). Of course one could argue that this point was always taken for granted before the twentieth century.

The case of Halley's comet has acquired some importance in the current debate among philosophers of science because philosopher Peter Lipton, a prominent advocate of predictivism, used this as his only historical example in an article in

Science magazine. But I have not found any documentation to support the view that astronomers or physicists actually did consider the successful prediction of the return of Halley's comet better evidence for the validity of Newton's theory of universal gravitation than, for example, Clairaut's earlier successful deduction of the already-observed motion of the moon's apsides—or any evidence that any scientist who had not already converted to the Newtonian theory did so because of the comet prediction. Sutton mentions that the abbé Nollet did use this example "to convince his readers of the efficacy of the law of universal gravitation," but those readers who were familiar with the other successful applications of the law may not have needed to be convinced. I did find one statement by a historian of science that casts doubt on the latter possibility: Valentin Boss, discussing Lomonosov's opposition to Newtonian gravity in 1756 (and later), says that this opposition is "rather extraordinary . . . since by that time there were few Cartesians left to withstand the general acceptance of the Newtonian theory."

Sutton argues that the favorable reception of Newtonian gravity among French scientists and intellectuals beyond the small group of Newtonian experts (Maupertuis, Clairaut, d'Alembert) was partly due to a famous textbook by the philosopher and writer François Marie Arouet de Voltaire, *Elements of the Philosophy of Newton* (1738). But he gives even more credit to the works of Gabrielle Émilie du Châtelet, a woman often identified only as Voltaire's mistress. In her *Institutions de Physique* (1740) she synthesized Newtonian physics with the philosophy of Leibniz, recognizing in particular (as physicists do now) that Leibniz's measure of motion, mv^2, is not incompatible with Newton's measure, mv. Physicists had previously disagreed about which quantity is conserved, in the so-called vis viva ("living force") controversy, but eventually decided that both are conserved in the appropriate situation. Thus Châtelet's compromise did not undermine Newtonian mechanics but enhanced it. Her position, writes Sutton,

> ironically, was one that placed the theoretical, the philosophical, and the mathematical before the empirical, the sensual, and the intuitive, embraced by Locke and Voltaire. It was of course this entirely theoretical, mathematical approach that Lagrange and Laplace would employ in the next generation to finish the solution of the solar system, a task that was not accomplished, despite the swaggering rhetoric of the participants in the voyages of exploration launched to discover the Newtonian world . . . The technical developments Mme du Châtelet witnessed and documented required a thorough integration of Leibniz's mathematical style and indeed of his theoretical mechanics, into Newton's physical system of universal gravitation.

Sutton remarks that her role in this has been obscured by those who can't bear to "attribute originality in science to women."

Most studies of the reception of Newtonian gravity implicitly agree that what Sutton calls the "solution of the solar system" by Lagrange and Laplace came too late to persuade scientists of the validity of Newton's theory. Thus the highly technical explanation of difficult problems such as the secular acceleration of the moon, the long inequality of Jupiter and Saturn, and the decrease in the obliquity of the ecliptic served mainly to support the conviction that the solar system is a clockwork mechanism in which all deviations from Kepler's laws (secular inequalities) are themselves cyclic and will not cause the system to run down. This view was sometimes called the Newtonian clockwork universe, despite the fact that Newton himself explicitly rejected it.

As regards the question whether prediction or explanation is more effective in persuading scientists to accept a new theory, these Newtonian reception studies tend to support explanation, with two controversial exceptions. There were indeed two spectacular confirmations of predictions from Newtonian theory: the oblate shape of the Earth and the time of return of Halley's comet. The first was announced in 1737; it may be argued that 50 years after the publication of the *Principia*, most physical scientists had already formed their opinions about the validity of Newton's theory. This argument is even more persuasive when applied to the situation two decades later when Halley's cometography was triumphantly confirmed. So until a more thorough study of the reasons for accepting Newton's gravitational theory has been done, I conclude that most physicists and astronomers accepted this theory because of its explanations rather than its predictions.

Similarly, the discovery of Neptune in 1846, thanks to astronomer Urbain Jean Joseph Leverrier's Newtonian analysis of anomalies in the motion of Uranus, was not needed to convince astronomers that Newton's theory of gravity was correct. One might, however, ask whether they would have abandoned that theory if a planet had *not* been found anywhere near the predicted position.

We can guess the answer from the subsequent history of Leverrier's work on planetary motion: when he found a discrepancy in the motion of Mercury, he proposed the existence of another unseen planet, close to the sun. Failure to observe it did not lead astronomers to reject Newton's theory—until Einstein *explained* the discrepancy from his general theory of relativity (Chapter 11).

2.4 DARWIN'S THEORY OF EVOLUTION BY NATURAL SELECTION

Before the eighteenth century, scientists generally believed that God had created humans, animals, plants, the earth, and the rest of the universe in more or less their present forms. During the next 150 years, there were several attempts to replace this creationist view with what came to be called evolutionary theories. The astronomers P. S. de Laplace and William Herschel succeeded with their nebular hypothesis for the origin of the sun and the solar system, while biologist Jean B.P.A. Lamarck advanced a view of biological evolution that was intermittently popular up to the early decades of the twentieth century. Scientists in the nineteenth century often adopted a schematic theory of cosmic evolution or evolutionary worldview, favoring a deterministic goal-directed irreversible progress from primeval chaos and simplicity to contemporary complex biological and physico-chemical organization. The evolutionary worldview replaced the Newtonian cyclic clockwork universe (Section 2.3), but did not establish the random undirected version of evolution that we now associate with Charles Darwin's theory.

Many books and articles have been published on the reception of Darwinism, and a reader unfamiliar with this literature might think that these publications deal primarily with the reception of Darwin's theory of evolution by natural selection. After all, the dictionary definition of Darwinism is "the biological theory of Charles Darwin concerning the evolution of species, etc., set forth especially in his works entitled *The Origin of Species by Means of Natural Selection, or the Preservation of*

Favoured Races in the Struggle for Life (1859) and *The Descent of Man and Selection in Relation to Sex* (1872)." But that is not good enough for today's scholars, who devoted an entire 114-page issue of a major journal to this question. As philosopher David Hull points out, scientists themselves do not agree on what Darwinism means; "a scientist could be a Darwinist without accepting natural selection as the major evolutionary mechanism," and conversely.

Most authors of works on the reception of Darwinism do not define what they mean by Darwinism, with the useful exception of historian Ronald Numbers, who flatly asserts that the word is "a term commonly used as a synonym for organic evolution."

If that is indeed the common usage, it is certainly confusing and paradoxical. It implies that Darwinism existed long before Darwin published *The Origin* in 1859; it implies that Lamarckism is one version of Darwinism; and, as it turns out, it means that Darwinism was generally accepted in the late nineteenth century despite the fact that what Darwin himself considered the primary mechanism for evolution was *not* accepted. It therefore means that when a historian writes that a particular scientist accepted Darwinism, the reader does not always learn whether that scientist accepted natural selection. Even for those who are interested only in the social aspects of evolutionary theory it causes confusion, since the concept of natural selection or survival of the fittest is generally understood to be central to social Darwinism even if it is not central to Darwinism.

Since I am primarily interested here in the reception of Darwin's theory of evolution by natural selection, I will avoid the vague term "Darwinism." Nevertheless, for information about the nineteenth-century reception I will rely on published studies by historians who do use that term, and try to learn what I can from their reports.

It is convenient to begin with the book by Ronald Numbers, *Darwinism Comes to America* (1998), because it gives the most direct answers to the questions: when and why was evolution accepted? Numbers reports that by the mid-1870s "most American naturalists who expressed themselves on the subject were favorable to evolution." Although only a few gave reasons for that belief, the most important reasons were the following: First, the theory was compatible with beliefs already held before 1859; naturalists had developed a "commitment to methodological naturalism" and a dislike for the invocation of supernatural causes in science (what we would now call creationism). This dislike for the religious viewpoint was "even more compelling than the positive evidence for evolution"; there was simply no *scientific* alternative to evolution. So the publication of *Origin of Species* "served more as the catalyst for the shift to organic evolution than as its cause." A similar argument can be made for the connection between Darwinism and a belief in progress, according to historian Michael Ruse.

The second important reason, according to Numbers, was Darwin's "skill in constructing a persuasive argument from familiar evidence"; this, rather than "the novelty of his views," was crucial. This remark points to the importance of rhetoric in science, a topic that has recently attracted a number of scholars, some of whom analyze Darwin's writings.

One of the reasons often brought up *against* evolution was the absence of missing links—transitional forms that should have existed when one species was changing to another. But that statement is itself a testable prediction, and the discovery of a new fossil species intermediate between two different ones could be considered to confirm a prediction of evolutionary theory.

The 1861 discovery of Archaeopteryx, apparently an intermediate between reptiles and birds, was regarded by some scientists as confirmation of Darwin's theory. Darwin himself had made a more specific prediction in a letter to geologist Charles Lyell in 1859: "If ever fossil birds are found very low in series, they will be seen to have a double or bifurcated wing. Here is a bold prophesy!" Biologist Gene Kritsky states that the appearance of the wings of Archaeopteryx was consistent with Darwin's prediction. But the prediction was not published before the discovery, and Kritsky notes that "in subsequent editions of *The Origin of Species*, Darwin downplayed the importance of Archaeopteryx as a transitional fossil."

Examples of compelling empirical evidence in support of evolution were contributed by American science and might have been expected to persuade Americans to accept evolution: In the early 1870s, paleontologist Othniel C. Marsh discovered fossil remains of toothed birds in Cretaceous rock from Kansas, and in 1876 he found a spectacular series of fossils showing different stages of the evolution of the horse, leading up to the modern horse. Darwin himself wrote to Marsh that his discoveries provide "the best support to the theory of evolution, which has appeared within the last 20 years." But, as Numbers points out, these discoveries came too late to account for the conversion of most American naturalists to Darwin, although they could have had an impact in other countries. So we have here an example of a theory that did make successful novel predictions, but it remains to be seen whether that success actually persuaded any scientists to accept the theory.

Another Darwinian prediction was that fossil evidence would show that humans originally evolved in Africa. This was confirmed beginning in the 1920s by anatomist and anthropologist Raymond Dart and others (discussed in more detail later in this chapter).

Numbers also makes effective use of historian and psychologist Frank Sulloway's analysis of the role of birth order, social attitudes, and age in the decision of a scientist to support a new theory. So we have three more reasons to support evolution for a person who (a) holds "liberal religious and political attitudes," (b) is younger than most of the established scientists, and (c) is not a firstborn child.

An earlier study of the reception of evolution in the United States by historian Edward J. Pfeifer yielded two other reasons for a favorable view: Variations of form or behavior within a species, as would be expected on Darwin's theory (natural selection needs variation in order to work); and, for opponents of slavery, it provided scientific support for the view that all human races had the same origin (monogenism).

Numbers, like other historians of biology, found that while American naturalists in the late nineteenth century generally accepted evolution, they did *not* accept natural selection as its major cause. Instead, in agreement with Pfeifer and with the thesis of historian Peter Bowler, there was actually a non-Darwinian revolution in which Lamarckian explanations of evolution again came to the fore. Like most historians who have studied the reception of Darwinism, Numbers does not discuss the revival and widespread acceptance of natural selection in the middle of the twentieth century, thereby leaving the impression that Darwin's *theory* died while Darwinism lived on.

In his study of the reception of evolution in England, historian M.J.S. Hodge reached conclusions somewhat similar to those of Numbers for the United States, but was more specific in presenting evidence that was considered to support Darwin's theory. Evolution explained the similarities of many organs in different animals for which there was no similarity of function or purpose: for example, why does the penguin's flipper have a pentadactyl (five-finger or five-part) structure like that of "the monkey's grasping hand, the mole's shoveling paw, and the sparrow's flying wing?" These are sometimes called rudimentary or vestigial organs, if they appear to serve no purpose at all. Evolution from a common ancestor was a plausible explanation. Also, evolution could explain why some ancestors persisted "long after descendant groups had sprung from them," for example the Archaeopteryx (a reptile-like fossil bird), which also seemed to be a transitional form filling the gap between reptiles and birds. Moreover, evolution could explain how species that migrated to a different area could be "competitively superior to the originals" in those areas. In this case natural selection is part of the explanation: "Successful colonizations by species of alien types were . . . frequently recognized as both an anomaly on the theory of independent creations and an uncontested factor in Darwin's favor." Hodge notes that "those who, like Hooker, Henry W. Bates, and Wallace, primarily concerned themselves with geographical —including what we would call ecological—questions about species and varieties were likely to insist on the importance of natural selection (not just allow it as a possibility, as Huxley did."

Historian Frederick Burkhardt, in a review of discussions in scientific societies in England and Scotland, confirmed that geographical variation provided important support for natural selection, and reported another kind of support: the mimicry of one species by another, apparently to gain protection against predators, seemed difficult to explain by any hypothesis other than natural selection.

Burkhardt also quotes Darwin's remark in an 1867 letter: "The belief in the descent theory is slowly spreading, even amongst those who can give no reason for their belief." If the scientists themselves can't explain why they choose one theory over another, how can the historian? Perhaps only by invoking unconscious predispositions to accept a new theory.

In a pathbreaking paper published in 1978, historian and philosopher David Hull and his colleagues Peter Tessner and Arthur Diamond raised the standard for reception studies to a new level. They subjected to critical scrutiny two widespread views: that Darwinism was accepted within a decade or two of the publication of *Origin*, and that younger scientists were more likely to accept it. The first view was expressed not only by Darwin himself (as quoted in the previous paragraph), but by many observers then and since. The second is an application of a general statement by the physicist Max Planck:

> A new scientific truth does not triumph by convincing its opponents and making them see the light, but rather because its opponents eventually die, and a new generation grows up that is familiar with it.

They refer to this as Planck's principle.

Hull's team identified 67 British scientists who were at least 20 years old in 1859 and lived until at least 1869, assuming that they would have "assimilated at least a little of the special creationist world view before being confronted by Darwin's theory"; thus they were (or were not) converted to Darwinism rather than being brought up with it. Darwinism was defined as belief in the evolution of species, but not necessarily by natural selection. They tabulated the age when the scientist first accepted evolution (based on available evidence) or age in 1869 for those who continued to reject it.

According to this study, 50 out of 67 scientists accepted evolution by 1869. The authors concluded that "the conversion of the British scientific community was not nearly as rapid nor as total as we have been led to believe." They found that age is a relevant factor: the average age of acceptors was 8½ years younger than that of rejectors (39.6 and 48.1 respectively), but that age explains less than 10% of the variation in acceptance.

The modesty and incompleteness of the results of Hull's team in this particular case is less important than the fact that they showed it possible to determine quantitatively the role of one significant factor in the reception of scientific theories. Their work has stimulated many other studies, some of which led to more surprising results.

The papers by Pfeifer, Hodge, and Burckhardt mentioned above were all published in the same book, *The Comparative Reception of Darwinism*, edited by Thomas F. Glick, based on a conference held at the University of Texas in 1972. Until more systematic studies of the kind advocated by Hull have been done, this book provides the best single source for our subject, since the contributors made some attempt to provide answers to the same questions about the reception in different countries. Unfortunately, they also agreed that reception should be studied mainly in the period 1859–1885, roughly corresponding to the interval between Darwin's announcement of his theory and his death. While many contributors expanded this period to include earlier and later developments, there was hardly any discussion of the fate of Darwin's theory after 1900, which seems to have been taken as the natural terminus for other reception studies as well. At the same time it was generally agreed that Darwin's work was largely responsible for the acceptance of evolution, even if not of his own explanation for its cause. Thus, despite the obvious importance of finding out whether evolutionists ever reached a consensus in favor of Darwin's own theory—whether they eventually agreed that natural selection is the primary factor in evolution—the reader is left completely in the dark on this point. The paucity of systematic reception studies for most of the twentieth century is one reason why I chose to take up the question myself (see Chapter 14).

Among the other countries discussed by contributors to Glick's book, I will mention first the two extremes, Russia and France. In Russia, according to historian Alexander Vucinich, there was a strong proevolution tradition before 1859, and Darwin's theory found a very friendly environment among the liberal antireligious intelligentsia (corresponding to the first and fourth reasons mentioned by Numbers). The Russian Darwinists were young, and their opponents were the older scientists who dominated the academies (reason 5). Biologists such as A. O. Kovalevsky, I. I. Mechnikov, and N. D. Nozhin concluded that striking similarities between the embryos of otherwise different lower and higher animals strongly suggested that the latter had evolved from the former. (This argument was discussed by Darwin and Huxley, but it is not clear whether it converted any British biologists who didn't already accept evolution for other reasons.)

In France, however, the early nineteenth-century discussion of evolution had produced a bias against what was called "transformisme." According to historian Robert Stebbins, the reaction to Darwin's theory was mostly silence. Although transformisme was accepted by 1900, Darwin's theory was never accepted and even in the late twentieth century was considered just one of several possible explanations for transformism. Unlike in other countries, Darwin does not even get credit for persuading French scientists to accept evolution or transformism.

According to Frank Sulloway, evolution (like other radical innovations) was supported mostly by later- borns, and among later-borns mostly by younger scientists. In most European countries in the mid-nineteenth century, later-borns outnumbered firstborns by a ratio of about 2.6 to 1, except in France where a demographic shift toward smaller families had begun in the late eighteenth century. "In 1859, French scientists had only 1.1 siblings compared with 2.8 siblings among scientists in other countries," Sulloway says. "This fact helps to explain why French scientists were so hostile to Darwinism compared with scientists in other countries."

In Germany, the pro-Darwinians were generally younger than the anti-Darwinians or those who accepted evolution but not Darwin's theory, according to historian William Montgomery. The evolutionists triumphed by 1870; they believed that organisms with similar forms (members of a systematic group) share a common ancestor, and were "impressed by the increasing tendency of later fossil forms to resemble modern ones, by the recapitulation of forms in embryonic development, and by rudimentary organs." For supporters of Darwin's theory, "the strongest proof of evolution lies in the anomalies of organic nature [e.g., rudimentary wings], anomalies that may be traced to the adaptive effects of Natural Selection." Some of them found evidence for Darwin's theory in the wide variety of biological phenomena, especially the wide variations in stages of growth, but they disagreed on whether these variations are caused deterministically or are random.

In the Netherlands , historian Ilse Bulhof found only two scientists who rejected Darwin in 1859, but listed 10 who supported him in the 1860s. The strongest supporter, Pieter Harting, had already favored a vaguely evolutionary view before 1859, and in his 1864 zoology textbook he presented evolution not as an established theory but as a hypothesis that may explain "forms that seem to belong to two species at the same time."

A book-length study of *Darwin in Italy* by historian Giuliano Pancaldi deals primarily with one zoologist, Giovanni Canestrini, and mentions some further reasons to accept evolution: Darwin's theory promised to improve the system of classifying species and varieties, and made it legitimate to explore new questions, such as why sea horses were losing the tail fin they had in common with other fish. It could be used to address the question of the origin of humans, currently of great popular interest (though sometimes treated cautiously by scientists) because of fossil discoveries in Italy and elsewhere.

Thomas Glick, whom we have to thank for stimulating and publishing so many valuable reception studies, is himself especially interested in Spain and the Hispanic world. He notes that Darwinian evolution entered Spain after the Revolution of 1868 as part of a general revival of scientific intellectual life. But the discussion of evolution involved mostly religion and philosophy: "Scarcely an original idea was uttered on either side, a reflection of the low level of scientific creativity (the neurohistologists were an outstanding exception) on the one hand, and the inane posturings of the orthodox, on the other. . . . It was as easy for anti-Darwinists to attack the mediocrity of the evolutionists as it was for Darwinists to attack the bigotry and obtuseness of the Catholics." Evolutionism simply went along with liberal religious and political attitudes. This seems to be the implicit conclusion one draws from several other reception studies: one can identify the political and religious affiliations of those who supported or opposed evolution, but it is hard to find good evidence of scientific reasons that motivated their views.

In summing up the results of the studies in the book edited by Glick, David Hull wrote:

> Among Western intellectuals, religious and philosophical considerations far outweighed matters more directly concerned with science in determining their *initial* reaction to evolutionary theory. (Their eventual acceptance or rejection of the theory is another story.)

Hull asserted that the eventual acceptance of Darwinian theory was determined by evidence and rational argument—but the above-mentioned reception studies don't demonstrate that, since they don't cover a long enough time period. Thus Glick, in a survey of reception studies, concluded that Darwinism (which, for him, is not the same thing as Darwinian theory as used by biologists) is socially constructed . (The issue of social construction of scientific theories will be discussed in Chapter 3.)

A more recent collection of reception studies for European countries, edited by philosopher and historian Eve-Marie Engels and Glick, fills in three gaps left by most of the studies mentioned above. First, it covers several more countries; second, in some cases it extends the time period into the twentieth century; third, in a few cases, it provides information about specific scientific reasons for accepting Darwinism. Conveniently, I can illustrate all three by summarizing a single chapter in this two-volume work.

That chapter, by historian and biologist Victoria Tatole, discusses the reception of Darwin's theory in Romania, a country often overlooked. It surveys works on Darwin's theory published from 1872 to 1939. In particular, it presents the views of three scientists who supported Darwinism for specific reasons:

1. In 1876, paleontologist Grigore Steanescu stated that 30,000 species of fossils are extinct "while many other new animals and plants . . . had no representation during the ancient times." This could only have happened because the ancient species were transformed into the modern ones, in accordance with the theory of evolution.
2. In 1903, zoologist Nicolae Leon invoked the biogenetic law of zoologist Ernst Haeckel, "ontogeny recapitulates phylogeny," and the existence of rudimentary organs which, he argued can only be explained by evolution. He introduced Darwin's theory in his course of general biology but also accepted Lamarckism.
3. In 1893, zoologist, anatomist, and cytologist Dimitrie Voinov reported his studies of abyssal marine fauna. He wrote that the existence of eyes in the larvae of the abyss of animals "is evidence that once these abyssal animals lived in an environment with light and that only after that they

> went down to the abyssal zone where the organ of light was useless, or it had to be transformed to make it useful." In 1903, Voinov declared that "the rudimentary organs are the strongest proofs that can be brought against finalism and in support of transformism."

I found very little discussion in these reception studies of the influence of confirmed Darwinian predictions, such as the intermediate stages in the evolution of the horse mentioned above, on the acceptance of his theory. At least three of them were tested only in the twentieth century.

First, in *Origin of Species* Darwin discussed the remarkable fact discovered by naturalist Thomas Vernon Wollaston that many of the beetles inhabiting the Madeira Islands in the Atlantic Ocean are wingless, unlike related beetles elsewhere; Darwin suggested that

> The wingless condition . . . is mainly due to the action of natural selection . . . each individual beetle which flew least, will have had the best chance of surviving from not being blown out to sea.

This was converted from an explanation to a confirmed (implied) prediction (see Section 1.3) in the 1930s (see Chapter 14).

The second prediction involved the complicated sex life of orchids. It did have an important influence on the acceptance of his theory by his contemporaries, according to an article by historian Richard Bellon. Bellon argues that naturalists were persuaded to accept evolution in the period 1863–1868 because of Darwin's application of his theory to "original problems, most notably orchid fertilization." In particular, "Darwinism received its triumphant settlement as the theory of the scientific establishment at the 1868 meeting of the British Association." In his presidential address, Joseph Hooker summarized Darwin's "very long and laborious" research on "the common Loosestrife (Lythrum Salicarico) which he showed to be trimorphic, this one species having three kinds of flowers . . . each flower has, further, three kinds of stamens . . . six kinds of pollen . . . and three distinct forms of style. To prove . . . that the coadaptation of all these stamens and pistils was essential to complete fertility, Mr. Darwin had to institute eighteen sets of observations, each consisting of twelve experiments, 216 in all . . . The results in this case . . . are such as the author's sagacity had predicted; the rationale of the whole was demonstrated, and he finally showed, not only how nature might operate in bringing these complicated modifications into harmonious operation, but how through insect agency she does this, and also why she does it."

According to biologist Duncan M. Porter, Darwin had previously discussed a trimorphic species and states that "we may safely predict the function of the three kinds of flowers." He does so, then shows that his predictions are correct in his 1864 Loosestrife paper. Porter concludes that this is what Hooker was referring to.

Third, Darwin made another prediction about orchids that was confirmed in the twentieth century. In his 1862 book on the fertilization of orchids by insects, he mentions

> the *Angraecum sesquipedale*, of which the large six-rayed flowers, like stars formed of snow-white wax, have excited the admiration of travellers in Madagascar.

> A whip-like green nectary of astonishing length hangs down . . . [several of them] eleven and a half inches long, with only the lower inch and a half filled with very sweet nectar . . . It is . . . surprising that any insect should be able to reach the nectar . . . in Madagascar there must be moths with probosces capable of extension to a length of between ten and eleven inches!

Darwin proposed a coevolution of the insect and the orchid. The moth postulated by Darwin was discovered in 1903 by Lionel Walter Rothschild and Karl Jordan; it was named *Xanthopan morganii praedicta*, and is informally called a hawk moth.

Fourth, in his book on the *Descent of Man* (1871), he considered where the precursors of humans might have lived. While we should not assume that the early progenitors were identical with our even closely resembled any modern ape or monkey, we can still guess that they probably lived in Africa since the gorilla and the chimpanzee, our closest relatives, live there now. That's where paleontologists should look for fossil remains linking apes and humans. In 1924 anatomist Raymond Dart obtained a fossilized skull blasted by workmen out of the face of a limestone cliff at Taung, in the Cape Province of South Africa. He identified it as belonging to a six-year-old child with features intermediate between apes and humans. I have found no evidence that any scientist was converted to a proponent of Darwinian evolution by the many confirmations of this prediction.

As we will see, for example, in the case of Einstein's prediction of light bending (Chapter 11), making a successful prediction may not count as evidence for a theory, at least not until it has been shown that no alternative theory could have produced the same prediction. This is what happened with Darwin's proposal that an orchid is fertilized by a moth with a long proboscis: in 1997, zoologist L. T. Wasserthal proposed an alternative explanation involving predators. This provoked controversy and more observations; we can't yet say who is right. Botanist L. Anders Nilsson doubted the validity of Wasserthal's theory but wrote:

> What is still needed after 135 years is quantitative observational data of *X. M. Praedicta pollinating A. Sesquipedale* in natural populations. This magnificent orchid grows epiphytically in the coastal rainforest of Madagascar, a disappearing unique habitat. To explore Darwin's prediction is indeed a race against time.

2.5 BOHR MODEL OF THE ATOM

By 1913, the existence of atoms (hotly contested in the late nineteenth century) was no longer in doubt, and quantitative estimates of their sizes, masses, and speeds were available. The success of Dmitri Mendeleev's periodic law (Chapter 5) showed that atoms of different elements exhibited systematic similarities and differences, suggesting some kind of internal structure. The electron, discovered at the end of the nineteenth century, was apparently one constituent. The discovery in Ernest's Rutherford's laboratory that an atom is mostly empty space, with most of its mass concentrated in a tiny nucleus, was an important clue. There was also some evidence that a particle much more massive than the electron, with opposite electric charge (soon to be named the proton), was also part of the atom.

At the same time, as described in the previous chapter, physicists were starting to take seriously the evidence that light has not only wave properties but, at least when interacting with matter, also has quantum properties, as described by Planck's hypothesis. Energy may be absorbed or emitted in units of hν, where ν is the frequency of the light.

The first successful attempt to imagine an atom with electrons orbiting a nucleus, like planets orbiting a star but with electric forces replacing gravity, was made by Niels Bohr in 1913. Bohr postulated that (contrary to Maxwell electrodynamics) I: the electron does not emit or absorb radiation as long as it stays in one orbit, but only when it moves from one orbit (#1) to another (#2). In that case the frequency of the radiation must be $\nu = (E_2 - E_1)/h$. If $E_2 < E_1$, radiation is emitted; if $E_2 > E_1$, radiation is absorbed. As noted by John Stachel, Bohr did not assume that light itself is composed of particles (photons).

The crucial step in Bohr's work was a rather strange procedure for determining the allowed orbits in which the electrons could move. In addition to contradicting Maxwell's electromagnetic theory, Bohr at one point suddenly changed the definition of the symbol *n*, so that his final result does not logically follow from his own assumptions (see notes to this chapter for details.) The result, which I call postulate II: for the hydrogen atom (one electron going around a proton) the energy of the *n*th state is

$$E_n = -C / n^2 \tag{1}$$

where $C = 2B^2 e^4 m_e k_e^2/h^2$

e = charge of electron

m_e = mass of electron

k_e = constant in Coulomb's law

h = Planck's constant

n = "principal quantum number" (integers from 1 to 4).

Note that the electron's energy E_n is negative, meaning that it is in a bound state. The lowest energy (ground state) is $E_1 = -C$, the first excited state is $E_2 = -C/4$, and so on. The radius of the orbit comes out to be

$$r = n^2 h^2 / 4B^2 e^2 m_e k_e$$

or

$$r = n^2 k_e e^2 / C.$$

If this were a single atom in an otherwise empty infinite universe, we could say that there are an infinite number of bound states squeezed into the finite energy interval between $-C$ and 0. If an electron absorbed enough radiation to make its energy greater than zero, it would no longer be bound to the nucleus and would move through the rest of the universe. As a free particle, its energy could be anywhere in the continuum of numbers from 0 to +4.

If postulates I and II are accepted, these results are equivalent to those from Newton's gravitational theory for a planet moving in a circular orbit around an infinitely massive sun, with gravitational force replaced by electrostatic force. So it was easy to generalize the results to give the sun a finite mass, and to allow the electron to move in elliptical as well as circular orbits.

The simplest empirical test of Bohr's theory is to compare C with the ionization energy of a hydrogen atom (the amount of energy needed to remove the electron from the ground state to the bottom of the continuum of unbound states). The theoretical result agreed with experiment, with an error of less than 1%.

A second test was to calculate the frequency of radiation emitted when an electron drops form an orbit with quantum number n" to one with a smaller quantum number n'. According to postulate I,

$$\Leftarrow (E'' - E')/h = (1/h)\,[-C/(n'')^2 + C/(n')^2] = C/h[1/(n')^2 - 1/(n'')^2] \quad (2)$$

This is equivalent to a well-known spectroscopic formula; for $n' = 2$ it corresponds to the Balmer series, discovered by Johann Jacob Balmer in 1885. The series for $n' = 1, 3, 4, 5$ are named after their respective discoverers Theodore Lyman, Friedrich Paschen, Frederick Brackett, and August Pfund.

At this point Bohr has provided a rather remarkable way to derive two separate known facts about hydrogen: its ionization energy and its spectrum. Bohr could also estimate the size of a hydrogen atom: the diameter of the smallest orbit is about 10^{-10} meters. This is just the order of magnitude estimated from the kinetic theory of gases, and provides further evidence in favor of the model.

Bohr was able to extend his theory's domain by applying it to singly ionized helium, He^+: a nucleus with an electric charge twice as large as that of hydrogen, and a single orbiting electron. Using the same procedure as for hydrogen, he could show that the frequencies are given by the formula

$$\nu_{He+} = (C/h)\left[1/(\tfrac{1}{2}n')^2 - 1/\tfrac{1}{2}n''^2\right] \quad (3)$$

Spectra corresponding to this formula had been reported by E. C. Pickering, but were believed to be produced by hydrogen, in which case they would provide exceptions to Eq. (2). Following a suggestion by Niels Bjerrum Bohr predicted that this spectral line should be found in a laboratory experiment with a mixture of helium and chlorine or another element that could pull an electron away from the helium atom. This prediction was confirmed in Rutherford's laboratory by E. J. Evans. According to Bohr's collaborator and biographer Leon Rosenfeld, this success "in no small degree contributed to its swift acceptance."

Bohr now had a theory that could predict previously unknown spectral lines for the hydrogen atom, singly ionized helium, and in general any atom with only one electron. One may ask whether Bohr's achievement was qualitatively different from that of Balmer. Balmer was able to guess the special case of Eq. 2 with $n' = 2$ from a knowledge of the frequencies for four hydrogen lines as determined by Anders Ångstrom, and from this to predict successfully the frequency of previously unknown lines.

The answer has to be affirmative, because even though Bohr's procedure included a certain amount of guesswork and arbitrary postulation (perhaps guided by some knowledge of the result he wanted to obtain), it opened a path to further research that Balmer's did not. The analogy of solar system dynamics suggested ways to attack problems such as the effect of electric and magnetic fields on the frequencies of the spectral lines (Stark effect and Zeeman effect), relativistic effects, two-body problems in which one body was not infinitely more massive than the other, and three or more body problems corresponding to neutral helium (nucleus + 2 electrons), the hydrogen molecule-ion H_2^+ (2 nuclei + 1 electron), and all the other atoms and molecules known to chemists.

The reception of the Bohr model has been discussed by historian Helge Kragh and others. Kragh's study is especially useful for my purposes because in several cases he gives the *reasons* why physicists accepted the model; he also identifies a date (end of 1915) when the community of experts on atomic physics generally accepted it.

Kragh lists 14 scientists who, in the period 1913–1915, were opponents or critics of the theory, 23 who were interested in the theory, and 20 who were supporters of the theory. According to Kragh, of the 20 supporters, nine physicists accepted Bohr's model because it explained the empirical formula for spectral lines, and gave an accurate calculation of Rydberg's constant from known values of physical constants. Three physicists, along with (unnamed) British physicists, mentioned the confirmed prediction of the He^+ spectral lines. Seven physicists and one chemist accepted the theory for other reasons, including the fact that it supported their own theory.

By 1916, Robert A. Millikan, in his presidential address to the American Physical Society, could also cite Arnold Sommerfeld's explanation and prediction of the fine structure of the hydrogen spectrum using his generalization of the Bohr model, to illustrate the "extraordinary success of the Bohr atom." Kragh reports that German physicists were "skeptical about atomic models in general" (quoting Sommerfeld) and did not have much interest in Bohr's theory. A significant exception was Einstein, who, according to a 1913 letter from George von Hevesy to Bohr, when told about the confirmation of Bohr's prediction that the Pickering-Fowler spectrum belongs to helium, not hydrogen, "was extremely astonished and told me . . . this an *enormous achievement*. The theory of Bohr must then be wright [sic]."

From these results we might conclude that explanation was more effective than prediction in persuading physicists to accept the theory—but this is not quite fair, since what was explained was a general formula and the numerical value of a previously arbitrary constant. What was predicted was one special case.

2.6 CONCLUSIONS AND GENERALIZATIONS

Thomas Glick, the leading performer and promoter of reception studies, has proposed a general model for the reception of scientific ideas. He stipulates four modes of reception:

1. thetic (positively asserted)—a new idea is absorbed into a given discipline;
2. antithetic—resistance "grounded in maintaining some preexistent ideological or disciplinary stance";
3. "a corrective logic takes place as a quasiscientific limitation or revision of the original thetic content of an idea"; and
4. an extensional logic takes shape through cultural appropriations in different political or social environments.

My own approach addresses primarily the first of these modes, with occasional ventures into the second. I am interested in finding out *how* a new idea is absorbed into a given discipline, and find it useful to consider three ways of accomplishing this: confirming novel predictions, providing new and convincing explanations of previously known but mysterious facts, and persuading influential scientists that the new idea is so beautiful that it must be true. Glick, as he himself states, is interested in both the scientific and popular reception of a new idea, and in practice his approach seems to give more emphasis to social and cultural factors. We would occupy common ground in considering the influence of scientists' religious or political views on the acceptance of a new scientific theory.

The receptions of the four theories reviewed in this chapter—those of Copernicus, Newton, Darwin, and Bohr—have only two common features, both negative. First, in each case there was no single empirical or theoretical argument that was decisive in persuading scientists to accept the theory. Second, successful predictions of previously unknown facts (novel predictions, as philosophers call them) played only a minor role. Each theory did indeed make successful predictions: the Copernican theory, especially when revised by Kepler, gave more accurate predictions of planetary motions; Newton's theory predicted the oblate shape of the Earth and the date when Halley's comet returned in the eighteenth century; Darwin's theory predicted the loss of wings by beetles on oceanic islands, the intermediate stages in the evolution of the horse, and the place where the earliest humans arose; Bohr's theory predicted spectral lines for atoms with one electron. But in general those confirmations came after leading scientists had already accepted the theory. Kuhn's statement—that astronomers accepted the Copernican theory before 1700 because it led to accurate predictions of planetary positions even though they could not yet accept the Earth's motion as physically real—suggests an important exception to this generalization. But he does not give direct evidence for his statement, and I have not found convincing support for it in other reception studies.

On the other hand, Kuhn's claim that the agreement of mathematical calculations with empirical data can, over time, lead scientists to accept the seemingly implausible theoretical assumptions on which those calculations are based, does have support from several other cases in the history of science. In addition to the success of absurd ideas like the Earth's motion and forces acting at a distance, we will see in Chapter 14 that natural selection overcame the opposition of many biologists because of the calculations of Ronald Fisher and others in the 1930s. It did not seem reasonable that a selective advantage too small to be observed by humans could, in a few generations, cause one form to displace another—yet that is what Fisher proved theoretically and others confirmed empirically.

Reception studies, as currently done by historians of science, may fail to uncover adequate information about the dynamics of theory change for at least two reasons: First, historians themselves have become more interested in social factors than technical factors, so they aren't looking for this information. As historian Robert Kohler recently noted, "when history of science was the history of ideas, philosophical issues of evidence and inference served as a general guide to our common labors," but this is no longer the case. Currently the alternative, at least for some historians, according to Kohler, is "the social mechanisms that turn information into acknowledged facts." Second, scientists often indicate their acceptance of a theory simply by using it, without making a public statement of their reasons, so one must use other means to infer those reasons.

According to historian and philosopher Theodore Arabatzis,

> Even though the relationship between the history of science and the philosophy of science has been discussed extensively, the focus of the discussion has been on the importance of the history of science for the philosophical understanding of science. To the best of my knowledge, there has been very little discussion of the ways in which the philosophy of science can enrich historiographical practice.

I would argue that reception studies, an established genre of the history of science, could be enriched by inclusion of philosophical issues such as prediction versus explanation. One should ask not just who accepted a new theory, but why.

Historians (and sociologists) of science do have access to the evidence needed to answer questions about the behavior of scientists as reflected in their published or unpublished statements; they have either collected the relevant data already or know how to find it quickly. But with a few exceptions, they have not systematically investigated the importance of empirical evidence and other factors in theory change. Yet I would insist that this is a central problem in the history of science.

Another suggestion is that historians of science should pay more attention to textbooks, which I have found to be one of the most useful sources of information about the reception of new ideas. Their authors are more likely than authors of scientific journal articles to explain *why* they use a new theory. Consecutive editions of a widely used book are especially useful in pinpointing *when* the author decided the theory was well enough established to be included.

Unfortunately, libraries generally do not systematically buy textbooks, and often discard an old one if it is replaced by a new edition. This is illustrated by the case of Mendeleev's periodic law (Chapter 5), which was initially introduced as a teaching aid and was not widely discussed in the research literature, so textbooks are the best source of evidence for its acceptance. But many chemistry textbooks that were used in American colleges before 1890 no longer exist in *any* American library whose books are accessible by interlibrary loan. Moreover, not enough series of consecutive editions have survived to allow the historian to determine whether the acceptance of the law was due primarily to the conversion of individual authors or to the replacement of these authors by younger authors who had already accepted the law.

My suggestion to look at textbooks is by no means original. Faithful readers of *Isis*, the flagship journal for professional historians of science, will find several good reasons in the articles by Marga Vicedo , Michael D. Gordin , Adam R. Shapiro, and David Kaiser in the March 2012 issue .

NOTES FOR CHAPTER 2

Section 2.1

030 *What reasons do they give* I mostly use published books and articles. Letters are occasionally used when available, but in most cases it would not be practical to search all the archives for all the scientists working in the relevant field at a particular time. (An exception is the Archive for History of Quantum Physics, copies of which are

deposited at a dozen or more institutions.) Moreover, published statements by a small group of respected experts can influence the acceptance of the theory by many other scientists. Inclusion in textbooks is an indication that a theory was considered part of established knowledge, at least at some time in the past.

030 *Using the electronic database* Although "reception" is not an index term in the *Isis Current Bibliography*, one can retrieve more than 100 publications with the word "reception" in the title from the History of Science, Technology, and Medicine database hosted by OCLC and available through its FirstSearch platform. HistSciTechMed contains all the data from the *Isis* bibliographies from 1974 to the present, supplemented by bibliographies of related subjects; see Stephen P. Weldon (ed.), *Isis Current Bibliography of the History of Science and its Cultural Influences*, 2009 (published for the History of Science Society by the University of Chicago Press). Additional citations can be retrieved using keywords such as "response" and "reaction." Unfortunately the only clue a few authors give in their title that they are presenting a reception study is the word "in" (e.g., "Darwin in China" or "Einstein in America").

030 *Cannon noted in 1976, there are no established standards* W. Faye Cannon, review of Glick, *Comparative Reception of Darwinism*, in *American Historical Review* 81 (1976): 559–561.

030 *philosophical distinctions like that between novel and non-novel predictions* Historians can't be expected to observe that distinction when scientists themselves sometimes don't do so; see Section 1.3.

030 *Together, those three account for more than half* My estimates are 20%, 18%, and 15% for evolution, relativity, and psychoanalysis respectively, and only about 3% for quantum theory. These figures are based on 169 publications retrieved by the HistSciTechMed search above. A more comprehensive search that included cultural impact studies, in connection with Section 2.4, turned up several more for Darwinian evolution, and a similar search would undoubtedly do the same for Freudian psychoanalysis.

030 *no published reception studies* See notes to Chapters 5, 6, 7, 9, 11, 12, 13, and 14.

031 *compendium of studies by 11 historians* The book edited by Thomas F. Glick, *The Comparative Reception of Relativity* (Boston: Reidel, 1987) was based on a double session of the Boston Colloquium for the Philosophy of Science held in 1983. Glick suggested guidelines for the authors, but admitted that they were not closely followed. In several cases the authors summarized or cited their own studies published at greater length elsewhere.

031 *only one serious attempt to construct a general theory* Thomas F. Glick and M. G. Henderson, "The Scientific and Popular Receptions of Darwin, Freud, and Einstein," in *The Reception of Darwinism in the Iberian World*, edited by T. F. Glick, M. A. Puig-Sampoer, and R. Ruiz (Dordrecht: Kluwer, 2001), pp. 229–238.

Section 2.2

031 *studied by a number of scholars* William Whewell, *History of the Inductive Sciences* (London, 1837; 3rd ed., 1857), Vol. I, Book V, Chapter III; Dorothy Stimson, *The Gradual Acceptance of the Copernican Theory of the Universe* (New York: Baker & Taylor, 1917); Francis R. Johnson, *Astronomical Thought in Renaissance England: A Study of the English Scientific Writings from 1500 to 1645* (Baltimore: Johns Hopkins University Press, 1937; reprinted, New York: Octagon Books, 1968); Jerzy Dobrzycki, editor, *The Reception of Copernicus' Heliocentric Theory: Proceedings of a Symposium Organized*

by the Nicolas Copernicus Committee of the International Union of the History and Philosophy of Science, Toru|, Poland, 1972 (Dordrecht: D. Reidel Publishing Company, and Warsaw: Ossolineum, Polish Academy of Sciences Press, copyright 1972). A useful collection is *The Copernican Achievement*, edited by Robert S. Westman (Berkeley: University of California Press, 1975). See also the following (other references can be found in the database HistSciTechMed, cited above): Hans Blumenthal, *Die Genesis der Kopernikansche Welt* (Frankfurt am Main: Suhrkamp Verlag, 1975), Part III; Jerzy Dobrzycki and Lech Szczuki, "On the Transmission of Copernicus's Commentariolus in the 16th Century," *Journal for the History of Astronomy* 20 (1989): 25–28; Henry Heller, "Copernican Ideas in Sixteenth Century France," *Renaissance and Reformation* 20 (1996): 5–26; James M. Lattis, *Between Copernicus and Galileo: Christoph Clavius and the Collapse of Ptolemaic Cosmology* (Chicago: University of Chicago Press, 1994); Victor Navarro Brotóns, "The Reception of Copernicus in Sixteenth-Century Spain: The Case of Diego de Zuñiga," *Isis* 86 (1995): 52–78; Isabelle Pantin, "New Philosophy and Old Prejudices: Aspects of the Reception of Copernicanism in a Divided Europe," *Studies in History and Philosophy of Sciences* 30 (1999): 237–262; Edward Rosen, *Copernicus and his Successors* (London: Hambledon Press, 1995); Robert S. Westman, "The Melanchthon Circle, Rheticus, and the Wittenberg Interpretation of the Copernican Theory," *Isis* 66 (1975): 165–193.

031 *Copernicus himself answered that objection* Nicolaus Copernicus, *On the Revolutions* [1543], edited by Jerzy Dobrzycki, translation and commentary by Edward Rosen (Baltimore: Johns Hopkins University Press, 1978), Book I, Chapter 9, p. 18. On the contemporary relevance of this argument (especially for science education) see S. G. Brush, "Anachronism and the History of Science: Copernicus as an Airplane Passenger," *Scientia Poetica, Jahrbuch für Geschichte der Literatur und der Wissenschaften* 8 (2004): 255–264.

032 *According to Thomas Kuhn* Thomas S. Kuhn, *The Copernican Revolution: Planetary Astronomy in the Development of Western Thought* (Cambridge, MA: Harvard University Press, 1957), pp. 185–186. For the debate about whether the Copernican and Ptolemaic systems were mathematically equivalent see Derek J. DeSolla Price, "Contra-Copernicus: A Critical Re-Estimation of the Mathematical Planetary Theory of Ptolemy, Copernicus, and Kepler," in *Critical Problems in the History of Science*, edited by Marshall Claggett (Madison, WI: University of Wisconsin Press, 1962), pp. 197–218; Norwood Russell Hanson, "Contra-Equivalence: A Defense of the Originality of Copernicus," *Isis* 55 (1964): 308–325; Imre Lakatos and Elie Zahar, "Why Did Copernicus' Research Program Supersede Ptolemy's?" in Westman, *Copernican Achievement*, 354–383, on pp. 360–364. For the debate about the role of precession see Curtis A. Wilson, "Rheticus, Ravetz, and the 'Necessity' of Copernicus' Innovation," in Westman, *Copernican Achievement*, 17–39. According to Arthur Berry, the reason why Reinhold's Copernican tables were better than the Ptolemaic Alfonsine Tables was that "Reinhold was a much better computer than the assistants of Alfonso" and that Copernicus had better mathematical tools than Ptolemy; their superiority "was only indirectly connected with the difference in the principles on which the two sets of tables were based." They nevertheless "had great weight in persuading astronomers to accept the Copernican system." Berry, *A Short History of Astronomy* (London: Murray, 1898), pp. 125–127.

032 *the Copernican system retained much of the complexity* In an *epicycle* orbit, the planet (or satellite) moves at constant speed in a small circle around a point, which itself moves in a circle around the central body (Earth or sun). In an *eccentric* orbit, the planet moves at constant speed in a circle around a point *O* in space that is near but

not at the center *C* of the central body. In an *equant* orbit, the planet moves in a circle around a point *D* that moves in a circle around a point *O* in space, while the motion of *D* is uniform as seen from another point *Q*. For details and diagrams see Gerald Holton and Stephen G. Brush, *Physics, the Human Adventure: From Copernicus to Einstein and Beyond* (New Brunswick, NJ: Rutgers University Press, 2001), pp. 11–15.

032 *arguably could have protected it from religious criticism* Bruce Wrightsman, "Andreas Osiander's Contribution to the Copernican Achievement," in Westman, *Copernican Achievement*, 213–243. Conversely, Giordano Bruno, who is now best known for having been executed by the Church for his advocacy of Copernicanism, did accept the Earth's motion but misrepresented almost every other aspect of Copernicus's system; see Ernan McMullin, "Bruno and Copernicus," *Isis* 78 (1987): 55–74.

032 *undermining the Aristotelian doctrine* Kuhn, *Copernican Revolution*, 206–208.

032 *Michael Mästlin's calculations* Robert S. Westman, "The Comet and the Cosmos: Kepler, Mästlin and the Copernican Hypothesis," in Dobrzycki, *Reception*, 7–30 discusses Michael Mästlin's calculations suggesting that the 1577 comet must have a heliocentric rather than a geocentric orbit, and the role of this conclusion in converting Kepler to Copernicanism. Westman, "The Astronomer's Role in the Sixteenth Century: A Preliminary Study," *History of Science* 17 (1980): 105–147. See also Navarro Brotóns, "Reception," p. 53.

032 *"would almost certainly have converted all astronomers"* Kuhn, *Copernican Revolution*, p. 219.

032 *Galileo Galilei's telescopic observations* Swerdlow, "Galileo's Discoveries with the Telescope and their Evidence for the Copernican Theory," in *The Cambridge Companion to Galileo*, edited by Peter Machamer, 244–270, and references cited therein (Cambridge: Cambridge University Press, 1998). Stimson, *Gradual Acceptance*, 68. Harold I. Brown, "Galileo on the Telescope and the Eye," *Journal of the History of Ideas* 46 (1985): 487–501. Samuel Y. Edgerton Jr., "Galileo, Florentine 'Disegno' and the 'Strange Spottednesse' of the Moon," *Art Journal* 44 (1984): 225–232. Mario Biagioli, "Replication or Monopoly? The Economics of Invention and Discovery in Galileo's Observations of 1610," *Science in Context* 13 (2000): 547–590.

032 *His proclamation of the principle of inertia* Johnson, *Astronomical Thought*, 164; Michael Wolff, "Impetus Mechanics as a Physical Argument for Copernicanism: Copernicus, Benedetti, Galileo," *Science in Context* 1 (1987): 215–256.

032 *phenomenon of ocean tides* Harold I. Brown, "Galileo, the Elements, and the Tides," *Studies in History and Philosophy of Science* 7 (1976): 337–351; Stillman Drake, "History of Science and Tide Theories," *Physis* 21 (1979): 61–69. Joseph C. Pitt, "The Untrodden Road: Rationality and Galileo's Theory of the Tides," *Nature and System* 4 (1982): 87–99. W. J. Shea, "Galileo's Claim to Fame: The Proof that the Earth Moves from the Evidence of the Tides," *British Journal for the History of Science* 5 (1970): 111–127. Francis Bacon, *On the Causes of the Tides*, as cited by Whewell, *History*, 273. Moesgaard, "Copernicanism . . . Denmark and Norway" in Dobrzycki, *Reception*, 137–138 (on C. N. Lesle).

032 *Johnson had concluded (contrary to Kuhn)* Johnson, *Astronomical Thought*, 291. Thomas Blundeville is a late sixteenth-century exception to Johnson's conclusion (Johnson, p. 207) and a good example of Kuhn's thesis: he praised the Copernican account of celestial motions but still insisted it was based on a false hypothesis. See also Russell, "Copernican System" in Dobrzycki, *Reception*, 196.

033 *which does largely agree with Kuhn's interpretation* Westman, "Melanchthon Circle." See also Owen Gingerich, "From Copernicus to Kepler: Heliocentrism as Model and as Reality," *Proceedings of the American Philosophical Society* 117 (1973): 513–522.

033 *Gellibrand, who wrote in 1635* Russell, "Copernican System in Great Britain," in Dobrzicki, *Reception*, 189–239, on p. 217.

033 *Some of the other historians content themselves* Stimson, *Gradual Acceptance*, 71–82. Johnson, *Astronomical Thought*, 110–119, 230. Whewell, *History*, 277 (only example given of the influence of Galileo's telescope discoveries is Bacon, who rejected heliocentrism); 288 (Riccioli "enumerated fifty-seven Copernican arguments" but Whewell mentions only two of them).

033 *importance of the textbook literature is noted* James M. Lattis, *Between Copernicus and Galileo: Christoph Clavius and the Collapse of Ptolemaic Cosmology* (Chicago: University of Chicago Press, 1994), p. xv.

033 *the Vatican had stopped actively enforcing* John L. Russell, "Catholic Astronomers and the Copernican System after the Condemnation of Galileo," *Annals of Science* 46 (1989): 365–386.

033 *a special commission . . . had found Galileo not guilty* The *Washington Post* began its report with the sentence: "The Earth revolves around the sun, even for the Vatican." See "Vatican Says Galileo Right After All," November 1, 1992, p. A40. See also Alan Cowell, "After 350 Years, Vatican Says Galileo Was Right: It Moves," *New York Times*, October 31, 1992, pp. 1, 4; Michael Segre, "Light on the Galileo Case?" *Isis* 88 (1997): 484–504, and review of Mario Artigas and Melchor Sánchez de Toca, *Galileo y el Vaticano*, in *Isis* 101 (2010): 249–250. Maurice Finocchiaro (ed.), *The Galileo Affair: A Documentary History* (Berkeley: University of California Press, 1989).

033 *not real but merely a useful hypothesis* Descartes, *Le Monde, ou Traité de la Lumière* (1664). According to Descartes, only motion relative to the surrounding matter is real, and all space is filled with matter, so one choose to consider the Earth as being at rest (relative to its own surrounding vortex) as well as the sun. So one can give Copernican explanations without assuming the Earth is really moving.

033 *Jean Dietz Moss . . . notes that the text* Jean Dietz Moss, *Novelties in the Heavens: Rhetoric and Science in the Copernican Controversy* (Chicago: University of Chicago Press, 1993), 53–54. Johnson (*Astronomical Thought*, 115) suggests that the Church was fooled by Osiander's preface and didn't realize Copernicus believed in the reality of the Earth's motion until Galileo's publications.

033 *"neither sufficiently absolute nor sufficiently pleasing to the mind" De hypothesibus motuum coelestium a se constitutis Commentariolus* (preprint, circa 1514, distributed to a few friends); translated with introduction and notes by E. Rosen in *Three Copernican Treatises* (New York: Dover, 1959). Contrary to the myth that Copernicus was offended by the excessive complexity of the Ptolemaic and proposed a drastic reduction in the number of epicycles with a great improvement in accuracy, Copernicus admitted that the geocentric system was "consistent with the numerical data" and had no objection to epicycles. Instead he objected to Ptolemy's use of the equant.

033 *"with uniform velocity neither on its deferent nor about the center" Commentariolus.* See Holton and Brush, *Physics*, p. 13.

034 *Moss also argues* Moss, *Novelties*, 53, citing Richard E. Young et al., *Rhetoric: Discovery and Change* (New York: Harcourt, Brace and World, 1970) for the introduction of Rogerian views into rhetoric.

034 *finds that birth order is an important factor* Frank J. Sulloway, *Born to Rebel: Birth Order, Family Dynamics, and Creative Lives* (New York: Pantheon Books, 1996), 39, 43, 185–187, 337–341; quotation from p. 339.

035 *Lakatos and Zahar asked* I. Lakatos and E. Zahar, "Why Did Copernicus' Research Program Supersede Ptolemy's?" in Westman, *Copernican Achievement*, 354–383, on 355.

035 *the Lakatos-Zahar assertion is based on a notorious historical myth* In view of the authors' obvious disinterest in historical accuracy, it is perhaps superfluous to point out that Copernicus did not predict the phases of Venus; see Neil Thomason, "1543—The Year that Copernicus Didn't Predict the Phases of Venus," in *1543 and All That: Image and Word, Change and Continuity in the Proto-Scientific Revolution*, edited by Guy Freeland and Anthony Corones (Dordrecht: Kluwer, 2000), 291–331. By Lakatos's original criterion, the Copernican theory (as distinct from its later modifications by Kepler and Newton) did not produce any confirmed novel predictions before the nineteenth century. Presumably Lakatos and Zahar still accept Popper's definition of prediction, according to which the prediction of another occurrence of a cyclic phenomenon does not count as evidence for a theory (see Section 1.2); this may explain why they don't give Copernicus any credit for making *better* predictions of planetary positions than Ptolemy could. Yet it is clear that sixteenth-century astronomers did see this as an advantage of the Copernican system. In addition to the evidence presented by Stimson, Johnson, Westman, and other historians, James Henderson reports a 1545 letter to Rheticus from Mathias Lauterwalt, who has just observed a conjunction of Mercury and Jupiter, and is pleased by the agreement with predictions based on the calculations of Copernicus, as compared with the geocentric *Alfonsine Tables* which predict it to occur a day too early. See "The Reception of Heliocentrism" in *The Nature of Scientific Discovery*, edited by Owen Gingerich, 393–429 (Washington, DC: Smithsonian Institution Press, 1975), on 432.

035 *Lakatos and Zahar claim that "several important facts"* Lakatos and Zahar, "Why Did," p. 376.

035 *remarks that it represents a significant step* away *from Popperian doctrine* Stephen Toulmin, "Commentary" in Westman, *Copernican Achievement*, 384–391.

035 *one* prediction *was not yet confirmed* N. S. Hetherington, "The First Measurements of Stellar Parallax," *Annals of Science* 28 (1972): 319–325; Dorrit Hoffleit, "The Search for Stellar Parallax," *Popular Astronomy* 57 (1949): 259–273; Otto Struve, "The First Stellar Parallax Determination," in *Men and Moments in the History of Science*, edited by H. M. Evans (Seattle: University of Washington Press, 1959), pp. 177–206.

Section 2.3

035 *proposed that a cannonball dropped from a tower* For Galileo's argument (using a thought experiment) see Holton and S. G. Brush, *Physics*, pp. 54–57.

036 *he accepted the axiom of René Descartes* "It is inconceivable, that inanimate brute matter should, without the mediation of something else, which is not material, operate upon, and affect other matter without mutual contact; as it must do, if gravitation, in the sense of Epicurus, be essential and inherent in it. And this is one reason, why I desired you would not ascribe innate gravity to me. That gravity should be innate, inherent, and essential to matter, so that one body may act upon another as a distance through a *vacuum*, without the mediation of anything else, by and through which their action and force may be conveyed from one to another, is to me so great an absurdity, that I believe no man who has in philosophical [i.e., scientific] matters a competent faculty of thinking, can ever fall into. Gravity must be caused by an agent acting constantly according to certain laws; but whether this agent be

material or immaterial, I have left to the consideration of my readers." Letter to Richard Bentley, February 25, 1693. For further discussion of Newton's views see Alexandre Koyré, *Newtonian Studies* (Cambridge, MA: Harvard University Press, 1965; reprint, University of Chicago Press, 1968), Appendix C. Andrew Janiak, *Newton as a Philosopher* (New York: Cambridge University Press, pp. 272–273.

036 *His famous motto "hypotheses non fingo"* Isaac Newton, *The Principia: Mathematical Principles of Natural Philosophy*, translated by I. Bernard Cohen and Anne Whitman (Berkeley: University of California Press, 1999), p. 943; see also the translators' comments on 274–77. On the problem of translating "fingo" see Koyré, *Newtonian Studies*, p. 35.

036 *Historians continue to debate what Newton really meant* Henry, "Gravity and *De Gravitatione*: The Development of Newton's Ideas on Action at a Distance," *Studies in History and Philosophy of Science* 42 (2011): 11–27, on p. 13, Warning: if you read this article, you can hardly avoid looking at three more articles on the subject in the same issue of this journal, by Steffen Ducheyne (pp. 154–159), Eric Schliessen (pp. 160–166), and Hylarie Kochiras (pp. 167–184).

036 *proposed incompatible hypotheses for the nature of light* René Descartes, *Discourse on Method, Optics, Geometry, and Meteorology* [1637] (Indianapolis: Bobbs-Merrill, 1965).

036 *Newton . . . never admitted attraction as a "physical" force* Koyré, *Newtonian Studies*, 16. Newton, *Principia*, Book I, Definition VIII. See also Gerd Buchdahl, "Gravity and Intelligibility: Newton to Kant," in *The Methodological Heritage of Newton*, edited by R. E. Butts and J. W. Davis (Toronto: University of Toronto Press, 1970), pp. 74–102. Buchdahl argues that Newton intends "only to give a mathematical notion of those forces, without considering their physical causes and seats; considering those forces not physically, but mathematically" (pp. 76–77, citing *Principia*, pp. 5–6). Leibniz wrote that matter cannot "naturally possess . . . attraction because it is impossible to *conceive* how this takes place, I. E., to explain it mechanically" (Buchdahl, 81, citing *New Essays Concerning Human Understanding*, translated by A. G. Langley, 1949, p. 61). What then happened, according to Buchdahl, is: although "gravity, regarded as 'action at a distance,' seemed intelligible . . . However, as time goes on, and Newtonianism gains in prestige, the temptation to meet the charge of unintelligibility via explanation tends to seem less and less attractive, though the charge itself is not abandoned. For this had to await the reconstruction of the concept without recourse to 'explanation'; an approach of which Kant's method . . . is a supreme example" (p. 84). The only other writers cited by Buchdahl are Clarke, Locke, Cotes, and Hume, indicating that Buchdahl is really discussing primarily the philosophical rather than the scientific reception of gravity.

036 *they proposed to explain gravity as the result of impacts* Huygens himself had done this in 1669, showing that bits of wax in a rotating cylinder filled with water would experience a force toward the center of the base of the cylinder when the rotation is stopped. Nollet described a more elaborate demonstration of this effect but had to admit that Huygens's Cartesian impact theory predicted a force toward the Earth's axis, not its center.

On the wax in water experiments see Geoffrey V. Sutton, *Science for a Polite Society: Gender, Culture, and the Demonstration of Enlightenment* (Boulder, CO: Westview (HarperCollins), 1995), 132, 234. For the objections to Newtonian attraction see: Henri Guerlac, *Newton on the Continent* (Ithaca, NY: Cornell University Press, 1981); E. J. Aiton, "The Vortex Theory in Competition with Newtonian Celestial Mechanics," in *The General History of Astronomy*, Volume 2, *Planetary*

Astronomy from the Renaissance to the Rise of Astrophysics, Part B, *The Eighteenth and Nineteenth Centuries*, edited by R. Taton and C. Wilson (New York: Cambridge University Press, 1995), pp. 3–21. See also Koyré, *Newtonian Studies*, Appendices A and B; Koffi Magio, "The Reception of Newton's Gravitational Theory by Huygens, Varignon, and Maupertuis: How Normal Science may be Revolutionary," *Perspectives on Science* 11 (2003): 135–169; Roberto de A. Martins, "Huygens's Reaction to Newton's gravitational theory," in *Renaissance and Revolution*, edited by J. F. Field and F. A. J. L. James (Cambridge: Cambridge University Press, 1993), pp. 203–213.

036 *"reduced the questions of physics to the art of solving equations"* Magio, "Reception," 154; for details see Aiton, in Taton and Wilson, *Planetary Astronomy*, 13–15, and *The Vortex Theory of Planetary Motions* (London: MacDonald, 1972), 196–200.

037 [The role of] Emilie du Châtelet Sutton, *Polite Society*, pp. 261–263. She recognized that Newton's mv is not incompatible with Leibnitz's mv^2.

037 *the first Frenchman . . . to declare himself a Newtonian* Aiton, *Vortex Theory*, 201; see also the works of P. Brunet he cited on p. 207. An updated summary of Aiton's book is in Taton and Wilson, *Planetary Astronomy*, pp. 3–21. See also A. R. Hall, *Isaac Newton: Eighteenth-Century Perspectives* (Oxford: Oxford University Press, 1999).

037 *"the Newtonian system . . . explained all the phenomena"* Aiton, *Vortex Theory*, 202, citing Maupertuis, *Discours sur les différentes Figures des Astres* (Paris, 1732), p. 11. Johann Bernoulli disliked attraction but admitted that the Cartesian system could not satisfactorily reconcile Kepler's Second and Third Laws (Aiton, 229).

037 *failure of the Cartesian theory to explain phenomena* Newton devoted Book II of the *Principia* to fluid mechanics, partly in order to show that the Cartesian vortex theory could not explain Kepler's laws. His somewhat primitive theory was corrected and extended (to solids as well as fluids) by Leonard Euler, the Bernoulli family of mathematicians, d'Alembert, Clairaut, Maupertuis and others. Despite the fact that Euler and Johann Bernoulli rejected Newtonian attraction, their contributions were merged into a science of continuum mechanics that was subsequently viewed as a mere application of Newton's laws. For a more accurate assessment of the historical importance of these contributions see C. Truesdell, "Reactions of Late Baroque Mechanics to Success, Conjecture, Error, and Failure in Newton's *Principia*," in his *Essays in the History of Mechanics* (New York: Springer-Verlag, 1968), pp. 138–183; "Rational Fluid Mechanics, 1687–1765," Editor's Introduction to *Leonhardi Euleri Opera Omnia*, series 2, vol. 12 (Zürich: Orell Füssli, 1954), pp. ix–cxxv; *The Rational Mechanics of Flexible or Elastic Bodies, 1638–1788*, editor's introduction to *Leonhardi Euleri Opera Omnia.*, series 2, vol. 10 and 11, in ibid., vol. 11, part 2 (Zürich: Orell Füssli, 1960). On Euler's continued opposition to Newtonian attraction see Valentin Boss, *Newton and Russia: The Early Influence, 1698–1796* (Cambridge, MA.: Harvard University Press, 1972).

037 *perhaps an instance of Planck's principle* See Section 2.4.

037 *"abandon the vortices with a clear conscience"* Aiton, in Taton and Wilson, *Planetary Astronomy*, pp. 19, 21.

037 *turned the Cartesian objection upside down* M. A. Finocchiaro, "Gravity and Intelligibility in Boscovich's Natural Philosophy," in *R. J. Boscovich*, edited by P. Bursill-Hall (Roma: Inst. Enc. Ital., 1993), pp. 149–167. In Boscovich's theory, Newtonian gravitational attraction is dominant only over a range of intermediate distances. At short distances it oscillates between attraction and repulsion several times; at very small and very large distances, it must be repulsive to avoid the collapse of point atoms that would otherwise result (Finocchiaro, p. 156).

037 *The mathematics of Newton and Leibniz* For further discussion of this point see Yves Gingras, "What Did Mathematics Do to Physics?" *History of Science* 39 (2001): 338–416.

037 *Geoffrey Sutton has traced the sequence* Sutton, *Polite Society.*

037 *The essays in Volume 2B of* The General History of Astronomy Taton and Wilson, *Planetary Astronomy.*

037 *"the adoption of the Newtonian system was well begun"* Sutton, *Polite Society.* 258; Seymour C. Chapin, "The Shape of the Earth," in Taton and Wilson, *Planetary Astronomy*, 22–34. For further references on this subject see Stephen G. Brush and Helmut Landsberg, *The History of Geophysics and Meteorology: An Annotated Bibliography* (New York: Garland Publishing, 1985), items cited under "Figure of earth" in Index.

037 *could not really be considered a crucial experiment* Harcourt Brown, "From London to Lapland: Maupertuis, Johann Bernoulli I, and La Terre Applatie, 1728–1738," in *Literature and History in the Age of Ideas*, edited by C. G. S. Williams (Columbus: Ohio State University Press, 1975), pp 69–96; R. Taton, articles on Cassini II and III in *DSB* 3 (1971): 104–106, 107–109. Volker Bialas, *Der Streit um die Figur der Erde: Zur Begründung der Geodasie im 17. Und 18. Jahrhundert* (Munich: verlag der Bayerischen Akademie der Wissenschaften, 1972).

037 *"could only have won over those already firmly leaning"* Sutton, *Polite Society*, 260.

038 *implies that this did not help to persuade scientists* Sutton, *Polite Society*, 86, 260, 270. Daniel Bernoulli [1740]; Descartes [1644]; Chatelet, *Principes Mathematiques de la Philosophie Naturelle* (Paris: Desaint & Saillant and Lambert, 1756). On the history of tidal theory see references in Brush and Landsberg, *History of Geophysics* (note 58), 274–277. Aiton offers evidence that Bouguer gave the explanation of tides as one reason to accept the Newtonian theory (*Vortex Theory*, 225).

038 *Galileo's claim that the tidal phenomena prove the Earth's motion* The major exception is E. J. Aiton, "The Contributions of Newton, Bernoulli and Euler to the Theory of the Tides," *Annals of Science* 11 (1955): 206–223. See also Joseph Proudman, "Newton's Work on the Theory of Tides," in *Isaac Newton, 1642–1727*, ed. W. J. Greenstreet, 87–95 (London: Bell, 1927); Hermann Thorade, "Ebbe und Flut in der Nordsee: ein geschichtlicher Rückblick," *Petermanns Mitteilungen Ergänzungsheft* 209 (1930): 195–206; *Probleme der Wasserwellen* (Hamburg: Grand, 1931).

038 *his theory explains many phenomena* Aiton, *Vortex Theory*, 225–226. A similar view is expressed by James Evans, "Fraud and Illusion in the Anti-Newtonian Rear Guard," *Isis* 87 (1996): 74–107.

038 *"came to accept the correctness of Newton's law"* Sutton, *Polite Society*, 269. See Craig Waff, "Clairaut and the Motion of the Lunar Apse," in Taton and Wilson, *Planetary Astronomy*, 35–46; *Universal Gravitation and the Motion of the Moon's Apogee: The Establishment and Reception of Newton's Inverse-Square Law, 1687–1749*, PhD dissertation, Johns Hopkins University, 1975. The latter work, while providing a very thorough technical account of Clairaut's work, offers no direct evidence for the conclusion that this work "played an important role in the establishment (in the eyes of the public) of the inverse-Square formulation of the gravitational law" (p. 222).

038 Clairaut's achievement also had important practical consequences: One of the two available methods for determining the longitude of a ship at sea depended on using lunar tables (the other relied on the development of a clock that could keep accurate time in the turbulent environment of a ship in the middle of an ocean). Tobias Mayer, using Euler's approximation method for the three-body problem,

produced tables that were printed in an almanac that could be carried on board ship. As a result he won (and his widow actually received) £3,000, part of the prize established by the British Parliament to encourage the development of a practical method. (John Harrison, inventor of the marine chronometer, got a larger share of the purse.) The story of the quest for the longitude was so dramatic that it formed the basis for a best-selling book two and a half centuries later (Dava Sobel, *Longitude* (New York: Walker, 1995). Astronomers are still arguing about whether Mayer or Euler deserved a bigger share: Haywood Smith Jr., "Mayer Earned the Lunar-Table Prize," *Physics Today* 63, no. 7 (July 2010): 9.

039 *the "making of the comet"* Schaffer, "Halley, Delisle, and the Making of the Comet," in *Standing on the Shoulders of Giants*, edited by Norman J. W. Thrower (Berkeley: University of California Press, 1990), pp. 254–298 on 287.

039 *"it was important that the result was given before the arrival of the comet"* Lalande, *Bibliographie Astronomique* (1803), quoted by Craig B. Waff, "The First International Halley Watch: Guiding the Worldwide Search for Comet Halley, 1755–1759," in Thrower, *Standing*, 373–411, on 395.

039 *used this as his only historical example* Peter Lipton, "Testing Hypotheses: Prediction and Prejudice," *Science* 308 (2005): 219–221; K. Stanger-Hall, D. Allchin, A. Aviv, S. G. Brush, J. Aach and G, M. Church, and P. Lipton, "Accommodation or Prediction," *Science* 308 (2005): 1409–1412.

040 *readers who were familiar with the other successful applications* Sutton, *Polite Society*, 273. See C. B. Waff, "Predicting the Mid-Eighteenth-Century Return of Halley's Comet," in Taton and Wilson, *Planetary Astronomy*, 69–86.

040 *"by that time there were few Cartesians left"* Boss, *Newton & Russia*, 194.

040 *the so-called* vis viva *("living force") controversy* See Thomas L. Hankins, "Eighteenth-Century Attempts to Resolve the *Vis-Viva* Controversy," *Isis* 56 (1965): 281–297; L. L. Laudan, "The *Vis Viva* Controversy: A Post-Mortem," *Isis* 59 (1968): 131–143; David Papineau, "The *Vis Viva* Controversy: Do Meanings Matter?" *Studies in History and Philosophy of Science* 8 (1977): 111–142; Keiko Kawashima, "La participation de Madame du Châtelet à la querelle sur les forces vives, *Historia Scientiarum* 40 (1990): 9–28; Brian Hepburn, "Euler, *Vis Viva*, and Equilibrium," *Studies in History and Philosophy of Science* 41 (2010): 120–127.

040 *"was one that placed the theoretical, the philosophical, and the mathematical before the empirical"* Sutton, *Polite Society*, pp. 271–272.

040 *those who can't bear to "attribute originality in science to women"* Sutton, *Polite Society*, p. 263.

040 *all deviations from Kepler's laws* Curtis Wilson, "The Problem of Perturbation Analytically Treated: Euler, Clairaut, d'Alembert" and "The Work of Lagrange in Celestial Mechanics," in Taton and Wilson, *Planetary Astronomy*, 89–107, 108–130; "Perturbations and Solar Tables from Lacaille to Delambre: The Rapprochement of Observation and Theory," *Archive for History of Exact Sciences* 22 (1980): 53–204; "The Great Inequality of Jupiter and Saturn: From Kepler to Laplace," *Archive for History of Exact Sciences* 33 (1985): 15–290. Bruno Morando, "Laplace," in Taton and Wilson, *Planetary Astronomy*, 131–150. On the secular acceleration of the moon, a problem that appeared to have been solved by Laplace at the beginning of the nineteenth century but was later found to be a deceleration whose analysis played an important role in theories of the origin of the moon, see S. G. Brush, *Fruitful Encounters: The Origin of the Solar System and of the Moon from Chamberlin to* Apollo (New York: Cambridge University Press, 1996), Chapter 4.2.

041 *ask whether they would have abandoned that theory* Greg Banford, "Popper and his Commentators on the Discovery of Neptune: A close Shave for the Laws of Gravitation," *Studies in History and Philosophy of Science* 27 (1996): 207–232. Jarrett Leplin, "The Historical Objection to Scientific Realism," *PSA 1982* (Proceedings of the Biennial Meeting of the Philosophy of Science Association) 1: 88–97.

SECTION 2.4

041 *a schematic theory of cosmic evolution* S. G. Brush, *Nebulous Earth: The Origin of the Solar System and the Core of the Earth from Laplace to Jeffreys* (Cambridge: Cambridge University Press, 1996), Part 1. Note that the original meaning of the word "evolution" is "unfolding," i.e., of a predetermined history.

041 *the dictionary definition of Darwinism* J. A. Simpson and E. S. C. Weiner, *Oxford English Dictionary*, 2d edition, Volume IV (Clarendon Press, Oxford, 1989), p. 257.

041 *devoted an entire 114-page issue of a major journal Studies in History and Philosophy of Biology and Biomedical Sciences*, Special Issue, *Defining Darwinism: One Hundred and Fifty Years of Debate, In Memoriam of David Hull (1935–2010)*, Guest Editor Richard G. Delisle, vol. 42, no. 1 (March 2011): 1–114.

042 *scientists themselves do not agree on what Darwinism means* David L. Hull, "Darwinism and Historiography," in *The Comparative Reception of Darwinism*, edited by Thomas F. Glick (Chicago: University of Chicago Press, 1974, 1988), pp. 388–402, on p. 398. See also Hull, "Darwinism as a Historical Entity: A Historiographic Proposal," in *The Darwinian Heritage*, edited by David Kohn (Princeton, NJ: Princeton University Press, 1985), pp. 773–812, p. 809. For an example of the confusion see G. E. Allen, "Thomas Hunt Morgan and the Problem of Natural Selection," *Journal of the History of Biology*, 1 (1968): 113–139, on p. 113 and note 1.

042 *who flatly asserts that the word is "a term commonly used"* Ronald L. Numbers, *Darwinism Comes to America* (Cambridge, MA: Harvard University Press, 1998), 1. According to Peter J. Bowler, "Darwinism" means "little more than the basic idea of evolution"; see *Evolution: The History of an Idea*, revised edition (Berkeley: University of California Press, 1989), p. 23. See also Marsha L. Richmond, "The 1909 Darwin Celebration: Reexamining Evolution in the Light of Mendel, Mutation, and Meiosis," *Isis* 97 (2006): 447–484, on p. 462.

042 *"most American naturalists who expressed themselves on the subject"* Numbers, *Darwinism*, pp. 1–2, 47–48. His sample contained 80 American naturalists elected to the National Academy of Sciences between 1863 and 1900: 34 geologists and paleontologists, 26 zoologists, 9 botanists, 8 specialists in anatomy, physiology or medicine, and 3 anthropologists. In his earlier book, *Creation by Natural Law* (Seattle: University of Washington Press, 1977), Numbers suggested that the widespread acceptance of the nebular hypothesis in the early nineteenth century broke down resistance to naturalistic explanations of the past and thereby prepared the way for the acceptance of organic evolution.

042 *connection between Darwinism and a belief in progress* M. Ruse, *Monad to Man* (Cambridge, MA: Harvard University Press, 1996).

042 rhetoric *in science, a topic that has recently attracted a number of scholars* John Angus Campbell, "Scientific Revolution and the Grammar of Culture: The Case of Darwin's Origin," *Quarterly Journal of Speech* 72 (1986): 351–376; Campbell, "Scientific Discovery and Rhetorical Invention: The Path to Darwin's *Origin*," in H. W. Simons (ed.): *The Rhetorical Turn* (Chicago: University of Chicago Press, 1990), pp. 58–90; Doreen A. Recker, "Causal Efficiency: The Structure of Darwin's

Argument Strategy in the *Origin of Species*," *Philosophy of Science* 54 (1987): 147–175; Linda S. Bergmann, "Reshaping the Roles of Man, God, and Nature: Darwin's Rhetoric in On the *Origin of Species*," in J. W. Slade and J. Y. Yee (eds.), *Beyond the Two Cultures* (Ames: Iowa State University Press, 1990), pp. 79–98; Matti Sintonen, "Darwin's Long and Short Arguments," *Philosophy of Science* 57 (1990): 677–689; Michael Spindler, "The *Origin of Species* as Rhetoric," *Nineteenth Century Prose* 91, no. 2 (1991/1992): 26–34; Martin J. S. Hodge, "Discussion: Darwin's Argument in the Origin," *Philosophy of Science*, 59 (1992): 461–464; Jeanne Fahnestock, "Series Reasoning in Scientific Argument: *Incrementum* and *Gradatio* and the Case of Darwin," *Rhetoric Society Quarterly* 26, no. 4 (1996): 13–40; Fahnestock, *Rhetorical Figures in Science* (New York: Oxford University Press, 1999); Keith Thomson, "Darwin's Literary Models," *American Scientist 98* (2010): 196–199.

043 *"If ever fossil birds are found"* Darwin to Lyell, October 14, 1859, as quoted by Gene Kritsky, "Darwin's Archaeopteryx Prophecy," *Archives of Natural History* 19 (1991): 407–410, on p. 409.

043 *"Darwin downplayed the importance of Archaeopteryx as a transitional fossil"* Kritsky, "Darwin's Archaeopteryx Prophecy," on p. 409.

043 *Marsh discovered fossil remains of toothed birds* Numbers, *Darwinism*, p. 44.

043 *"the best support to the theory of evolution"* Darwin to Marsh, August 31, 1880, quoted by Numbers, *Darwinism*, p. 44. T. H. Huxley, already an advocate of evolution, told his New York audience in 1876 that Marsh's discovery established the truth of evolution. See, e.g., Edward J. Pfeifer, "United States," in Glick, *Comparative Reception*, pp. 168–206, on p. 197. Peter J. Bowler also mentions this successful prediction, along with the discovery of *Archaeopteryx*, as a reason why evolution was accepted; see "Scientific Attitudes to Darwinism in Britain and America," in Kohn, *Darwinian Heritage*, pp. 641–681. But the story and its implications for the validity of Darwin's theory is more complicated; see Bowler, *The Non-Darwinian Revolution: Reinterpreting a Historical Myth* (Baltimore: Johns Hopkins University Press, 1988), 78–80.

043 *evidence would show that humans originally evolved in Africa* Dart, "*Australopithecus Africanus*: The Man-Ape of South Africa," *Nature* 115 (1925): 195–199.

043 *Sulloway's analysis of the role of birth order* Sulloway. *Born to Rebel: Birth Order, Family Dynamics, and Creative Lives* (New York: Pantheon, 1996), pp. 39, 43, 185–187.

043 *three more reasons to support evolution* Numbers, *Darwinism*, pp. 44–45.

043 *study . . . yielded two other reasons for a favorable view* Pfeifer, "United States," in Gleick, *Comparative Reception*, pp. 168–206.

043 *Lamarckian explanations of evolution again came to the fore* Bowler, *Anti-Darwinian Revolution: Reinterpreting a Historical Myth* (Baltimore: Johns Hopkins University Press, 1988); *The Eclipse of Darwinism: Anti-Darwinian Evolution Theories in the Decades Around 1900* (Baltimore: Johns Hopkins University Press, 1983); Numbers, *Darwinism*, p. 43; Pfeifer, "United States," pp. 198–202. Russia, where natural selection was regarded much more favorably, seems to be an exception to this statement; see Francesco M. Scudo and Michele Acanfora, "Darwin and Russian Evolutionary Biology," in Kohn, *Darwinian Heritage*, pp. 731–752. William B. Provine discusses British supporters of natural selection in "Adaptation and Mechanisms of Evolution after Darwin: A Study in Persistent Controversies," in Kohn, *Darwinian Heritage*, 825–866.

043 *"those who . . . concerned themselves with geographical . . . questions"* M. J. S. Hodge, "England," in Glick, *Comparative Reception*, pp. 3–31; see also "England: Bibliographical Essay," ibid. 75–80.

044 *confirmed that geographical variation provided important support* Fredrick Burkhardt, "England and Scotland: The Learned Societies," in Glick, *Comparative Reception*, pp. 32–74; Bowler, "Scientific Attitudes to Darwinism in Britain and America," in Kohn, *Darwinian Heritage*, pp. 641–681.

044 *"belief in the descent theory is slowly spreading"* Quoted in Glick, *Comparative Reception*, p. 60.

044 *raised the standard for reception studies to a new level* D. L. Hull, P. D. Tessner and A. M. Diamond, "Planck's Principle," *Science* 202 (1978): 717–723.

044 *a general statement by the physicist Max Planck* Max Planck, *Scientific Autobiography and other Papers* (London: Williams & Norgate, 1950), pp. 3–4.

045 *Their work has stimulated many other studies* Hull, *Science as a Process* (Chicago: University of Chicago Press, 1988), 302–303, 379–382.

045 *reception should be studied mainly in the period 1859–1885* Glick, preface to *Comparative Reception*, p. viii.

045 *a strong proevolution tradition before 1859* Alexander Vucinich, "Russia.. Biological Sciences," in Glick, *Comparative Reception*, pp. 227–268. For further discussion see Scudo and Acanfora, "Darwin and Russian Evolutionary Biology," in Kohn, *Darwinian Heritage; Daniel Todes, Darwin without Malthus: The Struggle for Existence in Russian Evolutionary Thought* (New York: Oxford University Press, 1989).

045 *a bias* against *what was called "transformisme"* Robert E. Stebbins, "France" and "France: Bibliographical Essay" in Glick, *Comparative Reception*, pp. 117–163, 164–167. The case of Gaston de Saporta is one exception to this statement; see P. Corsi, "Recent Studies on French reactions to Darwin," in Kohn, *Darwinian Heritage*, pp. 698–711; Cédric Grimoult, *L'Evolution biologique en France: Une Révolution scientifique, politique et culturelle* (Genève/Paris: Droz, 2001).

046 *evolution (like other radical innovations) was supported mostly by later-borns* Sulloway, *Born to Rebel*, pp. 33–36, quotation from p. 36.

046 *In Germany, the pro-Darwinians were generally younger* William M. Montgomery, "Germany," in Glick, *Comparative Reception*, pp. 91–116., quotations from pp. 97, 104. Other studies on Germany tend to concentrate on Haeckel, e.g., P. J. Weindling, "Darwinism in Germany," in Kohn, *Darwinian Heritage*, pp. 685–698.

046 *In the Netherlands, historian Ilse Bulhof found* Ilse Bulhof, "The Netherlands," in *Comparative Reception of Darwinism*, ed. Glick, 269–306, quotation from p. 280.

046 *Giuliano Pancaldi deals primarily with one zoologist* Guiliano Pancaldi, *Darwin in Italy: Science across Cultural Frontiers* (Bloomington: Indiana University Press, 1991), pp. 85–104. See also P. Corsi, "Recent Studies on Italian Reactions to Darwin," in Kohn, *Darwinian Heritage*, pp. 711–729.

046 *Evolutionism simply went along with liberal religious and political attitudes* Thomas F. Glick, "Spain," in Glick, *Comparative Reception*, pp. 307–345, quotation from p. 343.

047 *it is hard to find good evidence of scientific reasons* Articles by Roberto Moreno [1974/1988] on Mexico and by Nim A. Bezirgan [1974/1988] on the Islamic World, in Glick, *Comparative Reception*.

In a book of more than 200 pages entitled *Darwin in America: The Intellectual Response, 1865–1912* (San Francisco: Freeman, 1976), Cynthia Russett discussed in detail the response of only one scientist, C. S. Peirce, who liked natural selection (because of its statistical aspect) but thought it insufficient to explain evolution so he favored Lamarckism and large mutations.

In a 50-page section on "The Reception of the *Origin*" in her book *Darwin and the Darwinian Revolution* (London: Chatto & Windus, 1959), Gertrude Himmelfarb concentrated on those scientists who opposed Darwin's theory or should have done so because (she believes) it was fallacious. M.J.S. Hodge comments that "the radical defects of Himmelfarb's understanding of Darwinian science are too manifest to need emphasis"; she attributed Darwin's success mainly to "the enormous solidarity, camaraderie, and prestige enjoyed by the triumvirate of Lyell, Hooker, and Huxley" (Hodge in Glick, *Comparative Reception*, "Bibliographical Essay," p. 77) rather than to the merits of his theory.

047 *the* eventual *acceptance of Darwinian theory was determined by evidence and rational argument* Hull, "Darwinism and Historiography," in Glick, *Comparative Reception*, 388–402, quotation from pp. 391–392. I did not mention before now Hull's book *Darwinism and his Critics: The Reception of Darwin's Theory of Evolution by the Scientific Community* (Cambridge, MA: Harvard University Press, 1973) because it is not primarily a reception study but an extremely valuable compilation of primary sources, with useful commentaries.

047 *"The wingless condition . . . is mainly due to the action of natural selection"* Darwin, *On the Origin of Species by Means of Natural Selection: or, The Preservation of Favored Races in the Struggle for Life* (London: Murray, 1859), Chapter 5 on "effects of the increased use and disuse of parts."

047 *Darwinism . . . is socially constructed* Thomas F. Glick, "The Comparative Reception of Darwinism: A Brief History," *Science & Education* 19 (2010): 693–703, on p. 701.

047 *A more recent collection of reception studies for European countries* Eve-Marie Engels and Thomas F. Glick, editors, *The Reception of Charles Darwin in Europe*. (New York: Continuum, 2008).

047 *That chapter, by historian and biologist Victoria Tatole* Tatole, "Notes on the Reception of Darwin's Theory in Romania," Chapter 25 in Engels and Glick, *Reception*, pp. 463–479, 587–590.

047 *Grigore Stefanescu stated that 30,000 species of fossils are extinct* Stefanescu, "Locul omului în Natură," *Revista științifică* [Bucharest] (1876): 1–25, on p. 10, as cited by Tatole, "Notes," p. 469.

047 *Nicolae Leon invoked the biogenetic law* Leon, ""Generaționea spontanee și darinismul," *Convorbiri literare* [Bucharest] 37, no. 4 (1903): 343–364, p. 360, as cited by Tatole, "Notes," p. 471.

047 *can only be explained by evolution* Leon, "Organele rudimentare și dl. Prof. Paulesco," *Convorbiri literare* [Bucharest] 43, no. 3 (1909): 434–446, on p. 443, as cited by Tatole, "Notes," 471.

047 *the existence of eyes in the larvae of the abyss of animals* Voinov, "Noi cuceriri transformiste," *Literatură și știință* [Bucharest] 1 (1893): 99–109, on p. 109, as cited by Tatole, "Notes," p. 472.

047 *"the rudimentary organs are the strongest proofs"* Voinov, "Dovezile: Discuții cu prof. Paulescu," *Convorbiri literare* [Bucharest] 1907: 41, no. 8: 779–799, on p. 790, as cited by Tatole, "Notes," p. 473.

048 *an article by historian Richard Bellon* Bellon, "Inspiration in the Harness of Daily Labor: Darwin, Botany, and the Triumph of Evolution, 1859–1868," *Isis* 102 (2011): 393–420.

048 *"Darwinism received its triumphant settlement"* Bellon, "Inspiration," pp. 416–417.

048 *Joseph Hooker summarized Darwin's "very long and laborious"research* Hooker, "Presidential Address," *Report of the Thirty-eighth Meeting of the British Association for the Advancement of Science, Held at Norwich in August, 1868* (London: John Murray, 1869), pp. lviii–lxxv, on pp. lxvi–lxviii, as quoted by Bellon, "Inspiration," p. 417, and e-mail from Bellon to Stephen G. Brush, October 16, 2012.

048 *According to biologist Duncan M. Porter, Darwin had previously discussed* Darwin, "On the Existence of Two Forms, and on Their Reciprocal Relation, in Several Species of the Genus Linum," *Journal of the Proceedings of the Linnean Society (Botany)* 7 (1863): 69–83. Porter writes: "Darwin records finding a trimorphic species of Linum and states that 'we may safely predict the function of the three kinds of flowers.' He does so, then shows that his predictions are correct in his 1864 loosestrife paper. I'm sure that this is what Hooker was referring to." E-mail from Porter to SGB, December 12, 2011.

048 *Darwin discussed the remarkable fact discovered by naturalist Thomas Vernon Wollaston* See notes for Section 14.7.

048 *He identified it as belonging to a six-year-old child* "*Australopithecus africanus*: The Man-Ape of South Africa," *Nature* 115 (1925): 195–199.

048 *In his 1862 book on the fertilization of orchids* Charles Darwin, *On the Various Contrivances By Which British and Foreign Orchids Are Fertilised by Insects, and on the Good Effects of Intercrossing* (London: John Murray, 1862), pp. 197–198. In a letter to J. D. Hooker, January 30, 1862, Darwin wrote: "[James] Bateman has just sent me a lot of orchids with the Angraecum sesquipedale: do you know its marvellous nectary 11½ inches long, with nectar only at the extremity. What a proboscis the moth that sucks it, must have!" Darwin Correspondence Project, Letter 3421; www.darwinproject.ac.uk/entry-3421. According an essay "Was Darwin an Ecologist" by the editors of the Darwin Project, "the Impetus for *Orchids* was Darwin's perception that many of the strange morphological features of orchid flowers 'made sense' if looked at as part of a system to ensure intercrossing. . . . he argued that the insects that carried the pollen could, to some extent, be predicted by the form of the flower. . . . Darwin is interested not just in an organism's adaptation to a static or regularly changing environment but in the ever-changing coadaptation of different organisms to one another." See http://darwinproject.ac.uk/was-darwin-an-ecologist.

See also A. R. Wallace, "Creation by Law," *Quarterly Journal of Science* 4 (1867): 470–488.

A good introduction to this subject is the article by Gene Kritsky, "Darwin's Madagascar Hawk Moth Prediction," *American Entomologist* 37 (1991): 206–210. I thank Ariel Segal for the reference to this article and for locating the original sources.

049 *The moth postulated by Darwin was discovered in 1903* Rothschild and Jordan, "A Revision of the Lepidopterous Family Sphingidae," *Novitates Zoologicae* 9 (Suppl.) (1903): 1–972, on p. 32. I thank Ariel Segal for this reference.

049 *L. T. Wasserthal proposed an alternative explanation* Wasserthal, "The Pollinators of the Malagasy Star Orchids *Angraecum esquipedale, A. sororium*, and *A. Compactum* and the Evolution of Extremely long Spurs by Pollinator Shift," *Botanica Acta* 110 (1997): 343–359.

049 *"To explore Darwin's prediction is indeed a race against time"* Nilsson, "Deep Flowers for Long Tongues," *Trends in Ecology & Evolution* 13 (1998): 259–260, on p. 260. See also articles published under this title by Wasserthal, pp. 459–460; Mats G. E. Stevenson, Jens Rydell, and Jan Tove, p. 460; Michael J. Samways, p. 460; T. Jermy, *Trends in Ecology & Evolution* 14 (1999): 34.

Section 2.5

050 *Bohr did not assume that light itself is composed of particles* John Stachel, "Bohr and the Photon," in *Quantum Reality, Relativistic Causality, and Closing the Epistemic Circle*, edited by Wayne C. Myrvold and Joy Christian (n.p.: Springer, 2009), pp 69–83. Bohr did not use Einstein's light quantum hypothesis, but instead relied on Planck's second theory of radiation. Planck was not sure whether the quantum discontinuity occurred at the instant of absorption or emission, but in either case he rejected (until the 1920s) the idea that light in space is composed of particles (see Chapter 7).

050 *[A] rather strange procedure* The following is a summary of Bohr's argument in his paper "On the Constitution of Atoms and Molecules," *Philosophical Magazine* [series 6], 26 (1913): 1–25, 476–502, 857–875. The net force acting at any instant on the revolving electron, a centripetal Coulomb's law force of electric attraction directed to the central nucleus, is simply given (numerically) by

$$F_c = m_e v^2/r = k_e e \times e/r^2 \qquad (1)$$

where m_e = mass of electron, e = numerical charge on the electron and on the hydrogen nucleus, and r = distance between electron and nucleus.

But the speed of the electron is $v = 2Bfr$ for circular motion with frequency f. Substituting and rearranging terms in Eq. (1) yields

$$1/f^2 \equiv T^2 = (4\pi^2 m_e/k_e e^2) r^3. \qquad (2)$$

In short, $T^2 \propto r^3$, just Kepler's third law.

Bohr assumed that while classical mechanics and electricity could be used to discuss the motion of the electron in a stable and nonradiating orbit, "the passing of the systems between different stationary states cannot be treated on that basis."

Bohr suspected that there was probably some definite relation between the single frequency ν of the emitted light and the two different frequencies of the initial and final orbits, and that this relation involved the energies *E'* and *E"* of these orbits. He saw that what was needed was, first of all, a formula expressing the frequency of rotation in an orbit, *f*, in terms of the energy of that orbit, *E*. Such a formula could easily be obtained from classical theory: The energy of the orbiting electron is simply the sum of its kinetic energy and its potential energy:

$$E = \tfrac{1}{2} m_e v^2 - k_e e^2/r \qquad (3)$$

Since, according to Eq. (1),

$$\tfrac{1}{2} m_e v^2 = \tfrac{1}{2} k_e e^2/r$$

we can write the equation for the energy of the electron in its orbit of radius *r* as

$$E = -\tfrac{1}{2} k_e e^2/r$$

Rearranging, we obtain

$$r = -\tfrac{1}{2} k_e e^2/E$$

On substituting this formula for r into Eq. (2) and rearranging, we find the desired relation between orbital frequency and energy:

$$f = \sqrt{2}(-E)^{3/2} / Bk_e e^2 \sqrt{m}. \quad (4)$$

(Note that the energy E is negative for electrons bound in orbits around nuclei, so $-E$ will be a positive quantity.)

So far Bohr has not gone beyond classical physics, although in choosing to express frequency in terms of energy he has obviously been guided by Planck's hypothesis. But now he takes another bold step: he assumes that if the electron is initially at rest at a very large distance from the nucleus, so that its energy and frequency of revolution are both essentially zero, and if, when it is captured or bound by the atom, it drops into a final state with frequency and energy related by Eq. (4), *the frequency v of the emitted light is simply the average of the initial and final frequencies of orbital revolution.* If the initial frequency is zero and the final frequency, say, f, the average is just ½f. In a crucial paragraph Bohr writes: "Let us now assume that, during the binding of the electron, a homogeneous radiation is emitted of a frequency ν, equal to half the frequency of revolution of the electron in its final orbit; then, from Planck's theory, we might expect that the amount of energy emitted by the process considered is equal to $nh\nu$, where h is Planck's constant and n an entire [integer] number. If we assume that the radiation emitted is homogeneous, the second assumption concerning the frequency suggests itself, since the frequency of revolution of the electron at the beginning of the emission is zero."

Note that n means the *number of quanta* of energy $h\nu$ that are emitted, according to Planck's hypothesis, when an electron moves from rest at an infinite distance to an orbit with frequency f and energy E. Bohr proposes to set

$$E = \tfrac{1}{2} nhf$$

because he is asserting that the frequency of the emitted radiation is $\nu = \tfrac{1}{2}f$. Substituting this expression for E into Eq. (4), we can determine both ν and E in terms of the charge and mass of the electron, Planck's constant, and the integer n:

$$\nu = 2\pi^2 m_e e^4 k_e^{\,2} / n^3 h^3, \quad (5)$$

$$E = -2\pi^2 e^4 m_e k_e^{\,2} / n^2 h^2. \quad (6)$$

Note that all terms except m in these equations are known universal constants.

The theoretical result agreed with experiment with an error of less than 1%:

When Bohr substituted numerical values into these equations (see next note), he found that—provided that n was set equal to 1—he could obtain almost exactly the experimental values for the ionization energy of the hydrogen atom, and for the spectroscopic frequency corresponding to the capture of a free electron into the lowest energy level of a hydrogen atom. From Eq. (6) we find $E = -2.18 \times 10^{-18}$ J.

To put it another way, the electron must be given 2.18×10^{-18} joule of energy to allow it to move (against the force of attraction to the nucleus) from the first orbit to infinity, or to a large enough distance from the nucleus so that the remaining energy is virtually zero. But this corresponds to the ionization of the hydrogen atom; and thus Bohr's theory predicts that one must supply 2.18×10^{-18} J to a hydrogen atom to cause its ionization. Experimentally, it takes an accelerating voltage of 13.6 volts for the electron beam to induce a sudden change in conduction through or ionization in the gas tube. But with that accelerating voltage, each projectile, having the electronic charge *e*, carries an energy of

$13.6 \times 1.6 \times 10^{-19} = 2.17 \times 10^{-18}$ J—in excellent agreement with the calculation from Bohr's theory.

051 *Following a suggestion by Niels Bjerrum* Jammer, *Conceptual Development*, pp. 83–84.

051 *this success "in no small degree contributed to its swift acceptance"* Leon Rosenfeld, "Bohr, Niels Henrik David," *Dictionary of Scientific Biography* 2 (1970): 239–254, on p. 244. Jammer, *Conceptual Development*, p. 85.

052 *The reception of the Bohr model has been discussed* Kragh, "The Early Reception of Bohr's Atomic Theory (1913–1915): A Preliminary Investigation," *Reposs: Research Publications on Science Studies 9* (Aarhus, Denmark: Department of Science Studies, University of Aarhus, 2010); available online at http://css.au.dk/fileadmin/reposs/reposs-009.pdf.

Kragh, "Resisting the Bohr Atom: The Early British Opposition," *Physics in Perspective* 13 (2011): 4–35. Kragh, "Conceptual Objections to the Bohr Atomic Theory—Do Electrons Have a 'Free Will'?," *European Physical Journal H* 36 (2011): 327–352. Kragh, *Niels Bohr and the Quantum Atom: The Bohr Model of Atomic Structure 1913–1925* (Oxford: Oxford University Press, 2012). See also Jagdish Mehra and Helmut Rechenberg, *The Historical Development of Quantum Theory*, vol. 1 (New York: Springer-Verlag, 1982), pp. 193–257; Ulrich Hoyer, editor, *Niels Bohr Collected Works*, vol. 2 (Amsterdam: North-Holland, 1981).

052 *nine physicists accepted Bohr's model because it* explained *the empirical formula* H. Stanley Allen (1915c), Max Born (1914), Rudolf Seeliger (1914a), James Jeans (1913), Oliver Lodge (1913), Eduard Reicke (1915), Frederick Soddy (1914), Arnold Sommerfeld (1915b), Wilhelm Wien (1915). For details and references see Kragh, "Early Reception."

There is a significant ambiguity in the case of Jeans, who said in a discussion at the British Association meeting on September 12, 1913: "Bohr has arrived at a most ingenious and suggestive, and I think we must add convincing explanation of the laws of spectral lines." But on the next page he says that the theory explains "the series recently discovered by Fowler." This was just after E. J. Evans announced his preliminary results, indicating that the lines attributed to hydrogen by Fowler could be explained as due to He^{+} as suggested by Bohr, but before the disagreement about the source of these lines was completely resolved by Bohr's theoretical refinement and Evans's further experiments. Jeans's statement "the only justification at present put forward for [Bohr's] assumptions is the very weighty one of success" does not mention Bohr's confirmed *prediction* that the new spectral lines are due to helium, not hydrogen. See Jeans, in "Discussion on Radiation," *Report of the 83rd Meeting of the British Association for the Advancement of Science, Birmingham, September 10–17, 1913* (London: John Murray, 1914), pp. 376–386, on p. 379. Jammer, *Conceptual Development,* pp. 76–92 (includes references to original papers).

052 *Three physicists . . . mentioned the confirmed prediction* Friedrich Paschen, letter to Arnold Sommerfeld, February 24, 1915, in Michael Eckert and Karl Märker (editors), *Arnold Sommerfeld. Wissenschaftliche Briefwechsel*, Vol. 1, p. 500; Rudolf Seeliger (1914); Frederick Soddy (1914). For details and references see Kragh, "Early Reception."

052 *other reasons, including the fact that it supported their own theory* Samuel MacLaren (1913) and Rutherford (1913, 1914); H. G. Moseley (1913, 1914).

052 *By 1916, Robert A. Millikan . . . could also cite Sommerfeld's explanation* Millikan, "Radiation and Atomic Structure," *Science* 45 (1917): 321–330, on p. 326.

052 *Kragh reports that German physicists were "skeptical"* Kragh, *Niels Bohr*, p. 122 (quoting Sommerfeld).

052 *A significant exception was Einstein* Kragh, *Niels Bohr*, p. 95. Also, Hevesy wrote to Rutherford: "When I told him about the Fowler spectrum the big eyes of Einstein looked still bigger, and he told me 'Then it is one of the greatest discoveries'" (Kragh, *Niels Bohr*, p. 122).

052 *Seven physicists and one chemist accepted the theory for other reasons* Ernest Rutherford (1913), because it "complemented and justified his own earlier theory of the nuclear atom"; Samuel B. McLaren (1913), because it supported his own idea of an elementary magnetic quantity; Walther Kossel (1914) and Henry G. J. Moseley (1914) found it useful in discussing their results on X-ray absorption; Theodore Lyman (1914) because it was confirmed by his own observation of two new hydrogen spectral lines, "suggested by Bohr"; George Shearer (1915) based on X-ray data (ionization of hydrogen); the chemist was E. H. Buchner (1915), who liked Bohr's theory because it might explain analogies such as between ammonium ion and alkali ions.

Kragh concluded: "Two years after Bohr had announced his theory . . . it was widely accepted or at least seriously considered by physicists working with quantum theory and the structure of matter" ("Early Reception," pp. 81–82).

At least three other prominent physicists—Paul Ehrenfest, Max von Laue, and Otto Stern—were initially strongly opposed to the theory. According to science writer Louisa Gilder and other sources, they asserted that they would give up physics if Bohr proved to be right. All accepted it eventually. See Gilder, *The Age of Entanglement: When Quantum Physics was Reborn* (New York: Knopf, 2008), pp. 35, 38; Martin J. Klein, "Paul Ehrenfest, Niels Bohr, and Albert Einstein: Colleagues and Friends," *Physics in Perspective* 12 (2010): 307–337, on p. 308.

Section 2.6

052 *Thomas Glick . . . has proposed a general model* See Glick and Henderson, "Receptions."

053 *Kohler recently noted, "When history of science was the history of ideas"* Robert E. Kohler, "A Generalist's Vision," *Isis* 96 (2005): 224–229, on 226.

054 *there has been very little discussion of the ways in which the philosophy of science can enrich historiographical practice* Arabatzis, *Representing Electrons*, p. 2.

054 *articles by Marga Vicedo . . . in the March 2012 issue* Marga Vicedo, "Introduction: The Secret Lives of Textbooks," *Isis* 103 (2012): 83–87; Michael D. Gordin, "Translating Textbooks: Russian, German, and the Language of Chemistry," ibid., 88–98; Adam R. Shapiro, "Between Training and Popularization: Regulating Science Textbooks in Secondary Education," ibid., 99–110; Vicedo, "Playing the Game: Psychology Textbooks Speak Out about Love," ibid., 111–125. See also Stephen G. Brush, "How Theories Became Knowledge: Why Science Textbooks Should be Saved," in *Who Wants Yesterday's Papers? Essays on the Research Value of Printed Materials in the Digital Age*, edited by Yvonne Carignan et al. (Lanham, MD: Scarecrow Press, 2005), pp. 45–57, and works cited therein.

3

Prediction-Testing in the Evaluation of Theories

A Controversy in the Philosophy of Science

> In recent decades, mainstream history of modern science has been somewhat disinclined to treat seriously the history of scientific ideas and their philosophical implications . . . As a result, we have largely ceded this territory to the philosophers of science. It is time to take it back. If the history of science is to exist as a separate field and not be subsumed by general history, then we need to deal with "science" not only as a social and cultural phenomenon but also as an intellectual one.
>
> —LYNN K. NYHART, *President of the History of Science Society (2012).*

> Historians and sociologists study science as it is, whereas philosophers of science study science as it ought to be. Philosophy of Science is a *normative* discipline, its goal being to distinguish good science from bad, better scientific practices from worse.
>
> *—philosopher* ELLIOTT SOBER *(2008)*

3.1 INTRODUCTION

Why do scientists accept or reject theories? More specifically: why do they change from one theory to another? What is the role of empirical tests in the evaluation of theories? Do physical, biological, behavioral, and social scientists differ in the weight they give to empirical and other factors? If a scientific community is converted to a new theory, does that mean that the individual scientists are converted or (as Max Planck suggested) that the opponents just die out and the younger scientists grow up with the new theory? Have there been significant changes over time in the answers to these questions?

In this chapter I don't promise a comprehensive review of these issues, but will focus primarily on a very narrowly defined question: in judging theories, do scientists give greater weight (other things being equal) to successful *novel predictions* than to successful deductions of previously known facts?

It is primarily philosophers who seem to be seriously interested in that question, yet as I suggested in Chapter 1, it is crucial to the problem of whether or to what extent scientists follow the so-called scientific method. This problem in turn has important educational, financial, and legal ramifications: should creationism be given equal weight with evolution in public schools? Which scientific or medical or psychological approaches should receive government funding? What kind of expert testimony is admissible in court?

I begin by recalling some recent philosophical discussions about the weight given to predictions of novel facts (novel prediction or prediction in advance) compared to the weight given to the retrodiction of previously known facts (explanation). If philosophers are interested in the way scientists themselves deal with this question, they might turn to the literature on the history of science. However, until recently historians have not addressed the issue in a systematic way. There is a need for historical research designed specifically to answer such questions, and conversely there is a need for philosophers to apply the same critical scrutiny to historical arguments for their claims that they apply to logical arguments—or else, like Sober, to stipulate that they are discussing only what scientists *should* do, not what they actually do or have done.

At this point I should state that my "if" signifies a real choice. I don't claim that philosophy of science must be validated by scientific practice. Normative or logical studies of the relations between theory and evidence are perfectly legitimate activities for philosophers—as valuable as any other intellectual exercise. My remarks are addressed only to practitioners of naturalistic philosophy of science—to those who do claim that their analysis is informed by and relevant to the behavior of scientists.

Philosopher Imre Lakatos did claim to follow naturalism, to a limited degree. He postulated a "metacriterion":

> M_L: A rationality theory . . . is to be rejected if it is inconsistent with accepted "basic value judgments" of the scientific community.

But he also advocated rational reconstruction of the history of science: the philosopher need not be concerned with the personal details of why this scientist accepted that theory at a particular time. Instead he wanted to answer broader questions, such as whether it was rational for the physics community to accept quantum mechanics after but not before 1932. (This is a version of the question faced by physicist C. W. Oseen; see Section 9.2.) It seems to me a good question for a philosopher to ask, especially since many professional historians have avoided discussing the reasons why major long-term changes in science occur.

Of course, history of science is not an authority you can pull off the shelf to support your claims about how science works; it is a source of evidence that must be critically scrutinized, like any other kind of evidence.

3.2 NOVELTY IN THE PHILOSOPHY OF SCIENCE

> What did the President know and when did he know it?
>
> —*Senator* HOWARD BAKER *(1973)*

During the past three decades novel prediction has played a prominent role in three areas of the philosophy of science: (a) the quasihistorical methodology of scientific research programs developed by Imre Lakatos and his followers; (b) the problem of old evidence in Bayesian analysis; and (c) the miracle argument for scientific realism. (It should be noted that in the philosophy of science literature these are usually called Programmes when discussing the work of Lakatos, but the American spelling will be used here in general.)

The Lakatos Methodology of Scientific Research Programs

Lakatos introduced a temporal element into the assessment of theories: a series of theories or a research program should be judged progressive or degenerating according as later theories were more or less successful than earlier theories in the series. Success means, primarily, confirmed novel predictions.

But the application of the methodology to specific historical cases soon showed that it was unreasonable to claim that *only* novel predictions can provide evidence for a theory; scientists obviously do count the explanation of already-known facts in judging a theory. The Lakatos methodology was born refuted according to meta-criterion M_L, but in the spirit of the methodology, its followers tried to rescue it by changing the definition of "novel." Thus his junior colleague in London, philosopher Elie Zahar, argued that the "novelty of facts" should not be limited to *temporal* novelty, for then we should "have to give Einstein no credit for explaining the anomalous precession of Mercury's perihelion" and "we should have to say, contrary to informed opinion, that Michelson's experiment did not confirm Special Relativity and Galileo's experiments on free fall did not confirm Newton's theory of gravitation."

Zahar then proposed the following definition:

> N_Z: A fact will be considered novel with respect to a given hypothesis if it did not belong to the problem-situation which governed the construction of the hypothesis.

But, as Zahar immediately recognized, this "implies that the traditional methods of historical research are even more vital for evaluating experimental support than Lakatos has already suggested. The historian has to read the private correspondence of the scientist whose ideas he is studying; his purpose will not be to delve into the psyche of the scientist, but to disentangle the heuristic reasoning which the latter used in order to arrive at a new theory."

Zahar asserted in the same paper that Einstein's explanation from general relativity theory of the advance of Mercury's perihelion was a novel prediction by the definition N_Z because "Einstein did not use the known behaviour of Mercury's perihelion in constructing his theory."

With this statement Zahar planted a time bomb that exploded five years later, providing an even more devastating refutation of the Lakatos methodology. The American philosophers John Earman and Clark Glymour examined Einstein's unpublished correspondence, and found a 1915 letter to Arnold Sommerfeld which strongly suggested that Einstein did in fact use the known behavior of Mercury's perihelion in choosing his field equations for general relativity. Citing this example, philosopher Michael Gardner and others argue that Zahar's N_Z (and similar definitions of novelty that can be implemented only by detailed historical research in unpublished documents) cannot be used to explain the rational evolution of science;

after all, most other scientists could not have known at the time whether Einstein had used the Mercury behavior to construct his theory.

Other definitions of "novelty" have been proposed and used by philosophers to evaluate the empirical support of theories. But I have not yet seen any evidence that these definitions have any basis in the behavior of scientists, and there now seems to be considerable doubt whether the concept of novel prediction has any useful role to play in philosophical discussions of the history of science.

Bayesian Analysis

Thomas Bayes, an eighteenth-century minister and probability theorist, proposed a method for what is now called statistical inference: the estimation of causes from effects. In modern terms, suppose you have a hypothesis *H*, which initially has some plausibility on the basis for your previous knowledge, so that you can estimate the probability of its being correct to be prob (*H*). Now suppose you use it to predict some new fact *e*. Let prob (*e/H*) be the probability that *e* would be observed if *H* were true. Then according to Bayes' rule, the probability that *H* is true, given this *new* evidence *e*, is

$$prob\left(\frac{H}{e}\right) = prob\left(\frac{e}{H}\right)\left(\frac{prob\,(H)}{prob\,(e)}\right)$$

where prob (*e*) is the probability that the fact would have been observed regardless of the validity of *H*. (This corresponds to Popper's demand that the prediction should contradict what one would expect from already accepted theories.)

In 1980, philosopher Clark Glymour argued that Bayesian analysis implies "the absurdity that old evidence cannot confirm new theory." According to Glymour's interpretation of traditional Bayesian analysis, only novel predictions can test a theory (yielding new evidence *e*), contrary to actual scientific practice (in which estimating prob [*H*] on the basis of known facts may be more important). Hence Bayesian analysis is discredited by the Lakatos metacriterion M_L: it is inconsistent with accepted judgments of the scientific community.

Glymour's critique stimulated heated discussions about whether Bayesian theory actually does give no weight to old evidence (i.e., to the explanation of known facts), hence only novel predictions can confirm (or corroborate, in Popper's terminology) a theory; and, if so, whether it therefore fails to describe the behavior of scientists.

Philosophers have defended all four possible positions: Bayesian analysis is (1) valid because it favors novel predictions, (2) valid because it does not favor novel predictions, (3) invalid because it favors novel predictions, and (4) invalid because it does not favor novel predictions. As a nonphilosopher, I eagerly await the outcome of this controversy but don't wish to participate in it.

I am, however, interested in a byproduct of this debate: philosopher Colin Howson, who supports what he calls personalist Bayesianism, argues that the preference for novel predictions is based on a false assumption, the null-support thesis. This thesis states: "a theory that has been deliberately constructed to yield the [observed] effect" cannot derive any support at all from its (non-novel) prediction of that effect. This is because one can "generate arbitrarily many theories that are supported by any given piece of data." The falsity of the null-support thesis is a consequence of

> the fundamental principle of all inductive inference: evidence supports a hypothesis the more, the less it is explicable by any plausible alternative compared with its explicability by h. This is the criterion by which in practice we decide which data support a hypothesis and which do not, and it is a criterion that can just as easily by satisfied by a hypothesis constructed from that evidence as by one that was not.

I would add that while it may possible to "generate arbitrarily many theories that are supported by any given piece of data," there may not be *any* single theory that gives a consistent explanation of *all* the relevant data. That was the situation in atomic physics in 1924: the old quantum theory was a collection of different hypotheses, each of which could predict or explain some of the data (see below, Section 3.5 Part G, "Novelty Is Crucial"), but no one had melded them into a satisfactory self-consistent theory. When quantum mechanics was proposed by Heisenberg and Schrödinger it *explained* a remarkably large number of facts and was quickly accepted before any of its *predictions* of new facts had been confirmed (Chapter 8). This seems to refute the null hypothesis as an accurate account of the behavior of scientists in evaluating what now seems to be one of the most important theories of twentieth-century science.

The Miracle Argument for Scientific Realism

In 1978, philosopher Hilary Putnam wrote that "the typical realist argument against idealism is that it makes the success of science a *miracle*." How can theories of electrons, space-time, and DNA, he asked, "correctly predict observable phenomena if, in reality, there are no electrons, no curved space-time, and no DNA molecules"? If these entities don't exist, it is a miracle that a theory assuming their existence successfully predicts phenomena. This became known as the miracle argument for scientific realism, and has generally (but not always) been taken to mean that the success of *novel* predictions is incomprehensible unless one assumes that the theory generating the prediction refers to something in the real world.

One does not expect a philosopher to know that the nineteenth-century wave theory of light led to several successful predictions, despite the fact that it was based on an entity, the ether, that was considered not to exist after Einstein's theory of relativity was accepted. But one does expect a philosopher to know about—and avoid—the fallacy of affirming the consequent (see Section 1.2).

It takes only an elementary knowledge of the history of science—no detailed research in primary sources is required—to recognize that several successful novel predictions have come from theories now considered incorrect, and were later replaced by other theories that made the same (no-longer-novel) predictions. This was forcefully pointed out, with many examples, by philosopher Larry Laudan. In response to Laudan's critique, realists have retreated to weaker versions of the miracle argument or have argued that the false theories really refer to entities that do exist (according to current theory) rather than to what their authors said they did.

3.3 WHAT IS A PREDICTION? (REVISITED)

Before we can look into the way scientists use predictions to evaluate theories, we have to take account of an unfortunate linguistic confusion. In ordinary language "predict" means "foretell" or "prophesy," implying a statement about future events. But physicists (and many other scientists) currently use the word to mean "deduce from a theory," whether before or after the fact is known or has occurred. Occasionally one finds the phrase "predict in advance" used to specify prediction of a future event, and sometimes the context indicates that this is what is meant, but very little use is made of terms like "retrodiction" or "postdiction." The physicist-philosopher Henry Margenau made this quite explicit:

> The word *prediction*, as used in science, does not mean "forecast" in a temporal sense. *Pre* implies "prior to completed knowledge"; it does not contrast with *post* as does *ante*. The counterpart to *pre*fix is not postfix but *suf*fix. It is therefore unnecessary to coin a new word, *postdiction*, to denote what we should call prediction of the past. The use of this word, though it has been suggested, would seem a bit "*prepost*erous."

While a few philosophers have recognized this usage most now prefer to distinguish *novel* from other predictions. Since we have to name the distinction in order to discuss whether it is important, I support this as a general term, but I don't think much is to be gained by adopting one of the more restrictive definitions of novelty advocated in the discussion of Lakatosian research programs. There is still no suitable phrase for *non*-novel predictions. "Accommodation" seems too pejorative. I prefer "explanation," although I don't want to imply that deducing a known fact from a theory "explains" that fact in the sense of offering intellectual understanding. Understanding the explanation may come later, as scientists revise their views on how nature behaves in order to accept elliptical orbits and action at a distance. Most of us have not yet reached that level of understanding of the basic principles of relativity and quantum mechanics, even if we acknowledge that the novel predictions and explanations deduced from those theories have been astonishingly successful.

Another way to express this distinction has been suggested by philosopher Ronald N. Giere: "naturalists know that what counts as a scientific explanation changes over time"—for example, action at a distance was not a scientific explanation in the seventeenth century but after Newton's *Principia*, it was in the eighteenth century. Alternatively, one could say that the explanation did not at first convey understanding, but after repeated use it did.

3.4 DOES NOVELTY MAKE A DIFFERENCE?

The fact that physicists frequently do not distinguish novel prediction from retrodiction suggests that they don't ascribe much importance to novelty. Yet in their *popular* writings they often celebrate successful predictions such as Einstein's light bending and Mendeleev's new elements. Since "prediction" means "novel prediction" in popular discourse, and it's not always clear whether an author is using a word in the technical or popular sense, we cannot easily infer whether scientists believe the prediction was more important just because it was novel. In order to isolate the novelty factor, I try to find cases in which the same theory made both novel and non-novel predictions (preferably at the same time and of the same importance).

Novelty does have some benefits. The publicity generated by a successful novel prediction may lead scientists to *pursue* a theory that they would otherwise ignore; a novel prediction may stimulate experiments designed to test the theory and thus contribute more to the advance of knowledge than an explanation.

Conversely, explanation (retrodiction of known facts) does have some disadvantages: a theorist may be consciously or unconsciously influenced by knowing what the theory needs to explain, and almost any set of data can be fitted by a theory with enough adjustable parameters. (That's why some philosophers use the word "accommodation" rather than "explanation.")

It is useful to focus on technical reviews and science textbooks (rather than popular works or unpublished correspondence and interviews) in order to learn how a discipline articulates a public position on the evaluation of a theory, and how this position is passed on to students. Even then, it may be difficult to find an explicit statement by a scientist that a novel prediction counts more (or less) than an explanation in judging a particular theory. That's one reason why I study the more important cases, because they generate a large sample of published responses that can be examined. My conclusions may not be valid for more routine research where the validity of basic theories is not in doubt, but their application to specific situations may require testing a hypothesis by predicting unknown consequences (paradigm-governed rather than paradigm-changing, in Thomas Kuhn's terminology).

My results do challenge the views of those philosophers and educators who assert, with Popper, that a theory that provides satisfactory explanations but no confirmed novel predictions is unscientific. As philosopher and historian William R. Newman points out, in rejecting the view of philosopher Alan F. Chalmers that philosophical theories of matter—those that merely explain or accommodate known facts but don't predict ones—do not lead to new scientific knowledge and therefore are not scientific, he asserts:

> Needless to say, such a restrictive concept of "scientific knowledge" would present intractable problems even for the working scientists engaged today in high-energy physics, where much remains untested; it would also make it hard to count the explanation of physical anomalies as science.

In particular, Chalmers concludes that all atomistic theories before the late nineteenth or earlier twentieth centuries were philosophical, and hence unscientific.

"Needless to say" is of course just a rhetorical flourish—in fact it does need to be said that large areas of activity by modern physicists, chemists, astronomers, and biologists *are* scientific even if they do not involve testable novel predictions. Moreover, it also needs to be said that even by Chalmers's own criterion there were atomistic theories before the late nineteenth century that made testable predictions; a notable example is Maxwell's successful prediction that the viscosity of a gas is independent of density and increases with temperature (see Section 1.2).

3.5 EVIDENCE FROM CASE HISTORIES

> You can always make news with doomsday predictions, but you can usually make money betting against them.
>
> —JULIAN SIMON

I summarize here the results of my study of case histories in physical and biological science presented in detail in Parts Two through Four, but starting with one that I did not yet study in any depth.

Accepted Despite Failed Predictions

Is it possible for a theory to be accepted even though almost all of its predictions have been falsified? None of the cases I've studied myself fall into this category. (I don't include pseudosciences like astrology and creationism, because they are not accepted by most scientists.) But, according to economist Julian Simon, my late colleague at the University of Maryland, Malthusian theories about the effect of population increase on resources and the standard of living do fit this description. He argued that such theories continue to be supported by scientists even though they are inconsistent with known data and all their specific novel predictions have turned out to be wrong. He challenged population biologist Paul Ehrlich, a well-known proponent of Malthusian forecasts, to a $1,000 bet on Ehrlich's prediction that the prices resources would increase in price over a decade, as a result of population pressure.

According to science writer John Tierney,

> Ehrlich accepted and formed a consortium with two colleagues at Berkeley, John P. Holdren and John Harte. In 1980, they picked five metals and bet that the prices would rise during the next 10 years. By 1990, the prices were lower, and the Malthusians paid up, although they didn't seem to suffer any professional consequences. Dr. Ehrlich and Dr. Holdren both won MacArthur "genius awards" (Julian never did). Dr. Holdren . . . today serves as President Obama's science adviser.

Tierney decided to follow in the footsteps of Simon, who died in 1998 after having failed to get any other prominent doomsayers to take his bets. In 2005, Matthew R. Simmons, head of a Houston investment bank specializing in the energy industry, predicted that the price of oil, then about $65 a barrel, would increase in the next five years to more than $200, even after adjusting for inflation. Tierney and Simon's widow, Rita Simon, shared the bet of $5,000 against Simmons's prediction. The result: the 2010 average price was just under $80, around the equivalent of about $71 in 2005 dollars. Tierney and Rita Simon won the bet. So far, I've seen no indication that any Malthusians have abandoned their theory.

To avoid any misunderstanding, I want to state again that I am not trying to judge the behavior of scientists or the validity of their theories. In particular, I'm not saying that Ehrlich and his supporters are wrong or that Simon was right. (The basic premise that a finite earth cannot support an infinite number of people is a mathematical truism; in some of his pronouncements Simon gave the impression that he was denying this truism.) Rather, I'm trying to find out whether statements by philosophers about predictions are consistent with the way scientists do science. If Simon was right, this may be a flagrant example of the unimportance of prediction-testing.

Rejected Despite Confirmed Novel Predictions

Physicist and astrophysicist Hannes Alfvén developed an electromagnetic plasma theory of space and solar system physics from which he deduced several novel predictions. Most of them were confirmed but some were falsified or still remain in dispute. Even though he won the Nobel Prize in physics for his work, his theory has been rejected by most space scientists not because of its falsified predictions, but because its basic premises and procedures are considered unsatisfactory. When physicist Bibhas De used one of the confirmed phenomena to predict that the planet Uranus would be found to have rings, his paper was rejected for publication and was rejected again when it was resubmitted after the Uranian rings were discovered (Section 1.1). This is the example that originally led me to undertake this research on predictions.

Another example (not discussed in this book) has been suggested by Egbert Giles Leigh Jr.:

> Successful prediction, however, is not usually enough to win acceptance of a theory . . . The theory must also be reasonable . . . Wegener's theory of continental drift was rejected at first, despite a seemingly overwhelming array of successful "predictions," because there seemed to be no mechanism capable of moving continents about the planet. Since [such] forces were discovered . . . the theory has won general acceptance.

Acceptance Independent of Confirmation of Novel Prediction

In 1925 Werner Heisenberg proposed his matrix mechanics theory of the atom, and in 1926 Erwin Schrödinger proposed a wave mechanics theory; shortly thereafter the two theories were shown to be mathematically equivalent (at least with regard to their observational consequences). They are now considered different versions of a single theory, quantum mechanics. Widely acknowledged to be the foundation for theories of the properties of atoms, molecules, radiation, gases, liquids, solids, and metals, quantum mechanics was the most important new physical theory of the twentieth century.

As far as I can determine from a survey of books and articles published in the period 1927–1931, the confirmation of its novel predictions played essentially no role in the acceptance of quantum mechanics by physicists. In fact, my survey shows, in agreement with other studies, that most researchers active in atomic physics had already accepted the theory by 1928, although its interpretation and technical details continued to be debated for another decade. Because of this rapid acceptance, it seems that quantum mechanics was accepted before there was time to confirm any of its novel predictions.

Of course the new quantum mechanics was based on earlier theories—Einstein's light-quantum, Bohr's atomic model, and de Broglie's matter-waves—which did make successful novel predictions. But it was a much more general and powerful theory than its predecessors, and had philosophical consequences that had been only hinted at previously. Given the instrumentalist flavor of the writings of its advocates—especially Heisenberg and Bohr—one might have expected that the new theory would have been put forward with strong claims about its confirmed novel predictions—but that was not the case. Instead, they argued that quantum mechanics better accounted for the facts explained by the old theory, explained several anomalies that its predecessor had failed to resolve, and gave a single consistent method for doing calculations in place of a collection of ad hoc rules. Even after 1928, physicists did *not* ascribe any extra significance to the handful of experiments that could now be called confirmations of novel predictions (see Chapter 8).

According to historian Richard Staley, the special theory of relativity (proposed by Einstein in 1905) was accepted around 1911, before any of its predictions were confirmed. In fact, the most important empirical evidence for the theory was the null result of the Michelson-Morley experiment (no motion of the Earth relative to the ether or absolute space could be detected). The theory explained this result in a way that satisfied the handful of experts who understood it. In other words, it explained "nothing." Experiments on the change of mass of electrons at high speed were not accurate enough to show that Einstein's prediction was more accurate than those of rival theories, until after 1911. More important was the elegance and simplicity of a theory that succeeded in uniting mechanics and electromagnetism, and the aesthetic appeal of mathematician Rudolph Minkowski's four-dimensional space-time formulation. The theory seemed (to experts) so beautiful that it *must* be true.

Similarly, most of the biologists who accepted the Fisher-Haldane-Wright theory of natural selection in the 1940s and 1950s did so without basing their acceptance on its successful novel predictions. In fact, even when Popper and other philosophers alleged that Darwinism is not a scientific theory because it does not make testable predictions, Darwinists did not cite the testable (and confirmed) predictions that their theory had actually made. They argued instead that biology is different from physics and does not need to satisfy Popper's criterion in order to be considered scientific (see Chapter 14).

Retrodiction Counts as Much as Novel Prediction

The 1919 eclipse test of Einstein's general theory of relativity, by a British team led by Arthur Eddington, was an important event not only in the history of science but also in the history of the *philosophy* of science, because it led Karl Popper (according to his own account) to formulate his falsifiability criterion for distinguishing science from pseudoscience. He was impressed by the contrast between theories like psychoanalysis and Marxism, which could explain any given phenomenon but refused to make testable predictions, and Einstein's theory which made a novel quantitative prediction—the amount of bending of starlight by the sun's gravitational field—whose refutation would (or so Popper believed) have killed the theory.

The light-bending observation turns out to be a good case for examining the claim that novel prediction is a better test of a theory than retrodiction. There is a large amount of scientific and popular literature dealing with this case, so that one can hope to draw conclusions about the behavior of an entire scientific community, not just a handful of specialists.

Moreover, one can judge the weight ascribed to light bending by comparison with two other tests that were discussed at the same time: the advance of the perihelion of Mercury and the gravitational redshift of spectral lines. The former was a well-known discrepancy that theorists had failed to explain satisfactorily despite several decades of work; Einstein managed to calculate the observed effect within the observational error without introducing any arbitrary parameters. This was clearly not a novel prediction, even by the most generous definition of novel (Section 3.2).

The redshift was, like light bending, a prediction from general relativity theory, but its observational confirmation was still in doubt in the 1920s and remained so for several decades. I will therefore ignore it for now, and inquire simply whether scientists considered that light bending was better evidence than Mercury's orbit *because* it was a novel prediction.

Most of the published comments by physicists during the first two or three years after the 1919 eclipse test indicated that light bending and the Mercury perihelion advance counted equally strongly in favor of general relativity. If light bending was more important, that was not because it had been forecast in advance but because the data themselves seemed at first to be more definitive.

It later became clear to the experts that the Mercury effect was *stronger* evidence than light bending. In part this was because the observational data were, on further analysis, more accurate than those for light bending. It was very difficult to make good eclipse measurements, even with modern technology, and there was some suspicion that Eddington had cherry-picked the results that favored Einstein and discarded those that didn't. Moreover, the Mercury orbit calculation depended on a deeper part of the theory. The novelty of the light-bending prediction seems to count for little or nothing in these judgments.

The most interesting argument (though it was not often explicitly stated) is that rather than light bending providing better evidence because it was a novel prediction, it actually provides less secure evidence *for that very reason*. This is the case at least in the years immediately following the announcement of the eclipse result, because scientists recognized that any given empirical result might be explained by more than one theory. Because the Mercury discrepancy had been known for several decades, theorists had already had ample opportunity to explain it from Newtonian celestial mechanics and had failed to do so except by making implausible ad hoc assumptions. This made Einstein's success all the more impressive, and made it seem quite unlikely that anyone else would subsequently produce a better alternative explanation. Light bending, on the other hand, had not previously been discussed theoretically (with rare exceptions), but now that the phenomenon was known to exist one might hope that another equally—or more satisfactory—explanation would be found. (Recall the fallacy of affirming the consequent in Section 3.2.) It was only about 10 years later that Einstein's supporters could plausibly assert that no other theory could account for light bending.

Morgan's chromosome theory of heredity made two different kinds of confirmed novel predictions. One, nondisjunction, was regarded by geneticists as a proof of the validity of the theory, but was not usually identified as a novel prediction. The other, the ability to predict crossover frequencies from chromosome maps, was often mentioned as a successful prediction that supported the validity of the theory, but was not considered as important as the theory's ability to explain Mendelian genetics.

Acceptance After Novel Prediction Is Confirmed, But . . .

There are three well-known cases of new particles predicted from theories: the positron (from Dirac's relativistic quantum theory); the meson (from Yukawa's theory of nuclear forces); and the omega-minus, Ω^- (from Gell-Mann's SU3 or eight-fold way symmetry group theory). The first two are discussed in Chapter 9. In each case physicists accepted the theory within two or three years after the discovery of the predicted particle. But while the successful predictions certainly forced physicists to give serious consideration to those theories, there is no convincing evidence that within the context of such consideration a theory that merely retrodicted the particles would be any less acceptable, other things being equal. In each case, the original theory that predicted the particle was soon replaced by another theory.

Perhaps most remarkable is the case of the positron, which Dirac proposed as the antiparticle of the electron. The existence of an antiparticle followed directly from his equation, provided one accepted his interpretation of holes in a sea of negative-energy states as particles with the opposite charge and positive energy. Dirac's theory was later replaced by quantum electrodynamics, in which the existence of antiparticles was not deduced but simply postulated.

Also in this category is the big bang cosmology of Friedmann, Lemaitre, and Gamow. Alpher and Herman, and later Gamow, predicted from this theory the existence of a cosmic microwave background, with a Planck frequency distribution corresponding to a temperature a few degrees above absolute zero (see Chapter 12). The rival steady-state cosmology of Bondi, Gold, and Hoyle did not entail this consequence. Soon after the discovery of this background radiation in 1965, astronomers abandoned the steady state theory in favor of the big bang. Again, it is not clear whether the big bang theory, once it was revived by this successful prediction, got credit for novelty; steady-state advocates argued that if they could somehow retrodict the new phenomenon from their theory, they would be entitled to just as much credit for it. After they failed to do so, the big bang was accepted.

The cosmology case also shows the influence of Karl Popper's falsificationism on scientists. Bondi and Gold (but not Hoyle), citing Popper as a guide to scientific methodology, promised that they would abandon their theory if any reliable evidence were found that the large-scale properties of the universe were different in the past than they are now. When the cosmic microwave background provided evidence for a much hotter early state, they kept their promise. So falsificationism works, if you believe in it.

Novelty Does Count—A Little

In 1871, Dmitrii Ivanovich Mendeleev predicted from his periodic law, now known as the periodic table, the existence and properties of three elements needed to fill gaps in his table. Two of them, now called gallium and scandium, were discovered in the 1870s. The third, germanium, was found in 1886. Mendeleev also proposed that a few of the atomic weights assigned to known elements should be changed in order to fit them neatly into his table: beryllium from 14 to 9, yttrium from 60 to 88, cerium from 92 to 138, and uranium from 120 to 240. The new values were confirmed by experiments.

While chemists differed on the relative importance of prediction and explanation, it seems fair to approximate the consensus as follows. The reasons for accepting the periodic law are, in order of importance: (1) it accurately describes the correlation between physicochemical properties and atomic weights of nearly all known elements; (2) it has led to useful corrections in the atomic weights of several elements and has helped to resolve controversies such as those about beryllium; and (3) it has yielded successful predictions of the existence and properties of new elements.

I found that the best source of evidence for the acceptance of the periodic law was not research articles, but textbooks. (In part this was because the law was quite useful in teaching chemistry: it's easier for students to learn the properties of the elements by learning the patterns shown by the table rather than memorizing the properties of each element separately.) By 1890, the majority of chemistry textbooks and reference books published in the United States, the United Kingdom, France, and Germany (but also available in American libraries) included the periodic law, and many of them gave reasons for including it.

The early history of benzene (C_6H_6) theories (Chapter 6) shows that chemists were much more strongly committed to the Popperian scientific method than physicists or biologists. Chemist August Kekulé's 1865 model, with the six carbon atoms located at the vertices of a regular hexagon, each bonded to one hydrogen atom, successfully predicted how many different compounds could be produced by substituting 1, 2, 3, 4, 5, or 6 of the hydrogen atoms by (for example) chlorine atoms.

Unfortunately the model was inconsistent with a basic postulate adopted by organic chemists (including Kekulé himself): a carbon atom normally forms a total of four bonds with other atoms. In the model each carbon has only three bonds (with the carbons on each side and the hydrogen atom). One could assume that three (alternating) pairs of neighboring carbon atoms form double bonds, but that would make it very likely to react with other atoms, contrary to the fact that benzene is a very stable molecule. It would also contradict evidence that all six CC bonds are equivalent. So Kekulé also proposed that the molecule rapidly oscillates between the two possible forms (1=2-3=4-5=6-) and (1-2=3-4=5-6=). At this point he seemed to be resorting to ad hoc explanations to explain away faults in his original model, and other chemists jumped in with their own models, usually leading to predictions that could be tested.

Yet despite its defects, Kekulé's model was successful in correlating and predicting the structure of more complicated larger molecules formed from benzene. Only with the advent of quantum mechanics was it possible to give a satisfactory explanation of these aromatic (benzene and benzene-like) molecules.

In the period 1930–1970, two competing versions of quantum chemistry were developed: valence bond and molecular orbital. At the end of this period the molecular orbital theory had emerged victorious (though not for long), in part because of successful predictions and in part because it was easier to use in calculating properties of large molecules. Notable confirmed predictions were the properties of cyclobutadiene (C_4H_4), a smaller benzene-like molecule; the Woodward-Hoffmann rules for pericyclic reactions; and Fukui's frontier orbital theory (Chapter 10).

Novelty Is Crucial

Are there any cases in which a theory was accepted *primarily* because of its successful predictions of novel facts, and would not have been accepted if those facts had been previously known? This is the strong version of the Popper-Lakatos thesis; some textbook presentations of the scientific method imply that it is the norm in science. So far I have found only one possible example: the collection of hypotheses now known as the old quantum theory.

Max Planck originally proposed the quantum hypothesis in an attempt to *explain* the observed frequency distribution of black-body radiation, by using physicist Ludwig Boltzmann's theory of entropy. This led him to introduce the hypothesis that the energy of an atomic oscillator is divided into discrete parts, written as a multiple of frequency ν multiplied by Planck's constant h, for mathematical convenience rather than as a new physical hypothesis. Einstein was the first to introduce the assumption that electromagnetic energy in space behaves under some circumstances like a stream of actual particles (which still have wave properties).

Einstein's theory yielded, directly or indirectly, several confirmed novel predictions, the most important being the following:

1. **Albert Einstein**'s law of the photoelectric effect, based on his light quantum hypothesis: light cannot eject an electron from a metal unless its frequency is high enough that $h\nu$ exceeds the required amount of energy to liberate the electron from the metal (Nobel Prize 1921); confirmed by **Robert A. Millikan** (Nobel Prize 1923) and others.
2. **Niels Bohr's** prediction, based on his model of the atom as a massive nucleus orbited by light electrons (Nobel Prize 1922), predicted that electrons with energy E passing through a low-pressure gas produce no radiation until E is greater than the energy difference between the ground state and an excited state of the atom; confirmed by **James Franck** (Nobel Prize 1925) and **Gustav Hertz** (Nobel Prize 1925).
3. **Einstein**'s prediction, based on his light quantum hypothesis, that the specific heats of solids go to zero as absolute temperature goes to zero; confirmed by **Walther Nernst** (Nobel Prize 1920).
4. **Arthur Holly Compton**'s prediction that X-rays act like particles when they collide with electrons, so the result can be described by using the classical laws of conservation of momentum and energy; at the same time the X-rays can be treated as waves, and the change in their wavelength is a simple function of the angle between incident and scattered rays; confirmed by Compton himself (Nobel Prize 1927). He also predicted that a recoil electron should emerge with appropriate momentum and energy; this was confirmed by **C.T.R. Wilson** (Nobel Prize 1927).

The boldface names involved in these achievements were rewarded by seven Nobel Prizes (six in physics, the other, to Nernst, in chemistry) during the years 1920–1927, all for confirmed predictions. One was awarded for a successful explanation, to **Max Planck** (1918).

Unfortunately the old quantum theory, despite its many successful predictions, was not a coherent theory but a collection of inconsistent hypotheses, and by 1927 it had been replaced by quantum mechanics. As noted above, that theory was rapidly adopted because it immediately *explained* all the mysteries left unsolved by the old quantum theory, before there was any time to test the many predictions made by the new theory. There was, however, one remaining mystery, whose solution was not generally known until quite recently: why did it take so long for the creators of quantum mechanics to receive *their* Nobel Prizes? (See Chapter 9 for the answer.)

3.6 ARE THEORISTS LESS TRUSTWORTHY THAN OBSERVERS?

The reason why some philosophers and scientists want to give more credit to novel predictions is presumably their suspicion that theorists may be influenced in reaching their conclusions by knowledge of the phenomena to be explained. But have we forgotten so soon the phenomenon of theory-dependence of observations widely discussed a generation ago by philosophers of science? Isn't it just as likely that observers will be influenced in reporting their results by knowledge of theoretical predictions of those results? Doesn't the praise for novel prediction imply a double standard for theorists and observers, based on a discredited empiricist conception of science with its neutral observational language? Wouldn't it be just as reasonable, as Rutherford suggested, to give more weight to observations performed before rather than after a theoretical prediction, to avoid confirmation bias?

The assumption that observation provides an objective test of theory is especially precarious in the light-bending case. Arthur Eddington, the person primarily responsible for carrying out the eclipse observation project, was already convinced of the truth of Einstein's theory before making the observations. Several physicists who examined the results concluded that they did *not* provide convincing evidence for the theory, but Eddington selected a subset that did confirm it. One modern expert, physicist and historian Francis Everitt, writes that "this was a model of how not to do an experiment . . . It is impossible to avoid the impression—indeed Eddington virtually says so—that the experimenters approached their work with a determination to prove Einstein right. Only Eddington's disarming way of spinning a yarn could convince anyone that here was a good check of General Relativity." (Everitt completed in 2011 his own test of general relativity.)

3.7 THE FALLACY OF FALSIFIABILITY: EVEN THE SUPREME COURT WAS FOOLED

It's not only physics that fails to follow Popper's prescription. Most theories in astronomy and historical geology address phenomena that cannot be brought into the laboratory for controlled tests, but span large domains of space and time. How could one make *testable* predictions when the predicted events will not happen for millions of years? Of course there are exceptions, such as big bang cosmology and plate tectonics, which predict that events in the past have observable but previously unobserved effects in the present. But it seems foolish to adopt a definition of science that excludes most of astronomy and geology.

Popper himself was guilty of this foolishness when he declared that Darwinian evolution is not a scientific theory but only a speculative description of past events, unable to make testable predictions about what will evolve in the future. To his credit, he recognized his own mistake and retracted that statement, admitting that Darwinism is indeed a scientific theory if one adopts a more flexible definition of prediction—in fact going in the direction of the physicists' definition, which includes logical deductions of known facts.

But the damage had been done: creationists gleefully spread the word that a respected philosopher had said evolution was not scientific, and therefore it deserved no more support or time in public schools than their own doctrine. Paradoxically, creationism does qualify as a scientific theory according to Popper's definition—it leads to several predictions, all of which have been falsified.

The belief that scientific status depends on falsifiability is so widespread that some scientists have asserted that their theories have this virtue even when they clearly don't. Alan Gross points out that some papers on evolutionary taxonomy take advantage "of the impressiveness that an openness to falsification confers without taking the risks that such openness should entail." Thus they "give a mistaken impression of the strength of particular taxonomic claims." Taxonomists, as philosopher and historian David Hull says, "think that they can use his [Popper's] Principle of Falsifiability to show that *their* classifications are truly scientific, while those of their opponents are not." Moreover, it is doubtful whether even physics is a science in Popper's sense; in Popper's favorite example, the confirmation of general relativity by the light-bending observation, Gross points out (citing a book by Jeremy Bernstein) that "physicists were persuaded as much [by] the elegance of the theory as by its alleged resistance to falsification."

Other scientists have complained that the insistence on hypothesis-testing has made it difficult to get support for other kinds of research. According to biologist Richard Lewontin:

> NIH [National Institutes of Health] is getting more picky. They no longer want to support purely observational work. They want very well formulated hypotheses that can then be tested by very precise testing of statistical measurements—things which I do not believe in. We no longer have any research money because the NIH is not interesting in this continued kind of exploration of patterns of variation and constraint.

The Popperian version of scientific method has even been endorsed by the Supreme Court. In their 1993 ruling on the case of Daubert *v. Merrell Dow Pharmaceuticals*, the Court downgraded the 1923 Frye rule that "expert opinion based on a scientific technique is inadmissible unless the technique is 'generally accepted' as reliable in the relevant scientific community." Instead, the Court, relying on *Amici Curiae* briefs by the American Association for the Advancement of Science and the National Academy of Sciences, declared that a trial judge could determine "whether a theory or technique is scientific knowledge" based on "whether it can be (and has been) tested. Scientific methodology today is based on generating hypotheses and testing them to see if they can be falsified; indeed, this methodology is what distinguishes science from other fields of human inquiry."

In my opinion the *Daubert* decision is exactly wrong in canonizing Popper's definition of "scientific," and lawyers like David Faigman who applaud it as a scientific revolution in the law are misinformed about how science works. David Mercer argues that the use of Daubert's Popperian verdict has "inhibited the legal admissibility of various forms of medical and other expertise in ways generally favourable to large industrial and corporate interests. In general, plaintiff experts tend to rely on scientific claims that are more likely to be novel and less likely to have been subject to extended and often expensive testing."

Sociologist Steven Yearley, noting that legal commentators have applauded the use of Popperian falsifiability as a criterion for being scientific, argues:

> There are several grave and well-known problems with the notion of testability as a distinctive feature of the scientific method. The chief among these are revealed in the apparently straightforward logic of falsification advanced by Popper. . . . First, it is never easy to work out whether any particular experiment should count as a test of an idea. Perfectly legitimate and successful scientists appear to ignore lots of experimental tests which seem to falsify their theories because they assume the test was poorly done and was inconclusive. . . . Second, scientists typically respond to a negative test result by revising their theory rather than by rejecting it [as pointed out by Imre Lakatos].

The Frye rule wisely refrained from imposing a single methodology on all the sciences for all time. Why shouldn't we accept the fact that modern physics, chemistry, astronomy, geology, and biology have been successful by following the different methods that work best for them? If chemists solve chemical problems by using the scientific method, that's fine, but it doesn't mean physicists have to use the same method.

Why should we care about how the word "scientific" is defined? Because it carries so much prestige in our society that many people want to claim it for their own ideas and practices. If we let philosophers like Popper define it for us, we will *not* have a defense against the proliferation of junk science, astrology, and creationism. We *will* see that definition used to denigrate sciences that we consider legitimate, such as evolutionary biology.

One cannot expect to find a single scientific method followed by all scientists, since (as I hope to demonstrate in this book) different sciences, each undoubtedly successful, can use widely different methods.

3.8 CONCLUSIONS

The predictivist thesis—that novelty enhances confirmation—gains little empirical support from the history of science. Any attempt to rescue it by redefining novelty in terms of what the theorist knew, when he knew it, and what he did or could have done with the information puts the philosopher in the position of a Watergate investigator without a Deep Throat.

Novel predictions played essentially no role in the acceptance of the most important physical theory of the twentieth century, quantum mechanics (Chapter 8). Physicists quickly accepted that theory because it provided a coherent deductive account of a large body of known empirical facts, many of which could be explained only approximately or arbitrarily by earlier theories; the new experiments confirming the quantum-mechanical calculations of intensities of Stark lines, and of the scattering of identical particles, were rarely mentioned as crucial evidence for quantum mechanics.

As evidence for general relativity, light bending did not get more weight because of its novelty, though that feature did direct attention to the theory and may thus be considered a contributor to its ultimate acceptance. On the other hand, just *because* it was a new phenomenon, it could not immediately count as unequivocal evidence for relativity until rival theories had tried and failed to retrodict it. Similarly, the discovery of the cosmic microwave background predicted by the big bang theory did not clearly confirm that theory until advocates of the rival steady-state theory had tried and failed to retrodict it.

In several other cases my findings are less clearly against the predictiveness thesis, but still not very favorable to it. In any case, there seems to be no justification for the claim that it is unscientific to accept a theory that has not made successful novel predictions.

One case does give limited support to the predictiveness thesis: novel predictions did have some evidential value in the establishment of periodic law, though most chemists did not consider them as important as the success of the law in organizing knowledge about the known elements. Another case in which novelty may have made a difference is the old quantum theory (see Section 2.5).

NOTES FOR CHAPTER 3

Section 3.1

073 This section is based in part on my article "Dynamics of Theory Change: The Role of Predictions," in *PSA 1994, Proceedings of the 1994 Biennial Meeting of the Philosophy of Science Association*, edited by David Hull et al., Vol. 2 (East Lansing, MI: Philosophy of Science Association, 1995), pp. 133–145. For a more comprehensive treatment see Eric C. Barnes, *The Paradox of Predictivism* (New York: Cambridge University Press, 2008).

073 *"we need to deal with 'science' not only as a social and cultural phenomenon"* Nyhart, "*Wissenschaft* and *Kunde*: The General and the Special in Modern Science," *Osiris* 27 (2012): 250–275, on p. 251.

073 *"philosophers of science study science as it ought to be"* Elliott Sober, *Evidence and Evolution: The Logic behind the Science* (Cambridge: Cambridge University Press, 2008), p. xv. Philosophers of science are sometime tempted to claim that scientists (or at least "good" scientists) actually do behave the way philosophers say they should. Peter Lipton, a leading advocate of the predictivist thesis (that novel predictions are better than explanations), seemed to be making this claim about the successful prediction of the return of Halley's comet in the eighteenth century, but backed away from it (see Section 2.3).

073 *(as Max Planck suggested) that the opponents just die out* Planck, *Scientific Biography and other Papers* (London: Williams and Norgate, 1950), pp. 3–4.

074 *apply the same critical scrutiny to historical arguments* To assert that evidence must be critically scrutinized does not imply, reflexively, that a positivist or hypothetico-deductive method should be used to test methodological claims. This kind of criticism has been used by some philosophers to avoid giving serious consideration to a major project which does exactly what I am proposing; see,

e.g., the critiques by Colin Howson, "The Poverty of Historicism," *Studies in History and Philosophy of Science Part A* 21: 173–179 (1990), and Alan W. Richardson, "Philosophy of Science and its Rational Reconstructions. Remarks on the VPI Program for Testing Philosophies of Science," *PSA 1992* 1 (1992): 36–46 (1992) of Arthur Donovan, Larry Laudan and Rachel Laudan, *Scrutinizing Science* (Boston: Kluwer, 1988), and the changing assessments of Thomas Nickles, "Remarks on the Use of History as Evidence," *Synthese* 69 (1986): 253–266 and "Historicism and Scientific Practice," *Isis* 80 (1989): 665–669.

074 *practitioners of naturalistic philosophy of science* See, e.g., David Faust and Paul E. Meehl, "Using Scientific Methods to Resolve Questions in the History and Philosophy of Science: Some Illustrations," *Behavior Therapy* 23 (1992): 195–211.

074 *Lakatos . . . postulated a "metacriterion"* Imre Lakatos, "Popper on Demarcation and Induction," in *The Philosophy of Karl Popper*, edited by P. A. Schilpp (LaSalle, IL: Open Court, 1974), pp. 241–273, on p. 246. Lakatos himself did not accept this metacriterion (ibid., p. 248), but does not clearly explain why. Daniel Garber rejects the assumption (attributed to Larry Laudan) that "a theory of scientific change must make sense of at least some principal episodes of the history of actual scientific change"; his goal is to promote good science, not to explain the past. See Garber, "Learning from the Past: Reflections on the Role of History in the Philosophy of Science," *Synthese* 67 (1986): 91–114, on p. 92. According to Steve Fuller, "One quick-and-dirty way of trying to disqualify the falsification principle as a standard for judging scientific evidence is to show that it fails to capture how science is ordinarily practised. This is certainly true . . . The principle has attracted controversy in philosophy of science because its strict application would end up discrediting most normal science as well as what might be called 'junk science.' (Imre Lakatos liked to raise this point to show that Popper was more radical—and less correct—than his reputation would suggest)." Fuller, "Author's Response [to reviews of his book *Kuhn vs. Popper*], *Metascience* 14 (2005): 19–30, on p. 25.

Section 3.2

074 "*What did the President know*" *Presidential Campaign Activities of 1972, S. Res. 60. Watergate and Related Activities. Phase 1: Watergate Investigation. Book 8.* DOC-TYPE: Hearings—Digital Collection, HEARING-ID: HRG-1973-PCA-0016, July 31, Aug. 1, 2, 1973, 389 pp., LexisNexis Congressional Hearings Digital Collection. Page 3088.
See also Gretchen Morgenson's article "Seeing vs. Doing," which begins: " 'What did they know, and when did they know it?' Those are questions investigators invariably ask when trying to determine who's responsible for an offense or a misdeed." *New York Times, Sunday Business* (July 25, 2010), pp. 1, 6 on p. 1.

075 *Success means, primarily, confirmed novel predictions* Lakatos writes: "Thus the crucial element . . . is whether the *new theory* offers any novel, excess information compared with its predecessor and whether some of this excess information is corroborated"; Lakatos, "Falsification and the Methodology of Scientific Research Programmes," in *Criticism and the Growth of Knowledge*, edited by Lakatos and A. Musgrave (New York: Cambridge University Press, 1970), pp. 91–196, on p. 120. See also his discussion of novel predictions on pp. 123ff. He gives a list of his favorite novel predictions in "The Methodology of Scientific Research Programmes," *Philosophical Papers* (New York: Cambridge University Press, 1978), p. 184.

075 *we should "have to give Einstein no credit"* Elie Zahar, "Why Did Einstein's Programme Supersede Lorentz's?" *British Journal for the Philosophy of Science* 24 (1973): 95–123,

223–262, on p. 101; Zahar, *Einstein's Revolution* (LaSalle, IL: Open Court, 1989), pp. 13–14.

075 *"A fact will be considered novel"* Zahar, "Why did Einstein's . . . ", p. 103, emphasis in original.

075 *"The historian has to read the private correspondence of the scientist"* Zahar, "Why did Einstein's . . . " pp. 102–103; Zahar, *Einstein's Revolution*, p. 16. John Worrall points out: "if we have to look at the way that a theory was constructed in order to decide on confirmation, then things seem to become very messy. Wouldn't we need to have Einstein's psyche available for inspection in order to know how he arrived at his theory of relativity?" Worrall, "Scientific Discovery and Theory-Confirmation," in *Change and Progress in Modern Science*, edited by J. C. Pitt (Boston: Reidel, 1985), pp. 301–331, on p. 306.

075 *found a 1915 letter to Arnold Sommerfeld* John Earman and Clark Glymour, "Einstein and Hilbert: Two Months in the History of General Relativity," *Archive for History of Exact Sciences* 19 (1978): 291–308. The letter is now easily accessible, in both the original German and in English translation: see *The Collected Papers of Albert Einstein*, volume 8, edited by Robert Schulmann et al., Part A (Princeton, NJ: Princeton University Press, 1998), Document 153 (English translation in separate volume). In a follow-up letter to Sommerfeld, ibid., Doc. 161, Einstein wrote: "The result of the perihelion motion of Mercury gives me great satisfaction. How helpful to us here is astronomy's pedantic accuracy, which I often used to ridicule secretly!" The original German versions of the two letters, with a facsimile of the first one, were published in Armin Hermann (ed.), *Albert Einstein/Arnold Sommerfeld Briefwechsel* (Basel/Stuttgart: Schwabe & Co. Verlag, 1968), pp. 32–37.
Elie Zahar, *Einstein's Revolution* (LaSalle: Open Court, 1989), does not cite the Earman-Glymour paper, nor does Jarrett Leplin, who claims that the Mercury perihelion was a novel prediction of general relativity because "it was not an explanatory task of Einstein's program to account for empirical regularities which violate Newton's theory of gravitation"; Leplin, "The Historical Objection to Scientific Realism," *PSA 1982* 1 (1982): 88–97, see p. 94); nor does Ronald Giere, who says the Mercury data played no role in the construction of Einstein's theory—see *Understanding Scientific Reasoning*, 2nd ed. (New York: Holt, Rinehart & Winston, 1984), p. 161.

075 *Einstein did in fact use the known behavior of Mercury's perihelion* Zahar, "Why did Einstein's . . . ", p. 257, emphasis in original.

075 *Gardner and others argue that Zahar's* N_2 *. . . cannot be used* Michael Gardner, "Predicting Novel Facts," *British Journal for the Philosophy of Science* 33 (1982): 1–15.

076 *Other definitions of "novelty" have been proposed* Henry Frankel, "The Career of Continental Drift Theory: An Application of Imre Lakatos' Analysis of Scientific Growth to the Rise of Drift Theory," *Studies in History and Philosophy of Science* 10 (1979): 21–66; John Worrall, "Scientific Discovery and Theory-Confirmation," in *Change and Progress in Modern Science*, edited by J. C. Pitt (Boston: Reidel, 1985), pp. 301–331, and "The Value of a Fixed Methodology," *British Journal for the Philosophy of Science* 39 (1988): 263–275; Jarrett Leplin, "The Historical Objection to Scientific Realism," *PSA 1982* (Proceedings of the Biennial meeting of the Philosophy of Science Association) 1 (1982): 88–97; Nancey Murphy, "Another Look at Novel Facts," *Studies in History and Philosophy of Science* 20 (1989): 385–388; Alan Musgrave, "Logical vs. Historical Theories of Confirmation," *British Journal for the Philosophy of Science* 25 (1974): 1–23; Richard Nunan, "Novel Facts, Bayesian

Rationality, and the History of Continental Drift," *Studies in. History and Philosophy of Science* 15 (1984): 267–307, and "Heuristic Novelty and the Asymmetry Problem in Bayesian Confirmation Theory," *British Journal for the Philosophy of Science* 44 (1993): 17–36.

"Are physicists simply wrong?" asks Wesley Salmon, in saying that the Mercury perihelion effect, though clearly not novel, confirms general relativity at least as strongly as the novel effect, light bending" (see Chapter 11). "Of course not. The fault lies somewhere in our philosophical account of confirmation." Salmon, Review of *Bayes or Bust* by John Earman, *American Scientist* 82 (1994): 91–92, on p. 92.

076 *"the absurdity that old evidence cannot confirm new theory"* Clark Glymour, *Theory and Evidence* (Princeton, NJ: Princeton University Press, 1980), p. 86; John Earman, *Bayes or Bust? A Critical Examination of Bayesian Confirmation Theory* (Cambridge, MA: MIT Press, 1992).

076 *Glymour's critique stimulated heated discussions* Deborah Mayo reached the same conclusion for the opposite reason: scientists do (she claims) give greater weight to novel predictions but Bayesian analysis does not, hence Bayesian analysis is inadequate.

"To the extent that the historical record indicates the importance of novelty (in any of the senses mentioned here), it simultaneously counterindicates these statistical philosophies" (Bayesian and Likelihood). Mayo, "Brownian Motion and the Appraisal of Theories," in *Scrutinizing Science*, edited by Arthur Donovan et al., 219–243 (Boston: Kluwer, 1988), p. 233. See also Nunan, "Novel Facts."

Colin Howson argued that Bayesian analysis is valid because it *does* give "high supportive power" to "any novel facts it predicts" but nevertheless rejects the claim that old evidence counts less than new. Howson, "Bayesianism and Support by Novel Facts," *British Journal for the Philosophy of Science* 35 (1984): 245–251;

"Accommodation, Prediction, and Bayesian Confirmation Theory," *PSA 1988* 2 (1989): 381–392. "The Poverty of Historicism," *Studies in History and Philosophy of Science* 21 (1990): 173–179.

Daniel Garber and others have argued that old evidence does count in science, and assert that Bayesian analysis can be formulated in a manner that is consistent with this fact. Garber, "Old Evidence and Logical Omniscience in Bayesian Confirmation Theory," in *Testing Scientific Theories*, edited by John Earman (Minneapolis: University of Minnesota Press, 1983), pp. 99–131; I. Niiniluoto, "Novel Facts and Bayesianism," *British Journal for the Philosophy of Science* 34 (1983): 375–379; Roger Rosenkrantz, "Why Glymour is a Bayesian," in *Testing Scientific Theories*, edited by John Earman (Minneapolis: University of Minnesota Press, 1983), pp. 69–97; Ellery Eells, "Problems of Old Evidence," *Pacific Philosophical Quarterly* 66 (1985): 283–302; Allan Franklin, *The Neglect of Experiment* (New York: Cambridge University Press, 1986); Bas C. Van Fraassen, "The Problem of Old Evidence," in *Philosophical Analysis*, edited by D. F. Austin (Boston: Kluwer, 1988), pp. 153–165; Colin Howson and Peter Urbach, *Scientific Reasoning: The Bayesian Approach* (LaSalle, IL: Open Court, 1989); James A. Kahn, Steven E. Landsburg and Alan C. Stockman, "On Novel Confirmation," *British Journal for the Philosophy of Science* 43 (1992): 503–516; Robert J. Levy, "Another Day for an old Dogma," *PSA 1992* 1 (1992): 131–141; Kenneth F. Schaffner, *Discovery and Explanation in Biology and Medicine* (Chicago: University of Chicago Press, 1993).

077 *"the typical realist argument against idealism "* Hilary Putnam, *Meaning and the Moral Sciences* (London: Routledge & Kegan Paul, 1978), pp. 18–19.

077 *the theory generating the prediction refers to something in the real world* See the collection of articles edited by Jarrett Leplin, *Scientific Realism* (Berkeley: University of California Press, 1984) and especially Leplin's remark on page 217 which rejects the position, apparently taken by Putnam, that *any* empirical success supports realism. The novelty issue is discussed by Leplin, "The Historical Objection to Scientific Realism," *PSA 1982* 1 (1982): 88–97; Alan Musgrave, "The Ultimate Argument for Scientific Realism," in *Relativism and Realism in Science*, edited by R. Nola (Boston: Kluwer, 1988), pp. 229–252; Martin Carrier, "What is Wrong with the Miracle Argument," *Studies in History and Philosophy of Science Part A.* 22 (1991): 23–36; Paul E. Meehl, "The Miracle Argument for Realism: An Important Lesson to be Learned by Generalizing from Carrier's Counter-examples," *Studies in History and Philosophy of Science Part A 23* (1992): 267–282. An earlier example (without the word "miracle") in connection with Mendeleev's periodic law is in Benjamin Harrow, *The Romance of the Atom* (New York: Boni & Liveright, 1927), p. 40.

John Wright argues, similarly, that "at least part of a theory" must be "true, or close to the truth in some sense," if it makes successful novel predictions, but does not discuss even one example of such success: "Metaphysical Realism and the Explanation of the Success of Science," in *Realism and Anti-Realism in the Philosophy of Science*, edited by R. S. Cohen (Boston Studies in the Philosophy of Science, 169; Kluwer, 1996), pp. 227–243.

077 *several successful novel predictions have come from theories now considered incorrect* A. d'Abro, *The Evolution of Scientific Thought*, 2nd ed. (New York: Dover, 1950), p. 402); David Knight, *A Companion to the Physical Sciences* (New York: Routledge, 1989), p. 120.

077 *pointed out, with many examples, by philosopher Larry Laudan* Laudan, "A Confutation of Convergent Realism," *Philosophy of Science* 48 (1981): 19–49.

077 *retreated to weaker versions of the miracle argument* Leplin, "Truth and Scientific Progress," in *Scientific Realism*, ed. Leplin, 193–217 (Berkeley: University of California Press, 1984), pp. 204, 213; Martin Carrier, "What is Right with the Miracle Argument: Establishing a Taxonomy of Natural Kinds," *Studies in History and Philosophy of Science Part A* 24 (1993): 391–409.

077 *really refer to entities that do exist* Philip Kitcher, *The Advancement of Science* (New York: Oxford University Press, 1993), p. 144. This is sometimes known as the "causal theory of reference"; see John Zammito, *A Nice Derangement of Epistemes: Post-Positivism in the Study of Science from Quine to Latour*, (Chicago: University of Chicago Press, 2004), pp. 72–74; Steve Fuller, *Social Epistemology* (Bloomington: Indiana University Press, 1988), p. 70.

SECTION 3.3

078 *Margenau made this quite explicit* Henry Margenau, *The Nature of Physical Reality* (New York: McGraw-Hill, 1950), p. 105. Another physicist, Rudolf Peierls, asserted: "From the physicist's point of view there is . . . no difference in principle between prediction and post-diction . . . and it is not clear why one should have a different relation to explanation than the other"; *Atomic Histories* (New York: American Institute of Physics, 1996), p. 291. (This is a reprint of his 1965 review of N. R. Hanson's *Concept of the Positron*, 1963.) Biologists and geologists do not follow the practice of physicists here, and occasionally complain about the ambiguous use of the word "prediction"—see Ernst Mayr, "How Biology Differs from the Physical Sciences," in *Evolution at a Crossroads*, edited by D. J. Depew and B. C. Weber (Cambridge, MA: MIT Press, 1985), p. 49; Arthur N. Strahler, *Science*

and Earth History (Buffalo: Prometheus, 1987), p. 15). But I have found occasional instances in the social sciences of prediction or even forecast being used to deduce known facts; see Ted Robert Gurr and Mark Irving Lichbach, "Forecasting Internal Conflict: A Competitive Evaluation of Empirical Theories," *Comparative Political Studies* 19 (1986): 3–38..

078 *a few philosophers have recognized this usage* Hilary Putnam, "The 'Corroboration' of Theories," in *The Philosophy of Karl Popper*, edited by P. A. Schilpp (LaSalle, IL: Open Court, 1974, article dated 1969), p. 234; Ronald N. Giere, *Explaining Science* (Chicago: University of Chicago Press, 1988), pp. 199–200.

078 *in the sense of offering intellectual understanding* E. G. Boring, "The Validation of Scientific Belief: A Conspectus of the Symposium," *Proceedings of the American Philosophical Society* 96 (1952): 535–539; Philipp G. Frank, "The Variety of Reasons for the Acceptance of Scientific Theories," in *The Validation of Scientific Theories* (Symposium at AAAS meeting, 1953), edited by Frank, pp. 13–26 (New York: Collier, 1961). Hilary Putnam, *Reason, Truth and History* (New York: Cambridge University Press, 1981), pp. 184–185; Peter Caws, *The Philosophy of Science: A Systematic Account* (Princeton, NJ: Van Nostrand, 1965), p. 333.

078 *deducing a known fact from a theory "explains" that fact* Peter Achinstein, *Particles and Waves* (New York: Oxford University Press, 1991), p. 6. Some philosophers, influenced by cognitive science, seem willing to use a broader definition; see Giere, *Explaining Science*, pp. 104–105. I use the word explain in the sense that (I assume) Giere uses it in the title of his book.

078 *Most of us have not yet reached that level of understanding* "I can safely say that nobody understands quantum mechanics" Richard P. Feynman, *The Character of Physical Law* (Cambridge, MA: MIT Press, 1967), p. 129. "We cannot make the mystery [of atomic behavior] go away by "explaining" how it works. We will just *tell* you how it works" (Feynman, Robert B. Leighton and Matthew Sands, *The Feynman Lectures on Physics* (Reading, MA: Addison-Wesley, l965), vol. 3, p. 1–1. "The new theories . . . are built up from physical concepts which cannot be explained in terms of things previously known to the student, which cannot even be explained adequately in words at all"; P. A. M. Dirac, *The Principles of Quantum Mechanics* (Oxford: Clarendon Press, 1930), p. v.

078 *Another way to express this distinction has been suggested* Giere, "Critical Hypothetical Evolutionary Naturalism," in C. Hayes and D. L. Hull (eds.), *Selection Theory and Social Construction* (Albany, NY: SUNY Press, 2001), pp. 53–70, on p. 55. Stephen R. Grimm has reviewed the usage of the words "explain" and "understand": Grimm, "The Goal of Explanation," *Studies in History and Philosophy of Science*, 41 (2010): 337–344.

Section 3.4

079 *influenced by knowing what the theory needs to explain* See the Krugman quotation in Section 1.2.

079 *almost any set of data can be fitted* Christopher Hitchcock and Elliott Sober, "Prediction versus Accommodation and the Risk of Overfitting," *British Journal for the Philosophy of Science* 55 (2004): 1–34.

079 *As philosopher and historian William R. Newman points out* Newman, "How Not to Integrate the History and Philosophy of Science: A Reply to Chalmers," *Studies in History and Philosophy of Science* 41 (2010): 203–213; *Atoms and Alchemy: Chemistry and the Experimental Origins of the Scientific Revolution* (Chicago: University

of Chicago Press, 2006). See Alan F. Chalmers, The Scientist's Atom and the Philosopher's Stone (Dordrecht: Springer, 2009); "Boyle and the Origins of modern Chemistry: Newman Tried in the Fire," *Studies in History and Philosophy of Science* 41 (2010): 1–20.

Section 3.5

080 *"You can always make news with doomsday predictions"* Quoted by John Tierney, "Economic Optimism? Yes, I'll Take That Bet," *New York Times/Science Times* (December 28, 2010), pp. D1, D3, on D3.

080 *Malthusian theories about the effect of population* Julian L. Simon, "Resources, Population, Environment: An Oversupply of False Bad News," *Science* 208 (1980): 1431–1437; *The Ultimate Resources* (Princeton, NJ: Princeton University Press, 1981); *Population Matters* (New Brunswick, NJ: Transaction Publishers, 1990).

080 *"the Malthusians paid up, although they didn't seem to suffer any professional consequences"* Tierney, "Optimism?" p. D3.

080 *the price of oil . . . would increase in the next five years* Tierney, "Optimism?" p. D3.

081 *from which he deduced several novel predictions* S. G. Brush, "Alfvén's Programme in Solar System Physics," *IEEE Transactions on Plasma Science* 20 (1992): 577–589.

081 *"Wegener's theory of continental drift was rejected"* Egbert Giles Leigh Jr., "Ronald Fisher and the Development of Evolutionary Theory. 1. The Role of Selection," *Oxford Surveys in Evolutionary Biology* 3 (1986): 187–223, on p. 189. The point is stated clearly by Marilyn Vos Savant, in response to a question from Scott Morris: how could the ancient Babylonians accurately predict eclipses if they didn't know that the earth actually travels around the sun rather than the reverse? She wrote: "They didn't need to know *why* the eclipse was occurring"; they just collected and analyzed a lot of data. "*This is an excellent example of how prediction—widely accepted by scientists as the truest test of the accuracy of a theory—is utterly inadequate*" (emphasis in original). See "Ask Marilyn," *Parade*, August 10, 1997, pp. 4–5.

082 *According to historian Richard Staley. Einstein's Generation: The Origins of the Relativity Revolution* (Chicago: University of Chicago Press, 2008), Part III.

082 *The 1919 eclipse test of Einstein's general theory of relativity* Popper, *Conjectures and Refutations* (ref. 25), pp. 33–37; "Autobiography of Karl Popper," in *The Philosophy of Karl Popper*, edited by P. A. Schilpp (LaSalle, IL: Open Court, 1974), pp. 3–181, on pp. 23–33; *Unended Quest: An Intellectual Autobiography* (LaSalle, IL: Open Court, 1976), pp. 31–44.

086 *some textbook presentations of the scientific method* Biologists seem to think it is the norm in physics. See, e.g., Leigh, "Ronald Fisher," who writes: "In physics, major theories have won acceptance . . . because they yielded a few precise, detailed, distinctive predictions that were so completely and unexpectedly in accord with observation that each of these theories 'could not but be near the truth.' So it was with Newton's celestial mechanics . . . quantum mechanics and the special theory of relativity" (pp. 188–189). As will be seen, at least two of these three examples are wrong.

Section 3.6

087 *to give more weight to observations performed before* See Section 1.4. On confirmation bias see M. J. Mahoney, *Scientist as Subject* (Ballinger, 1976); Ryan D. Tweney, Michael E. Doherty and Clifford R. Mynatt, *On Scientific Thinking* (New York: Columbia University Press, 1981).

087 *"this was a model of how not to do an experiment"* C.W.F. Everitt, "Experimental Tests of General Relativity: Past, Present and Future," in *Physics and Contemporary Needs*, vol. 4, edited by Riazuddin (New York: Plenum, 1980), pp. 529–555, on pp. 533–534. See also John Earman and Clark Glymour, "Relativity and Eclipses: The British Eclipse Expeditions of 1919 and their Predecessors," *Historical Studies in the Physical Sciences* 11 (1980): 49–85; Steven Weinberg, *Dreams of a Final Theory* (New York: Pantheon, 1992), pp. 96–97, and other criticisms cited in Chapter 11.

087 *Everitt completed in 2011 his own test of general relativity* Dennis Overbye, "52 Years and $750 Million Prove Einstein was Right," *New York Times*, May 5, 2011, p. A17; C. W. F. Everitt et al., "Gravity Probe B: Final Results of a Space Experiment to Test General Relativity," *Physical Review Letters* 106 (Published 31 May 2011): 221101 DOI:10.1103/PhysRevLett.106.221101. The experiment confirmed the predicted geodetic and frame-dragging effects.

Section 3.7

087 *he declared that Darwinian evolution is not a scientific theory* Karl Popper, "Autobiography," pp. 133–143.

088 *he recognized his own mistake and retracted that statement* Popper, "Natural Selection and the Emergence of Mind," *Dialectica* 32 (1978) 339–355; "Evolution," *New Scientist* 87 (21 August 1980): 611

088 *creationism* does *qualify as a scientific theory* See *Reports of the National Center for Science Education* and its predecessor, *Creation/Evolution*. For other examples see Joseph Agassi, "The Nature of Scientific Problems and their Roots in Metaphysics," in *The Critical Approach to Science and Philosophy*, edited by Mario Bunge (London: Free Press, 1964), pp. 189–211. But intelligent design, at least as presented by Michael Behe, is not falsifiable, as he admitted in response to my question after his lecture at the University of Maryland several years ago: "when philosophers of science testify as expert witnesses in trials concerning creationism and intelligent design [they] often make their case on the basis of demarcation between science and nonscience. The appeal of such a courtroom strategy is obvious. Show that creationism or intelligent design is not science and the case is closed. Judges [e.g., Judge John E. Jones III in the Dover case] like such clean criteria. So it's no surprise that the strategy has worked. But the problem with the demarcationist strategy was shown already more than 20 years ago, when, after the Arkansas case, Larry Laudan (1982) argued that each of the criteria of demarcation adduced by Michael Ruse (1982) in his expert testimony is violated by some well-accepted scientific theory and that many of the same criteria are satisfied by 'creation science.'" See articles by Ruse and Laudan in *Science, Technology and Human Values* 7, no. 40 (1982): 72–78 and no. 41, pp. 16–19. But the defects of demarcationism were "well known among philosophers of science" by then, so "why then, is [it] presented to the public as representing settled wisdom among philosophers of science?" Don Howard, "Better Red than Dead—Putting an End to the Social Irrelevance of Postwar Philosophy of Science," *Science & Education* 18 (2009): 199–220, on p. 215. See also Larry Laudan, "The Demise of the Demarcation Problem," in *Physics, Philosophy and Psychoanalysis*, edited by R. S. Cohen and L. Laudan (Boston: Reidel, 1983), pp. 111–127.

088 *Gross points out that some papers on evolutionary taxonomy* Alan Gross, *The Rhetoric of Science* (Cambridge, MA: Harvard University Press, 1990), 40–41, 47. David Hull, "The Principles of Biological Classification: The Use and Abuse of Philosophy," *PSA*

1976 (Proceedings of Biennial Meeting of Philosophy of Science Association), vol. 2, pp. 130–153, on p. 142; *Science as a Process: An Evolutionary Account of the Social and Conceptual Development of Science* (Chicago: University of Chicago Press, 1988), see index entry for "falsifiability."

088 *Gross points out (citing a book by Jeremy Bernstein)* Jeremy Bernstein, *Einstein* (New York : Viking Press, 1973) pp. 141–146.

088 *insistence on hypothesis-testing has made it difficult* R. C. Lewontin, interview in *Thinking about Evolution*, edited by R. S. Singh et al., vol. 2, pp. 22–61 (New York: Cambridge University Press, 2001), on p. 52. A similar problem has been recognized in science fairs, where students are forced to fit their research into the mold of scientific method even when it is not appropriate: see Valerie Strauss, "Science-Fair Hypothesis Fraying," *Washington Post*, February 20, 2001, p. A9.

088 *In their 1993 ruling on the case of* Daubert "William Daubert, *et ux., etc., et al.*, Petitioners, v. Merrell Dow Pharmaceuticals, Inc. No. 92–102. Argued March 30, 1993. Decided June 28, 1993." *West's Supreme Court Reporter Interim Edition*, volume 113B, Cases Argued and Determined in the Supreme Court of the United States, October Term, 1992 (St. Paul, MN: West Pub. Co., 1993), pp. 2786–2800. Most of the quotations come from the majority opinion delivered by Justice Harry A. Blackmun. The quotation "Scientific methodology today . . . " is taken by Blackmun from Green, "Expert Witnesses and Sufficiency of Evidence in Toxic Substances Litigation: The Legacy of *Agent Orange* and Bendectin Litigation," *Northwestern University Law Review* 86 (1992): 643–699. Blackmun also cites C. Hempel, *Philosophy of Natural Science* (1966) and Popper, *Conjecture and Refutations* (5th ed., 1989), for falsifiability as a criterion for being scientific. Chief Justice William H. Rehnquist, concurring in part and dissenting in part, wrote: "I defer to no one in my confidence in federal judges; but I am at a loss to know what is meant when it is said that the scientific status of a theory depends on its 'falsifiability,' and I suspect some of them will be, too" (p. 2800). Sheila Jasanoff argues, contrary to the view that the *Daubert* decision urged judges to "think like scientists," rather that it enhanced the autonomy of judges and their power over juries and scientists. It "codified . . . the naive sociology of science that was already an established component of judicial practice"—favoring evidence based on "objectivity—quantification, instrumental readings . . . the 'repertoire of empiricism' "—; Jasanoff, "Science and the Statistical Victim: Modernizing Knowledge in Breast Implant Litigation," *Social Studies of Science* 32 (2002): 37–69, on p. 51.

088 *lawyers like David Faigman who applaud it* David L. Faigman, "Is Science Different for Lawyers?," *Science* 297 (2002): 339–340.

088 *Mercer argues that the use of* Daubert's *Popperian verdict* David Mercer, "A parting shot at Misunderstanding: Fuller vs. Kuhn," *Metascience* 14 (2005): 3–7, on p. 6.

089 *Yearley, noting that legal commentators have applauded the use of Popperian falsifiability* Yearley, *Making Sense of Science: Understanding the Social Study of Science* (London: Sage, 2005), pp. 148–158, quotation from p. 155. He cites especially Bert Black, Francisco J. Ayala and Carol Saffran-Brinks, "Science and the Law in the Wake of *Daubert*: A new Search for scientific Knowledge," *Texas Law Review* 72 (1994): 715–802.

4

The Rise and Fall of Social Constructionism 1975–2000

> Science as something already in existence, already completed, is the most objective, impersonal thing that we humans know. Science as something coming into being, as a goal, is just as subjectively, psychologically conditioned as are all other human endeavors.
>
> —Albert Einstein *(1932)*

> Study of the history, philosophy, and/or sociology of science places one in a better position to judge what makes something a science than simply study in one of the sciences themselves.
>
> —*sociologist* Steve Fuller *(2009)*

> The sociologist knows less than the natural scientist, while the sociologist of science knows still less. Those engaged from day to day with the problem of reflexivity would, if they could achieve their aims, know nothing at all. We might say that SSK [Sociology of Scientific Knowledge] has opened up new ways of knowing nothing.
>
> —*sociologists* H. M. Collins and Steven Yearley *(1992)*

In this chapter I discuss the proposal that scientific knowledge, rather than being created as new ideas are developed and accepted on the basis of the success of predictions or explanations of phenomena, is *constructed* by some kind of social process. According to this proposal, scientists do not discover a preexisting reality; rather, whatever is constructed—including science itself—is then accepted as being a description of reality.

4.1 THE PROBLEM OF DEFINING SCIENCE AND TECHNOLOGY STUDIES

During the last quarter of the twentieth century there arose a new academic discipline known as science and technology studies (STS). Sometimes this discipline is called simply science studies, even if technology is still considered part of it. Since many practitioners of STS—here called "STSers"—advocated a very definite answer to the question "How did theories become knowledge?", we need to consider this discipline and its relation to history and philosophy of science.

STS used to be described as a combination of several disciplines: history of science, philosophy of science, sociology of science, history of technology, and so on. But as it now defines itself and as it is perceived by the rest of the academic world, STS has greatly diminished the role of history and philosophy of science and is now dominated by a particular version of the sociology of science called sociology of scientific knowledge (SSK). It has also tried to distinguish itself from another movement that shares the same initials, science, technology, and society.

This situation was described authoritatively (and, quite frankly, almost brutally) by historian and STSer Mario Biagioli in the introduction to his anthology The Science Studies Reader (1999). Biagioli states that the essays in his book, and the field of science studies generally, "focus on modern and contemporary science, mostly physical and biological" while "philosophy of science is largely absent." Medicine and technology, on the other hand, are excluded from the reader only because they would require additional volumes; mathematics is not mentioned at all. History of science is included not for its own sake, but because it offers useful material for sociological analysis.

At about the same time, STSer Sheila Jasanoff was complaining that while the Society for Social Studies of Science (4S), of which she was president, has always welcomed historians of science and "showered on them most of the prizes and honours at our Society's disposal," historians have not reciprocated. Their "attitude toward 4S's intellectual programme bespeaks . . . considerably greater wariness," she said. "A similar jitteriness about being caught out in risqué company marks the hiring practices of our major history of science departments. . . . to quote a number of historians of science, the constructivist strain that threads through so much contemporary work in S&TS seems to be a source of puzzlement and profound disquietude. Historians of science often seem to share with scientists the suspicion that the firm ground of reality will dissolve into the quagmire of make-believe if social constructivists are allowed to have their way with science and technology."

What is this constructivist strain—hereafter to be called social constructionism , or SC— that is allegedly responsible for the split between history of science and STS , even though some historians initially supported it, and many STSers later abandoned it? If scientific knowledge is indeed socially constructed, that factor would have to be included in any discussion of the reception of new ideas, along with prediction and explanation. I have therefore examined the claims of SC in some detail.

4.2 THE RISE OF SOCIAL CONSTRUCTIONISM

> No one can describe the coherence of a thought system better than an anthropologist.
>
> —*anthropologist and* STSer Bruno Latour *(2010)*

> Through the first nine decades of the twentieth century there were few scholarly books addressing themselves to the act of lying, and among those all but a very few presented the act as simply wrong. Suddenly, however, in 1990

> there opened a gusher of scholarly publications with titles chosen to make it clear that their frame of reference turned modernity's upside down. Instead of taking truth telling as the norm and the rule . . . now lying is presented as a usual, acceptable, even commendable practice . . . the new legitimacy of lying . . . effectively differentiates . . . postmodernity from . . . modernity.
>
> —*historian* Paul Forman *(2012)*

For the purposes of this chapter it is convenient to use as our definition of social constructionism historian Jan Golinski's definition of constructivism:

> Scientific knowledge is a human creation . . . rather than simply the revelation of a natural order that is pre-given and independent of human action.

Consider the assertion by Supreme Court Justice Robert Jackson that "we are not final because we are infallible, but we are infallible only because we are final." The SC counterpart is something like "scientists do not accept a statement because it is true; rather, a statement is true because scientists accept it."

The subfield of STS that advocates SC is often known as sociology of scientific knowledge (SSK). It is sometimes considered part of the more general cultural movement called postmodernism. That would suggest an origin in the 1970s or after, but according to historian Mary Jo Nye, the basic principles of SC were discussed by philosophers and scientists in the 1930s. She discusses physical chemist Michael Polanyi, who, reflecting on resistance by other scientists to his surface theory of adsorption and his work on X-ray diffraction in cellulose, turned to "sociological explanation, rather than logical explanation, for the mechanism by which scientific priority and recognition are accepted within the structure of scientific authority." But, she writes, Polanyi never doubted that scientific research uncovers truth about the nature of the world. So he did not accept the antirealist position that came to be associated with SC in the 1970s.

In this section I will summarize the contributions to SC of Kuhn, Forman, Bloor, Latour, Collins, Pickering, Shapin, and Schaffer. The following section will sketch how SC was undermined or substantially revised by some of these scholars.

Social constructionists often trace the origin of their movement to historian and philosopher of science Thomas S. Kuhn's *Structure of Scientific Revolutions* (1962). In his book he argued that there is no objective criterion—no appeal to a real world out there, independent of ourselves—that can justify the decision to switch from one paradigm to another. This is because successive paradigms are "incommensurable": they do not assess theories by the same criteria, so there is no higher court than the collective decision of the scientific community. Thus social factors must be involved, at least in a revolution, though Kuhn did not say precisely what they are or how they work.

In 1971, Kuhn's student Paul Forman published his study of Weimar culture and indeterminism, often invoked by social constructionists as a canonical case. Kuhn was very impressed by this study Forman claimed that Werner Heisenberg's indeterminacy principle, and more generally the randomness associated with the standard interpretation of quantum mechanics, were accepted by mathematicians and physicists as part of their response to criticisms of science in German-speaking countries in the 1920s. Although (in my opinion and as shown in detail by philosophers

Robert Nola and Howard Sankey) Forman did not prove that the change in physics was *caused* by changes in the cultural environment, he did provide a large amount of evidence to suggest a *correlation* between these changes—a correlation that exemplifies a series of connections between science and culture in the nineteenth and twentieth centuries. Moreover, while the Forman thesis can be and has been criticized as an inaccurate account of an episode in the history of science, at least it can be *tested* by seeking better historical evidence.

A very influential version of social constructionism, called the strong program was developed in the 1970s by David Bloor, Barry Barnes, and other sociologists at Edinburgh. They rejected the tradition of explaining the development of *accepted* ideas as the uncovering and elucidation of true facts about the world, while explaining stubborn adherence to *rejected* ideas by invoking social or personal factors that caused scientists to deviate from the correct path of science. Instead, Bloor introduced a symmetry principle: the same causes should be used to explain truth and error. Or rather, the sociologist should not presume to judge what *is* the truth about nature—that's the job of the scientist—but only try to explain how scientists come to *decide* what the truth is. (Up to this point, my own agenda in this book agrees with Bloor's.) Instead of asserting that scientists accept a theory because it is true, one should say that a theory is true because scientists accept it. (Here is where I start to disagree.) *All* knowledge, they seem to say, is socially constructed; it is the task of the sociologist to construct a sociology of scientific knowledge (following in the steps of Karl Mannheim's sociology of knowledge).

Bloor frequently protests that, contrary to critics, he does not say social forces completely determine scientific knowledge; other factors are always present too. But he does seem to reject the possibility that some sciences, or some scientists, could be more objective and less influenced by social factors than others. As has often been remarked, social scientists liked Kuhn's theory for just this reason: Kuhn did not support the conventional premise that the physical sciences are "harder," their results more reliable, than the behavioral and social sciences. That is just the sort of premise that the followers of sociologist Robert K. Merton in the previous generation were willing to test, and sometimes confirm. For example, studies of the peer review system by Merton and his colleague sociologist Harriet Zuckerman showed that referees for a major physics journal were not likely to give more favorable reports on papers submitted by higher-status authors, whereas studies by other sociologists showed that a leading sociological journal was more likely to publish papers submitted by higher-status authors even when their referees judged them to be equally good.

If (according to Mertonian research) sociologists are more biased than physicists against low-status authors, does that undermine the validity of sociological SC research? This is known as a reflexive objection to SC (its claims apply to itself). The answer is that social constructionists are willing to admit that their own results are socially constructed, but only because they assert that *everyone's* results are socially constructed. They simply ignore the contrary findings of Zuckerman and Merton.

Advocates of the strong program and other social constructionists attempted to become knowledgeable observers of (and sometimes even participants in) the scientific community in order to find out how discoveries and ideas are developed and accepted; another group took the approach of the anthropologist who can observe only the *behavior* of another culture, without really understanding it. The most famous example of this approach is a book by Bruno Latour and sociologist Steve Woolgar, titled (in its first edition) *Laboratory Life: The Social Construction of Scientific Facts* (1979). As summarized by sociologist Pierre Bourdieu, Latour and Woolgar claim that products of science are "artificial in the sense of manufactured" and thus are "fictitious, not objective, not authentic." To "produce this effect of 'derealization,' the authors . . . highlight the very important role of *texts* in the *fabrication of facts as fiction*. They argue that the researchers they observe during their ethnography at the Salk Institute did not investigate things in themselves; rather they dealt with 'literary inscriptions' produced by technicians working with recording instruments . . . the researchers' naïvely realist belief in a reality external to the laboratory is a pure illusion."

Similar statements are made by sociologist Karin Knorr Cetina, another leader in the field of laboratory studies. She wrote:

> Constructionism is one of the major, perhaps the major, outcome of laboratory studies . . . [c]onstructionism holds reality not to be given but constructed . . . There are, for constructionism, no initial, dissimulatable "facts": neither . . . scientific objectivity, nor reality itself.

That's clear enough, but then two pages later:

> Constructionism did not argue the absence of material reality from scientific activities, it just asked that "reality," or "nature," be considered as entities continually retranscribed from within scientific and other activities.

And 11 pages after that Cetina notes that a:

> Frequent criticism of laboratory studies is that they assert that material objects become real as a result of the inquiry. They are snapped into existence through the accounts produced by science, which most readers find "wildly implausible."
>
> However, there is a more sympathetic reading of these claims according to which what does indeed come into existence, within, usually, a longer term process, when science "discovers" a microbe or a subatomic particle, is a specific reality distinguished from other entities (other microbes, other particles) and furnished with a name, a set of descriptors, and a set of techniques in terms of which it can be produced and handled. In other words, some part of a preexisting material world becomes specified and thereby real . . .

I don't understand how the material world can exist but is not real. Do we go back to Aristotle's potential reality, or do we assume that it was like Schrödinger's cat before we opened the box to observe whether it is dead or alive?

Sociologist Harry Collins analyzed several controversies in modern physics. His claims are generally less antirealist that those of Latour and Woolgar and Knorr Cetina, but he became notorious for a single statement. In describing what he called the relativist program, he wrote that "the natural world has a small or nonexistent role in the construction of scientific knowledge." This seems to mean that scientists do not discover true facts about the world; they just make them up to suit their ideological or social interests. Scientists who read this statement protested that Collins is trying to debunk science. Collins and his defenders claim that the quote is taken out of context (I have given a more extended quotation in the notes to this chapter, so you can judge for yourself).

In other publications at the same time Collins asserted that he did not advocate ontological or epistemological relativism. Rather, he was recommending that the sociologist should adopt a relativistic or nonjudgmental attitude to the scientific theories and phenomena being studied, in constructing an account of the process of scientific research.

At this point we encounter a fundamental ambiguity in the meaning of the phrase "social construction." For Collins and some others, it sometimes does not mean that the scientist is constructing knowledge (rather than discovering it), but rather that the sociologist is constructing an explanation of how scientists do science, without making any claims about whether scientific knowledge corresponds to reality. We might call this sociological constructionism.

Forman, on the other hand, went beyond this position, arguing that social forces actually influenced the content of science in the early days of quantum mechanics. Physicist and historian James Cushing argued that the acceptance of the (acausal) Copenhagen interpretation of quantum mechanics by physicists was "historically contingent": David Bohm's rival causal interpretation might have been accepted instead if certain historical events had been different or in a different sequence. Thus the influence of Weimar culture might have been partly responsible for the outcome.

Collins's original statement that nature plays little or no role in our knowledge about it sounds like social determinism. For a nonsociologist like me, it is often hard to tell whether a particular claim is to be interpreted as sociological construction, social influence (which might be identified with Bloor's weak program), or social determinism.

Another major work from the mid-1980s was sociologist Andrew Pickering's *Constructing Quarks*. This is an excellent account of the origin of the modern theory, highly praised by historian Yves Gingras and physicist and historian Sylvan S. Schweber in their review. Here the social constructionism is only implicit through most of the account, and becomes blatant only in the final chapter where the author proclaims:

> On the view advocated in this chapter, there is no obligation upon anyone framing a view of the world to take account of what twentieth-century science has to say.

The statement is literally correct: that *is* the view he adopts in that chapter. Again, a scientist reading the statement would get the message that science does not discover true facts about the world but just makes them up. (For the context of this quotation see the longer quote in the notes.)

Finally, I note the very influential book *Leviathan and the Air-Pump*, by sociologist and historian Steven Shapin and historian Simon Schaffer. The authors argue that the debate between philosopher Thomas Hobbes and physicist and chemist Robert Boyle, over the legitimacy of experiments as a way of establishing truth, was ultimately won by Boyle not because he was right but because he mobilized more effectively the social forces active in seventeenth-century England. Boyle succeeded in constructing the experimental method as the correct way to do science; Hobbes's objections, which Shapin and Schaffer consider valid, were rejected for political rather than scientific reasons.

4.3 THE FALL OF SOCIAL CONSTRUCTIONISM

> Without the belief that it is possible to grasp reality with our theoretical constructions, without the belief in the inner harmony of the world, there could be no science. This belief is and always will be the fundamental motive for all scientific creation.
>
> —Albert Einstein *(1938)*

> Nature, apparently vanquished by some of the early "relativists," seems to be making a comeback as an explanatory factor. . . . The constructivist paradigm that is the very core of STS, according to some theorists, is considered moribund, dead, or a grave error by others.
>
> —Sal Restivo *(1995)*

> The leaders of science studies themselves bemoan a lack of rigor, even a crisis, in their field. . . . What has happened to the fizz and feistiness of science studies? *Où sont les programmes forts d'antan?*
>
> —Lorraine Daston *(2009)*

> Deconstructing science is a fool's game. In the '90s, literary critics used to try. They'd argue that science is a system of metaphors, complete with a style and an ideology, rather than the royal road to the truth. They were laughed at as cultural relativists, posers high on Gauloises and nut jobs who didn't believe in gravity.
>
> Science writers play rough. They like hoaxes, humiliations and Oxbridge-style showdowns that let them use words like "claptrap" and "gibberish." There's a reason people don't call themselves deconstructionists and pick fights with science anymore. The old battle is won: books called "The Science of X" fly off the shelves, while "The Culture of" books are remaindered.
>
> —Virginia Heffernan *(2010)*

Although some of its advocates claim that social constructionism is still alive and well (except in Paris), there are unmistakable signs that it has been declining since the 1980s. The famous Sokal hoax, though not a major cause of this decline, was a telling indication that the movement was in trouble. It was part of a counterattack against SC by those who considered it antiscience.

In the early 1990s, a few scientists became aware of SC publications and related postmodernist writings in social science and the humanities that denied that science uncovers objectively valid knowledge about the world. At a time when science seemed to be under attack from several powerful groups in society—budget-cutting politicians, polluting corporations, religious fundamentalists—these scientists felt threatened by the criticism of professors on the other side of the campus.

Biologist Paul Gross and mathematician Norman Levitt struck back in 1994 with a book titled Higher Superstition: *The Academic Left and its Quarrels with Science.* The controversy started (or at least aggravated) by this book became known as "science wars." In 1996 a special issue of the postmodernist journal *Social Text* was devoted to the topic. (According to postmodernism, everything is a text.) None of the articles appeared to support the science side of the wars, but one by physicist Alan Sokal turned out to be a clever parody of the postmodernist SC view of science. It included the claim that the value of B (ratio of circumference to diameter of a perfect circle) has no objectively fixed value, but varies from one culture to another. The editors of *Social Text* were so completely ignorant of science that they failed to realize how ridiculous some of Sokal's statements were, nor did they bother to have his paper refereed by a scientist. When Sokal revealed the hoax (it might more appropriately be called a sting), it was widely publicized as an exposure of the incompetence of the editors and an embarrassing defeat for constructionist STS—even though the editors and most of the contributors were not leaders of that discipline.

Less publicized but more destructive was the fact that two of SC's founding fathers—Kuhn and Latour—disinherited it, while other leaders seemed to lose their enthusiasm. Although in 1970 Thomas Kuhn had admitted that "some of the principles deployed in my explanation of science are irreducibly sociological," by 1983 he could see the unintended consequences of *Structure*. In his remarks after accepting the Bernal Award of the Society for Social Studies of Science (known as 4S), he insisted that there was a large gap between his concerns and those of 4S. One cause was

> the background of the people being recruited into the social study of science. If the gap is to be bridged, that will have to be done by people with sufficient specialized training to read the technical papers of the scientists they study.

Another cause was the careless application of models from other areas of sociology to the sociology of science, and especially an impoverished concept of interests that excluded the motivations of scientists themselves.

Eight years later, Kuhn complained that the situation had become even worse. The strong program in the sociology of science seemed to claim "that power and interest are all there are. Nature itself, whatever that may be, has seemed to have no part in the development of beliefs about it." Such absurd claims represented "deconstruction gone mad."

How did the social constructionists respond to Kuhn's attack? Some simply refused to believe that Kuhn meant what he said. Thus David Bloor wrote: "despite some uncharacteristically ill-focused remarks of his own to the contrary, Kuhn is properly called a 'social constructivist.'" Others deny that they have made the extreme claims ascribed to them by Kuhn, or complain that their frequently quoted phrases have been taken out of context. Only a few, like "Paul Forman, have pursued their own research agendas " without accepting or rejecting the SC label.

In the meantime, the term "paradigm" (which Kuhn himself abandoned in his later writings) has been degraded to a label for any idea or thesis that one wants to use without taking the trouble to justify. For example, Alberto R. Gonzales, appointed in 2005 as attorney general in the administration of George W. Bush, wrote the following in a 2002 memorandum about how the Geneva Convention should apply to the treatment of captured members of Al Qaeda and the Taliban:

> The nature of the new war places a high premium on . . . factors, such as the ability to quickly obtain information from captured terrorists and their sponsors in order to avoid further atrocities or war crimes . . . In my judgment, this new *paradigm* renders obsolete Geneva's strict limitations on questioning of enemy prisoners.

Pet owners may also experience paradigm shifts, with better evidence than Gonzales seemed to need. Veterinary scientist Linda Lord examined the common opinion that cats (unlike dogs) dislike wearing collars and may rip them off or injure themselves trying to do so. After studying 538 collared cats for six months, she concluded that "cats will tolerate wearing a collar , and this could be a new paradigm shift in thinking."

What about the other founders of social constructionism? David Bloor, when he received the Bernal Award, gave a speech whose title "Remember the Strong Program?" revealed his fear that no one does. Or rather, as the text explained, that what the science studies community remembered as being the strong program was not what Bloor himself meant by it.

Sociologist Stephen Cole challenged the SC advocates to produce just one example of social influence on the *substantive content* of the core of accepted scientific knowledge, as opposed to the direction and rate of progress of research. In response, Bloor pointed to Andrew Warwick's study of the reception of relativity at Cambridge University. But this example, which concerns only the years *before* British physicists had accepted Einstein's theory, shows only that social influences affect how science works at the frontier (before the new theory is accepted)—which even Cole accepts. (See Chapter 11 for further discussion of the reception of relativity.)

In a comprehensive survey of science studies published in 1997, anthropologist David J. Hess concluded:

> By the 1990s there had been several developments beyond the sociology of scientific knowledge, so that SSK had come to occupy a rearguard position. A major factor was the explosion of feminist, anthropological, and cultural studies of science and technology, which often began with SSK insights but moved beyond the concerns of the SSK analysts. The second was the turn to technology, policy, the environment, and public understanding of science issues. . . . feminist science studies . . . had emerged in the 1980s but went largely ignored by most of the constructionist research programmes.

He pointed out that STS must justify the existence of its academic programs in a world where college budgets are being cut. The best way (politically) to do this is to show the value of STS in guiding policy decisions that involve science. But that "requires an exit from the neutrality and relativism of the constructivist period."

Similarly, in 2005, Karyn L. Freedman wrote:

> The Strong Programme is old school SSK [sociology of scientific knowledge], and from the perspective of those working on the vanguard of the Science Studies of today it arguably holds little more than historical value.

When Latour and Woolgar published the second edition of their book *Laboratory Life* in 1986, they pointedly omitted the word "social" from its subtitle. In the meantime Latour and sociologist Michel Callon had started to develop their actor-network theory as a replacement for social constructionism. In a 1992 article, "One More Turn After the Social Turn," Latour asserted that "social studies of science are at a standstill." Either the research is too microscopic to account for worldviews and global features of society, or too macroscopic to come to grips with the detailed content of science (quoted at length in the notes to this chapter).

Latour's solution was to apply Bloor's symmetry principle more symmetrically than Bloor himself had done. Rather than explaining truth with nature and error with society, Bloor had wanted to explain both truth and error with society. Latour proposed to explain *both* nature and society in the same terms. Otherwise one would be in the odd situation of claiming that society is more real than nature (if society is invoked to construct nature). In 1999 he wrote that full-blown social constructionism is a position by which one could not be convinced for "more than three minutes. Well, let's say an hour, to be fair."

Latour's coauthor Steve Woolgar has also abandoned SC (though not as loudly) because it did not go far enough. In 1983 he wrote:

> It is perhaps especially timely to ask whether social studies are now in danger of being bogged down by constructivism. Will the repeated application of the constructivist formula prevent us from yielding the full potential of our acquaintance with the details of scientific practice? . . . When applied to the social study of science. . . it seems to me that constructivism does us a grave disservice. For unless we are to be guided by some political motive for the "demystification" of science, constructivism can only be a distraction from any attempt to come to terms with the fundamentals of knowledge production.

In a 1988 book, Woolgar presented what seemed to be a strong argument in favor of SC. But at the end he undermined it, by using the reflexive objection in a somewhat different way that did Bloor (Section 4.2). According to the sociology of scientific knowledge, all the results of science are socially constructed. But sociology claims to be a science. Hence this result of SSK must itself be socially constructed, rather than a discovery about the real nature of science. So science is *not* really socially constructed.

In a 2004 article, Latour confessed his feelings of guilt for having undermined the authority of science by his constructivist writings. Quoting a Republican strategist who proposed to stave off any remedial action by stressing the lack of scientific certainty about global warming, Latour admitted:

> I myself have spent some time in the past trying to show "*the lack of scientific certainty*" inherent in the construction of facts. . . . But I did not exactly aim at fooling the public by obscuring the certainty of a closed argument—or did I? After all, I have been accused of just that sin. . . . In which case the danger would not longer be coming from an excessive confidence in ideological arguments posturing as matters of fact . . . but from an excessive *distrust* of good matters of fact disguised as bad ideological biases! . . . And yet entire PhD programs are still running to make sure that good American kids are learning the hard way that facts are made

> up, that there is no such thing as natural, unmediated, unbiased access to truth . . . and so on, while dangerous extremists are using the very same argument of social construction to destroy hard-won evidence that could save our lives. Was I wrong to participate in the invention of this field known as science studies? . . . a certain form of critical spirit has sent us down the wrong path, encouraging us to fight the wrong enemies and, worst of all, to be considered as friends by the wrong sort of allies.

To make matters even worse, Latour continues, science studies did not even achieve its original goal of giving "a social explanation of scientific facts."

> At first we tried . . . to use the armaments handed to us by our betters and elders to crack open . . . religion, power, discourse, hegemony. But fortunately (yes, fortunately!) . . . we witnessed that the black boxes of science remained closed and that it was rather the tools that lay in the dust of our workshop, disjointed and broken.

Six years later, Latour seemed to revert to his earlier position, arguing (without giving any new evidence) that "construction and truth become synonymous," so we don't have to choose between them. But for Latour, "construction" does not necessarily mean *social* construction. Indeed, "social constructivism obliges us to have as many misconceptions about its presumed tireless laborer as it does about the entities it mobilizes."

At the same, he announced that the Latour-Callon actor-network theory had been superseded by a new approach:

> We seek to replace the black and white television image of the "actor-networks" with a color picture, by fabricating enough analyzers to register the principal contrasts that seem important to the "actors-themselves," our only masters.

Harry Collins, as we have seen, no longer claims that scientists construct knowledge rather than discover it, his sociological constructionism (the methodological relativism mentioned above), like Kuhn's bland disciplinary matrix, seems unobjectionable and hardly worth fighting a science war about. Instead, he now claims that he can be an expert on theoretical physics (e.g., the relativistic theory of gravitational waves) without understanding the mathematics on which it depends.

Bloor and his Edinburgh colleagues Barry Barnes and John Henry also renounce that claim , asserting that

> sociologists should be willing to acknowledge the existence and the causal relevance of the physical environment when they study the growth of knowledge. And having acknowledged this, they should acknowledge also the ability of individual human beings to monitor the physical environment and learn about it.

After all, they point out, *animals* learn from experience and adjust their behavior (e.g., pushing levers) to rewards and punishments; "it would be perverse to insist that what rats manage to accomplish in this context, human beings cannot hope to emulate."

One has to wonder *whom* they are accusing of being perverse enough to deny such an obvious fact—if not their social constructionist rivals. The charge seems directed against Collins, and they explicitly reject the methodological relativism he used to justify his 1981 statement that scientific knowledge is constructed with little or no input from the world. They also refuse to go along with the antirealism expressed by other Social Constructionists. Barnes, Bloor, and Henry don't explicitly reject social constructionism; rather (if their index is accurate), they never mention it at all.

Rather than defend his views against such attacks, Collins simply declared victory and left the battlefield. In 2002 he proclaimed, in a paper with Robert Evans, a third wave of science studies to build on the success of social constructionism, now called Wave 2. Wave 1, in the 1950s and 1960s, wanted only to explain and reinforce "the success of the sciences, rather than questioning their basis." Wave two stressed social construction, but noted that

> by emphasizing the ways in which scientific knowledge is like other forms of knowledge, sociologists have become uncertain about how to speak about what makes it different . . . they have become unable to distinguish between experts and nonexperts.

But now we (sociologists of scientific knowledge) should claim that *we are* "experts in the field of knowledge itself." Thus while Wave 2 replaced Wave 1, Wave 3 builds on the success of Wave 2. Having shown that "science and technology are much more ordinary than we once thought," we must now explain how they *do* have a special place because of expertise and experience—but "with enough delicacy" to avoid admitting that we were wrong, which would "shatter the whole edifice" of relativism. If sociology of scientific knowledge is not going to be completely useless in public policy, we must adopt a position that "seems incompatible with much that the authors of this paper have previously argued." But not to worry; all we need is a little compartmentalization.

Collins and Evans did not enjoy an enthusiastic reception from the STS community. Some initial reactions by Sheila Jasanoff , Brian Wynne, and Arie Rip, with a response from Collins and Evans, were published in 2003. Jasanoff characterized the third wave as a retrenchment from the "unbridled social constructivism" of the second wave. Collins and Evans protested that it was not a "drawing back" but a "drawing away from the exclusive concentration on the construction of knowledge that was vital in the early days of Wave Two . . . 3-wave moves forward to new questions to do with the making of policy in real time, before the knowledge has been socially constructed."

It is historians of pre-twentieth-century science, whose concerns had been effectively excluded from Wave 2 (see Section 4.1), who are most enthusiastic about the study of expertise. Historian Eric H. Ash edited a volume of *Osiris*, the annual publication of the History of Science Society, on the topic "Expertise: Practical Knowledge and the Early Modern State."

While citing Collins as one of the instigators of the current scholarly interest in expertise, Ash does not accept his view that "expertise is to be treated as real and as more than an attribution by others." He writes:

> Much of Collins's concern stems from a sort of backpedaling from the stronger version of the social constructivist argument in science and technology studies (STS). His concern is that as STS has systematically called into question the privileged status of scientific knowledge as 'truth' about nature, it has had a deleterious effect upon public, democratic debates touching on issues of science and technology, because it has undermined the authority of the scientific community and hindered its ability to bring its greater expertise to bear. . . . However, I would argue that expertise remains a negotiated concept, one that relies on a broad consensus for its meaning and efficacy.

Over the past few decades Collins has in fact tried to make himself an expert on the experimental search for gravitational waves, by observing and interacting with physicists. In his book *Gravity's Ghost* (2011) he describes his current view:

> Wave Three makes it explicit that in spite of the logic of Wave Two, which shows how sciences [sic] claim to true knowledge can be "deconstructed," science is still the best thing we have where knowledge about the natural world is concerned.

He makes it clear that the existence or nonexistence of gravity waves is an objective fact about the world, not a social construction. There is still a place for a sociology of physicists based, for example, on observations of how they decide to accept or reject a particular piece of data; thus

> the "proximate cause" of the first discovery of a gravitational wave will be, not a physical event, but a vote . . . A vote is a sure sign that something sociological is going on.

But Collins apparently no longer considers himself an STSer; he rejects

> the popular "Machiavellian" approaches to science found in the field of Science and Technology Studies, which . . . see the achievement of scientific success as essentially a political process.

At the same time his advocacy of *expertise* translates into a defense of science against the attacks of postmodernists (who were affiliated with social constructionists in the science wars of the 1990s):

> The attack on experts and expertise that has come with the academic movement known as postmodernism actually does the same work as the attack on the professions begun by Margaret Thatcher and Ronald Reagan and take the same view on experts as religious fundamentalists . . . The arms of this grotesque pincer are squeezing science from three sides.

Andrew Pickering, when faced with severe criticism of his *Constructing Quarks* by philosophers of science in 1990, did not strongly defend it but changed his position in a way that seems to me to water down social constructionism. In his reply and at greater length in his book *Mangle of Practice*, he describes the resistance of the material world to scientists' practice. With this change (which he regards as only a refinement or elaboration of his original theory) he can deny that he is talking about mere construction. He also denies that he is rejecting realism, but I have great difficulty understanding his position. He calls his picture of scientific practice "a realist one . . . It points to a constitutive role for reality—the material world—in the production of knowledge, but it carries no necessary connotation of correspondence (or lack of correspondence) for the knowledge produced." I gather that the Edinburgh sociologists are not very enthusiastic about this position either.

The process of testing constructionist claims by historical evidence (rather than just complaining about antiscience bias) was already underway. In 1998, philosopher Noretta Koertge edited a collection of essays criticizing in considerable detail several social constructionist works, under the title *A House Built on Sand: Exposing Postmodernist Myths about Science.*

The response from the social construction establishment was quite revealing. In 1999, Sheila Jasanoff reviewed it for *Science, Technology & Human Values*, the official organ of the Society for Social Studies of Science. In a six-page essay she did not attempt to rebut any of the specific criticisms against the accuracy of the writings of her colleagues. Instead, she chose to inquire into the book's "intellectual and cultural origins . . . to ask here and now what made it the way it is." This approach is reminiscent of the strategy used by some Freudians and Marxists: respond to an attack not by showing that your critic's statement is wrong, but by explaining how his damaged psyche or false consciousness has prevented him from accepting the truth you have revealed. In the same spirit, Jasanoff socioanalyzed her opponents rather than engaging them in argument. She noted the authors of *House* attacked the factual accuracy of specific science studies accounts (including those by Latour, Collins, Shapin and Schaffer) but, rather than defend those accounts, she complained that the critics haven't shown they are representative of the field or how they "figure within the overall tapestry of science studies." The critics fail to notice how significant and respected within science studies are the authors they attack (as if this should make them immune from criticism). The critics are "obsessed with the desire to show that social factors do not influence the core of science"—but rather than contesting that point, Jasanoff changes the subject to "science's place in society." She acknowledges the "grave charges" of Meera Nanda that the cultural relativism associated with science studies supports "the religious, military, and economic fundamentalisms that fuel the repressive regimes of the developing world." But Jasanoff answers Nanda's argument that science and rationality are "the only effective weapons" against this fundamentalism, with the weak rejoinder that science and rationality themselves have been misused to "perpetuate authoritarian regimes"—and, moreover, don't blame our *methodological* relativism if others use it to conclude that all knowledge claims are equal. But, near the end of her essay, Jasanoff admits "that science studies as a field has not been especially effective thus far in challenging the monopoly of reading [i.e., how to interpret scientific works] claimed by its critics." In other words, we social constructionists are losing the science wars—but we shouldn't give up, because we possess the truth.

That seems to be the attitude of Wiebe Bijker, as expressed in his contribution to a 2001 survey *Visions of STS* edited by Stephen H. Cutcliffe and Carl Mitcham. According to the editors, his essay "makes a strong brief for what is called the social constructionist view." But he doesn't actually make *any* case for that view—he simply states it:

> ***Scientific facts are not found***, literally dis-covered, in nature, but they are actively construed by scientists. Readings from instruments do not speak for themselves but need to be constructed into scientific facts by researchers. . . . nature does not dictate scientific facts.

That's it. No evidence is given for this assertion, and no reply to the many criticisms of social constructionism. It seems that SC is not expected to produce objective *knowledge* about how science works, only perspectives and insights.

The most authoritative account of the current position of STS experts is the third edition of *The Handbook of Science and Technology Studies*. According to its preface,

> This *Handbook* was produced under the aegis of the Society for Social Studies of Science [4S]. The Society selected the proposal by the editorial team and constituted the Handbook Advisory Board to monitor and assist in the process. . . . During 4S annual meetings, consecutive steps for developing the Handbook were presented by the editors and discussed with 4S members. . . . It is, then with conviction and pride that 4S grants its imprimatur to this book.

Chapter 1, by Sergio Sismondo, begins by endorsing SC: "STS looks to how the things it studies are constructed." But that may not be obvious to a casual reader, because "the metaphor of 'construction' or 'social construction' was so ubiquitous in the 1980s and 1990s that now authors in STS bend over backward to avoid using the term." This may explain why the casual reader looking in the index may conclude that, aside from Sismondo's 14-page article, only four pages in a book more than 1,000 pages long discuss "constructivist approach to science and technology" (99–100 and 726–727).

Try "Strong Programme" (or "strong program") which has several index entries, though they don't lead to substantive conclusions about it. Try "relativism," because, Sismondo reminds us, "Collins's methodological relativism asserts that the natures of materials play no role in the resolution of controversies"; the index has six entries for relativism. But Collins is apparently no longer interested in relativism; he now wants to talk about "expertise" (see Section 4.3). And this fits in quite well with a new trend celebrated by Sismondo, one that is "less concerned with understanding science and technology in and of themselves, and more with making science and technology accountable to public interests." If STSers can become experts on science as Collins and Evans claim, so that scientists no longer have privileged access to the truth, they can help the public to make well-informed decisions: "Collins and Evans assume that expertise is real and that it represents genuine knowledge."

How can this be? Doesn't it require STSers to abandon the claim that all scientific knowledge is socially constructed? Or is this where they say "that was just a metaphor"? In any case, one may ask why the public would accept scientific advice from scholars who have previously made that claim.

After wading through this muddle, it's a relief to read Park Doing's trenchant article "Give Me a Laboratory and I Will Raise a Discipline." He reviews laboratory studies by Latour and Woolgar, Lynch, Knorr Cetina, Collins, Pinch, and others. They all produced interesting information about what goes on in a scientific lab, but fail to take the last and really crucial step: to show how a specific scientific fact was constructed. He concludes:

> The first thing any new lab study should do is go directly for what laboratory studies have missed—a particular fact—and wrestle with how its endurance obtains within the "*in situ*" world of practice. Let's make detectable the dark matter in STS lab studies and get the books straight. I do not know just what such accounts will look like, but I do know that they should not begin with the ironic line, "Laboratory studies have shown" . . . perhaps we should say, at least for now, that we did not really do what we said.

And remember: 4S grants its imprimatur to this article.

To get a semiquantitative idea of the rise and fall of SC, I have checked in Google Scholar the number of publications with the phrase "social construction of scientific knowledge" in the title, for five-year periods from 1971 to 2006–2010. The count starts at zero for 1971–1975, rises to 17 for 1981–1985, then drops to only two for each of the period 1996–2000, 2001–2005, and 2006–2010. For the shorter phrase "social construction," Google Scholar shows a peak of 1,750 in the period 1996–2000. The frequency of the longer phrase in the texts of books (Google's Books Ngram Viewer) follows a similar pattern; the peak of the curve is around 1999. (Both sources were accessed in April 2011.) These results should not be taken too seriously, but are roughly consistent with the evidence presented here, and with Latour's 1992 pronouncement (quoted in the notes to this chapter) that *social* studies of science collapsed around 1989, taking account of the normal time lag of a few years between composition and publication of a book or article. Of course this should not be taken to indicate anything more than the use of the *words* "social construction"—it is certainly quite possible that the *spirit* of social construction permeates science studies so thoroughly that no one needs to use that phrase explicitly anymore, as suggested by Sismondo.

Another confirmation of the fall of SC is found in a 2011 article by sociologists Marion Blute and Paul Armstrong. They analyzed the publications of 10 sociologists or sociologically-minded philosophers of science, whom they considered prominent creators of grand theories of the scientific and scholarly process. Seven of them were also interviewed. Blute and Armstrong indicated, among other things, whether their subjects were now constructionists. This label was found not to fit any of them; three (Steve Fuller, Bruno Latour, and Donald McKenzie) were described as "yes, but," meaning that the person "later either backed off that position or claimed to have been misunderstood."

Like many movements in sociology, social constructionism was a fad that was extremely popular for a time but disappeared without leaving any trace of an obvious reason for either its rise or fall. Before memories fade, I would like to suggest some of those reasons.

4.4 POSTMORTEM

Benefits of Constructionism

Even if social constructionism is not quite dead yet, this seems to be a good time to assess its successes and failures. Again I emphasize that I'm doing this from the perspective of a historian of science, and I rely in part on the (favorable) account by Jan Golinski in his *Making Natural Knowledge*.

Liberation from Whiggery

SC reinforces the efforts of professional historians of science to replace the whig interpretation of the history of science by a historicist or contextual viewpoint. Historian Herbert Butterfield, in his 1931 book *The Whig Interpretation of History,* criticized those who described British history as a tale of progress toward the victory of the Whig party, now known as the Liberal party. Historians of science have adapted this phrase to criticize those who judge past science by the extent to which it represents progress toward present (more nearly correct) knowledge. Instead, historians now prefer to interpret past science on its own terms, looking at the state of knowledge, instruments, philosophical viewpoints, and so on at a particular time and place. Many of the social constructionist case studies do this very well.

Laboratory Life

SC also reinforces the movement by nonconstructionist historians (e.g., Allan Franklin) to focus on what happens in the laboratory, rather than concentrating on theoretical developments. By looking closely at scientific practice and the interactions among individual scientists, social constructionists have helped dispel the myth of a single scientific method unique to successful science, and have shown how scientists use the same reasoning skills as other intelligent people.

Valid Claims

In many cases there *is* good evidence for social influence on the content of science, independent of sociological construction. Most of these cases are in the biological or behavioral-social sciences, which social constructionists seem reluctant to examine too closely (in spite of their claim to adopt a reflexive attitude in applying SC to SC). Perhaps this is because they would have to defend the scientific legitimacy of theories widely regarded by scientists as false.

The best-known example is Lysenko's theory in biology. As philosopher and sociologist Peter Slezak remarks, according to SC "the purges of orthodox geneticists [during the Lysenko era] must count as an instance of successful science." (As Thomas Kuhn wrote—see Section 4.2—there is no court higher than the consensus of the scientific community, so if you can eliminate your opponents your theory will win.) Similarly, "the political success of fundamentalism [creationism] would *ipso facto* constitute scientific success."

In his book addressing the question of whether evolution is a social construction, Michael Ruse argues that the answer evolved over time. Erasmus Darwin's theory was a speculation heavily influenced by social views; after 1859 objective scientific evidence became more important. By the mid-twentieth century the balance shifted from mostly social to mostly epistemic. Whether or not one agrees with Ruse's thesis, it is much more fruitful to ask and try to answer the question of whether the amount of social construction in this particular field decreases as scientists do more research than simply to assume that all scientific knowledge is always socially constructed.

There is another example of SC for which its advocates might want to claim some credit. Historian Carl Degler has argued that the swing away from hereditarian toward environmentalist theories in the behavioral and social sciences in the United

States during the middle decades of the twentieth century was primarily motivated by ideology rather than empirical evidence. I have not seen Degler's book cited in any science studies publication, though one might suppose his results, if valid, would bolster the credibility of social constructionism. (Degler won the Pulitzer Prize in History for an earlier book.)

Mathematician Richard C. Brown notes that SSK "seems not to be very interested" in analyzing the behavioral-social sciences, perhaps because "SSK has a political dislike of the hard sciences and therefore prefers to find ideology at work in areas like physics, astronomy, or paleontology."

Psychologists Thomas Sturm and Gerd Gigerenzer , though not identifying themselves as social constructionists, suggested two examples that might be considered to support the strong program. First, the adoption of inferential statistical techniques by psychology departments in leading American universities in the 1950s, and second, the popularity of computer models of the mind, could have been due primarily to "the institutionalization of psychological methods and . . . the public standing of psychology" rather than to the needs of their own research programs. Psychologists, in order to get public respect and government funding, wanted to be regarded as scientists, using the latest mathematical and experimental methods. This would be consistent with Bloor's assertion:

> Much that goes on in science can be plausibly seen as a result of the desire to maintain or increase the importance, status and scope of the methods and techniques which are the special property of the group.

However, Sturm and Gigerenzer seem reluctant to admit that these cases actually do support the strong program, arguing that there were more substantive reasons for adopting inferential statistics and the computer model of their mind. Thus the subtitle of their article mentions the *weaknesses* of the strong program (emphasis added). They assert that "it has become too fashionable in recent decades to say that something or other (nature, childhood, the subject, population statistics, madness . . .) is socially constructed." On the other hand, they remark that in the science wars "defenders of the authority of science have criticized certain historians of science in unfair ways."

Valuable Scholarship

The books by Shapin and Schaffer and by Pickering mentioned earlier include sound historical research, mostly unaffected by the constructionist viewpoint. This is shown by the many favorable published remarks about these books by authors who themselves are not social constructionists.

Gender Bias

An example of construction by gender stereotypes is presented by Emily Martin. She shows that biologists frequently describe the sperm as active and the egg as passive in the process of human reproduction, contrary to observational evidence. Her research is frequently cited by feminist critics of science who argue that "established knowledge" in reproductive biology would be significantly different if more women were doing the research.

Feminist scholars have also shown that the descriptions of behavior of male and female primates given by male primatologists—descriptions that influenced and may have been influenced by some of our stereotypes about gendered human behavior—were grossly inaccurate, until a new generation of female primatologists started asking different questions. Despite chivalrous bows to the importance of feminist science studies, most social constructionists ignore these examples when they need to adduce support for their position.

I would add another example of the social construction of sex differences (or nondifferences, in this case). When psychologist Lewis M. Terman administered an American version of the French psychologist Alfred Binet's intelligence test to California children, he found that the girls earned higher scores (on average) than the boys. This result contradicted the well-known fact that men are more intelligent than women. So he adjusted the test by increasing the proportion of questions that boys answered correctly more often than girls, so that the average IQ of both sexes is 100 *by definition*. Later, psychologists found that blacks got lower scores than whites (on average), but since this was consistent with cultural stereotypes in America they did not adjust the tests to equalize the average scores. Therefore the statements often made in textbooks and popular books, that men and women have the same average IQ but whites have higher IQs than blacks, are both social constructions.

Physical Construction

The thesis of social influence or even social determinism is not so alien to scientific thinking as both sides (pro- and anti-SC) seem to assume. The idea that reality does not exist apart from human observation, and that the properties of the world must be actively constructed by experiments, is congenial to physicists who adopt the Copenhagen interpretation of quantum mechanics. Physics has its own strictly circumscribed version of social constructionism for experiments in the atomic realm: an electron does not have an objectively real position or speed, but you can make it have either property by choosing to measure it. Wave or particle? It depends on what experiment you choose to perform. Few social constructionists have tried to explore this connection.

Open Minds

The social constructionists were remarkably open to criticism (at least before the science wars started in the mid-1990s) and reflexively skeptical of the truth of their own views; some of the most devastating critiques of social construction can be found in their own publications (look for the catch-phrase "epistemological chicken"). This situation conveyed, even to an outsider like myself, an impression of vigorous controversy and excitement.

Entertaining Writing

A final word of praise: Violating our American cultural stereotypes about turgid sociologese, some of the SC literature is actually quite well written and fun to read. This is one reason why the defection of Bruno Latour from social constructionism has weakened it. I enjoy most of his writings, but I especially recommend his piece on the sociology of the door-closer.

Faults of Social Constructionism

What Is It?

There is still much confusion and ambiguity about the definition of social constructionism, which I have very roughly summarized by distinguishing the three versions: sociological constructionism, social influence, and social determinism. Many social constructionists have a tendency to make statements that sound like social determinism, then—when attacked—claim they have been misquoted or misunderstood, they really meant something like sociological constructionism.

Where's the Evidence?

Social constructionists often refuse to provide *evidence* that scientific knowledge is socially constructed, or even refuse to admit that any evidence is needed; rather it is treated as an axiom. Nadine Weidman expressed this view quite bluntly in the preface to her book on Lashley:

> I did not intend to treat social constructivism as a scientific hypothesis, to be tried and tested on historical subjects [and rejected] if found lacking. Rather, I believed it was a way of looking at the world, a way of making sense of science, its history, and its relationship to society.

Logical Fallacy

Others considered SC a necessary consequence of the alleged "underdetermination of theory by evidence." Thus Shapin , in response to Cole's challenge to produce just one example of core scientific knowledge that has been socially constructed, wrote:

> A quite typical form of social constructivist case-study involves the examination of scientific controversy. How is one to account for variation in scientific judgment when both parties to a controversy have access to the same evidence and, presumably, to the same canons of right reasoning? Here social constructivists have argued that empirical evidence has a causal role but not a discriminating role. If nature is one and the same, then one has to look elsewhere for variation in belief and judgment. It is primarily for this reason that methodological—not ontological—relativism has recommended itself to sociologists of scientific knowledge.

Collins makes a similar assertion: since the scientific method cannot "account adequately for the generation of scientific knowledge," we must admit that "the application of rules of scientific method is always mediated through the social."

Ullica Segerstrale, reviewing The Golem (1993) by Collins and Pinch, writes:

> This book represents a point near the exhaustion of a paradigm of science-bashing that has flourished since the early 1970s or so. This paradigm has yielded important insights, but it is now at a point where sweeping claims are backed up by a relatively small set of selected case studies . . . it is high time to turn the question from What's wrong with science? To How come science works at all?

Yves Gingras, reviewing Collins's Gravity's Shadow, asserts that "far from being a good methodological principle, the idea of treating nature as though it did not affect our perception of it, introduces a major bias since it a priori excludes what should in fact be an empirical question to be decided case by case."

The reasoning of Shapin and Collins is akin to the creationist argument that if evolutionary theory does not explain all biological facts, creationism must be invoked, as if there were no other alternatives. Both arguments assume an effect has only two possible causes, so if cause A cannot completely account for the effect, cause B must also (if not exclusively) be invoked. But it is not hard to think of third, fourth, and subsequent causes that may be invoked.

Fallacious Claim to Be Scientific

In 1981, philosopher Larry Laudan attacked Bloor's strong program on several grounds, one of which is quite relevant to an issue discussed in this book. Bloor claims that his program is scientific, which means he has to specify criteria for demarcating science from pseudoscience. Laudan shows that Bloor's criteria would make quantum mechanics and statistical mechanics nonscientific, because their basic principles are not causal (deterministic). Any other theory that does not explain all events by the *same* causal mechanism would also be unscientific, by another criterion. (Laudan showed in a paper published in 1983 that all attempts to demarcate science from pseudoscience have failed.) In his reply to Laudan, Bloor added another criterion: "the most important aspects of science concern the making and testing predictions." Yet he presents no example of a confirmed prediction made from his strong program.

Why Sociology But Not Psychology?

Some critics, while not rejecting the importance of social factors, have criticized what historian Mara Beller calls "the flattening of the cognitive to the social." Psychological factors may also include the influence of an individual scientist's age or birth order on the decision to accept or reject a theory (as in the reception of Darwinian evolution; see Section 2.4).

Historical Errors and Distortions

As mentioned above, several of the case studies used by social constructionists have been criticized as inaccurate by historians and scientists. Going beyond factual errors, one may make a more fundamental objection to social constructionism. The American historian V. Betty Smocovitis recognized Shapin and Schaffer's *Leviathan* as a "landmark . . . in the history of science," but charged that it violated an important precept of modern historiography by failing to "understand the perspective of the historical actors." In the new contextualist historiography of science, "this feature of *Leviathan and the Air Pump* grew to become an unacceptable historical method." Its authors

> remained firmly inside the scientific and positivistic sociology of science. Thus, whereas they could argue persuasively for the social construction of scientific knowledge, they were unwilling to apply the same sociology of knowledge to their own practices as historian-analysts. Shapin and Schaffer's argument, and other

> such attempts to argue for the social construction of scientific knowledge, therefore [faced] a serious contradiction: while they argued against simple-minded scientific empiricism, they argued for similar simple-minded historical empiricism; and at the same time they acted to 'deprivilege' the knowledge-making claims and positions of scientists, they also served to privilege their own knowledge-making claims and positions as historian-analysts of science.

Science Is Not Just Local

Social constructionism has not solved what Golinski calls the "problem of construction": if all knowledge is constructed at a specific time and place, how can it be found to be valid everywhere else? This is a problem for postmodernism in general, which, according to historian Lewis Pyenson, "maintains the denial of universals, whether with regard to human values or truths about nature." Golinski thinks the problem can be solved by postulating diffusion from one place to another. The most obvious objection to that solution is the phenomenon of multiple discovery: the same idea or discovery appears within a few months or even a few days at widely separated places, with no evidence of any direct communication between the scientists involved.

Big Picture Is Missing

> Social constructionism does not try to explain the big picture of the progress of science over several centuries, because it deals only with science at one time and place. (This criticism has been very effectively made by Margaret Jacob.) It shares this defect with the older contextualist historiography, but there the problem was not as severe because many historians denied that they had any responsibility to explain historical change at all—they just describe it. The social constructionists often reject the need for a narrative account of the history of science, but at the cost of making their program seem quite inadequate.

Is Social Constructionism Anti-Science?

> Though I welcome the turn to the external history of science as addressing a balance which has long been seriously askew, its new popularity may not be an unmixed blessing. One reason it now flourishes is undoubtedly the increasingly virulent anti-scientific climate of these times.
>
> —Thomas S. Kuhn *(1971)*

> SSK [Sociology of Scientific Knowledge] has been animated by a goal . . . to undermine the authority of science by destroying the epistemological privilege that has traditionally justified that authority.
>
> —Gary Bowden *(1995)*

> The study of scientific knowledge is primarily seen [by SC] to involve an investigation of how scientific objects are produced in the laboratory rather than a study of how facts are preserved in scientific statements about nature . . . nowhere in the laboratory do we find the "nature" or "reality" which is so crucial to the descriptivist interpretation of inquiry.
>
> —Karin Knorr-Cetina *(1983)*

> The result [of field studies of laboratory practices] . . . was that nothing . . . "scientific" was happening inside the sacred walls of these temples (Knorr, 1981). After a few years of studies, however, our critics would be right in raising again the naive but nagging question: if nothing scientific is happening in laboratories, why are there laboratories to begin with and why, strangely enough, is the society surrounding them paying for these places where nothing special is produced? . . . Scientific fact is the product of average, ordinary people and settings.
>
> —Bruno Latour *(1983)*

The social constructionists often appear to be *debunking* science. This is not surprising given the origin of science studies in the radical movements of the 1960s, and the influence on some of its founders of earlier critics of science. Indeed, Steve Fuller argues that as social scientists one would have expected them to become even more radical critics of the scientific-technological establishment, but they were deflected from that path by Kuhn's influence. Nevertheless, they are now perceived by scientists as part of a dangerous antiscience movement that includes also postmodernism, creationism, alternative medicine, and other varieties of pseudoscience. This charge was made by Gross and Levitt in their book *Higher Superstition* (1994), and by other scientists in recent years, including the authors of a book edited by Gross, Levitt, and Lewis, *The Flight from Science and Reason* (1995). Historian Helge Kragh faults social SC not only for misreading the historical record but for undermining science education by reinforcing popular antiscientific attitudes.

In fact social constructionists, like postmodernists and multiculturalists, seem to tolerate creationism along with parapsychology, cold fusion, and levitation; they refuse to give the scientific community the right to define what is science and what is pseudoscience. Sociologist Steve Fuller (quoted at the beginning of this chapter) went even further, appearing as a witness for the intelligent design (ID) movement at the Dover, Pennsylvania, trial on including creationism in public school biology courses. Robert Pennock has recently pointed out the surprising connection between postmodernism and intelligent design creationism: ID "is the bastard child of Christian fundamentalism and postmodernism."

Brian Martin and Evelleen Richards explained why social constructionists often seem to support ideas that scientists regard as pseudoscience or antiscience: in discussing controversies, constructionists assume that "the 'truth' or 'falsity' of scientific claims is considered as deriving from the interpretations, actions, and practices of scientists rather than as residing in nature." In practice, this means that the constructionist approach "gives more credibility to opponents of orthodoxy . . . and thus provides de facto support for the opponents" (e.g., more support for creationists, global warming deniers, tobacco companies).

Barry Barnes explained in 1974 why SSK is antiscience:

> Real changes in the institutional base of science lie at the root of the matter. Since its purchase by government, industry, and the military, it has become impossible to combine hostility to existing major institutions, or to capitalism as a whole with an entirely positive evaluation of science. A minority opposed to science has thereby almost automatically been generated.

Richard Brown gave a similar explanation from a somewhat different perspective:

> The disdain of postmodernists in general for ordinary concepts of truth and rationality . . . is a natural continuation of the antiintellectualism of the radical student movement of the Vietnam war era. . . . More recently . . . the continuing power and intensity of this generation's opinions reflect . . . rage and jealousy at the growing power and influence of technoscience and mathematics . . . at the expense of the humanities . . . academic intellectuals in the humanities continue to feel deprived of political power just as they had been as students by groups they regard as their intellectual inferiors (university administrators then, scientists and corporate types now.

Brown also suggested a psychological role played by SSK and PIS, saying that they

> are excellently designed to transform feelings of inadequacy *vis a vis* science into feelings of superiority . . . [their theses] are quite simple and may be mastered without the need of learning the details of a science.

Literary scholar George Levine, in his contribution to the *Science Wars* collection, admitted that "many of us are in the humanities because we were bad at or turned off by science; the ethos we joined tended, conventionally, toward antiscience." Philosopher Ian Hacking, in his extensive analysis of SC, concludes: "most people who use the social construction idea enthusiastically want to criticize, change, or destroy some X that they dislike in the established order of things."

Sociologist Stephen Kemp argued that Bloor's strong program has two antiscientific features. First, it is guilty of "weak idealism," meaning that it treats "scientific discourse as free-floating and unrelated to the world of things." Second, constructionist approaches (including the strong program) "challenge "the credibility of the scientific knowledge that they analyse." These views of science, if taken seriously by scientists, would undermine their enterprise:

> In order to be committed to science as an enterprise, scientists must believe that, at least some of the time, the scientific community gives credibility to a theory because there are convincing arguments in support of it, arguments that call on the results of observation, experiment, and testing. . . . By contrast, social constructionists deny that these kinds of reasoning processes provide an adequate explanation of how differences in credibility are achieved.

Many social constructionists indignantly protest the antiscience label, claiming that they are just trying to understand how scientists behave and don't contest the scientists' judgment about what is true or false; this is where the methodological stance I called sociological construction comes in handy. But their protest is disingenuous; as professed experts on the behavior of scientists, they ought to realize that telling scientists they are constructing facts rather than discovering them will provoke an angry reaction. After all, if that were true, why should anyone fund scientific research? And if scientists themselves believed it, what motivation would they have for doing research, especially in a society that does not reward most scientists with high salaries or public approval?

Latour, like some other science studies scholars, has a tendency to make statements that, taken out of context, can easily be interpreted as antiscience. For example, in 1998 he wrote: "The transformation of society by science has produced, to be sure, many beautiful ruins, but not a better society." Elsewhere in the same essay he defines science as an old-fashioned cold, objective, dogmatic enterprise disconnected from society, which he says is already being replaced by research, a tentative, passionate, socially involved activity. In other words, he criticizes what science used to be and praises what it is becoming. One wonders if the members of the American Association for the Advancement of Science, which commissioned this essay as one of a series celebrating its sesquicentennial, appreciate Latour's point.

Latour and his colleague Michel Callon stated, more forthrightly:

> The field of science studies has been engaged in a moral struggle to strip science of its extravagant claim to authority . . . we wish to attack scientists' hegemony.

Historian Naomi Oreskes, reviewing Latour's book *Politics of Nature*, also addresses Callon's assumption about science's hegemony:

> While Latour frequently insists on his respect, even admiration, for scientists, the language he chooses often seems to imply nefarious intent. . . . Latour also credits science with being far more powerful than it nowadays seems to be. Climate scientists have argued at considerable length that global warming is a fact . . . but this has not stopped politicians, economists, business executives and others from contesting it.

Blurring the Distinction Between Honesty and Fraud

If scientific knowledge is socially constructed, how would one distinguish between knowledge that is based on logical reasoning and empirical evidence, and knowledge that has been made up in order to win prestige and prizes, or to support political or religious views? The issue has been raised by (among others) Peter Slezak. He states that social constructionists like Pinch and Collins

> maintain a scrupulous neutrality concerning the substantive merits of different theories such as those of orthodox science compared with astrology and parapsychology. . . . they have no grounds to distinguish honourable scientists from unscrupulous charlatans . . . Deception is just the way the game is played . . . truth is what you can get away with. . . . Latour and Woolgar espouse essentially the same idea by attempting to eradicate and distinction between truth and falsehood.
>
> Since all that matters is what you can persuade people to accept,
>
> The sociology of scientific knowledge could have offered no principled objection to teaching the racial theories of *Mein Kampf* when they were believed by a majority.

Similarly, science writer Horace Freeland Judson writes that according to the doctrine of social construction,

> all science must appear fraudulent. . . . Curiously, perhaps not coincidentally, social constructionism reached its apogee just in the period when cases of scientific fraud arose and multiplied.

The Natives Are Restless—and Dangerous

Since social constructionists sometimes see themselves as anthropologists studying an exotic society of scientists, an analogy from that field may be helpful. In his discussion of the Mead-Freeman controversy about the behavior of adolescent girls in Samoa, Hiram Caton points out that the modern Samoans are angered by Mead's famous book *Coming of Age in Samoa* because she, as an agent (however unwitting) of colonialism, has told the world that their mothers and sisters are sluts. They are not pacified by the explanation that the anthropologist practices cultural relativism and therefore gives equal respect and validity to Samoan and Western value systems; "the native who accepts this cosmopolitanism has by that very fact forsaken the old ways." The Samoans do not accept that their own religion and worldview are "just as valid" as those of the Westerners; they know they are superior. The scientists who are told by social constructionists that the scientific worldview is just as valid as creationism, mysticism, and pseudoscience are equally angry—but unfortunately for the social constructionists, the power relations are reversed here, and the scientists have the ability to do considerable damage to them. Latour, historians M. Norton Wise and Paul Forman, and perhaps others have experienced, at least indirectly, the backlash against social constructionism.

Science and technology studies leader Sheila Jasanoff reported in 2010 that "STS remains weakly institutionalized in the upper reaches of academia . . . there are few fully fledged STS departments in the United States and even fewer in Europe." One reason is that the science wars of the 1990s "called into question whether constructivist approaches fairly portray progress in science or advances of technology. Although difficult to document, such worries about the field's intellectual soundness and descriptive accuracy, coming at a time when universities were becoming increasingly dependent on their links to science-based industries, may have inhibited the institutionalization of STS in the upper reaches of academia in several Western countries. The widely decried hostility toward science during the US presidency of George W. Bush, coupled with a growing perception that scientific progress and technological innovation are crucial for economic growth, may also have undermined institutional support for scholarship seen as questioning the authority of science."

But long before that, SSK had failed to find, or lost, a home in sociology departments, and was being criticized by a prominent sociologist of science. In 1981, Joseph Ben-David wrote:

> Sociology of science began to be recognized as a specialty within sociology in the 1960s . . . this changed drastically during the seventies as a result of changes in the location and uses made of sociology of science and in the characteristics of the practitioners in the field.

In the 1970s, interdisciplinary programs combining the sociology, economics, and politics of science

> attracted to the field a number of younger scholars trained in experimental science, philosophy, or history of science. . . . a significant part of sociology of science in Europe—especially in Britain, has been practiced in interdisciplinary units rather than departments of sociology.

The younger scholars criticized the preoccupation of older sociologists with the sociology of the scientific profession rather than the content of science. But by adopting the phrase "sociology of knowledge," they inherited the stigma attached to a tradition—identified with the writings of Karl Marx, Émile Durkheim, and Karl Mannheim—"unacceptable to the majority of sociologists. The sociology of knowledge tradition assumed that all ideas, irrespective of their truth, were socially conditioned."

This assumption was rejected by the older sociologists of science:

> The empirical evidence on covariation between social base and the structure of knowledge was never satisfactorily established, and because none of the theories of sociology of knowledge contained a satisfactory explanation of how, by means of what mechanisms, knowledge is determined by the social base (Merton, 1973a). The rejection of sociology of knowledge seemed also preferable from the point of view of the epistemologies prevailing among sociologists, according to which "rationality" as conceived by scientists "is a sufficient basis for the acceptance or rejection of theories, requiring no further social underpinning.

Ben-David then attacks the validity of the strong program, which assumes that "causal explanation of the same sociological kind is required of the emergence and acceptance of any belief irrespective of 'truth or falsity.'" This is refuted by a statement of another founder of the strong program, Barry Barnes, who admits that all beliefs are *not* sociologically equivalent: "there are limits to the possibility of human thought." This "implies the admission of some standards which are not socially determined," so one cannot give "the same causal explanation" for all beliefs.

Ben-David concludes that "no success can be claimed for the new Marxian-Mannheimian attempts to find a systematic . . . relationship among macro-social location, ideology, and scientific theory."

He seems to approve of "the attempt to investigate the similarities between science and other belief systems," without claiming a causal relation, and of placing "scientific developments in the context of general cultural and social history" (Brush, 1967). But so far, this work has not been integrated into sociology of science in any systematic way.

Evidence Ignored

As mentioned above, social constructionists seem to ignore the best evidence for their own claims (primatology, Lysenko, environmentalism in the behavioral and social sciences). In particular, what Donna Haraway calls "malestream science studies" has refused to take seriously the feminist research that supports social influence without adopting the constructionist rhetoric or methodology.

Symmetry Ignored

As Latour pointed out, the social constructionists ignore their own symmetry principle and risk absurdity by treating social forces as more real than natural forces. But Latour goes to the opposite extreme, treating not only animals but inanimate objects as actors (not just as a source of resistance to human action, as Pickering would have it).

Politically Incorrect

In an issue of *Social Studies of Science* devoted to the politics of sociology of scientific knowledge, social constructionists complained that they often end up on the politically wrong side when they get involved in public controversies. For example, they would have to support the tobacco companies against the charge that since, from their own research, they knew that smoking kills decades before this fact was firmly established by the scientific community, they committed a crime. For a social constructionist, it was not a fact until the scientific community accepted it, hence the tobacco companies can't be held responsible for ignoring their own knowledge.

In some countries, social constructionism, by its attacks on the objectivity of Western science, undermines efforts to discredit repressive regimes, according to Meera Nanda. The same spirit of controversy and lack of consensus that made science studies articles so entertaining to read ultimately accelerated the disintegration of social constructionism. But rather than recognizing their failure to establish any general propositions about the nature of science, constructionists, suggests historian Scott Montgomery, seem to have projected the disunity of their own movement onto science, rejecting the universalistic picture that science-historians have drawn.

Projection

In psychology, projection is described as ascribing one's own feelings, thoughts, or attitudes to others. There is some evidence suggesting (not proving) that sociologists of science think physical and biological knowledge is socially constructed because they believe that to be true of their own field. In 1987, three sociologists—Von Bakanic, Clark McPhail, and Rita J. Simon—published a study of the process by which manuscripts were submitted, refereed, and published or rejected by a leading journal in their field, *American Sociological Review (AJR)*. They found that author characteristics (title, sex, age, prestige of the institution that granted their PhD or currently employed them) did *not* affect the referees' ratings of the manuscripts. (The identities and institutions of authors had been removed from the manuscripts, although in some cases the referee might have accurately guessed them). But they *did* affect the publication decision of the *AJR* editors, who did know them. Authors with higher rank (e.g., full professor) at more prestigious institutions were more likely to have their manuscripts accepted (even if the referees' evaluations were equally favorable). This is one way in which social factors may influence what knowledge is accepted by the profession.

Daryl E. Chubin, a science policy analyst, and STSer Edward J. Hackett, in their 1990 book Peerless Science, called this study "strongly suggestive evidence of particularistic decision-making—favoritism, in plain language." But Bakanic, McPhail, and Simon avoided explicitly stating this conclusion, probably because they were submitting their paper to *ASR* and did not want to criticize its editors, especially since Simon and McPhail themselves "were editor and deputy editor, respectively, *during* the period from which the data were drawn."

Finally, I note that Chubin and Hackett, who are themselves involved in STS research, stated near the beginning of their book, "in the process of doing research, scientists do not merely operate on reality, they *construct* it, trying to persuade others to accept those constructions" (emphasis in original). This may be true for research in sociology and STS. But as noted in Section 4.2, two other sociologists, Harriet Zuckerman and Robert Merton, analyzed referee reports on manuscripts submitted to a leading physics journal, *Physical Review*, and found no evidence of favoritism in the publication process.

Changing Zeitgeist

On another level of analysis, we might argue that SC is an aspect of postmodernism, and thus was fated to go out of favor when that cultural movement was replaced by another one. It is too soon to know exactly what that successor movement will be; one candidate is postpositivistic realism, as developed by Paula M. L. Moya, Satya P. Mohanty, and others; another is postmodern production, according to Wilhelm E.S.J. Fudpucker.

Just Plain Silly

> It would be about as helpful to argue that the blood had no pressure before insurance companies began pushing the use of sphygomanometers as it would be to claim there was no mortality before there were actuaries.
>
> —Theodore M. Porter *(2000)*

Apparently in Galileo's time there were people who thought scientific facts were socially constructed. In his *Dialogue concerning the two chief World Systems*, Galileo wrote about people who "so firmly believe the earth to be motionless" that when they hear "that someone grants it to have motion" they "imagine it to have been set in motion when Pythagoras (or whoever it was) first said that it moved, and not before." They "believe it first to have been stable, from its creation up to the time of Pythagoras, and then made movable only after Pythagoras deemed it to be so." Similarly, in 1923 physicist Felix Auerbach, in a book on the history of modern physics, asserted that "X-rays are not a 'natural phenomenon,'" they didn't exist until Röntgen invented them . In 1927, philosopher Martin Heidegger wrote: "before Newton 's Laws were discovered, they were not true."

In February 2010, physicists Robert A. Harris and Leo Stolsky published a letter in *Physics Today* stating that they had not been given proper credit for their discovery and explanation in the early 1980s of what is now called the quantum decoherence effect, just because they had not used that term (which had not yet been coined at the time). They continued: "The situation is amusingly reminiscent of the case of Ramses II. Apparently, some people claim the pharaoh couldn't have died of tuberculosis because the disease wasn't discovered until the nineteenth century." "Some people" turns out to be Bruno Latour, in a 1998 article.

For these scientists, social constructionism was not a threat, only a joke.

4.5 CONSEQUENCES FOR SCIENCE STUDIES

> We should not mourn too much the demise of SSK/PIS. When it was fashionable it satisfied real needs: it provided a potent psychological therapy for the frustrations and *ressentiment* of its practitioners. On a more mundane level it helped to secure jobs, tenure, grants, administrative positions, and some influence in the scientific bureaucracy. One can ask for little more in the humanities.
>
> —*mathematician* Richard C. Brown *(2009)*

Now that the phrase "social construction of scientific knowledge" seems to be disappearing from the writings of sociologists of science (it scarcely appears in recent issues of *Social Studies of Science*), historians and others can once again discuss social and other influences on science without having to make any commitment for or against the ambiguous program once called by that name. Moreover, we can learn much from the detailed research done under the banner of social constructionism but not contaminated by its dubious philosophical or methodological relativism. What we need is to specify clearly what is meant by a social influence, and what criteria should be used to determine whether or not it is present. For example, one should not simply assert that a social factor is important in a particular case without positive documentary evidence—here Paul Forman's Weimar culture paper shows the kind of work that has to be done. One should demonstrate that when one varies the social factor by going to a different society, the scientific knowledge also varies, as Loren Graham argues in his comparisons of the former USSR and Western countries.

One should also agree on what kind of evidence would show that social factors are probably *not* important, for example when a scientist accepts a result contrary to his or her own previously stated beliefs or results without any indication that the relevant social factors have also changed. One of the most impressive works on the influence of ideology on the acceptance of scientific ideas, Degler's *In Search of Human Nature*, might seem less convincing if such criteria were applied—Degler doesn't vary the social factor much (he deals only with the United States, though he does discuss the effects of the changing demography of the academic profession), and he fails to recognize that several of the behavioral scientists he identifies as left-wing environmentalists were previously hereditarians but changed their views because of new empirical evidence they couldn't explain away.

As a guide to the kinds of research that might be fruitful in twenty-first-century science studies, I recommend David Hull's paper "Studying the Study of Science Scientifically." Whether or not you agree with the plausibility of the hypotheses he discusses, at least he shows the willingness to submit those hypotheses to empirical test that any scientist—even a sociologist—should display.

NOTES FOR CHAPTER 4

099 *According to this proposal, scientists do not discover a preexisting reality* For a comprehensive survey of the issues mentioned in this chapter see John H. Zammito, *A Nice Derangement of Epistemes: Post-Positivism in the Study of Science from Quine to Latour* (Chicago: University of Chicago Press, 2004). I also recommend the book by Richard C. Brown, *Are Science and Mathematics Socially Constructed?*

A Mathematician Encounters Postmodern Interpretations of Science (Singapore: World Scientific, 2009).

099 *"Science as something already in existence"* Einstein's address to students at UCLA, February 1932, in *Builders of the Universe* (Los Angeles: US Library Association, 1932), p. 91. Quoted in *The Ultimate Quotable Einstein*, collected and edited by Alice Calaprice (Princeton, NJ: Princeton University Press, 2011), p. 384.

099 *"a better position to judge what makes something a science"* Steve Fuller, letter to the Editor, *Isis* 100 (2009): 115–116, on 115.

099 *"The sociologist knows less than the natural scientist"* H. M. Collins and Steven Yearley, "Epistemological Chicken," in *Science as Practice and Culture*, edited by Andrew Pickering (Chicago: University of Chicago Press, 1992), pp. 301–326, on p. 302.

Section 4.1

100 *a combination of several disciplines* Stephen H. Cutcliffe, *Ideas, Machines, and Values: An Introduction to Science, Technology, and Society Studies* (Lanham, MD: Rowman & Littlefield, 2000). As the title of this book suggests, the letters STS can also stand for "science, technology, and society," but this phrase usually refers to an undergraduate (or occasionally graduate) educational program. The professors in such programs are generally involved in research in science and technology studies, and I will use STS in the latter sense unless otherwise specified (see the next note).

For historians of science like me, it is difficult to know whether we are automatically part of a field called science and technology studies, or whether we belong to it only if we study science from a particular viewpoint that may not be historical in the traditional sense. This can be a very important question if you want to get a grant to support your research. According to the US National Science Foundation, science and technology studies is a research field that simply grew out of history and philosophy of science (HPS) by adding sociologists and other social scientists. Moreover, the NSF has ruled that the entire field is part of the social sciences, despite the protests of some of us who consider history part of the humanities and recall the disastrous effects of the social science affiliation on our funding in the 1970s and early 1980s (Margaret W. Rossiter, "The History and Philosophy of Science Program at the National Science Foundation," *Isis* 75 [1984]: 95–104). But we would be even worse off if we had to rely for funding on the National Endowment for the Humanities. At the end of the twentieth century, scholars in HPS still constituted about two-thirds of the panel members in the STS program and received a reasonable share of its grants, according to Ronald Overmann, "Where's Tolstoy when you Need him? Personal Perspectives on HPS to STS," Colloquium, Committee on the History and Philosophy of Science, University of Maryland, College Park, February 24, 2000.

100 *another movement that shares the same initials* "Earlier, more activist, and less academically specialist programs in 'science, technology, and society' were responsive to the apparent corruption of science and technology by industry and government. The more recent . . . graduate programs in 'science and technology studies' . . . sought to distance [themselves] from the old STS, with its emphasis on undergraduate education, its association with scientists turned STS scholars, and its societal orientation. Cornell University even modified the acronym [sic] to S&TS in case anyone didn't get the point. The ampersand figured as the most visible element in its logo, seemingly advertising inclusion but in fact noting the field's repudiation of its past

for success as a normal discipline with journals, peer-reviewed government funding, and graduate students." William T. Lynch, Review of *Visions of STS*, edited by S. H. Cutcliffe and C. Mitcham, in *Science, Technology, and Human Values* 23 (2003): 326–331. Brian Martin recalled a similar trend: STS became more insular and removed from the real world, less willing to cite radicals like Bob Young or Rose and Rose; see "The Critique of Science Becomes Academic," *Science, Technology, and Human Values* 18 (1993): 247–259. For further discussion of the depoliticization of STS as it became S&TS, see Steve Fuller, "Why Science Studies Has Never Been Critical of Science." Some recent Lessons on how to be a helpful Nuisance and a harmless Radical," *Philosophy of the Social Sciences* 30 (2000): 5–32.

100 *his anthology* The Science Studies Reader Published by Routledge (New York, 1999). Note the word "the" and the fact, announced on the title page, that it has been prepared "in consultation with Peter Galison, Donna J. Haraway, Emily Martin, Everett Mendelsohn, Sharon Traweek"—five of the best-known scholars in the field.

100 *"philosophy of science is largely absent"* But some influential writings on Science Studies do give considerable importance to philosophy of science: David J. Hess, *Science Studies: An Advanced Introduction* (New York: New York University Press, 1997); Steve Fuller, *Thomas Kuhn: A Philosophical History for our Times* (Chicago: University of Chicago Press, 2000).

100 *historians have not reciprocated* Jasanoff, "Reconstructing the Past, Constructing the Present: Can Science Studies and the History of Science Live Happily Ever After?" *Social Studies of Science* 30 (2000): 621–631. For a response from a leading historian of science see Lorraine Daston, "Science Studies and the History of Science," *Critical Inquiry* 35 (2008–2009): 798–813; for a rejoinder see Peter Dear and Sheila Jasanoff, "Dismantling Boundaries in Science and Technology Studies," *Isis* 101 (2010): 759–778. Neither of these two papers has much to say about SC except to comment on whether Bruno Latour still accepts it. I have learned that it is a waste of time to argue about what Latour really believes.

100 *History of science is included not for its own sake* How did it happen that history and philosophy of science have been marginalized in a field they originally dominated? In February 1975 an academic journal announced it was changing its name. After publishing four volumes (1971–1974) under the title *Science Studies: Research in the Social & Historical Dimensions of Science & Technology*, the journal would henceforth be called *Social Studies of Science: An International Review of Research in the Social Dimensions of Science & Technology*, omitting the word "historical." The editors remained the same but were now listed in the reverse order: David O. Edge (Science Studies Unit, Edinburgh) and Roy M. MacLeod (History & Social Studies of Science, Sussex) instead of MacLeod and Edge. The publisher was no longer Macmillan (an established publisher of scientific journals) but SAGE, a newer company that specialized in the social sciences. According to the editors, all of these changes implied no change in editorial policy, so they simply reprinted their editorial content from the first issue of *Science Studies*. The journal would continue to publish "original research, whether empirical or theoretical, which brings fresh light to bear on the concepts, processes and consequences of modern science. It will be interdisciplinary in the sense that it will encourage appropriate contributions from political science, sociology, economics, history, philosophy, psychology, social anthropology, and the legal and educational disciplines." Roy MacLeod and David Edge, "Editorial," *Science Studies* 1 (1971): 1–2, on p. 2.

In retrospect it seems that the change in title did portend a significant change in emphasis. *Social Studies of Science* became one of the major outlets for articles on science and technology studies (S&TS) As if to reassure readers that the omission of the word "historical" from the subtitle of *Social Studies of Science* was not intended to exclude articles on the history of science and technology, the first issue of the renamed journal started with an article, by historian Harold Burstyn, on the Christmas Island phosphate industry, 1886–1914. But the emphasis did shift away from the kind of historical research that was usually found in journals like *Isis, Annals of Science*, and *The British Journal for the History of Science* before 1975. In its place appeared the sociology of scientific knowledge, often known as SSK.

According to Adrian Johns, a "real weakness in today's sociological history of science [is that] . . . its SSK branch in particular—has become parochial. It has become more and more technically adept, dealing with smaller issues, less historically, for a shrinking audience of specialists. As it does so, it effectively abandons the historical field to popular science writers." Johns, author's response to a review of his book in *Metascience* 10 (2001): 14–22.

Also in 1975, the international Society for Social Studies of Science (known as 4S) was founded; it now (2012) has a membership of about 1300 (according to Wesley Shrum), compared to about 2200 for the History of Science Society (according to Jay Malone).

At the three-society meeting in Cleveland, November 2011, the History of Science Society had 614 registered attendees, Society for History of Technology had 344, and 4S had 1115. These figures are from Michael D. Gordin and Matthew L. Jones, "History of Science Society Annual Meeting, 2011. The Program," *Isis* 103 (2012): 356–357.

100 *What is this constructivist strain . . . that is allegedly responsible for the split between history of science and STS . . .* Lorraine Daston, "Science Studies and the History of Science," *Critical Inquiry* 35 (2009): 798–803. See response by Peter Dear and Sheila Jasanoff, "Dismantling Boundaries in Science and Technology Studies," *Isis* 101 (2010): 759–774.

Section 4.2

100 *"No one can describe the coherence of a thought system"* Bruno Latour, *On the Modern Cult of the Factish Gods* (Durham, NC: Duke University Press, 2010), p. 50. On p. 20 he redefined "science studies" as "the anthropology of sciences."

101 *"Instead of taking truth telling as the norm"* Paul Forman, "On the Historical Forms of Knowledge Production and Curation: Modernity entailed Disciplinarity, Postmodernity Entails Antidisciplinarity," *Osiris* [series 2] 27 (2012): 156–197, on p. 76.

101 *Jan Golinski's definition of constructivism* Golinski, *Making Natural Knowledge: Constructivism and the History of Science* (New York: Cambridge University Press, 1998). As far as I can determine the first use of the phrase "social construction" in reference to a scientific concept appeared in the paper by psychologists Beth L. Wellman and George D. Stoddard, "The IQ: A Problem in Social Construction," *The Social Frontier* 5 (1939): 151–152.

David Bloor reports an earlier use of the term "social constructiveness" by psychologist Frederic Bartlett. The example discussed by Bloor involves the construction not of *scientific* knowledge but of *technology*: the development of aircraft detection devices (using sound-location, before the invention of radar) by the cooperation of several engineers, with somewhat different results in different countries. See Bartlett, *Remembering: A Study in Experimental and Social Psychology* (Cambridge: Cambridge University Press, 1932); Bloor, "Whatever Happened to 'Social Constructiveness'?" In *Bartlett, Culture and Cognition*, edited by Akiko Saito (London: Psychology Press, 2000), pp. 194–215. In my opinion some of the objections to social construction of science do not apply to social construction of technology.

Contrary to Golinski's usage, I will avoid the word "constructivism." This word is widely used by science educators to denote a process by which children learn or can be taught about the natural world. As we will see, it is already difficult enough to agree on a clear definition of social construction as a method by which professional scientists establish knowledge, without the additional confusion associated with an ongoing debate about pedagogy. Science studies should indeed be able to contribute something to science education, but it is by no means obvious that social constructionism would assist a constructivist program in science education; in fact a strong argument can be made that it would do just the opposite. For those interested in understanding the relations between these ideas I recommend the review article by Steven Turner and Karen Sullenger, "Kuhn in the Classroom, Lakatos in the Lab: Science Educators Confront the Nature-of-Science Debate," *Science, Technology, and Human Values* 24 (1999): 5–30; Helge Kragh, "Social Constructivism, the Gospel of Science, and the Teaching of Physics." *Science & Education* 7 (1998): 231–243; Michael R. Matthews, *Constructivism in Science Education* (Dordrecht: Kluwer, 1998); Matthews, "Editorial," *Science & Education*, 9 (2000): 491–505.

101 *the assertion by Supreme Court Justice Robert Jackson* The Jackson quotation comes from John Yoo, in a *New York Times* op-ed piece, November 25, 2000, p. A31. See Latour, *Science in Action* (Cambridge, MA: Harvard University Press, 1987), p. 12, "Janus's Fourth Dictum."

101 *the more general cultural movement called postmodernism* Mathematician Richard C. Brown uses the acronym PIS to designate the Postmodern Interpretation of Science. He argued that although SSK and PIS hold many of the same views, PIS is more clearly antiscience. SSK often claims to be proscience. Sociology calls itself a science, and STS sometimes call itself "the science of science," so if you accept those claims you would have to say that SSK is modern, not postmodern. Brown, *Are Science and Mathematics Socially Constructed? A Mathematician Encounters Postmodern Interpretations of Science* (London: Imperial College Press, 2000), p. 15.

101 *"sociological explanation, rather than logical explanation"* Mary Jo Nye, *Michael Polanyi and his Generation: Origins of the Social Construction of Science* (Chicago: University of Chicago Press, 2011), pp. 86, 222.

101 *no higher court than the collective decision of the scientific community* Thomas S. Kuhn, *The Structure of Scientific Revolutions* (Chicago: University of Chicago Press, 1962, 2nd ed. 1970), p. 94. (The third edition, from 1996, apparently differs from the second only by the addition of a one-page index.) I abstain from attempting to define "paradigm" here. Kuhn later insisted that in his view scientists do use empirical evidence, logic and mathematics in order to persuade others to accept their paradigms, and that critics such as Israel Scheffler are wrong to say that "adoption of a new scientific theory is an intuitive or mystical affair, a matter for psychological description rather than logical or methodological codification" (quoting Scheffler, *Science and Subjectivity*, Indianapolis: Bobbs-Merrill, 1967, p. 18); Kuhn, *The Road Since Structure* (Chicago: University of Chicago Press, 2000), p. 157.

Full disclosure: My first and only academic course on history of science was a graduate seminar on nineteenth-century thermodynamics taught by Kuhn when I was a junior at Harvard in spring 1954. He suggested the kinetic theory of gases as a topic for my seminar paper. This developed into a series of published articles, collected into a book *The Kind of Motion We Call Heat* (Amsterdam: North-Holland, 1976), although I continued my research in statistical mechanics until 1965 and did not hold a full-time position as an historian of science until 1968. Kuhn was one of my references for that position at the University of Maryland.

101 *Forman published his study of Weimar culture* Paul Forman, "Weimar Culture, Causality, and Quantum Theory, 1918–1927: Adaptation by German Physicists and Mathematicians to a hostile intellectual Environment," *Historical Studies in the Physical Sciences* 3 (1971): 1–115.

Kuhn was very impressed by this study. "I remember when I first read that. I was at Princeton, I went and put a note on the bulletin board of the department office, saying, 'I have just read the most exciting piece that I have read since I discovered Alexandre Koyré!"; Kuhn, *Road Since Structure*, p. 304.

"In his early work Forman produced one of the most successful examples in the sociology of knowledge, forever connecting the rise of uncertainty in physics with the particularities of Weimar culture. Despite years of criticism, the Forman thesis continues to infuse historical research. What Forman's critics never appear to notice is the great remorse he articulated about the role of Weimar culture in the production of a new kind of physics. One might easily read Forman as a scholar being dragged kicking and screaming to a conclusion he found uncomfortable and undesirable"; M. A. Dennis, "Postscript: Earthly Matters: On the Cold War and the Earth Sciences," *Social Studies of Science* 33 (2003): 809–819.

101 *philosophers Robert Nola and Howard Sankey* Nola and Sankey, "A Selective Survey of Theories of Scientific Method," in *After Popper, Kuhn and Feyerabend*, edited by Nola and Sankey (Dordrecht: Kluwer, 2000), pp. 1–65, on pp. 41–42. Forman's thesis is called an example of the Strong program's "flawed causal methodology." He should show that (1) A = physicists' *awareness* of the hostile culture milieu accompanies B = their *belief* in acausality; (2) P, an internalist story based in *physics* about how, on the basis of issues in science alone, acausality came to be believed, accompanies B; (3) "P does not cause B"; (4) A causes B. Forman shows (1) and (2) but not (3) or (4). They quote J. Hendry, "Weimar Culture and Quantum Causality," *History of Science* 18 (1980): 155–180, who writes: "Forman has succeeded in demonstrating that physicists and mathematicians were generally aware of the values of the milieu" but contrary to his claim "there were strong internal reasons for the rejection of causality" (p. 160).

102 *a series of connections between science and culture* Stephen G. Brush, *The Temperature of History: Phases of Science and Culture in the 19th Century* (New York: Franklin, 1978); "The Chimerical Cat: Philosophy of Quantum Mechanics in Historical Perspective," *Social Studies of Science* 10 (1980): 393–447. A more detailed account of the personal interactions involved in the formulation of the Indeterminacy Principle is provided by Mara Beller, *Quantum Dialogue: The Making of a Revolution* (Chicago: University of Chicago Press, 1999), Chapter 4.

102 *the strong program was developed in the 1970s* Bloor, "Wittgenstein and Mannheim on the Sociology of Mathematics," *Studies in History and Philosophy of Science* 4 (1973): 173–191; *Knowledge and Social Imagery* (London: Routledge & Kegan Paul, 1976); Barry Barnes, *Interests and the Growth of Knowledge* (London: Routledge,

1977); Zammito, *Nice Derangement*, 131–147. See comments on Bloor by David Kaiser, "A Mannheim for all Seasons: Bloor, Merton, and the Roots of the Sociology of Scientific Knowledge," *Science in Context* 11 (1998): 51–87.

102 *he does not say social forces* completely *determine scientific knowledge* Is there a "weak program" to compare with the "strong program"? According to George Couvalis, the Edinburgh group would apply that term to the claim that "*sometimes* scientists will accept or reject theories merely because of social influences." Couvalis, *The Philosophy of Science: Science and Objectivity* (London: SAGE, 1997), p. 143. For example, a traditional sociologist could criticize the claim that women are innately inferior as biased and inaccurate in the light of actual data, but a follower of Bloor "would hold that the claim that women are innately intellectually inferior is part of current scientific knowledge, though it may not be true." (It may be neither true nor false.) So, as I understand Bloor's position, there is no *objective* argument a feminist could make against this view unless she had enough *power* to make scientists take it seriously; within the Strong program there is no place to make a case that some established theories are refuted by *evidence.*

102 *referees for a major physics journal were not likely to give more favorable reports* H. Zuckerman and R. K. Merton, "Patterns of Evaluation in Science: Institutionalization, Structure and Functions of the Referee System," *Minerva* 9 (1971): 66–100; reprinted in Merton's *The Sociology of Science* (Chicago: University of Chicago Press, 1973). Daryl E. Chubin and Edward J. Hackett, *Peerless Science: Peer Review and US Science Policy* (Albany, NY: SUNY Press, 1990), pp. 90, 104.

102 *did not support the conventional premise that the physical sciences are "harder"* Kuhn, *Structure*, p. 208.

102 *a book by Bruno Latour and sociologist Steve Woolgar Laboratory Life: The Social Construction of Scientific Facts* (Beverly Hills, CA: SAGE, 1979).

103 *Similar statements are made by sociologist Karin Knorr Cetina* Knorr Cetina, "Laboratory Studies: The Cultural Approach to the Study of Science," in *Handbook of Science and Technology Studies*, edited by S, Jasanoff et al. (Thousand Oaks, CA: Sage Publications, 1995), pp. 140–166, on pp. 141, 147–148, 149, 160–161. Similarly, Lorraine Daston writes: "On the constructionist view, scientific objects are eminently historical, but not real." Daston, "The Coming into Being of Scientific Objects," introduction to *Biographies of Scientific Objects*, edited by Daston (Chicago: University of Chicago Press, 2000), pp. 1–14, on p. 3.

103 *he became notorious for a single statement* H. M. Collins, "Stages in the Empirical Programme of Relativism." *Social Studies of Science* 11 (1981): 3–10, on p. 3. Here is a more complete quotation:

"Modern philosophy of science has allowed an extra dimension—time—into descriptions of the nature of scientific knowledge. Theories are now seen as linked to each other, and to observations, not by fixed bonds of logic and by correspondence, but by a network, each link of which takes *time* to be established as consensus emerges and each link of which is potentially revisable—given *time*. [Endnote cites Toulmin, Quine, Hesse, Popper, Lakatos, Kuhn, Feyerabend, Fleck, Ravetz, Ziman.] Many contributors to this new model intend only to make philosophy of science compatible with history while maintaining an epistemological demarcation between science and other intellectual enterprises. One school, however, inspired in particular by Wittgenstein and more lately by the phenomenologists and ethnomethodologists, embraces an explicit relativism in which *the natural world has a small or nonexistent role in the construction of scientific knowledge*. [Endnote cited

Wittgenstein, Winch, B. Wilson, Bloor, Barnes, Collins and Cox, McHugh, Berger and Luckman, Garfinkel, Mulkay, Edge.] Relativist or not, the new philosophy leaves room for historical and sociological analysis of the processes which lead to the acceptance, or otherwise, of new scientific knowledge. One set of such analyses is gathered in this issue of *Social Studies of Science*" (emphasis added).

104 *he did not advocate ontological or epistemological relativism* In another paper in the same issue he wrote: "The existence of (hf) gravity waves is now literally *incredible*. My claim is not that sociology can bring them back, but that their demise was a social (and political) process. Where Weber . . . distinguished between the physics and politics of experiment, I have tried to show that they are not so easily distinguishable. To do this requires that, *at least for the purpose of constructing the account, a relativistic attitude is taken to the scientific phenomenon under investigation. To press the account forward required that it be taken that the phenomenon itself does not dictate the outcome of the debate*, otherwise the failure of the defeated party—the incredibility of the discredited phenomenon—will seem so natural as not to require an explanation at all. The appropriate attitude is to assume that 'the natural world in no way constrains what is believed to be.' I hope that the detailed empirical work found in this paper, and in other papers in the same tradition, will bear out the claims made for the relativistic approach and encourage its adoption as a methodological prescription, even by those to whom it is epistemologically distasteful." Collins, "Son of Seven Sexes: The Social Destruction of a Physical Phenomenon," *Social Studies of Science* 11 (1981): 33–62, on p. 54 (emphasis added).

In a 1998 e-mail to me, Collins stated that he still believes that, for the purpose of pursuing methodological relativism, the natural world should be treated *as if* it does not determine the outcome of the scientific debate.

104 *We might call this* sociological *constructionism* See also the discussion by Arthur Fine, who proposes "methodological constructivism" to replace what he calls "a great deal of nonsense" written by constructivists. Fine, "Science Made up: Constructivist Sociology of Scientific Knowledge," in *The Disunity of Science: Boundaries, Contexts, and Power*, edited by Peter Galison and David J. Stump (Stanford, CA: Stanford University Press, 1996), pp. 231–254, 481–484, on p. 232.

104 *the influence of Weimar culture might have been partly responsible* James T. Cushing, *Quantum Mechanics: Historical Contingency and the Copenhagen Hegemony* (Chicago: University of Chicago Press, 1994). *Weimar Culture and Quantum Mechanics: Selected Papers by Paul Forman and Contemporary Perspectives on the Forman Thesis*, edited by Catherine Carson, Alexei Kozhevnikov, and Helmuth Trischler (London: Imperial College Press; Singapore and Hackensack, NJ: World Scientific, 2011.).

104 *Pickering's* Constructing Quarks Pickering, *Constructing Quarks: A Sociological History of Particle Physics* (Chicago: University of Chicago Press, 1984). Gingras and Schweber, *Social Studies of Science* 16 (1988): 372–383.

104 *"what twentieth-century science has to say"* Pickering, *Constructing Quarks*, p. 413. Here is an extended quotation, which indicates that Pickering's statement was framed as a response to one by Polkinghorne: "To summarise, the overall conclusions to be drawn from the history of HEP [high energy physics] are these. The quark-gauge theory picture of elementary particles should be seen as a culturally specific product. The theoretical entities of the new physics, and the natural phenomena which pointed to their existence, were the joint products of a historical process—a process which culminated in a communally congenial representation of

reality. I have analysed that process in terms of the conditioning of scientific judgments by the dynamics of practice. I offered a simple model for that dynamics—'opportunism in context' was the slogan—and I have shown how it can illuminate major historical developments. The model is so simple that the query might legitimately arise: how could science be otherwise; what else might one expect? But the 'scientist's account' is deeply rooted in commonsense intuitions about the world and our knowledge of it. Many people do expect more of science than the production of a world congenial to social understanding and future practice. Consider the following quotation: 'Twentieth-century science has a grand and impressive story to tell. Anyone framing a view of the world has to take account of what it has to say . . . It is a nontrivial fact about the world that we can understand it and that mathematics provides the perfect language for physical science: that, in a word, science is possible at all.' Such assertions about science are commonplace in our culture. In many circles they are taken to be incontestable. But the history of HEP suggests that they are mistaken. It is *unproblematic* that scientists produce accounts of the world that they find comprehensible: given their cultural resources, only singular incompetence could have prevented members of the HEP community producing an understandable version of reality at any point in their history. And, given their extensive training in sophisticated mathematical techniques, the preponderance of mathematics in particle physicists' account of reality is no more hard to explain than the fondness of ethnic groups for their native language. *On the view advocated in this chapter, there is no obligation upon anyone framing a view of the world to take account of what twentieth-century science has to say.* The particle physicists of the late 1970s were themselves quite happy to abandon most of the phenomenal world and much of the explanatory framework which they had constructed in the previous decade. There is no reason for outsiders to show the present HEP worldview any more respect. In certain contexts, such as foundational studies in the philosophy of science, it may be profitable to pay close attention to contemporary scientific beliefs. In other contexts, to listen too closely to scientists may be simply to stifle the imagination. World-views are cultural products; there is no need to be intimidated by them." (emphasis added)

(Pickering's note 14 is: "Polkinghorne (1983). Professor Polkinghorne was for many years a leading HEP theorist at the University of Cambridge. In 1979 he left physics for the Church.") Pickering describes Kuhn's influence on his work in "Reading the Structure," *Perspectives on Science* 9 (2001): 499–510.

104 *The statement is literally correct* For twenty-first-century readers this statement is reminiscent of the defense offered by Condoleezza Rice for President George W. Bush's claim, in his 2003 State of the Union Address, that "the British government has learned that Saddam Hussein recently sought significant quantities of uranium from Africa." Rice, at that time National Security Adviser, said on Fox News that "the statement that he made was indeed accurate. The British government did say that." *New York Times*, July 14, 2003, p. A7. How many people died as a result of this conclusion?

104 *Hobbes's objections, which Shapin and Schaffer consider valid* Steven Shapin and Simon Schaffer, *Leviathan and the Air Pump* (Princeton, NJ: Princeton University Press, 1985).

Section 4.3

105 *"Without the belief that it is possible to grasp reality"* Albert Einstein and Leopold Infeld, *The Evolution of Physics* (New York: Simon & Schuster, 1938), p. 313.

105 *"Nature, apparently vanquished"* Sal Restivo, "The Theory Landscape in Science Studies: Sociological Traditions," in Jasanoff, *Handbook*, pp. 115–113, on p. 95.

105 *"The leaders of science studies themselves bemoan"* Lorraine Daston, "Science Studies," p. 799.

105 *"Deconstructing science is a fool's game"* Virginia Heffernan, "Unnatural Science," *New York Times Magazine* (August 1, 2010), p. 167.

105 *SC publications and related postmodernist writings* Brown frequently lumps together sociology of scientific knowledge (SSK) and postmodern interpretations of science (PIS) but argues that SSK is not a variety of PIS because SSK thinks of itself as a "science of science" and is thus "modern" rather than "postmodern" (*Are Science*, p. 15).

105 *struck back in 1994 with a book titled* Higher Superstition P. R. Gross and N. Levitt, *Higher Superstition: The Academic Left and its Quarrels with Science* (Baltimore: Johns Hopkins University Press, 1994).

106 *According to postmodernism, everything is a text* Leroy White and Ann Taket, "The Death of the Expert," *Journal of the Operational Research Society*, 45, no. 7 (1994): 733–748, on p. 733; Massimo Pigliucci, *Nonsense on Stilts: How to Tell Science from Bunk* (Chicago: University of Chicago Press, 2010), pp. 286–291.

106 *an embarrassing defeat for constructionist STS* A. D. Sokal, "Transgressing the Boundaries: Toward a Transformative Hermeneutics of Quantum Gravity," *Social Text* nos. 46/47 (1996): 217–252; "A Physicist Experiments with Cultural Studies," *Lingua Franca* 6, no. 4 (1996): 62–64. When the "Science Wars" issue of *Social Text* was reprinted as a book, the Sokal article that made this issue famous was omitted; see *Science Wars*, edited by Andrew Ross (Durham, NC: Duke University Press, 1996). This article and several pieces commenting on the hoax were reprinted in *The Sokal Hoax: The Sham that Shook the Academy*, edited by the editors of *Lingua Franca* (Lincoln: University of Nebraska Press, 2000). For a more extensive statement of Sokal's views see Sokal and Jean Bricmont, *Fashionable Nonsense: Postmodern Intellectuals' Abuse of Science* (New York: Picador, 1998). Detailed references for the Sokal hoax and its early repercussions may be found in the article by Nick Jardine and Marina Frasca-Spada, "Splendours and Miseries of the Science Wars," *Studies in History and Philosophy of Science* 24 (1997): 219–235.

Before chastising the editors of *Social Text* for accepting the statement that π is socially constructed, recall that American society blithely constructed the twentieth century to be only 99 years and ridiculed the scientists who objected that 2000 is the last year of the twentieth century, not the first year of the twenty-first. (It was generally agreed at the time that 1901, not 1900, was the first year of the twentieth century.)

Editors of scientific journals have also been fooled by hoaxes. See for example L. M. Krauss, "A Fifth Force Farce," *Physics Today* 61, no. 10 (October 2008): 53–55; Kelli Whitlock Burton (ed.), "Fiddle without Fear," *Science* 323 (2009): 693.

106 *Kuhn had admitted that "some of the principles"* Quotations from "Reflections on my Critics," in *Criticism and the Growth of Knowledge*, edited by I. Lakatos and A. Musgrave, 231–278 (New York: Cambridge University Press, 1970), on pp. 237–238.

106 *a large gap between his concerns* "Reflections on Receiving the John Desmond Bernal Award." *4S Review* 1, no. 4 (1983): 26–30. In the latter he also wrote: "Interests remain the dominant factor that practitioners of the new field employ in explanation, and the interests they deploy remain predominantly socioeconomic. To me the result often seems disaster. Almost by definition, resort to the term 'interest' has excluded the special cognitive interests inculcated by scientific training and required, as an

ingredient in explanations, by any sociology that claims to absorb or displace the traditional concerns of internalist studies. We shall not, in my view, account at all for scientific knowledge without recognizing the determining role in scientific practice of just such special interests: love of truth (fear of the unknown, if one prefers); fascination with puzzle solving (compulsion to exhibit special skills). . . . *Structure*, after all, was intended to suggest that the status of knowledge is in no way reduced when knowledge is seen as social." See also T. S. Kuhn, *The Trouble with the Historical Philosophy of Science* (Rothschild Lecture, 1991: Cambridge, MA: Department of the History of Science, Harvard University, 1992): pp. 8–9; reprinted in Kuhn, *Road since Structure*. In the preface to a collection of his essays, *The Essential Tension*, (Chicago: University of Chicago Press, 1977), Kuhn rejects the views of "sociologists who, drawing on my work and sometimes informally describing themselves as 'Kuhnians' " (p. xxi).

106 *"deconstruction gone mad"* Here is a longer quotation, which includes the sentences in the text: "the form taken by studies of 'negotiation' has, as I've indicated, made it hard to see what else may play a role as well. Indeed, the most extreme form of the movement, called by its proponents 'the strong program,' has been widely understood as claiming that power and interest are all there are. *Nature itself, whatever that may be, has seemed to have no part in the development of beliefs about it.* Talk of evidence, of the rationality of claims drawn from it, and of the truth or probability of those claims has been seen as simply the rhetoric behind which the victorious party cloaks its power. What passes for scientific knowledge becomes, then, simply the belief of the winners.

I am among those who have found the claims of the strong program absurd: an example of deconstruction gone mad. And the more qualified sociological and historical formulations that currently strive to replace it are, in my view, scarcely more satisfactory. These newer formulations freely acknowledge that observations of nature do play a role in scientific development. But they remain almost totally uninformative about that role—about the way, that is, in which nature enters the negotiations that produce beliefs about it." Kuhn, *The Trouble with the Historical Philosophy of Science* (Rothschild Lecture, 1991: Cambridge, MA: Department of the History of Science, Harvard University, 1992) (emphasis added); reprinted in Kuhn, *Road since Structure*.

106 *"despite some uncharacteristically ill-focused remarks of his own to the contrary"* David Bloor, "The Conservative Constructivist," *History of the Human Sciences* 10 (1997): 123–125, on p. 124.

106 *Paul Forman . . . pursued [his] own research agenda[s]* Having linked Kuhn with Forman in the previous section, I should note that Forman has not renounced SC but instead tried to put it in a larger culture perspective while defending it against attacks. See, e.g., "Truth and Objectivity," *Science* 269 (1995), 565–567, 707–710.

106 *"paradigm" . . . has been degraded* "Any sociologist who feels so inclined will declare himself the bearer of a 'new paradigm,' a new ultimate theory of the world"; Bourdieu, *Science of Science*, p. 8. The Gonzales quote is from "Excerpts from Gonzales's Legal Writings, *New York Times*, November 11, 2006, p. A27. A brief but well-documented critique of the Gonzalez paradigm is given by Vincent Iacopino, Scott A. Allen, and Allen S. Keller, "Bad Science Used to Support Torture and Human Experimentation," *Science* 331 (2011): 34–35. For further examples from the Bush administration see Eric Lichtblau, *Bush's Law: The Remaking of American Justice*, New York: Pantheon Books/Random House, 2008, pp. xi, xiii, 8, 43.

107 *Bloor, when he received the Bernal Award, gave a speech* Bloor, “Remember the Strong Program?” *Science, Technology & Human Values* 22 (1997): 373–385.

107 *“cats will tolerate wearing a collar”* quoted by Sindya N. Bhanov, “Helping Cats to Make their Way Back Home,” *New York Times/Science Times*, September 14, 2010, p. D3.

107 *Stephen Cole challenged the SC advocates* Cole, *Making Science: Between Nature and Society* (Cambridge, MA: Harvard University Press, 1992); “Voodoo Sociology,” in *The Flight from Science and Reason*, edited by P. R. Gross, N. Levitt and M. W. Lewis, pp. 274–287, New York Academy of Sciences, 1996.

107 *“SSK had come to occupy a rearguard position”* Hess, *Science Studies: An Advanced Introduction* (New York: New York University Press, 1997), p. 84.

107 *STS must justify the existence of its academic programs* Hess, *Science Studies*, pp. 148–149.

107 *“The Strong Programme is old school SSK”* Freedman, “Naturalized Epistemology, or What the Strong Programme Can’t Explain,” *Studies in History and Philosophy of Science*, 36 (2005): 135–148, on p. 137.

108 *Bloor pointed to Andrew Warwick’s study* Warwick, “Cambridge Mathematics and Cavendish Physics: Cunningham, Campbell and Einstein’s Relativity 1905–1911,” *Studies in History and Philosophy of Science* 23 (1992), 625–656; 24 (1993), 1–25. In a related case, historian Loren Graham argues that the Russian physicist V. A. Fok’s theory of gravity was affected by social influences; see “Do Mathematical Equations Display Social Attributes?” *Mathematical Intelligencer* 22, no. 3 (2000): 31–36.

108 *they pointedly omitted the word “social”* Bruno Latour and Steve Woolgar, *Laboratory Life: The Construction of Scientific Facts*, 2nd ed. (Princeton, NJ: Princeton University Press, 1986), p. 281: “A misunderstanding which has been more consequential” . . . concerns the use of the word ‘social.’ Given our explicit disavowal of ‘social factors’ in the first chapter, it is clear that our continued use of the term was ironic

108 *“social studies of science are at a standstill”* “Like Antony, I could say to philosophers, uncertain whether to stone or welcome the young domain of social studies of science, ‘I come to bury those studies, not to praise them.’ After years of swift progress, social studies of science are at a standstill. Cornered in what appears to be a blind alley, its main scholars are disputing with one another on where to go next.” What led social constructionism to its quandary, in Latour’s view, was the inadequacy of the resources of sociology and anthropology to explain the practice of science: “All the efforts at using macrosociology to account for the microcontent of science are fraught with difficulties; only very broad features like styles, worldviews, and cultures have been explained. The only research programs that have been successful are those which put to use a fine-grained sociology . . . The problem with those programs is that they, indeed, account nicely for the details of scientific practice but entirely lose track of the main goals of macrosociology—that is, an account of what holds the society together. . . . It seems that either the social science is subtle enough to explain the content of science but the making of a global society is left in the dark, or that macrosociology is back in but the details of science disappear from view. . . . most of the so-called social studies of science are largely *internalist* studies. They did not appear so to the English-speaking world because of the very abstract way in which philosophy of science had been carried on in the Anglo-American tradition before we began to work. When, for instance, Harry Collins added to the gravitation waves animals such as replication, negotiation, styles, core sets, and authority, philosophers of science mistook that zoo for the social (and so did Collins). Viewed,

however from a Continental point of view, most of the 'sociological' points could have been made—and indeed had been made—by *internalist* philosophers informed by the history of scientific practice, like Duhem, Mach, Bachelard, or Canguilheim. Social studies of science were not adding society to science but were adding some historical flesh to the often barren English-speaking philosophy of science."

When did social studies of science collapse? The year 1989 might not be a bad date for its demise, Latour suggested: "since it is the very same year that witnessed, on the one hand, the dissolution of socialism and, on the other, the dissolution of naturalism. . . . The fall of the Berlin Wall and the first conferences on global warming all point to the same transformation as the one I have outlined here: it is impossible to dominate nature and to dominate society separately. . . . We do not have to . . . recant constructivism . . . but . . . the whole domain of science studies, once its "social" attire has been set aside, becomes an exciting domain." Bruno Latour, "One More Turn after the Social Turn," in *The Social Dimensions of Science*, edited by Ernan McMullin, pp. 272–294 (Notre Dame, IN: University of Notre Dame Press, 1992). Quoted from the reprint in Biagioli, *Science Studies Reader*, pp. 287–288.

108 *a position by which one could not be convinced Pandora's Hope: Essays on the Reality of Science Studies* (Cambridge, MA: Harvard University Press, 1999), p. 125 as cited by Yearley, *Making Sense*, p. 97

108 *coauthor Steve Woolgar has also abandoned SC* Woolgar, "Irony in the Social Study of Science," in *Science Observed: Perspectives on the Social Study of Science*, edited by Karin Knorr-Cetina and Michael Mulkay (London: SAGE, 1983), pp. 239–266, on pp. 239, 262.

108 *But at the end he undermined it Science: The Very Idea* (Chichester, London & New York: Ellis Horwood and Tavistock, 1988), Chapter 6. Later he wrote: "I am not especially interested in proving or defending the position of social constructivism, although I am keen to use it as a provocation . . . My disappointment with social constructivism is that it has tended not to take on board this more ambitious agenda . . . in the end merely substituting a social for a natural realism, and thereby missing the more difficult task of challenging . . . our realist auspices per se." Woolgar, "Re Thinking Agency: New Moves in Science and Technology Studies," *Mexican Journal of Behavior Analysis* 20 (1994): 213–240, on 216 and 217.

108 *Latour confessed his feelings of guilt* "Why Has Critique Run Out of Steam? From Matters of Fact to Matters of Concern," *Critical Inquiry* 30 (2004): 225–248. The quotations come from the text posted in JSTOR, pages 2–3, 16. See www.jstor.org.proxy-um.researchport.umd.edu/stable/full/10.1086/421123. See also Judith Warner, "Fact-Free Science: How the Right Is Using Tactics Learned from the Left to Discredit Climate Change," *New York Times Magazine*, February 27, 2011, pp. 11–12, and the article by Michael Berubé mentioned therein.

109 *Six years later, Latour seemed to revert* Latour, *Modern Cult*, pp. 22, 23.

109 *actor-network theory had been superseded* Latour, *Modern Cult*, p. 51 and note 58, on p. 137.

109 *he now claims that he can be an expert on theoretical physics* Collins, "Mathematical Understanding and the Physical Sciences," *Studies in History and Philosophy of Science* 38 (2007): 667–685.

109 *Bloor, and his Edinburgh colleagues . . . also renounce that claim* Barry Barnes, David Bloor and John Henry, *Scientific Knowledge: A Sociological Analysis* (Chicago: University of Chicago Press, 1996), pp. 76–77.

110 *They also refuse to go along with the antirealism* "Radical antirealists are liable . . . to find themselves employing a realist strategy, and drawing upon the copious resources

of the realist mode of speech in just the way they find objectionable in those they criticize. Sociologists opposed to realism seem to have recognized this. Thus, those who 'deconstruct' scientific discourse and show how its theoretical entities are human inventions, rather than revelations of 'what is really there,' [sic] generally accept that their deconstructions are themselves constructions, invented accounts of their own rather than revelations of 'what a theoretical entity really is.' In effect they accept that deconstruction itself amounts to a realist strategy, sustained by appropriate use of a realist mode of speech. Their major aim appears to be to weaken the authority of accounts of what is real, and they have no need of any particular authority for their own accounts. They can allow that all speech, including their own, is a creature of a day. Nonetheless, this seems tantamount to an acceptance that is not realism as such, or the use of the realist mode of speech, which merits criticism, but rather a particular attitude to realist accounts." Barnes et al., *Knowledge*, p. 87.

110 *a third wave of science studies* H. M. Collins and Robert Evans, "The Third Wave of Science Studies: Studies of Expertise and Experience," *Social Studies of Science* 32 (2002): 235–296, on 239, 240. Stephen Turner seems to agree with Collins and Evans in his article "The Third Science War," *Social Studies of Science* 32 (2003): 581–611. But others have a competing definition for third wave, e.g., "a third generation [of] science studies," advocated by Geoffrey C. Bowker and Susan Leigh Star, that "derives from the 'pragmatist turn' which provides for a very strong form of social constructivism." D. Turnbull, review of *Sorting Things Out: Classification and its Consequences*, by Bowker and Star, in *Metascience* 11 (2002): 56–60.

In a 2010 book Collins barely mentions SC and then only in the past tense: *Tacit and Explicit Knowledge* (Chicago: University of Chicago Press, 2010), p. 166. The Collins-Evans wave 3 should not be confused with Jon Agar's wave 3: "Wave 3 was marked by an orientation toward the self, in diverse ways. Modern science studies is a phenomenon of Wave Three." Agar, "What Happened in the Sixties?" *British Journal for the History of Science* 41 (2008): 567–600, on p. 567.

Agar's wave 3 is somewhat similar to what John M. Nicholas called the "new wave," defined as "part of a movement in the sociology of science which insists that the locus of social control in science operates on the esoteric scientific artefacts and activities of scientists, rather than simply at the larger institutional level. Thus the movement eschews the Mertonian research programme which, it is alleged, attends only to scientific values which are 'content neutral' in that they are independent of the particular scientific artefacts and procedures with which the researching scientist is embroiled. With a little qualification, their views resemble Thomas Kuhn's thesis that the technical problem solving 'exemplar' is the embodiment and implicit vehicle for scientific values and criteria of judgment and choice. In this chapter I shall use the term more restrictively for the radical wing, in particular, Barry Barnes and David Bloor, and their cohorts." Nicholas, "Scientific and Other Interests," in *Scientific Rationality: The Sociological Turn*, edited by J. R. Brown (Dordrecht: Reidel, 1984), pp. 265–294, on p. 288.

Perhaps sociologists could feel more scientific if they imitated chemists and astronomers, who have established international committees to approve and register proposed names of elements and compounds or heavenly bodies, respectively.

110 *reactions by Sheila Jasanoff, Brian Wynne, and Arie Rip* Jasanoff, "Symposium. Breaking the Waves in Science Studies: Comment on H. M. Collins and Robert Evans, 'The Third Wave of Science Studies,'" *Social Studies of Science* 33 (2003): 389–400; Wynne, "Seasick on the Third Wave? Subverting the Hegemony

of Propositionalism: Response to Collins and Evans (2002)," ibid. 401–417; Rip, "Constructing Expertise: In a Third Wave of Science Studies?," ibid. 419–434; H. M. Collins and Robert Evans, "King Canute Meets the Beach Boys: Responses to the Third Wave," ibid., 435–452.

110 *the third wave as a retrenchment* Jasanoff, "Comment," p. 390.

110 *not a "drawing back" but a "drawing away"* Collins and Evans, "King Canute," p. 443.

111 *Pickering, when faced with severe criticism* Paul Roth and Robert Barrett, "Deconstructing Quarks," *Social Studies of Science* 20: 579–632. "Reply: Aspects of Sociological Explanation," *Social Studies of Science* 20 (1990): 729–745. In his contribution to this discussion, Thomas Nickles faulted Roth and Berrett for attacking *Constructing Quarks* (which I thought was the assigned topic for the discussion, SGB) instead of discussing Pickering's more recent, mostly unpublished, views, which lean away from "Edinburgh holism" toward a "weak" realism. Nickles, "How to Talk with Sociologists (or Philosophers)," *Social Studies of Science* 20 (1990): 633–638. On the relations between philosophy and SC see Martin Kusch, "Philosophy and the Sociology of Knowledge," *Studies in History and Philosophy of Science* 30 (1999): 651–685; also André Kukla, *Social Constructivism and the Philosophy of Science*, New York: Routledge, 2000.

112 *He calls his picture of scientific practice "a realist one"* Andrew Pickering, "Knowledge, Practice, and Mere Construction." *Social Studies of Science* 20 (1990): 682–729; *The Mangle of Practice: Time, Agency, and Science*, Chicago: University of Chicago Press, 1995.

112 *the Edinburgh sociologists are not very enthusiastic* Barnes et al., *Scientific Knowledge*, p. 205.

112 *Noretta Koertge edited a collection of essays* Koertge, editor, *A House Built on Sand: Exposing Postmodernist Myths about Science* (New York: Oxford University Press, 1998).

112 *Sheila Jasanoff reviewed it* Jasanoff, "Review of *A House Built on Sand*, edited by N. Koertge," *Science, Technology, and Human Values* 24 (1999): 495–500.

112 *She did not attempt to rebut any of the specific criticisms* Jasanoff, "Review of *A House Built on Sand*," p. 499. Collins followed a similar approach in his review of this book, "The Science Police," *Social Studies of Science* 29 (1999): 287–294, resorting to name-calling (as in the title of his review), complaining that Franklin's article failed to mention Collins's earlier reply in the text though it was listed in the bibliography, and that most of the authors don't explain the positions of their opponents; but in an eight-page review he found no space to give a substantive critique of the main arguments of the book. Similarly, Elisabeth Lloyd in her review alleges that Gross, Levitt, Holton, Wolpert, and other critics of SC are themselves the "real enemies" of science. A. I. Tauber, reviewing Norman Levitt's *Prometheus Bedeviled* (Rutgers University Press, 1999), rejects Levitt's attack on SC but gives no specific evidence for SC's validity, simply asserting that one major result of science studies was that science is socially constructed.

112 *reminiscent of the strategy used by some Freudians and Marxists* Alexei Kozhevnikov traces the origin of SC to a 1918 brochure *Socialism of Science* by the Marxist writer Aleksandr Bogdanov; it was transmitted to the West by Boris Hessen at the 1932 London Congress on History of Science. See Kozhevnikov, "The Phenomenon of Soviet Science," *Osiris* 23 (2008): 115–135. According to Sal Restivo, "Marxism is at the root of most of what is innovative about science studies"; in Jasanoff, *Handbook*, p. 103.

Similarly, David Bloor replied to critics of his Strong program by asserting that they treat science as if it is sacred, and are alarmed when it is scrutinized by sociology, which is *not* considered a science but rather belongs to "the sphere of the profane." *Knowledge and Social Imagery*, 2nd ed. (Chicago: University of Chicago Press, 1991), p. 48. (According to the preface, this section has not been significantly changed from the 1976 original.)

112 *his essay "makes a strong brief"* Stephen H. Cutcliffe and Carl Mitcham, editors, *Visions of STS: Counterpoints in Science, Technology and Society Studies* (Albany, NY: SUNY Press, 2001), "Introduction: The Visionary Challenges of STS," pp. 1–7, on p. 4.

113 *"Scientific facts are not found"* Bijker, "Understanding Technological Culture through a Constructivist View of Science, Technology, and Society," in *Cutcliffe and Mitcham, Visions of STS*, pp. 19–34, on p. 24.

113 *The most authoritative account of the current position of STS experts The Handbook of Science and Technology Studies*, 3rd ed., edited by Edward J. Hackett, Olga Amsterdamska, Michael Lynch, and Judy Wajcman, published in cooperation with the Society for Social Studies of Science (Cambridge, MA: MIT Press, 2008), page xi. I submitted an early version of this chapter for publication in the Handbook; it was rejected because it "does not fit well with the *Handbook* as it has developed." Letter to SGB from Ed Hackett. July 27, 2004.

113 *begins by endorsing SC* Sismondo, "Science and Technology Studies and an Engaged Program," *Handbook*, 3rd ed., pp. 13–33, on p. 14.

113 *"Collins's methodological relativism asserts" Handbook*, 3rd ed., p. 15.

113 *"less concerned with understanding science and technology" Handbook*, 3rd ed., p. 18.

113 *scientists no longer have privileged access Handbook*, 3rd ed., p. 23.

113 *"assume that expertise is real and that it represents genuine knowledge" Handbook*, 3rd ed., p. 23. See Collins and Evans, "Third Wave." The term "STSers" for practitioners of science and technology studies is used by Park Doing in *Handbook*, 3rd ed., p. 285. "Expertise" has also become a fashionable topic for historians of science: see Eric H. Ash, editor, "Expertise: Practical Knowledge and the Early Modern State," *Osiris* 25 (2010): 1–259.

113 *Park Doing's trenchant article "Give me a Laboratory"* "Give Me a Laboratory and I will Raise a Discipline: The Past, Present, and Future Politics of Laboratories in STS," in *Handbook*, 3rd ed., pp. 279–295. Latour and Woolgar, *Laboratory Life: Social Construction; Michael Lynch, Art and Artifact in Laboratory Science: A Study of Shop Work and Shop Talk in a Research Laboratory* (London: Routledge & Kegan Paul, 1985); Karin Knorr Cetina, *The Manufacture of Knowledge: An Essay on the Constructivist and Contextual Nature of Science* (Oxford: Pergamon Press, 1981); H. M. Collins, *Changing Order: Replication and Induction in Scientific Practice* (London: Sage, 1985); Trevor Pinch, *Confronting Nature: The Sociology of Solar Neutrino Detection* (Dordrecht, The Netherlands: D. Reidel, 1986).

114 *"we did not really do what we said"* Doing, "Give Me," p. 292.

114 *Latour's 1992 pronouncement* Latour, "One more turn." Zammito writes that "In 1992 . . . science studies hit a wall" with the epistemological chicken debate and a concurrent argument between Latour and Bloor (*Nice Derangement*, p. 203).

114 *a 2011 article by sociologists Marion Blute and Paul Armstrong* "The Reinvention of Grand Theories of the Scientific/Scholarly Process," *Perspectives on Science* 19 (2011): 391–425. The definition of "yes, but" is on page 418.

114 *Like many movements in sociology* James B. Rule, *Theory and Progress in Social Science* (New York: Cambridge University Press, 1997).

114 *popular for a time but disappeared* Lindsay Waters, executive editor for humanities at Harvard University Press, in answer to the question "What areas do you think are overpublished?" replied: "The time is up for Social Constructionism and New Historicism. The 'construct' part of that phrase looks too mechanical and reductionistic now. Those many thought were the coming gods in the 1990s are going to sink into well-deserved obscurity"; Leila Salisbury et al., "The Future of Scholarly Publishing," *Chronicle of Higher Education*, June 12, 2009, pp. B12–B13, on B13.

Section 4.4

115 *SC reinforces the efforts of professional historians* For the classic statements of the antiwhig doctrine see George W. Stocking Jr., "On the Limits of 'Presentism' and 'Historicism' in the Historiography of the Behavioral Sciences," *Journal of the History of the Behavioral Sciences* 1 (1965): 211–218; Thomas S. Kuhn, "The History of Science," in *International Encyclopedia of the Social Sciences*, vol. 14 (New York: Macmillan/Free Press, 1968), pp. 74–83. Peter S. Alagona is one writer who clearly associates SC with antiwhiggism: see his review of *James Lovelock: In Search of Gaia* by John Gribbin and Mary Gribbin, in *Isis* 101 (2010): 253–254.

115 *nonconstructionist historians (e.g., Allan Franklin)* Franklin, *The Neglect of Experiment* (Cambridge: Cambridge University Press, 1986); *Experiment, Right or Wrong*, ibid., 1990; "Experimental Questions," *Perspectives in Science* 1 (1993): 127–146; "Discovery, Pursuit, and Justification," ibid.1 (1993): 252–284; *Can That Be Right? Essays on Experiment, Evidence, and Science* (Dordrecht: Kluwer, 1999).

115 *have shown how scientists use the same reasoning skills* Golinski, *Making Natural Knowledge*, p. 31.

115 *"an instance of successful science"* Slezak, "Sociology of Scientific Knowledge and Science Education," in *Constructivism in Science Education: A Philosophical Examination*, edited by Michael R. Matthews (Dordrecht: Kluwer, 1998), pp. 159–188; previously published in *Science & Education* 3 (1994): 265–294.

115 *Michael Ruse argues that the answer evolved over time* Ruse, *Mystery of Mysteries: Is Evolution a Social Construction* (Cambridge, MA: Harvard University Press, 1999). See also Gregory Radick, "Cultures of Evolutionary Biology," *Studies in History and Philosophy of Biological and Biomedical Sciences* 34 (2003): 187–200. Instead of asking to what extent a scientific theory or fact is socially constructed and to what extent it reflects the truth about an objective real world, the constructionist asserts there is no such thing as the truth apart from society; society, Radick writes, fixes "the rules that determine how truth and falsehood are to be assessed in the first place" (p. 198).

115 *Degler has argued that the swing away from hereditarian* Degler, *In Search of Human Nature: The Decline and Revival of Darwinism in American Social Thought* (New York: Oxford University Press, 1991).

116 *Brown notes that SSK "seems not to be very interested"* Brown, *Are Science*, p. 286.

116 *Thomas Sturm and Gerd Gigerenzer . . . suggested two examples* Stern and Gigerenzer, "How Can We Use the Distinction between Discovery and Justification? On the Weaknesses of the Strong Programme in the Sociology of Science," in *Revisiting Discovery and Justification: Historical and Philosophical Perspectives on the Context Distinction*, edited by Jutta Schickmore and Friedrich Steinle (Dordrecht: Netherlan ds: Springer, 2006), pp. 133–158, on pp. 146–149.

116 *"Much that goes on in science"* Bloor, "The Strengths of the Strong Programme" in *Scientific Rationality: The Sociological Turn*, edited by J. R. Brown (Dordrecht: Reidel, 1984), pp. 75–94, on p. 80.

116 *She shows that biologists frequently describe* Martin, "The Egg and the Sperm: How Science has Constructed a Romance based on stereotypical male and female roles," *Signs* 16 (1991): 485–561. Paul Gross points out that the active role of the egg, stressed by Martin, was recognized by biologists as early as 1919. That does not necessarily contradict Martin's evidence from more recent publications. Gross, "Bashful Eggs, Macho Sperm, and Tonypandy," in *A House Built on Sand*, edited by N. Koertge (New York: Oxford University Press, 1998), pp. 59–70.

117 *most social constructionists ignore these examples* Golinski, *Making Natural Knowledge*, pp. 65–66, 162, 183–184. Eulalia Pérez Sedeño, "Gender: The Missing Factor in SRS," in *Visions of STA*, edited by S. H. Cutcliffe and Carl Mitcham (Albany, NY: SUNY Press, 2001), pp. 123–138.

117 *men and women have the same average IQ but whites have higher IQs* Lewis M. Terman, *Measurement of Intelligence* (Boston: Houghton Mifflin/Cambridge, MA: Riverside Press, 1916), pp. 68–72; Terman and Maud A. Merrill, *Measuring Intelligence* (Boston: Houghton Mifflin, 1937, pp. 22, 34.) In the 1970s the verbal part of the Scholastic Aptitude Test, based originally on the IQ test, was adjusted to increase the proportion of questions that boys answered correctly more often than girls. Previously girls got higher average scores on the verbal part while boys did better on the mathematics part; now boys get higher average scores on both parts. See Phyllis Rosser, *The SAT Gender Gap* (Washington, DC: Center for Women Policy Studies, 1989). This is a socially constructed fact that has serious consequences for girls who apply for admission and scholarships at selective colleges; see S. G. Brush, "Women in Science and Engineering," *American Scientist* 79 (1991): 404–419, and reply to letters, ibid. 80 (1992): 6, 8.

117 *Butterfield . . . criticized those who described British history as a tale of progress* See however his later book *The Englishman and his History* (Cambridge: Cambridge University Press, 1944, pp. 3–4.

117 *physics has its own strictly circumscribed version* See, e.g., Karen Barad, "Agential Realism: Feminist Intervention in Understanding Scientific Practices," in Biagioli, *Science Studies Reader*, 1–11. Social constructionists use the example of a baseball game: balls and strikes acquire reality only when the umpire calls (constructs) them. I have heard physicists use this metaphor to explain how quantum properties of subatomic particles acquire reality only when they are observed. Stanley Fish, "Professor Sokal's Bad Joke," *New York Times* (May 19, 1996): A23; Frederick Grinnell, "The Practice of Science at the Edge of Knowledge," *Chronicle of Higher Education* (March 24, 2000): B11–B12; E. J. Dionne Jr., "Deconstruction Zone," *Washington Post Magazine* (April 19, 2000): 9.

117 *his piece on the sociology of the door-closer* Jim Johnson [i.e., Bruno Latour], "Mixing Humans and Non-Humans Together: The Sociology of a Door-Closer," *Social Problems* 35 (1988): 298–310.

118 *have a tendency to make statements that sound like social determinism* Bourdieu, *Science of Science*, pp. 26–27 (on Latour); Peter Slezak, "Sociology of Scientific Knowledge and Scientific Education, Part 1," in *Constructivism in Science Education*, edited by M. R. Mathews (Dordrecht: Kluwer, 1998), pp. 159–188.

118 *Weidman expressed this view quite bluntly* Weidman, *Constructing Scientific Psychology: Karl Lashley's Mind-Brain Debates* (New York: Cambridge University Press, 1999), p. xiii. Guyatri Chakravorty Spivak expressed a similar view: Postmodern does not give arguments or evidence for itself, it is "a transformation of consciousness—a changing mind set." *The Postcolonial Critic* (New York: Routledge, 1991), p. 20.

Alexei Kozhevnikov wrote "my obligatory college-level training in Marxist philosophy became very useful as an introduction to the more recent Anglophone literature on social constructionism" in his "Introduction [to issue on science in Russian contexts]: A New History of Russian Science," *Science in Context* 15 (2002): 177–182, on pp. 179–180. Against the position of Weidman and others, Latour writes: "the project of science studies is not to state *a priori* that there exists 'some connection' between science and society, because *the existence of this connection depends on what the actors have done or not done to establish it.* Science studies merely provides the means of tracing this connection *when it exists.*" *Pandora's Hope: Essays on the Reality of Science Studies* (Cambridge, MA: Harvard University Press, 1999), pp. 86–87. David Gooding casts doubt on both sides: "individuals cannot always articulate the rules they are following. Sociologists use this fact to support the relativisation of knowledge to conceptual schemes and these, in turn, to practices. However, there are no hard facts that support relativism or any other epistemological position" ; "Review Symposium, What Can Particle Physicists Count On?" in *Metascience* 8 (1999): 361–367, on 362.

118 *the alleged "underdetermination of theory by evidence"* See Larry Laudan, "Demystifying Underdetermination," in *Scientific Theories*, edited by C. W. Savage (Minneapolis: University of Minnesota Press, 1990), pp. 267–297. The underdetermination is partly valid; see the fallacy of affirming the consequent discussed in Section 1.2.

118 *Shapin, in response to Cole's challenge* Steven Shapin, "Mertonian Concessions" (review of Cole, *Making Science*) *Science* 259 (1993): 839–841.

118 *Collins makes a similar assertion* H. M. Collins, "Philosophy of Science and SSK: Reply to Koertge," *Social Studies of Science* 29 (1999): 785–790.

118 *Ullica Segerstrale, reviewing* The Golem *(1993)* Segerstrale, "Science By Worst Cases," *Jurimetrics Journal* 35 (1995): 361–364.

119 *Yves Gingras, reviewing Collins's* Gravity's Shadow "Everything You Did Not Necessarily Want to Know About Gravitational Waves. And Why," *Studies in History and Philosophy of Science* 38 (2007): 268–282.

119 *Laudan attacked Bloor's strong program* Laudan, "The Pseudo Science of Science," in *Scientific Rationality: The Sociological Turn*, edited by J. R. Brown (Dordrecht: Reidel, 1984), pp. 41–73; reprinted from *Philosophy of the Social Sciences* 11 (1981): 173–198. For a review of philosophical objections see Dimitris P. Papayannakos, "Philosophical Skepticism not Relativism is the Problem with the Strong Programme in Science Studies and with Educational Constructivism," *Science & Education* 17 (2008): 573–611.

119 *all attempts to demarcate science from pseudoscience* Laudan, "The Demise of the Demarcation Problem," in *Physics, Philosophy and Psychoanalysis*, edited by R. S. Cohen and L. Laudan (Boston: Reidel, 1983), pp. 111–127.

119 *Bloor added another criterion* Bloor, "The Strengths of the Strong Programme," in *Scientific Rationality: The Sociological Turn*, edited by J. R. Brown (Dordrecht: Reidel, 1984), pp. 75–94.

119 *"the flattening of the cognitive to the social"* Beller, *Quantum Dialogue* (Chicago: University of Chicago Press, 1999), p. 310.

119 *the influence of an individual scientist's age or birth order* Frank Sulloway, *Born to Rebel: Birth Order, Family Dynamics, and Creative Lives* (New York: Pantheon Books, 1996).

119 *case studies used by social constructionists have been criticized* Jacob, "Science Studies"; Noretta Koertge, "The Zero-Sum Assumption and the Symmetry Thesis," *Social Studies of Science* 29 (1999): 777–784; Roger G. Newton, *The Truth of Science: Physical Theories and Reality* (Cambridge, MA: Harvard University Press, 1997); Rose-Mary Sargent, *The Diffident Naturalist: Robert Boyle and the Philosophy of Experiment* (Chicago: University of Chicago Press, 1995); Margaret Osler, in *Rethinking the Scientific Revolution*, edited by M. Osler, (Cambridge: Cambridge University Press, 2000), pp. 8, 19; Jamie C. Kassler, "On the Stretch: Hobbes, Mechanics and the shaking Palsey," in *1543 and All That*, edited by G. Freeland and A. Corones (Dordrecht: Kluwer, 2000), pp. 151–187; Roger G. Newton, *The Truth of Science: Physical Theories and Reality* (Cambridge, MA: Harvard University Press, 1997), p. 40; Barbara J. Shapiro, *A Culture of Fact: England, 1550–1720* (Ithaca, NY: Cornell University Press, 2000). Cassandra L. Pinnick, "What Is Wrong with the Strong Programme's Case Study of the 'Hobbes-Boyle Dispute'?" in *A House Built on Sand*, edited by N. Koertge (New York: Oxford University Press, 1999), pp. 227–239.

Sheila Jasanoff dismisses these criticisms in one sentence: "Historians of science often seem to share with scientists the suspicion that the firm ground of reality will dissolve into the quagmire of make-believe if social constructivists are allowed to have their way with science and technology." See her article "Reconstructing the Past, Constructing the Present: Can Science Studies and the History of Science Live Happily Ever After?" *Social Studies of Science* 30 (2000): 621–631, on p. 623. Note the implication in her title that History of Science is not part of Science Studies.

119 *it violated an important precept of modern historiography* Smocovitis, *Unifying Biology: The Evolutionary Synthesis and Evolutionary Biology* (Princeton, NJ: Princeton University Press, 1996), pp. 80–82.

120 *"maintains the denial of universals"* Lewis Pyenson, "The Enlightened Image of Nature in the Dutch East Indies: Consequences of Postmodernist Doctrine for Broad Structure and intimate Life," *Historical Studies in the Natural Sciences*, 41, no. 1 (2011): 1–40, on p. 2, note 3.

120 *the problem can be solved by postulating diffusion* Golinski, *Making Natural Knowledge*, p. 33.

120 *the phenomenon of multiple discovery* Robert K. Merton, "Singletons and Multiples in Scientific Discovery: A Chapter in the Sociology of Science," *Proceedings of the American Philosophical Society* 105 (1961): 470–486.

120 *criticism has been very effectively made by Margaret Jacob* Jacob, "Science Studies."

120 *social constructionists often reject the need for a narrative account* Golinski, *Making Natural Knowledge*, p. xii. Keith M. Parsons, *Drawing out Leviathan: Dinosaurs and the Science Wars* (Bloomington: Indiana University Press, 2002).

120 *the origin of science studies in the radical movements* Stephen H. Cutcliffe, *Ideas, Machines, and Values;* David Kaiser, "A Mannheim for all Seasons: Bloor, Merton, and the Roots of the Sociology of Knowledge," *Science in Context* 11 (1998): 51–87.

120 *they were deflected from that path by Kuhn's influence* Fuller, *Thomas Kuhn: A Philosophical History for our Times* (Chicago: University of Chicago Press, 2000); "Why Science Studies Has Never Been Critical of Science: Some Recent Lessons on How to Be a Helpful Nuisance and a Harmless Radical," *Philosophy of the Social Sciences* 30 (2000): 5–32.

120 *"Though I welcome the turn to the external history of science"* Kuhn, "The Relation Between History and the History of Science," *Daedalus* 100 (1971): 271–304,

reprinted in Kuhn, *The Essential Tension* (Chicago: University of Chicago Press, 1977), pp. 127–161.

120 *"SSK has been animated by a goal"* Bowden, "Coming of Age in STS: Some Methodological Musings," in Jasanoff, *Handbook*, pp. 64–79, on p. 71.

120 *"The study of scientific knowledge is primarily seen"* Knorr-Cetina, "The Ethnographic Study of Scientific Work: Towards a Constructivist Interpretation of Science," in *Science Observed: Perspectives on the Social Study of Science*, edited by Karin Knorr-Cetina and Michael Mulkay (London: SAGE, 1983), pp. 115–140, on p. 119.

120 *"nothing . . . 'scientific' was happening"* Latour, "Give Me a Laboratory and I Will Raise the World," in Knorr-Cetina and Mulkay, *Science Observed*, pp. 141–170, on p. 141.

121 *a dangerous antiscience movement* While creationists attack science from the (ideological) right and postmodernists (including SC) attack from the left, the two antiscience camps sometimes join forces. See S. G. Brush, "Postmodernism versus Science versus Fundamentalism: An Essay Review," *Science Education* 84 (2000): 114–122; Deborah Dysart, "Truce-Making in the Science Wars," *Metascience* 10 (2001): 66–69; Alasdair MacIntyre, *Three Rival Versions of Moral Enquiry* (Notre Dame, Ind.: University of Notre Dame Press, 1990); Levitt, *Prometheus Bedeviled*, pp. 183–186. Anthropologist Marvin Harris notes that "disparagement of Western science and technology" is "the most prominent and important" theme of postmodernism; see *Theories of Culture in Postmodern Time* (Walnut Creek, CA: Altamire, 1998), p. 153.

121 *a book edited by Gross, Levitt, and Lewis* Paul R. Gross and Norman Levitt, *Higher Superstition: The Academic Left and its Quarrels with Science* (Baltimore: Johns Hopkins University Press, 1994), 2nd ed. 1998; Gross, Levitt and M. W. Lewis (eds.) *The Flight from Science and Reason* (New York: New York Academy of Sciences, 1996); Gerald Holton, *Science and Anti-Science* (Cambridge, MA: Harvard University Press, 1993); R. G. Turner, "The History of Science and the Working Scientist," in *Companion to the History of Modern Science*, edited by R. C. Olby et al. (New York: Routledge, 1990), pp. 23–31.

121 *for undermining science education* Kragh, "Social Constructivism." See also Mercé Izquierdo-Aymerich and Agustín Adúriz-Bravo, "Epistemological Foundations of School Science," *Science & Education* 12 (2003): 27–43, p. 28.

121 *they refuse to give the scientific community the right* Barnes et al., *Scientific Knowledge*; M. Cartmill, "Oppressed by Evolution." *Discover* 19, no. 3 (1998): 78–83; H. M. Collins and T. J. Pinch, "The Construction of the Paranormal: Nothing Unscientific is Happening." *Sociological Review Monograph* 27 (1979): 237–270.; Andrew Pickering, "Science, Technology, and the Mangling of the Social," Colloquium, Committee on History & Philosophy of Science, University of Maryland, College Park, April 10, 1997; B. Simon "Undead Science: Making Sense of Cold Fusion after the (Arti)fact." *Social Studies of Science* 29 (1999): 61–85. See also the quotation from Steve Fuller at the beginning of this chapter. Fuller is not a full-fledged constructionist but a sociological provocateur.

121 *appearing as a witness for the intelligent design (ID) movement* Steve Fuller, "A Step Toward the Legalization of Science Studies," *Social Studies of Science* 36 (2006: 827–838. According to Norman Levitt, "Fuller's connection with the ID crowd is a rather old one. He signed on as a fellow-traveler as early as 1998, embracing Intelligent Design theory as a ploy in his more general campaign to challenge the hegemony of standard science . . . But despite his long familiarity with ID and its rationales, Fuller utterly failed to sway Judge Jones . . . Jones's decision refers frequently to Fuller's testimony, but only to find further damning evidence that ID is, root and branch, a theological imposture lacking scientific status. . . . Fuller, as 'expert' witness for the

defense, proved to be one of the plaintiff's most effective exhibits." Levitt, "Steve Fuller and the hidden Agenda of Social Constructivism," www.talkreason.org/articles/Fuller.cfm, posted February 19, 2006.

121 *Pennock has recently pointed out the surprising connection* Robert T. Pennock, "The Postmodern Sin of Intelligent Design Creationism," *Science & Education* 19 (2010): 757–778. See also Brush, "Postmodernism."

121 *explained why social constructionists often seem to support ideas* Martin and Richards, "Scientific Knowledge, Controversy, and Public Decision Making," in Jasanoff, *Handbook*, pp. 506–526, on p. 514.

121 *explained in 1974 why SSK is antiscience Scientific Knowledge and Sociological Theory* (London: Routledge & Kegan Paul, 1974), as quoted by Brown, *Are Science*, p. 260.

122 *Brown gave a similar explanation Are Science*, pp. 260–261.

122 *Brown also suggested a psychological role Are Science*, p. 262.

122 *"many of us are in the humanities because we were bad at or turned off by science"* Andrew Ross, ed., *Science Wars* (Durham, NC: Duke University Press, 1996), article by Levine. (This is described as an expanded version of the science wars issue of *Social Text*, but it omits the Sokal hoax article that made that issue famous.) Ross himself made clear his antipathy to science when he dedicated an earlier book, *Strange Weather: Culture, Science, and Technology in the Age of Limits* (London: Verso, 1991), to "all the science teachers I did not have, it could only have been written without them" (quoted b y Brown, *Are Science*, p. 259).

122 *Kemp argued that Bloor's strong program* Kemp, "Saving the Strong Programme? A Critique of David Bloor's Recent Work," *Studies in History and Philosophy of Science* 36 (2005): 706–719, on pp. 707, 715. See the replies by Bloor and Márta Fehér and the rejoinder by Kemp, ibid., 38 (2007): 210–253.

122 *"most people who use the social construction idea enthusiastically"* He writes: "Social construction work is critical of the status quo. Social constructionists about *X* tend to hold that:

(1) *X* need not have existed, or need not be at all as it is. *X*, or *X* as it is at present, is not determined by the nature of things; it is not inevitable.

Very often they go further, and urge that:

(2) *X* is quite bad as it is.

(3) We would be much better off if *X* were done away with, or at least radically transformed.

A thesis of type (1) is the starting point: the existence or character of *X* is not determined by the nature of things. *X* is not inevitable. *X* was brought into existence or shaped by social events, forces, history, all of which could well have been different. Many social construction theses at once advance to (2) and (3), but they need not do so. One may realize that something, which seems inevitable in the present state of things, was not inevitable, and yet is not thereby a bad thing. But *most people who use the social construction idea enthusiastically want to criticize, change, or destroy some X that they dislike in the established order of things.*" Hacking, *Social Construction*, pp. 6–7 (emphasis added).

122 *telling scientists they are constructing facts* Latour, *Pandora's Hope*, 2–18. But later in the same book he writes: when science studies "showed the constructivist character of science . . . it [was] as if we were undermining science's claim to truth" (p. 115). Having recognized the reason why scientists object to science studies, Latour still refuses to take their objection seriously and pushes it aside by fancy wordplay. Steve Fuller writes that STSers were surprised and confused by the science wars "even though it should have been perfectly evident that the diminished status STSers

assigned to institutionalized science . . . would eventually meet with resistance by the scientific community." Fuller, *Thomas Kuhn: A Philosophical History for Our Times* (Chicago: University of Chicago Press, 2000), p. 355.

122 *has a tendency to make statements* Latour, "From the World of Science to the World of Research?" *Science* 280 (1998): 208–209, on p. 209. Latour, *Pandora's Hope*, pp. 20, 115. Letters to Editor by R. B. Gallagher et al., "Passionate Science," *Science* 282 (1998): 1821.

122 *"engaged in a moral struggle to strip science of its extravagant claim"* Callon and Latour, "Don't throw the Baby out with the Bath School!: A Reply to Collins and Yearley," in *Science as Practice and Culture*, edited by A. Pickering (Chicago: University of Chicago Press, 1992), pp. 343–368, on p. 346.

122 *also addresses Callon's assumption about science's hegemony* Review of *Politics of Nature: How to Bring the Sciences into Democracy* (Cambridge, MA: Harvard University Press, 2004), *Science* 305 (2004): 1241–1242.

122 *social constructionists like Pinch and Collins "maintain a scrupulous neutrality"* Slezak, "Sociology," p. 163. T. J. Pinch and H. M. Collins, "Private Science and Public Knowledge: The Committee for the Scientific Investigation of the Paranormal and its use of the Literature," *Social Studies of Science* 14 (1984): 521–546.

122 *"Latour and Woolgar espouse essentially the same idea"* Slezak, "Sociology," pp. 163–164, citing Latour and Woolgar, *Social Construction*, p. 284.

122 *"could have offered no principled objection"* Slezak, "Sociology," p. 164.

124 *"all science must appear fraudulent"* H. F. Judson, *The Great Betrayal: Fraud in Science* (New York: Harcourt, 2004), p. 41. See also Brown, *Are Science*, p. 225, note 17: "Assume that science is merely a matter of propaganda and political success which is either independent of or actually creates 'nature.' Then it is difficult to make sense of the notion of 'cheating.' The possibility of cheating seems to imply the existence of 'facts' which are independent of the scientist's volition. If science is *only* a social construction, why not 'cheat' if it helps in the victory of our construction. We would not be falsifying anything for there is nothing to falsify."

124 *"the native who accepts this cosmopolitanism"* Hiram Caton, editor, *The Samoa Reader: Anthropologists take Stock* (Lanham, MD: University Press of America, 1990), p. 9. Conversely, those who do want to liberate themselves from oppression need the ideal of objective science; the ethnoscience favored by constructionists always support authority, according to Meera Nanda, "The Epistemic Charity of the Social Constructivist Critics of Science and Why the Third World Should Refuse the Offer," in *A House Built on Sand*, edited by N. Koertge (New York: Oxford University Press, 1998), pp. 286–311.

124 *the scientists have the ability to do considerable damage* The social constructionists point to two events at the Princeton Institute for Advanced Study (IAS) as evidence of their own powerlessness relative to natural scientists. In 1990 the IAS School of Social Science wanted to appoint Bruno Latour, at that time the most influential advocate of SC. The appointment was voted down by the physicists and mathematicians who dominate the IAS on the grounds that Latour's writings on science were inaccurate. (D. Berreby, " . . . That damned elusive Bruno Latour," *Lingua Franca*, vol. 4, no. 6 [1994]: 22–32, 78). About three years later the proposed appointment of M. Norton Wise, a respected historian of science who had defended STS against the attacks of scientists, was rejected (L. McMillen, "The Science Wars Flare at the Institute for Advanced Study," *Chronicle of Higher Education*, May 16, 1997: A13).

A more recent episode shows the intensity of some scientists' anger at their postmodern critics. The book review editor of *Science* magazine, Katherine Livingston, was reprimanded by the editor of *Science* and decided to retire because of harsh criticism of Paul Forman's negative review of anti-SC book *The Flight from Science and Reason*. Some readers, including a few contributors to the book, objected to what physicist Robert Park called Forman's "nasty diatribe against science itself." Park, *What's New* (electronic newsletter published by American Physical Society) Friday, 14 November 1997) http://bobpark.physics.umd.edu/WN97/wn111497.html; J. D. Robinson et al., "Deconstructing science," *Science* 276 (1997): 1953–1955.

124 *"STS remains weakly institutionalized"* Jasanoff, "A Field of its Own: The Emergence of Science and Technology Studies," in *Oxford Handbook of Interdisciplinarity*, edited by Robert Froderman et al. (Oxford: Oxford University Press, 2010), pp. 191–204, on pp. 202, 198.

124 *"Sociology of science began to be recognized"* Ben-David, "Sociology of Scientific Knowledge," in *The State of Sociology: Problems and Prospects*, edited by James F. Short (Beverly Hills, CA: Sage, 1981), pp. 40–59, on p. 40.

125 *"attracted to the field a number of younger scholars"* Ben-David, "Sociology," pp. 40–41.

125 *"The sociology of knowledge tradition assumed that all ideas"* Ben-David, "Sociology," p. 42.

125 *This assumption was rejected by the older sociologists of science* Ben-David, "Sociology," p. 43. R. K. Merton, "Paradigm for the Sociology of Knowledge," in The Sociology of Science: *Theoretical and Empirical Investigations*, edited and introduced by N. W. Storer (Chicago: University of Chicago Press, 1973), pp.?? [get page numbers]

125 *"'rationality' as conceived by scientists"* Ben-David, "Sociology," pp. 7–41.

125 *"causal explanation of the same sociological kind is required"* Ben-David, "Sociology," p. 44; Bloor, *Knowledge and Social Imagery* (London: Routledge & Kegan Paul, 1976), pp. 4–5.

125 *"there are limits to the possibility of human thought"* Ben-David, "Sociology," p. 46; Barry Barnes, *Scientific Knowledge and Sociological Theory* (London: Routledge & Kegan Paul, 1974), p. 43.

125 *This "implies the admission of some standards"* Ben-David, 'Sociology" p. 46. See also Mary Hesse, *Revolution and Reconstruction in the Philosophy of Science* (Brighton: Harvester Press, 1980), p. 56 (as cited by Ben-David).

125 *Ben-David concludes that "no success can be claimed"* Ben-David, "Sociology," p. 52.

125 *"the attempt to investigate the similarities"* Ben-David, "Sociology," pp. 53, 54. S. G. Brush, "Thermodynamics and History: Science and Culture in the 19th Century," *The Graduate Journal* 7 (1967) 477–565; expanded version published as *The Temperature of History: Phases of Science and Culture in the Nineteenth Century* (New York: Burt Franklin & C., 1978). See also Brush, "The Chimerical Cat: Philosophy of Quantum Mechanics in Historical Perspective," *Social Studies of Science* 10 (1980): 393–447.

125 *what Donna Haraway calls "malestream science studies"* Haraway, "Modest Witness: Feminist Diffractions in Science Studies," in *The Disunity of Science*, edited by Peter Galison and David J. Stump (Stanford, CA: Stanford University Press, 1996), pp. 428–441, 521–526, on p. 437.

125 *has refused to take seriously the feminist research* Mark Elam and Oskar Juhlin, "When Harry met Sandra: An alternative Engagement after the Science Wars," *Science as*

Culture 7 (1998): 95–109. Judy Wajcman, "Reflections on Gender and Technology Studies: In What State Is the Art?" *Social Studies of Science* 30 (2000): 447–64. Eulalia Pérez Sedeño, "Gender: The Missing Factor in STS," in Cutcliffe and Mitcham, *Visions of STS*, pp. 123–138.

126 *the tobacco companies can't be held responsible* M. Ashmore, "Ending Up on the Wrong Side: Must the Two Forms of Radicalism Always Be at War?" *Social Studies of Science* 26 (1996): 305–322.

126 *undermines efforts to discredit repressive regimes* Nanda, "The Science Wars in India," *Dissent* (Winter 1997), reprinted in *The Sokal Hoax*, edited by the editors of *Lingua Franca* (University of Nebraska Press, 2000), pp. 205–213; "Against Social (de) Construction of Science: Cautionary Tales of the Third World," *Monthly Review* 48, no. 10 (1997): 1–20; *Prophets Facing Backwards: Postmodern Critiques of Science and Hindu Nationalism in India* (New Brunswick, NJ: Rutgers University Press, 2003). See also Carlos Eduardo Sierra, "Post-Modernism as an Epistemological Obstacle in the Teaching of the History of Science: The Latin-American Case," *Education Forum* (newsletter for the Education Section of the British Society for the History of Science) no. 39 (Feb. 2003): 5–6; Aijaz Ahmad, *In Theory: Classes, Nations, Literatures* (London: Verso, 1992).

126 *constructionists, suggests historian Scott Montgomery, seem to have projected* Montgomery, review of books by Gerald Holton, *Isis* 91 (2000): 577–578.

126 *projection is described as ascribing one's own feelings Random House Dictionary of the English Language*, 2nd ed., 1987, p. 1546.

126 *the process by which manuscripts were submitted, refereed, and published or rejected* Bakanic, McPhail and Simon, "The Manuscript Review and Decision-Making Process," *American Sociological Review* 52 (1987): 631–642.

126 *in their 1990 book* Peerless Science Chubin and Hackett, *Peerless*, pp. 104–106.

127 *found no evidence of favoritism in the publication process* Zuckerman and Merton, "Evaluation."

127 *was fated to go out of favor* Scott McLemee, "Think Postpositive: A Latina cultural theorist wrestles with notions of identity and experience," *Chronicle of Higher Education* (February 13, 2004): A12–A14. Fudpucker, "Postmodern Production and STS Studies: A Revolution Ignored," in Cutcliffe and Mitcham, *Visions of STS*, pp. 139–154.

127 *"It would be about as helpful to argue that the blood had no pressure"* Theodore M. Porter, "Life Insurance, Medical Testing, and the Management of Mortality," in *Biographies of Scientific Objects*, edited by Lorraine Daston (Chicago: University of Chicago Press, 2000), pp. 226–246, on p. 246.

127 *they "imagine it to have been set in motion" Galileo Galilei, Dialogue Concerning the Two Chief World Systems*, translated with revised notes by Stillman Drake (Berkeley: University of California Press, 1967), pp. 188–189.

127 *X-rays . . . didn't exist until Röntgen invented them Auerbach, Entwicklungsgeschichte der moderne Physik: Zugleich eine Übersicht ihrer Tatsachen, Gesetze und Theorien* (Berlin: Springer, 1923), p. 5, as translated by H. Otto Sibum, "Science and the Changing Senses of Reality Circa 1900," *Studies in History and Philosophy of Science* 39 (2008): 295–297, on p. 296. Also quoted in Sibum, "Experience-Experiment: The Changing Experential Basis of Physics," in *Aurora Torealis: Studies in the History of Science and Ideas in Honor of Tore Frängsmyr*, edited by Marco Beretta, Karl Grandin, Svante Lindqvist (Sagamore Beach, MA: Science History Publications, 2008), pp. 181–191, on p. 181.

127 *"before Newton's Laws were discovered, they were not true"* Heidegger, *Being and Time* (New York: Harper & Row, 1962), p. 269 (translated from *Sein und Zeit*, 1927), as quoted by Babette Babich in *The History of Continental Philosophy*, vol. 3 (Durham: Acumen, 2010), p. 284. Like some of the other notorious sentences quoted in the text, this one deserves a little context, although perhaps only a German philosopher could really understand it.

"Constituted by disclosedness, Da-sein is essentially in the truth. Disclosedness is an essential kind of being of Da-sein. *'There is' ['gibt es'] truth only insofar as Da-sein is and as long as it is.* Beings are discovered only *when* Da-sein is, and only *as long as* Da-sein *is* are they disclosed. Newton's laws, the law of contradiction, and any truth whatsoever, are true only as long as Da-sein *is*. Before there was any Da-sein, there was no truth; nor will there be any after Da-sein is no more. For in such a case truth as disclosedness, discovering, and discoveredness *cannot* be. Before Newton's laws were discovered, they were not 'true.' From this it does not follow that they were false or even that they would become false if ontically no discoveredness were possible any longer. Just as little does this 'restriction' imply a diminution of the being true of 'truths.' The fact that before Newton his laws were neither true nor false cannot mean that the beings which they point out in a discovering way did not previously exist. The laws became true through Newton, through them beings in themselves accessible for Da-sein. With the discoveredness of beings, they show themselves precisely as the beings that previously were. To discover in this way is the kind of being of 'truth.'" Martin Heidegger, *Being and Time, A Translation of Sein und Zeit* translated by Joan Stambaugh (Albany, NY: SUNY Press, 1996, from the 7th German edition (Tübingen: Max Niemeyer Verlag, 1953). According to the Author's Preface to the Seventh German Edition (p. xvii), "The present reprint . . . is unchanged with respect to the text, but have been newly revised with regard to quotations and punctuations." This may mean that he has added the scare quotes around the word "true" after the second mention of Newton's laws.

If you are still unsure what Heidegger means, he advises you to read his *Einführung in die Metaphysik* which "presents the text of a lecture course delivered in the summer semester of 1935"—which happens to be the year in which the even more notorious (but "true," contrary to its authors' expectations) Einstein-Podolsky-Rosen paradox was published (see any book on the philosophy of quantum mechanics).

127 *Robert A. Harris and Leo Stolsky published a letter* "Anachrony in Decoherence," *Physics Today* 63, no. 2 (February 2010): 10. For "the pharaoh couldn't have died" they cite J. R. Searle, *New York Review of Books*, 56, no. 4 (2009): 88, who cites Latour, "Ramses II est-il mort de la tuberculose?" *La Recherche* (March 1998) pp. 84–85. Latour gives a somewhat better explanation in "On the Partial Existence of Existing and Nonexisting Objects," Daston, *Biographies*, pp. 247–269. If any evidence were needed that disease-causing organisms existed before they were discovered in the nineteenth century it has been found in a fourteenth-century graveyard in London, where researches identified the bacterium that caused the Black Death; see *Science* 334 (2011): 297. See also Brown, *Are Science*, p. 121; Nick Tosh, "Science, Truth and History, Part II. Metaphysical Bolt-Holes for the Sociology of Scientific Knowledge," *Studies in History and Philosophy of Science* 38 (2007): 185–209, on pp. 202–204; Jouni-Matti Kuukkanen, "I Am Knowledge. Get Me Out of Here! On Localism and the Universality of Science," ibid. 42 (2011): 590–601, on pp. 598–599.

Section 4.5

128 *"We should not mourn too much"* Brown, *Are Science*, p. 290.

128 *Forman's Weimar culture paper . . . Loren Graham argues* Forman, "Weimar Culture"; Loren Graham, *What have We Learned about Science and Technology from the Russian Experience?* (Stanford, CA: Stanford University Press, 1998).

128 *agree on what kind of evidence* See Gerald Holton, "R. A. Millikan's Struggle with the Meaning of Planck's Constant," *Physics in Perspective* 1 (1999): 231–237; Chapter 7 and S. G. Brush, "Dynamics of Theory Change in the Social Sciences: Relative Deprivation and Collective Violence," *Journal of Conflict Resolution* 40 (1996): 523–545.

128 *Degler doesn't* vary *the social factor* Degler, *In Search of Human Nature* (New York: Oxford University Press, 1991).

128 *I recommend David Hull's paper* David L. Hull, (1998, pub. 1999) "Studying the Study of Science Scientifically," *Perspectives on Science* 6 (1998, publ. 1999): 209–231.

PART TWO

Atoms, Molecules, and Particles

5

Mendeleev's Periodic Law

The greatest genius is he who sees furthest beyond the facts.

—*chemist* IRA REMSEN *(1889)*

Dr. Moody, 56, recalled that the periodic table had 104 elements when he was in high school. At the time, chemists thought the list was about finished, he said. He added that he recently spoke about his work to some high school students and found them fascinated. To them the periodic table "is an icon," he said. "The fact that it can change and it can be added to, I think, is a novel idea for younger people."

Not so for older people. Dr. Moody said he does not talk much about his work at parties "because people don't generally invite you back."

—*Associated Press, interview with chemist* KEN MOODY, *codiscoverer of elements 114 and 116 (2011)*

Historical studies of the reception of scientific discoveries have concentrated on a few major theories such as Darwin's evolution and Einstein's relativity. Other theories, including some of considerable importance to science, have been neglected. In particular, the reception of the periodic law of the elements, which is fundamental to modern chemistry and foreshadowed the physical theory of atomic structure, has attracted only a handful of historians, although the story of its discovery and verification is well known.

The results reported here will not be surprising to historians of chemistry. They gain their significance mainly in the context of the other cases presented in this book; this was the first case I found in which scientists gave substantial weight to evidence *predicted in advance* rather than treating such evidence as being no more important than similar evidence already known and deduced from or accommodated by the new theory.

In starting a discussion of the establishment of modern atomic theories, one might begin with the origins of atomism in ancient times. Aside from the fact that I have nothing new or interesting to say about that topic, philosopher Alan Chalmers offers a plausible reason for omitting it, consistent with the purpose of my book: before the mid-nineteenth century, atomism offered no testable predictions or convincing quantitative explanations of any empirical facts. In his 2009 book *The Scientist's Atom and the Philosopher's Stone*, he argues that atomism did not really become scientific until the late nineteenth century:

> Given the state of affairs in 1860 there was no guarantee that physical and chemical atoms would not be banished from science in the way that the aether came to be.

Before 1860, Chalmers declares, atomism was philosophical, not scientific.

Chalmers bases this statement on his own demarcation principle, similar but not identical to Popper's (Section 1.4). The atomic theories of the Greeks were "not confirmed by observational evidence." Robert Boyle's atomistic mechanical philosophy "involves mere accommodation to evidence"—as he himself claims it is so flexible (it can ascribe any desired size and shape to atoms) that it can explain any phenomenon, but "was not capable of predicting any phenomena and so not capable of guiding experimental investigation." Newton 's atomistic matter theory was "accommodated to rather than confirmed by observation and experiment." Dalton's atomic theory was a little better; it "made contact with experiment insofar as it predicted and explained why substances combine in constant proportions by weight." But "the considerable advances in nineteenth century chemistry did not owe much to Dalton's atomism."

These quotations suggest that Chalmers accepts Popper's criterion: to be scientific, a theory must make testable (novel) predictions. But Chalmers makes an exception to that statement: he wants to distinguish "between the confirmation of a theory by data and mere accommodation of a theory to data . . . a theory is confirmed to the extent that it predictions *or explains* a range of phenomena that follow naturally from it in con junctions with independently testable hypotheses" (emphasis added). So it seems that he wants to use the word "prediction" more or less as physicists do: to mean logical entailment, not necessarily novel prediction (see Section 1.3). And while Popper defines the alternative to science as pseudoscience, for Chalmers, it is philosophy.

I leave it to the philosophers to debate whether philosophy is unscientific. What concerns me is the assumption that science includes *only* the construction and experiment testing of theories by making predictions (novel or not) that can be tested by experiments and observations.

On the contrary, science now and in the recent past includes many other activities. These include, on the empirical side, the collection of accurate data and the construction of instruments to make measurements; and on the theoretical side, the construction of idealized models and the development of mathematical techniques to calculate the properties of those models.

An example of the latter (with which I was personally involved) is the attempt to answer the question: would a fluid composed of particles, interacting with repulsive forces only, freeze into a solid crystal? (The answer is yes.) This is certainly science, even though it involves no new experiment and no claim that the theory accurately predicts or explains other properties of any real substance.

5.1 MENDELEEV AND THE PERIODIC LAW

Three events, each the culmination of decades of scientific (not just philosophical) research, occurred around 1860 and led to the periodic law of chemical elements.

1. The chemist Stanislao Cannizzaro published in 1858 a short "Sketch of a Course of Chemical Philosophy" in which he revived Avogadro's distinction between atoms and molecules, and his hypothesis that every gas at the same pressure, volume, and temperature contains the same number of *molecules*. Together with Gay-Lussac's law of combining volumes for gaseous chemical reactions, this implies that the number of molecules is proportional to the volume and provides a consistent way of determining the atomic weights of elements. After copies of his sketch were distributed after an international chemical congress at Karlsruhe in 1860, the Avogradro-Cannizzaro system was gradually adopted. This meant, for example, that a molecule of water consists of *two* atoms of hydrogen and one of oxygen—denoted by H_2O—instead of HO as proposed by Dalton. The atomic weight of oxygen (relative to hydrogen) is then 16 (rather than about 8, as Dalton believed).
2. In the late 1850s, physicists Rudolf Clausius and James Clerk Maxwell revived the kinetic theory of gases (proposed by Daniel Bernoulli in 1738, but incomplete then because the absolute temperature scale had not yet been established). Among other things, this provided a theoretical basis for Avogradro's hypothesis, and provided a method used by Josef Loschmidt to calculating Avogadro's number and the size of a molecule.
3. Physicist G. R. Kirchhoff and chemist R. W. Bunsen established in 1859–1860 the basic principles of spectrum analysis. Each element has a characteristic set of wavelengths (or their inverse, frequencies) at which it may absorb or emit radiation. This made it much easier to determine (at least qualitatively) the chemical composition of a sample of material, and led directly to the discovery of several elements whose spectra did not match those of known elements.

The periodic law emerged in the 1860s, following the general agreement of chemists to accept Stanislao Cannizzaro's method for determining atomic weights. The spectroscopic discovery of two new alkali metals, cesium and rubidium, in 1860 and 1861 respectively, helped to emphasize the periodicity of physical as well as chemical properties for this monovalent family of elements. With this convergence of historical events allowing several chemists to see new regularities, the periodic law appeared as an example of multiple simultaneous discovery. Credit for the discovery has been claimed for at least six scientists, given here in alphabetical order: Alexandre-Emil Béguyer de Chancourtois, Gustav Hinrichs, John Newlands, Dmitrii Ivanovich Mendeleev, Lothar Meyer, and William Odling. Mendeleev is now generally recognized as *the* discoverer, the names of the others being known only to historians of chemistry.

For example, as philosopher and historian Eric Scerri has argued, Lothar Meyer should be recognized for his prediction of the element now called germanium, published in 1864. His table shows a gap in the column of quadrivalent elements between silicon (atomic weight 28.5) and tin (117.6). Interpolating would give the unknown element an atomic weight of 73.1, "which when discovered," writes Scerri, "was found to have an atomic weight of 72.3. This prediction of the element germanium, which was first isolated in 1886, is usually attributed to Mendeleev, even though it was clearly anticipated by Lothar Meyer in this early table of 1864."

Mendeleev was familiar with the ideas of European chemists, having worked with Robert Bunsen at Heidelberg in 1859–1860, and having attended the famous International Chemical Congress at Karlsruhe in 1860. When he began to prepare his own textbook, he was forced him to think about how best to organize and present to students a large quantity of chemical information. As historian Nathan M. Brooks writes, "Mendeleev succeeded in producing a viable system, where his predecessors had failed, precisely because he was writing a textbook and, thus, he was looking at the issue from a different perspective." Mendeleev formulated his periodic law in March 1869, and published a series of papers on it beginning in that year. The discovery in 1875 of gallium, with density and other properties in agreement with those predicted by Mendeleev, aroused considerable interest in the periodic law. By 1890 Mendeleev's confirmed predictions of the properties of two other new elements, and his successful corrections of the atomic weights of several known elements, had brought him widespread recognition and helped his system to displace competitors.

5.2 NOVEL PREDICTIONS

As discussed in Chapter 3, while there have been many successful predictions in modern physics, astronomy, and biology, there is not much evidence that the theories generating those predictions were more effective than good explanations in persuading scientists to accept those theories. If the old quantum theory is an exception, it must be noted that this theory was soon replaced by quantum mechanics with its superior explanatory power (Chapter 8).

Turning to chemistry, we may quote in this connection one of the best-known works on scientific method written by a chemistry professor, E. Bright Wilson Jr.:

> Successful [novel] prediction is usually considered stronger support for a hypothesis than the explanation of an equal quantity of observation known to the creator of the hypothesis at the time of its creation. This is not hard to justify on purely logical grounds and appears valid as a result of experience.

5.3 MENDELEEV'S PREDICTIONS

The periodic law (as formulated in the nineteenth century) states that when the elements are listed in order of atomic weight, properties such as valence will recur periodically—for example, after seven elements. Mendeleev and other advocates of the periodic law made numerous suggestions that new elements would be discovered to fill the gaps in their various sequences and tables. In most cases only the atomic weight and chemical family of the new element were specified. But in 1871 Mendeleev described in quantitative detail the properties of three unknown elements:

Eka-boron, atomic weight 44
Eka-aluminium, atomic weight 68, density 6.0, atomic volume 11.5; should come below aluminum in Group III
Eka-silicon, atomic weight 72, density 5.5, atomic volume 13

Mendeleev also predicted the following elements:

eka-manganese, atomic weight 100
eka-niobium, atomic weight 146
eka-caesium, atomic weight 175
tri-manganese, atomic weight 190
dvi-caesium, atomic weight 220
eka-tantalum, atomic weight 235

In addition, he proposed changes in the accepted atomic weights of several elements in order to fit them into his table, and in some cases where chemists disagreed on the atomic weight he supported one value over another for the same reason:

beryllium had been assigned either 9 or 14; Mendeleev chose 9 so that it would fit in his table with valence 2
yttrium should be changed from 60 to 88
cerium should be changed from 92 to 138 because it has valence 4
uranium should be changed from 120 to 240
tellurium's atomic weight should be changed from 128 to 125 since it comes before iodine (127)

All of these may be called novel predictions. To state that a previously accepted property must be incorrect is obviously different from stating a property of a previously unknown entity, and perhaps deserves a different name—say, *contraprediction*. But it fits very well the requirement that a prediction based on a new hypothesis should disagree with what one would expect on the basis of the knowledge available before the hypothesis was proposed. One might even argue that if successful novel predictions are better evidence than retrodictions, successful contrapredictions are better yet. Moreover, they cast doubt on the orthodox view that theories are tested by observations; observations themselves are subject to test by contrapredictions.

Were all these confirmed predictions equally important? According to philosopher Michael Akeroyd, the predictions about gallium and scandium were less important than those about germanium and uranium. Both gallium and scandium could have been predicted on the basis of the earlier Triad (natural family) method, proposed in 1829 by chemist Johann Wolfgang Döbereiner. Akeroyd argues that LeCoq de Boisbaudran actually did predict gallium, based on evidence from his notebooks, even though he did not publish the prediction. Akeroyd also suggests that chemist Per Teodor Cleve predicted scandium, though he had not found any positive evidence for this. Of course in either case other scientists would not give credit to LeCoq de Boisbaudran or Cleve if the prediction was not published. But Akeroyd argues that it would have been more difficult to predict germanium, or an atomic weight of 240 for uranium, from the Triad method, and therefore these predictions count more heavily in favor of Mendeleev's periodic law.

5.4 RECEPTION BY WHOM?

Normally a new scientific theory has to prove its value as a guide to ongoing research, and only after it does that will it be incorporated into textbooks as part of the established knowledge of a science. Although the reception of the periodic law followed this pattern to some extent, it should be recalled that the law was originally developed as an aid to teaching, and its pedagogical value was recognized very quickly. Teachers saw an urgent need to show their students that chemistry is not just a huge collection of facts to be memorized, but that, in the words of chemist M. M. Pattison Muir, "law and order pervade [its] vast domain." Once the student can connect the atomic weight of an element with its position in a general classification scheme, "he gains a basis on which he may rest the superstructure of facts as they are presented to him."

The number of explicit references to the periodic law to be found in late nineteenth-century chemistry journals is small and fluctuates irregularly from one year to the next. The number in American and British journals increases from 6 in the period 1871–1875, to 30 in 1876–1880, then 39 in 1881–1885 and 41 in 1886–1890. Of these, the number that mentioned the confirmation of Mendeleev's predictions about new elements was 1, 21, 11, and 8 respectively. There is a similar pattern in other countries, with a sharp increase around 1875–1881, a slow decline thereafter, and a small jump around 1889–1890.

From these data alone it would be difficult to judge whether the periodic law was actually accepted by the entire chemical community or was merely an exotic concept, of interest only to a few specialists. But by looking at textbooks and comprehensive chemistry reference works, we can get a much better idea of the extent of the law's diffusion. If a majority of chemistry textbooks published in a country present the periodic law—as is the case for the United States and Great Britain by 1890—it is reasonable to conclude that the law was generally accepted in that country.

Two important cautions must be placed against such a conclusion: (1) the textbooks published at a given time, especially the most popular ones, were generally written several years earlier, as indicated by the dates of their prefaces and copyrights—thus most *authors* had accepted the periodic law by 1885 or earlier; (2) many textbooks are no longer available in American libraries, even though we know they were used during the period before 1890. I think the survivors are numerous enough to support firm conclusions about the reception of the periodic law in Great Britain and the United States, but not about the reception in other countries.

These statements are based on an attempt to examine all the chemistry textbooks published in the period 1871–1890, now available in American libraries. Unfortunately many of these textbooks are no longer held in any library that I could search. Apparently many libraries discarded a textbook when a newer edition was acquired, thus destroying some of the evidence that an historian needs to track the *changes* made in response to new advances in the science. Also, some texts were printed on acidic paper, which has now disintegrated.

Nevertheless we must use incorporation into textbooks and reference works as a primary measure of the reception of the periodic law, because the number of chemists who actually mention it in their publications is only a small fraction of the total number of chemists.

5.5 TESTS OF MENDELEEV'S PREDICTIONS

I consider here only the first three discoveries of elements whose properties were predicted by Mendeleev, because as will be seen the periodic law was generally accepted soon after the third discovery. I do not discuss anomalies like tellurium, which could be resolved only in the twentieth century with help of the new science of nuclear physics.

Gallium

On August 27, 1875, between 3 and 4 p.m., the French chemist Paul Émile Lecoq de Boisbaudran discovered spectroscopically a new metal, which he named gallium, supposedly in honor of his native France. Mendeleev immediately suggested that gallium is the element eka-aluminium, which he had predicted from his periodic law. In his next report, Lecoq de Boisbaudran stated that he had not known about Mendeleev's description of the supposed properties of his hypothetical metal.

Lecoq de Boisbaudran's first density determination gave the value 4.7 at temperature 15E. He considered his result to be a confirmation of his own theoretical prediction (based on spectral lines) that the new element's properties are midway between those of aluminum and indium (which would give 4.7), but did not mention Mendeleev's prediction, 5.9 or 6.0. He soon recognized the importance of that prediction (which had now become a contraprediction) and, suspecting that his sample may have contained cavities filled with water, he redetermined the density after heating the metal and solidifying it in a dry atmosphere. He was thus able to obtain a density of 5.935, confirming Mendeleev's views.

Scandium

Early in 1879 the Swedish chemist Lars-Frederik Nilson discovered a new element, which he named *scandium* in honor of Scandinavia. At first he estimated its atomic weight to be between 160 and 180, and suggested that it might fill the gap in group IV between tin (118) and thorium (234); no mention was made of Mendeleev's periodic law or his predictions. Another Swedish chemist, Per-Teodor Cleve, isolated enough of the new element to determine its atomic weight fairly accurately; he reported a much lower value, about 45. He pointed out that the properties of scandium seemed to correspond to Mendeleev's ekaboron. Nilson confirmed this by his own further experiments and remarked on the success of Mendeleev in predicting the atomic weight and other properties of two elements, gallium and scandium.

Germanium

On February 6, 1886, the German chemist Clemens Alexander Winkler discovered a new element, which he called *germanium*. Even before determining its atomic weight he guessed that it would fill the gap between antimony and bismuth in the periodic table, identifying it with Mendeleev's predicted ekastibium. Mendeleev himself thought it more likely to be ekacadmium, an element between cadmium and mercury. Victor von Richter and Lothar Meyer argued that it should be identified with ekasilicon, atomic weight 72 or 73 (as Meyer had predicted, see Section 5.1), and this view was accepted by Winkler and Mendeleev.

Atomic Weights

Three of Mendeleev's revisions of the atomic weights of known elements are of special interest in assessing the reception of his theory: those of uranium, beryllium, and tellurium.

Before 1870 *uranium* was commonly assigned atomic weight 120 (sometimes 60). Mendeleev gave it the value 240, making it the heaviest of the known elements. This was accepted by many chemists without objection, and definitely confirmed by Clemens Zimmerman in 1882.

There were two possible values for the atomic weight of *beryllium*, 9 and 14, depending on whether one regarded this element as bivalent or trivalent respectively; in the early 1870s the lower value was favored by most chemists. Mendeleev chose 9 because he needed beryllium to fill the space in Group III. After considerable debate and further research, chemists eventually decided, in 1884, that Mendeleev's value 9 is correct.

Following Mendeleev's suggestion that the atomic weight of *tellurium* should be 125 rather than the previously accepted value 128, so that it could be placed before iodine (127), several attempts were made to determine this value more accurately. W. L. Willis redetermined it as 126.83 ± 0.198, slightly greater than the most recent values for iodine and therefore contrary to Mendeleev's assignment. In 1883 Bohuslav Brauner reported values between 124.94 and 125.4, but later determined the atomic weight to be 127.64. As of 1889, Mendeleev's contraprediction seemed to have been refuted, but Brauner still hoped to rescue the periodic law by showing that tellurium is not a pure substance.

5.6 BEFORE THE DISCOVERY OF GALLIUM

Before 1876 most of the references to the periodic law in chemistry journals—only two or three a year—were in the papers of Mendeleev, Lothar Meyer, William Odling, and other advocates of competing classification systems. Rudolph Fittig may have been the first nontheorist to mention it in a publication; in an encyclopedia article on the atom published in 1873 he briefly noted that Mendeleev, Lothar Meyer, and Bauhauer had discussed regularities in the atomic weights. In the same year, E. F. von Gorup-Besanez presented in his textbook two rows of a periodic series (lithium to fluorine and sodium to chlorine) to illustrate the regularities in atomic weights that had recently been discussed by (unnamed) chemists.

As far as I know, the first Western-language chemistry textbook to present Mendeleev's table was the one by the German chemist Karl Rammelsberg, published in 1874. Rammelsberg also noted Mendeleev's predictions of the properties of yet-undiscovered elements such as eka-silicon. I have found no English-language textbooks that used Mendeleev's table before those of Ira Remsen and Thomas Edward Thorpe in 1877, but readers of the ninth edition of the *Encyclopaedia Britannica* could find that table in H. E. Armstrong's article on Chemistry.

According to the extensive survey by Ludmilla Nekoval-Chikhaoui, the first French textbook to introduce the periodic law was published by Paul Schützenberger in 1880.

5.7 THE IMPACT OF GALLIUM AND SCANDIUM

At the 1988 biennial meeting of the American Philosophy of Science Association, philosopher Patrick Maher defended the predictivist thesis: "if a piece of evidence for a theory was known at the time the theory was proposed, then it does not confirm the theory as strongly as it would if the evidence had been discovered after the theory was proposed." He asserted that "scientific practice accords well with the predictivist thesis," but in support of that assertion he presented only a single example: Mendeleev's periodic table. In 1871 Mendeleev predicted the existence and properties of three elements needed to fill gaps in the table. Two of them, gallium and scandium, were discovered in the 1870s. Scientists, previously skeptical, now acclaimed Mendeleev's periodic law, and in 1882 the Royal Society awarded him its Davy Medal. According to Maher, since the number of elements in the table had increased by only a relatively small number—from 62 to 64—we must conclude that scientists were more impressed by the novel predictions of two elements than by the larger number of elements accommodated (explained) by the table. As another philosopher, Peter Lipton, expressed it, using the same example to support his version of the predictivist thesis, "sixty accommodations paled next to the two predictions."

The Maher-Lipton proposition gives us a semiquantitative estimate of the evidential superiority of novel prediction over explanation: predicting a new element is worth more than 31 times as much as explaining a known one. Unfortunately Maher and Lipton give no documentation for their proposition. Did the Royal Society award the Davy Medal to Mendeleev because of his successful predictions? Perhaps some evidence for this claim can be found, but the most obvious source refutes it. The citation for the award does not even mention those predictions. Instead, it discusses the foundations of a general system of classification of the elements, lists the atomic weights of 15 elements *not* including gallium or scandium, remarks on "the marvellous regularity with which the differences of property" of the sequence from lithium to fluorine are reproduced in the sequence from sodium to chlorine, and mentions the graphs of physical properties such as atomic volume plotted against atomic weight. Moreover, the Davy Medal was awarded jointly to Lothar Meyer, who was responsible for the atomic volume graph but did not successfully predict new elements.

Thus, if we are to believe the public announcement, the Davy Medal was awarded for explanation, not for prediction. (The Davy Medal had been awarded to Lecoq de Boisbaudran in 1879 for the discovery of gallium, with a citation that did mention Mendeleev's prediction of that element.)

Maher and Lipton may still be correct about the periodic law in a broader sense. As noted in sections 1.6 and 1.7, chemists do seem to give more weight to novel predictions than do physicists, at least in the handful of cases I have studied. My survey of chemistry textbooks and articles in the late nineteenth century suggests that many chemists did give some credit for novelty; they considered that, other things being equal, the prediction of a new element and its properties counted more than fitting a known element into the table. But not 31 times as much.

No American or British textbooks mentioned the periodic law at all until after the discovery of gallium; of the four that did mention it in the period 1876–1880, two authors (Tidy and Watts) noted that successful prediction, but the other two (Remsen and Thorpe) did not. In the following decade, most of the authors who introduced the periodic law also mentioned the prediction of new elements; a similar pattern was followed by German and French authors, although my sample is much smaller.

Similarly, in France, as historian Nekoval-Chikhaoui's survey shows, the periodic law was not mentioned in any scientific publication before Mendeleev in 1875 claimed the credit for predicting the properties of the new element discovered by LeCoq de Boisbaudran. She concludes that the law undoubtedly owed its success to its predictive character.

As in other cases, such as the confirmation of Einstein's general relativity theory by the 1919 eclipse observations, we have to distinguish between the value of a successful novel prediction in *publicizing* a theory, thus forcing scientists to give it serious consideration, and its value as *evidence* for that theory in comparison to other evidence. It is certainly true that the number of published discussions of the periodic law increased substantially after 1875 when Lecoq de Boisbaudran discovered gallium, and especially after 1876 when he admitted that Mendeleev's predicted density (5.9) was more accurate than his own first experimental determination (4.7). Similarly, when Nilson discovered scandium, he called attention to Mendeleev's successful predictions of the properties of that element and of the properties of gallium.

Conversely, the publicity and credit accorded to Mendeleev for his predictions led him to be increasingly recognized as the principal—if not the only—founder of the periodic law. After the mid-1880s, other founders such as Lothar Meyer and John Newlands were ignored or given only secondary status.

Mendeleev stressed the importance of predictions when he reviewed the status of his periodic law in 1879:

> No natural law acquires any scientific importance unless it introduces, so to speak, some practical conclusions, or, in other words, unless it admits of logical conclusions capable of explaining what has before remained unexplained, and, above all, unless it raises questions which can be confirmed by experience.

Such questions include the properties of presently unknown elements. Later he declared that he regarded Lecoq de Boisbaudran and Nilson, as well as Clemens Winkler, who discovered the predicted element germanium in 1886, "as the true corroborators of the periodic law. Without them it would not have been accepted to the extent that it now is."

Several statements published in the decade 1876–1885 by the British chemists William Crookes and M. M. Pattison, German chemists Carl Hell and August Michaelis, the Swedish chemist Lars-Frederik Nilson, American chemists Josiah Parsons Cooke, A. P. Lothrop, A. L. Hodges, and F. W. Clarke, and the French chemist Édouard Grimaux all affirmed, in one way or another, that the confirmation of predictions like those of Mendeleev provides strong evidence for the periodic law (see the notes to this chapter).

5.8 THE LIMITED VALUE OF NOVEL PREDICTIONS

Nevertheless, almost every discussion of the periodic law in nineteenth-century chemistry textbooks, including Mendeleev's, gives much more attention to the correlations of properties of the known elements with their atomic weights than to the prediction of new ones. Frequently the reader is given the impression that the periodic law is established by these correlations, and then *applied* to make predictions.

The reason why the predicted elements were less important to chemists than the known elements that were initially correlated by the periodic law is the same reason why they had not already been discovered before 1869: their abundance at the Earth's surface is very small. Of course this doesn't mean that rare elements are necessarily unimportant (think of radium). But in late nineteenth-century chemistry books these elements get little space; often it is said that they are of interest mainly because of the fact that Mendeleev predicted them. They were of no use except to confirm the periodic law.

While chemists differed on the relative importance of prediction and accommodation, it seems fair to approximate the consensus as follows. The reasons for accepting the periodic law are, in order of importance:

1. it accurately describes the correlation between physicochemical properties and atomic weights of nearly all known elements;
2. it has led to useful corrections in the atomic weights of several elements and has helped to resolve controversies such as those about beryllium;
3. it has yielded successful predictions of the existence and properties of new elements.

Although the periodic law was generally presented in chemistry textbooks after 1890, it was not as a rule introduced at the beginning or used as an organizing principle for those books, even by Mendeleev himself. This final stage of acceptance had to wait for the twentieth century and a detailed understanding of how periodicity is related to atomic structure.

Some chemists seemed reluctant to admit that the validity of the periodic law was implied by its successful predictions. The American chemist Ira Remsen, the British chemist Henry Roscoe, and the German chemist Wilhelm Ostwald were impressed by Mendeleev's predictions but did not seem to think that a few such successes by themselves demonstrated that reliable knowledge had been gained. Even in the case of gallium, in spite of Mendeleev's striking prediction, its discoverer noted that there are still some discrepancies between theory and observation.

Chemist Adolphe Wurtz, in his influential book on the atomic theory, argued that while the discovery of gallium shows that the periodic law must be taken seriously in any consideration of the classification of the elements, it does not confirm that law; gallium's atomic weight is "sensibly different" from Mendeleev's predicted value. Moreover, if cobalt had not already been known it could not have been predicted from the periodic law since its atomic weight is nearly the same as that of nickel.

Brauner recalled that Robert Bunsen, who trained many of the world's leading chemists at the University of Heidelberg, did not mention the periodic law in his lectures of 1878–1879. When Brauner pointed how well the atomic weights of the rare earths—as determined by Brauner himself and by Bunsen's student Norton—confirmed the values proposed by Mendeleev, he answered: "Leave these conjectures alone."

The strongest criticism of the periodic law came from chemist Marcellin Berthelot. In his 1885 book on the origins of alchemy, Berthelot argued that the successes of the Law should be credited to the previously recognized regularities of the atomic weights in families of elements, along with the "convenient trick" of creating a net so fine-grained (successive terms differing by no more than two units) that "no new body, whatever it may be, can fall outside the meshes of the net." Moreover, he claimed, in trying to fill the gaps in their tables, proponents of the periodic law "have interpolated elements already known which are manifestly strangers to the family: such as Mo inserted between Se and Te; W and U added in like manner to the series. To the series of Li = 7, H = 1 has been placed at the head and at the end Cu - 63, Ag = 108 and Au = 198." By such "fanciful" means, Berthelot alleged, regularities have been artificially invented rather than discovered in nature.

Berthelot was a well-known antiatomist, and one might suppose that he would have opposed the periodic law for that reason. However, Nekoval-Chikhaoui points out that (contrary to the views of G. Urbain and others) there is not a simple correlation between views on atomism and views on the periodic law. In fact, both atomists and antiatomists incorporated the periodic law into their teaching: it appealed to the positivist respect for empirical facts, while at the same time offering new opportunities for theoretical speculation.

5.9 IMPLICATIONS OF THE LAW

Berthelot's timing was bad. Winkler's discovery of germanium, the following year, brought further endorsements of the periodic law. Even Wilhelm Ostwald, a leading critic of atomism, found it difficult to maintain his skepticism about the law, conceding that it had now "stood the test both in the prediction of the properties of elements at the time undiscovered, and in the indication of errors in the atomic weights previously accepted." By the late 1880s, the periodic law was widely accepted.

The discussion now turned to the theoretical inferences that might be drawn from the validity of such a law. Mendeleev himself repudiated the idea that his law was based on any assumption about the compound nature of elementary atoms or the "unity of matter." But other chemists, as early as 1875, concluded that such an assumption was not only permissible but perhaps even a necessary consequence of the periodic law.

In modern times the best-known example of the unity of matter was William Prout's hypothesis that the elements are compounds of hydrogen, based on the empirical generalization that all atomic weights are integer multiples of that of hydrogen. At the time when Mendeleev proposed his periodic law it was believed that Prout's hypothesis had been refuted by the more accurate atomic weight determinations of Jean Servais Stas. In his 1869 presidential address to the Chemical Society of London, Alexander W. Williamson said: "the question whether our elementary atoms are in their nature indivisible, or whether they are built up of smaller particles, is one upon which I, as a chemist, have no hold whatever, and I may say that in chemistry the question is not raised by any evidence whatever."

But new research in the 1870s and 1880s persuaded some chemists that the hypothesis might be valid after all. The suggestion that the periodic law was analogous to the classification of organic compounds in homologous series amounts to an indirect endorsement of the hypothesis that elements are compounds. Mendeleev, though skeptical about that hypothesis, suggested that the weight of a compound atom would not have to be precisely equal to the sum of its constituents since the energy of motion and energy of mass could be interconvertible, the law of conservation of mass being only a special case of the law of conservation of energy.

During the same period, the speculations about atomic dissociation of astronomer J. Norman Lockyer were attracting much attention. Lockyer noted that the spectrum of a compound changes when it breaks into elementary atoms, for example in a Bunsen burner; changes observed in atomic spectra at high temperatures, for example in the sun, should therefore be ascribed to dissociation of atoms into smaller atoms (i.e., those of lower atomic weight). His dissociation hypothesis was first proposed at a meeting of the British Association in 1873.

The expectation that the elements would eventually be shown to be composite, and to be decomposable or combined into other elements, was widespread enough that Rutherford's newer alchemy of radioactive transmutation encountered little resistance when it was proposed less than two decades later.

For some chemists, the success of Mendeleev's predictions meant that their science had at last attained the respected status long enjoyed by astronomy and physics. Astronomy's spectacular 1846 triumph in predicting and discovering a new planet was still a vivid memory for the older generation and a well-known historical event for the younger; gallium, scandium and germanium were compared to Neptune as demonstrations of the capacity of mature exact science. (No one remarked, however, that astronomers already accepted Newtonian celestial mechanics before 1846 and did not need this successful novel prediction to persuade them.)

5.10 CONCLUSIONS

After spending considerable time perusing the crumbling pages of late nineteenth-century chemistry journals and textbooks, I have confirmed the traditional account: Mendeleev's periodic law attracted little attention (at least in the United States and Great Britain) until chemists started to discover some of the elements needed to fill gaps in his table and found that their properties were remarkably similar to those he had predicted. The frequency with which the periodic law was mentioned in journals increased sharply after the discovery of gallium; most of that increase was clearly associated with Mendeleev's prediction of the properties of the new element. By the late 1880s, most textbooks published in the United States and Great Britain discussed the periodic law to some extent.

As in other cases such as the discovery of gravitational light bending (Chapter 11.1), the confirmation of a novel prediction forced other scientists to pay attention to the theory that had yielded the successful prediction. Also, as in other cases, the new theory was expected to do much more than foretell an exotic new phenomenon or substance: it had to prove its value by providing a superior way of organizing existing knowledge. But, unlike other cases I have examined, novelty was accorded a significant (though not a dominant) role in weighing the evidence for the theory: predicting the properties of one previously unknown element did count more heavily, for many chemists, than explaining the properties of one known element. Hardly anyone argued that Mendeleev's periodic law should not get all the credit for that element if another theory could (retrospectively) predict its existence.

Could it be true that chemists give more weight to novel predictions than practitioners of physics—that, as suggested in Chapter 1, chemistry is more scientific (in Popper's sense) than physics? I am not aware of any other comparative studies of scientific practice that have suggested this conclusion. In fact, one philosopher of science, Eric Scerri, has argued just the opposite: that chemistry is a *less* predictive science that physics, and that the periodic law was established because of its explanations rather than its predictions.

Scerri also points out that Mendeleev made several unsuccessful predictions. He lists 16 firm predictions of elements to which Mendeleev gave provisional names, such as eka-boron and dvi-tellurium. Of these, three were found before 1890, and five were found later, the most recent being francium in 1937. So, as of 2007 when Scerri's book was published, Mendeleev's success rate was only 50%—yet his "unsuccessful predictions do not seem to have counted against the acceptance of the periodic system."

One might argue that Mendeleev's predictions could be confirmed by discovering the predicted element, but they could not be refuted by the failure to discover it, since there is always the possibility that it might turn up sometime in the future. So the predictions are not falsifiable in Popper's sense, yet the three that were confirmed before 1890 certainly influenced the acceptance of the periodic law.

The weight given to novel prediction helps to explain why other scientists who claimed to have discovered the periodic law have generally been forgotten. For example, Lothar Meyer, who seemed to have a rather strong claim to partial credit, left gaps in his table for undiscovered elements, but did not boldly predict the quantitative properties of those elements in the way that Mendeleev did.

In the twentieth century, the general success and occasional failure of Mendeleev's Law (i.e., elements that did not fit their place in the periodic table) was explained in terms of atomic physics. But the role of novel predictions in the original acceptance of his law has been somewhat exaggerated, while the correlation of chemical properties of known elements and the correction of some atomic weights has been taken for granted and almost forgotten. For philosophers of science who overlook the failures, Mendeleev's prediction of the properties of gallium, scandium, and germanium has been useful as an illustration of the proper application of the scientific method. Until we look at the reasons why scientists themselves accept or reject theories in other cases, we will not know whether this case is typical or exceptional.

NOTES FOR CHAPTER 5

157 *Mendeleev's Periodic Law* This chapter is based on my article "The Reception of Mendeleev's Periodic Law in America and Britain," published in *Isis* 87 (1996): 595–628, by permission of The History of Science Society; this article should be consulted for further details and references to publications before 1996.

For general background on Parts Two and Three see Gerald Holton and Stephen G. Brush, *Physics, the Human Adventure: From Copernicus to Einstein and Beyond* (New Brunswick, NJ: Rutgers University Press, 2001).

157 *"The greatest genius"* Ira Remsen, "The Chemistry of To-Day," *Popular Science Monthly* 34 (March 1889): 591–597, on p. 595. A more complete version is: "The greatest genius is he who sees furthest beyond the facts, and with the aid of his imagination is able to bring together into a harmonious whole those facts which seem least connected. But, it must be remembered, it is the imagination of the thoroughly trained mind, kept in subjection by profound knowledge, that leads to great results."

157 *"people don't generally invite you back.'"* "The Periodic Table Expands Once Again," *New York Times* (June 9, 2011), p. A17.

157 *the story of its discovery and verification is well known* In addition to the publications cited in Brush, "Reception," see Eric R. Scerri, *The Periodic Table: Its Story and its Significance* (Oxford: Oxford University Press, 2006); Scerri et al., Special issue on the Periodic System of the Elements, *Foundations of Chemistry* 3, no. 2 (2001): 57–195, www.wkap.nl/journalhome.htm/1386-4238. On Mendeleev see Michael D. Gordin, *A Well-Ordered Thing: Dmitri Mendeleev and the Shadow of the Periodic Table* (New York: Basic Books, 2004).

Additional publications may be located in the *Isis Cumulative Bibliographies* for 1913–1965, 1966–1975, 1976–1985, 1986–1995, and annual "Current Bibliographies" published as supplements to *Isis*; for the online version see Stephen Weldon, Kim Rudolph, and Sam Spence, "Searching Smartly in the HistSciTechMed Database," *History of Science Society Newsletter* 39, no. 2 (April 2010): 26–40.

158 he *argues that atomism did not really become scientific* Chalmers, *The Scientist's Atom and the Philosopher's Stone: How Science Succeeded and Philosophy Failed to Gain Knowledge of Atoms* (Dordrecht: Springer, 2009), on p. 189.

158 *Newton's atomistic matter theory was accommodated* Newton showed that *if* one assumes gases are composed of particles with repulsive forces, *then* one can deduce Boyle's law by stipulating that the forces are inversely proportional to the distance from the particle, but act only on their nearest neighbors: "Whether elastic fluids do really consist of particles so repelling each other is a physical question. We have here demonstrated mathematically the property of fluids consisting of particles of this kind, that hence philosophers may take occasion to discuss that question." *Sir Isaac Newton's Mathematical Principles of Natural Philosophy and His System of the World*, translated by A. Motte, revised and annotated by F. Cajori (Berkeley: University of California Press, 1934), p. 302. In modern terminology this would be considered a contribution to science, not philosophy (see next note).

158 *interacting with repulsive forces only* B. J. Alder and T. E. Wainwright, "Phase Transition for a Hard Sphere System," *Journal of Chemical Physics* 27 (1957): 1208–1209. S. G. Brush, H. L. Sahlin, and E. Teller, "Monte Carlo Study of a One-Component Plasma. I" *Journal of Chemical Physics* 45 (1966): 2102–2118. These two calculations consider the extremes of very short-range and very long-range forces, respectively. The particles are governed by Newtonian mechanics; no quantum effects are included. For possible astrophysical applications see Brush, *Statistical Physics and the Atomic Theory of Matter* (Princeton, NJ: Princeton University Press, 1983) pp. 166–167. Whether computer simulations of this kind are really theory or experiment is apparently a cause of confusion for sociologists, though not to scientists: see Mikaela Sundberg, "Cultures of Simulations vs. Cultures of Calculations? The Development of Simulation Practices in Meteorology and Astrophysics," *Studies in History and Philosophy of modern Physics* 41 (2010): 273–281 and publications cited therein.

Section 5.1

159 *in which he revived Avogadro's distinction between atoms and molecules* Henry M. Leicester, "*Cannizzaro, Stanislao,*" *Dictionary of Scientific Biography*, vol. III, edited by C. C. Gillispie (New York: Charles Scribner's Sons, 1971), pp. 45–49.

159 *revived the kinetic theory of gases* S. G. Brush, *The Kind of Motion We Call Heat: A History of the Kinetic Theory of Gases in the 19th Century* (Amsterdam: North-Holland Publishing Company, 1976).

159 *the basic principles of spectrum analysis* L. Rosenfeld, "Kirchhoff, Gustav Robert," *Dictionary of Scientific Biography*, vol. VII (1973), pp. 379–383. Susan G. Schacher, "Bunsen, Robert Wilhelm Eberhard," *Dictionary of Scientific Biography*, vol. II (1970), pp. 586–590. William McGucken, *Nineteenth-Century Spectroscopy: Development of the Understanding of Spectra, 1802–1897* (Baltimore: Johns Hopkins University Press, 1970). Hans Kangro, editor, *Untersuchungen über das Sonnenspectrum und die Spectrum der chemischen Elemente, und weitere ergänzende Arbeiten aus den Jahren 1859–1862* (Osnabrück: Otto Zeller, 1972). Clifford L. Maier, *The Role of Spectroscopy in the Acceptance of the internally structured Atom* (New York: Arno Press, 1981).

160 because *he was writing a textbook* Nathan M. Brooks, "Dmitrii Mendeleev's Principles of Chemistry and the Periodic Law of the Elements," in *Communicating Chemistry*, edited by A. Lundgren and B. Bensaude-Vincent, pp. 295–309 (Canton, MA: Science History Publications, 2000), on p. 296.

160 *Mendeleev was familiar with the ideas of European chemists* According to Michael D. Gordin, "he turned to a prior exemplar of effective classification: the type theory of the organic chemist Charles Fréderic Gerhardt," but was hostile to organic structure theory and valence theory. Gordin, "The organic roots of Mendeleev's Periodic Law," *Historical Studies in the Physical and Biological Sciences* 32 (2002): 263–290, on p. 263. See also Gordin, *A well-ordered Thing: Dmitrii Mendeleev and the Shadow of the Periodic Table* (New York: Basic Books, 2004); Nathan M. Brooks, "Developing the Periodic Law: Mendeleev's Work during 1869–1871," *Foundations of Chemistry* 4 (2002): 127–147; Igor S. Dmitriev, "Scientific Discovery *in statu nascendi*: The Case of Dmitrii Mendeleev's Periodic Law," *Historical Studies in the Physical and Biological Sciences* 34 (2004): 233–275.

Section 5.2

160 *"Successful [novel] prediction is usually considered stronger support"* E. B. Wilson Jr., *An Introduction to Scientific Research* (New York: McGraw-Hill, 1952).

Section 5.3

160 *properties such as valence will recur periodically* Thus the first two rows of Mendeleev's original table would be:

Li = 7 Be = 9.4 B=11 C=12 N=14 O=16 F=19

Na=23 Mg=24 Al=27.4 Si=28 P=31 S=32 Cl=35.3

Each element is similar in its chemical properties to the one above or below it, but different from the ones in the same row before and after it. The valence of Li and Na is 1, meaning that each can combine with a single atom of hydrogen; the valence of carbon and silicon is 4, etc. Mendeleev's 1872 table, reproduced in Holton and Brush, *Physics*, p. 300, incorporates these and other changes.

At that time the noble gases (He, Ne, etc.) had not been investigated chemically (helium was discovered in the solar spectrum in 1868, but not available for laboratory experiments until the 1890s). It was eventually determined that they should all be considered to have valence zero because they do not normally react with any other elements in the periodic table.

159 *the properties of three unknown elements* "Das periodische Gesetzmässigkeit der chemische Elemente," *Liebigs Annalen der Chemie und der Pharmacie* (Supplement) 8 (1871): 133–229. According to Scerri, *Periodic Table*, p. 96, the existence of germanium was predicted before Mendeleev by Lothar Meyer in 1864.

161 *a prediction based on a new hypothesis should disagree* See Brush, *Reception*, note 7.

161 *According to philosopher Michael Akeroyd* Akeroyd, "The philosophical Significance of Mendeleev's successful Predictions of Properties of Gallium and Scandium," *Foundations of Chemistry* 12 (2010): 117–122.

Section 5.4

162 *reception of the periodic law followed this pattern to some extent* Masanori Kaji, "Social Background of the Discovery and the Reception of the Periodic Law of the Elements: Recognizing the Contribution of Dmitri Ivanovich Mendeleev and Julius Lothar Meyer," in *Chemical Explanation: Characteristics, Development, Autonomy*, edited by Joseph E. Earley Sr. (New York: *Annals of the New York Academy of Sciences*, vol. 988, 2003), pp. 302–305. This work is a collection of articles from a conference proceeding, published by the NYAS in their Annals.

162 *explicit references to the periodic law to be found* See Brush, "Reception," pp. 619–623.

162 *"law and order pervade [its] vast domain"* M. M. Pattison Muir, "On the Teaching of Chemistry," *Nature* 36 (1887): 536–538, on p. 537.

162 *chemistry textbooks published in the period 1871–1890* See Brush, "Reception," pp. 624–628.

162 *we can get a much better idea of the extent* Anders Lundgren and Bernadette Bensaude-Vincent, *Communicating Chemistry: Textbooks and their Audiences 1789–1939* (Canton, MA: Science History Publications, 2000); José Ramon Bertomeu-Sánchez, Antonio Garcia-Belmar, and Bernadette Bensaude-Vincent, "Looking for an Order of Things: Textbooks and Chemical Classification in Nineteenth Century France," *Ambix* 49 (2002): 227–250; Bernadette Bensaude-Vincent, J. R. B. Sanchez and A. G. Belmar, "Chemistry Textbooks in the 19th Century: A Genre of Scientific Literature," *Chemical Heritage*,22, no. 4 (Winter 2004–2005): 10–11, 27–31; William B. Jensen, "Textbooks and the Future of the History of Chemistry as an Academic Discipline," *Bulletin for the History of Chemistry* 31 (2006): 1–8; José Ramon Bertomeu-Sánchez, Anders Lundgren and Manolis Patiniotis, "Introduction: Scientific and Technological Textbooks in the European Periphery," *Science & Education* 15 (2006): 657–665; Santiago Alvarez, Joaquim Sales, and Miguel Seco, "On Books and Chemical Elements," *Foundations of Chemistry* 10 (2008): 79–100.

Section 5.5

163 *he named* gallium, *supposedly in honor of his native France* Lecoq de Boisbaudran, "Caractères chimiques et spectroscopiques d'un nouveau métal, le gallium, découvert dans un blende de la mine de Pierrefitte, Vallée d'Argelès (Pyrénées)," *Comptes Rendus Hebdomadairesdes Sciences de l'Académie des Sciences, Paris* 81 (1875): 493–495; "Sur un nouveau métal, le gallium," *Annales de Chimie* (series 5), 10 (1877): 100–141. It was later alleged by cynics that Lecoq de Boisbaudran had chosen this name because the Latin word *gallus* means "coq" in French ("cock," male bird, in English),

thereby glorifying himself. This derivation is now found in dictionaries, and was cited as a precedent when it was proposed to name element 106 *seaborgium*, contrary to the custom that an element is never named after a living person. The name was approved before the death in 1999 of Glenn Seaborg, codiscoverer of several elements. See Seaborg's letter quoted in *Chemical Heritage* 13, no. 1 (Winter 1995–96): 30; *The Shorter Oxford English Dictionary* (Oxford: Clarendon Press, 3rd ed. with corrections and addenda, 1955), p. 771; *The Random House Dictionary of the English Language* (New York: Random House, second edition, 1987), p. 784; Isaac Asimov, *Asimov's Biographical Encyclopedia of Science and Technology* (Garden City, NY: Doubleday, 1964), p. 331. Geoff Rayner-Canham and Zheng Zheng, "Naming Elements after Scientists: An Account of a Controversy," *Foundations of Chemistry* 10 (2008): 13–18.

163 *He considered his result to be a confirmation of his own theoretical prediction* "He had some years earlier produced his own classification of the elements, based on spectral lines. Using these regularities, he made a prediction of the atomic weight of an analog of aluminum that was actually fairly close to Mendeleev's value (and closer to today's accepted valued)." Gordin, *Mendeleev*, p. 38; see also Gordin's note 86 on p. 269.

Section 5.7

165 *Patrick Maher defended the predictivist thesis* Maher, "Prediction, Accommodation, and the Logic of Discovery," in *PSA 1988 [Proceedings of the 1988 Biennial Meeting of the Philosophy of Science Association]* (East Lansing, MI: Philosophy of Science Association, Vol. 1, (1988): pp. 273–285, on pp. 273, 174.

165 *"sixty accommodations paled next to the two predictions"* Lipton, "Prediction and Prejudice," *International Studies in the Philosophy of Science* 4 no. 1 (1990): 51–65.

165 *does not even mention those predictions* William Spottiswood, "President's Address," *Proceedings of the Royal Society of London* 34 (1882): 302–329.

166 *Mendeleev stressed the importance of predictions* Dmitrii Mendeleev, "The Periodic Law of the Chemical Elements," *Chemical News* 40 (1879): 231 ff, on p. 292.

166 *he regarded Lecoq de Boisbaudran and Nilson* Mendeleev, *The Principles of Chemistry*, 2nd English ed. (New York: Collier, 1902), Vol. 2, Pt. 3, p. 26 note 13.

166 *Several statements published in the decade 1876–1885* See Brush, "Reception," pp. 610–612.

Section 5.8

167 *the periodic law is* established *by these correlations* Brush, "Reception," note 47; see also Simon Schaffer's review of Michael D. Gordin's *A Well-Ordered Thing* (2004), in which he writes: "It's not entirely clear whether Mendeleev's colleagues were much more moved by his prophecies of new elements than by his remarkable skill in ordering what they already knew": *Chemical Heritage* 24, no. 3 (Fall 2006): 45–48, on p. 48.

167 *"Leave these conjectures alone"* Bohuslav Brauner, "D. I. Mendeleef," *Collection of Czechoslovak Chemical Communications* 2 (1930): 219–243, on p. 227.

168 *"no new body, whatever it may be, can fall outside"* Marcellin Berthelot, *Les Origins de l'Alchemie* (Paris: Steinheil, 1885), p. 311, quotation from the translated extract in F. P. Venable, *The Development of the Periodic Law* (Easton, PA: Chemical Publishing, 1896), pp. 111–112.

168 *"stood the test both in the prediction of the properties of elements"* *Grundriss der allgemeinen Chemie*, 2nd ed. (Leipzig: Engelmann, 1890), pp. 34–37, 184–187; the quotation is from the translation by James Walker, *Outlines of General Chemistry* (London/New York: Macmillan, 1890), p. 186.

168 *Mendeleev himself repudiated the idea that his law was based on any assumption* According to Bernadette Bensaud-Vincent, for Mendeleev "the individuality of the elements was an irreducible fact"; he did not need atoms. "Lessons in the History of Science," *Configurations* 8 (2000): 201–214, on p. 211.

168 *theoretical inferences that might be drawn* See M. A. El'iasevich, "The Mendeleev Periodic Law, Atomic Spectra, and Atomic Structure: On the History of the Physical Interpretation of the Periodic Table of the Elements," *Soviet Physics Uspekhi* 13 (1970): 1–23; Helge Kragh, "The first Subatomic Explanations of the Periodic System," *Foundations of Chemistry* 3 (2001): 129–143; Matteo Leone and Nadia Robotti, "Are the Elements Elementary? Nineteenth-Century Chemical and Spectroscopic Answers," *Physics in Perspective* 5 (2003): 360–383.

168 *"the question whether our elementary atoms are in their nature indivisible"* A. W. Williamson, "On the Atomic Theory," *Journal of the Chemical Society [London]*, 2nd series, 7 (1869): 328–365, on p. 365.

168 *predicting the properties of one previously unknown element* "Predictions (novel facts) . . . played a crucial role in the development of the periodic table"; Mansoor Niaz, Mario A. Rodriguez, and Angmary Brito, "An Appraisal of Mendeleev's Contribution to the Development of the Periodic Table," *Studies in History and Philosophy of Science* 35 (2004): 271–284, on p. 271.

Section 5.9

169 *Mendeleev, though skeptical about that hypothesis* Mendeleev, "The Periodic Law of the Chemical Elements. V. On the Use of the Periodic Law for the Correction of Atomic Weights," *Chemical News & Journal of Physical Science* 41 (1880): pp. 93–94.

169 *the speculations about atomic dissociation* W. H. Brock, "Lockyer and the Chemists: The First Dissociation Hypothesis," *Ambix* 16 (1969): 81–99.

Section 5.10

169 *periodic law was established because of its explanations* E. Scerri and J. Worrall, "Prediction and the Periodic Table," *Studies in History and Philosophy of Science* 32A (2001): 407–452; for discussion of this argument see F. Michael Akeroyd, "Prediction and the Periodic Table: A Response to Scerri and Worrall," *Journal for General Philosophy of Science* 34 (2003): 337–355; "Philosophy of Science and History of Science: A Non Troubling Interaction," *Journal for General Philosophy of Science* 33 (2002): 159–162; E. Barnes, "On Mendeleev's Predictions: Comment on Scerri and Worrall," *Studies in History and Philosophy of Science* 36A (2005): 801–812; E. R. Scerri, "Response to Barnes's Critique of Scerri and Worrall," *Studies in History and Philosophy of Science* 36A (2005): 813–816; J. Worrall, "Prediction and the 'Periodic Law': A Rejoinder to Barnes," *Studies in History and Philosophy of Science* 36A (2005): 817–826; S. G. Brush, "Predictivism and the Periodic Table," *Studies in History and Philosophy of Science* 38 (2007): 256–259. Samuel Schindler, "Use-Novel Prediction and Mendeleev's Periodic Table: Response to Scerri and Worrall (2009)," *Studies in History and Philosophy of Science* 39 (2008): 265–269.

6

The Benzene Problem 1865–1930

> The benzene problem offers an "excellent example of the scientific method."
>
> —C. K. INGOLD *(1953)*

Are chemists more likely than other scientists to follow the hypothetico-deductive method and favor theories that have yielded successful predictions?

My studies of historical cases in the physical sciences suggest that physicists and astronomers generally do not give greater weight to novel facts (evidence not known before it was predicted by a hypothesis) than to comparable "old" facts—novelty per se often has no value. But in the case of Mendeleev's periodic law (previous chapter), the discovery of a new element such as gallium, having the predicted physical and chemical properties, did count more than the explanation of a known element. Some authors have claimed that the history of the benzene problem offers an "excellent example of the scientific method" because chemists accepted or rejected models of its structure on the basis of empirical tests of novel predictions based on those models. In the following chapter I investigate this claim.

6.1 KEKULÉ'S THEORY

In the 1870s chemists had only recently begun to embrace the hypothetico -deductive method. During this period they were offered another chemical theory, as famous as Mendeleev's periodic law and at least as important to the development of modern chemistry: August Kekulé's theory of the structure of the benzene molecule. His theory, which was immediately generalizable to the large and rapidly growing class of aromatic compounds, may be defined for our purposes as consisting of the following postulates and hypotheses:

Postulate I: The numbers of atoms of different elements in a molecule is determined in accordance with chemical atomic theory, using Stanislao Cannizzaro's system to compute the atomic weights. In particular the atomic weight of carbon is about 12.

Postulate II. The carbon atom has valence 4.

These postulates, so designated because they were not considered subject to falsification by experiments on benzene, had been proposed only a few years before Kekulé published his benzene theory in 1865. P-II worked so well in explaining the properties of other organic compounds that no one seems to have seriously considered deviating from it in the case of benzene, although chemists became accustomed to using diagrams for aromatic compounds that simply ignored the fourth valence of carbon.

Hypothesis K-1: The six carbon atoms in benzene are located at the vertices of a planar hexagon or are spaced equally around a circle. These two versions, both called the benzene ring, were used interchangeably in later years since there was no way to determine whether the valence bonds are straight or curved lines. Each carbon atom is bonded to one hydrogen atom (usually assumed to be located in the same plane as the carbon atoms).

Hypothesis K-2: In addition to the three valence bonds linking a carbon atom to its two neighbors in the ring and to a hydrogen atom, there is a fourth valence bond directed toward *one* of its neighbors, in such a way that single and double carbon-carbon bonds alternate around the ring.

Hypothesis K-Osc: In 1872 K-2 was revised to read: the fourth valence of each carbon atom oscillates between its two neighbors, synchronously with all the other fourth valences, so that the structure switches rapidly between the two structures in Figure 6.1. This means that all the carbon-carbon bonds are equivalent; a compound formed by replacing H^1 and H^2 with chlorine atoms, for example, cannot be distinguished experimentally from one formed by replacing H^1 and H^6.

I have presented these hypotheses as they were discussed in the literature of late nineteenth-century chemistry, not as Kekulé originally formulated them.

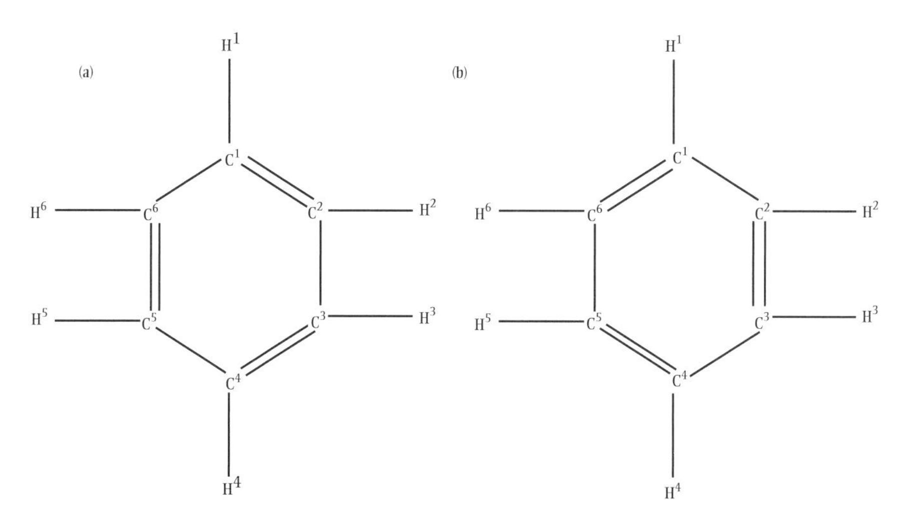

Figure 6.1 Kekulé's structure for benzene. According to his oscillation hypothesis the arrangement of double and single bonds between carbon atoms changes rapidly between (a) and (b).

In this chapter I do not discuss how Kekulé arrived at his benzene theory or the possible influence of the ideas of other chemists—issues that have been argued at great length by others. Instead I focus on how chemists have changed their theories of benzene since 1865, and especially on how those changes have resulted from experimental discoveries and theoretical reasoning. Were theories accepted (or rejected) because of empirical confirmation (or refutation) or for other reasons?

To answer these questions I have used not only published research articles but also reviews and textbooks (as explained in Section 5.4). However, textbooks are not as useful here as they were in studying the reception of the periodic law, because some of the arguments are rather technical and were probably considered too difficult for students to understand.

I can tell only the first half of the story in this chapter, because historically the benzene problem could not be resolved until the advent of quantum mechanics. So I will finish the story in Chapter 10.

6.2 THE FIRST TESTS OF KEKULÉ'S THEORY

The early history of Kekulé's theory has been extensively described by several chemists and historians so only a brief summary of the *outcomes* of this theory is needed. The scientific significance of this work is suggested by the claim of Gustavus Hinrichs in 1874: the "*phenomena of isomerism* [of organic compounds] *are* the best possible evidence of the reality of chemical atoms, and in themselves quite sufficient to establish the atomic theory as firmly as any other physical theory is established."

> **Prediction 1**. Hypothesis K-1 implies that all six carbon atoms are equivalent and all six hydrogen atoms are equivalent. Thus if one hydrogen is replaced by another atom such as chlorine, or by five such atoms, there is only one isomer of the resulting compound.

As Kekulé pointed out, a plausible alternative hypothesis that regarded the six carbon atoms as divided into three groups of two each would imply two different kinds of hydrogen atoms, inside or outside, and would thus predict two isomers of a monosubstitution product. Thus a research program to synthesize all the possible substitution products of benzene could help to choose between these two hypotheses.

Prediction P-1 seemed to be refuted by the existence of two isomers of benzoic acid, reported by Hermann Kolbe, and two isomers of pentachlorobenzene, reported by Émile Jungfleisch (working in Berthelot's laboratory) in his prizewinning dissertation, and confirmed by Robert Otto. Both reports were found to be erroneous, thereby leaving Kekulé's prediction unrefuted. The first case (there is only one isomer of benzoic acid) does not count as a novel prediction, since the second isomer of benzoic acid was shown not to exist and Kolbe withdrew his result before Kekulé published his prediction. Jungfleisch's second isomer of pentachlorobenzene remained an anomaly until 1874, when Albert Ladenburg repeated the research and found that the second isomer does not exist; that does count as a confirmed novel prediction by Kekulé.

While Ladenburg himself later mentioned the pentachlorobenzene case as a success of Kekulé's theory, and Victor Meyer praised Kekulé's "chemical feeling" (as distinct from deduction) that led him to deny the existence of isomers that would contradict his theory, it quickly disappeared from the literature of organic chemistry.

The opportunity to use this successful novel prediction as support for Kekulé's benzene theory was not seized by Kekulé himself or other chemists. Ladenburg's 1874 paper was cited as a proof that all six carbon atoms in benzene are equivalent.

Prediction 2. According to K-2, when two hydrogen atoms are replaced by other atoms such as bromine, there should be four isomers of the resulting compound, shown in Figure 6.2. As was shown by Wilhelm Körner, we can determine whether the replaced hydrogens were bonded to adjacent carbon atoms—defined as "ortho"—to carbons with one in between, such as C^1 and C^3, defined as "meta"—or to carbons at opposite sides of the ring such as C^1 and C^4, defined as "para." All meta isomers (positions 1-3, 2-4, 3-5, 4-6, 5-1, 6-2) are indistinguishable: there is only one. Similarly, all para isomers (1-4, 2-5, 3-6) are indistinguishable. But because of the alternating double and single bonds, there are two ortho isomers (1-2 and 1-6).

Körner found that when a third bromine atom is substituted, one of the diderivatives yields a single triderivative (Br at 1-2-4) as would be expected for a para derivative (1-4); another one yields three triderivatives (1-2-4 or 2-4-6 or 2-3-4), as would be expected for a meta diderivative (2-4); and the other yields two triderivatives (1-3-4 and 2-3-4), as would be expected for the ortho diderivative (3-4).

It soon became clear that there are only three double-substituted or diderivative isomers, not four, and that there is only one ortho diderivative, not two. Thus prediction P-2 is falsified.

Prediction 3. According to hypothesis K-2, benzene should easily react with other chemicals because of its double bonds; it is "unsaturated."

In fact it was already well known that benzene is a very stable compound; although the hydrogen atoms can be replaced by other monovalent atoms or groups, the benzene ring itself remains unchanged, and the double bonds, if they are really present, do not seem to represent an unsaturation (potential for bonding with additional atoms), such as one finds in double-bonded compounds such as C_2H_4. So P-3 was also falsified.

Chemists did not express much concern about the anomalous stability of benzene until decades later, but Kekulé himself soon recognized that the validity of K-2 was endangered by the failure of P-2. He therefore proposed the oscillation hypothesis, K-Osc. This not only allowed him to deduce the correct number of diderivative isomers (3), but also the correct numbers of all other multiderivatives:

Prediction 4. If all carbon-carbon bonds are equivalent, then the number of isomers of benzene with

one hydrogen substituted is	*one*;
two . . .	*three*;
three . . .	*three*;
four . . .	*three*;
five . . .	*one*
six . . .	*one*.

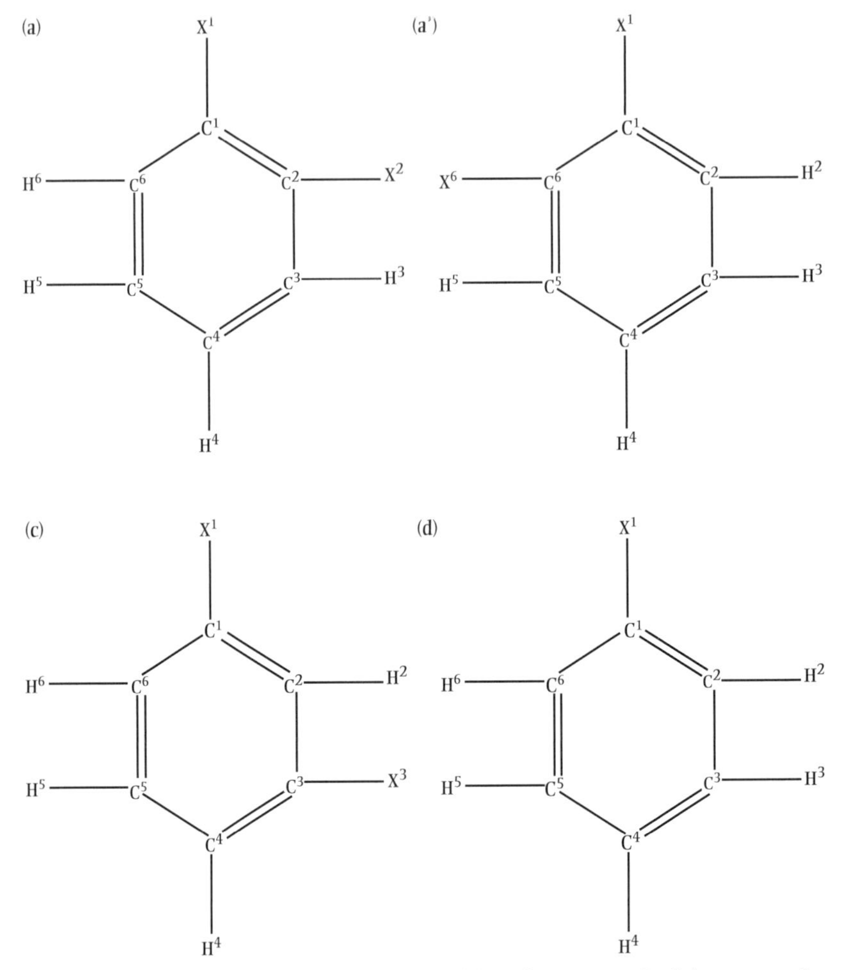

Figure 6.2 Isomers of diderivatives of benzene: (a) Ortho isomer of a diderivative of benzene, according to Kekulé's structure; the hydrogen atoms bonded to adjacent carbon atoms (labeled 1 and 2), which are them-selves connected by a double bond, are replaced by another atom such as bromine. (a') Second ortho isomer of a diderivative of benzene; the hydrogen atoms bonded to adjacent carbon atoms 6 and 1, which are themselves connected by a single bond, have been replaced by another atom, According to Kekulé's theory (a') is different from (a) unless the double and single bonds oscillate so that all CC bonds are effectively equivalent.
(b) Meta isomer of a diderivative of benzene: the hydrogen atoms bonded to carbon atoms 1 and 3 have been replaced by another atom. According to Kekulé's theory, there is only one such isomer whether or not the double and single bonds oscillate.
(c) Pure isomer of a diderivative of benzene: hydrogen atoms bonded to carbon atoms 1 and 4 have been replaced by another atom. According, to Kekulé's theory there is only one such isomer.

This prediction was confirmed by Beilstein and Kurbatow in 1878.

According to the strict rules of scientific method K-Osc is an *ad hoc hypothesis*; Kekulé did not claim that it led to any new testable predictions, other than those that could be deduced directly from K-1 by simply ignoring the fourth valence of carbon.

The agreement between K-Osc (and other hexagonal formulas in which all carbon atoms and carbon-carbon bonds are equivalent) and Körner's research on isomers led chemists to *define* the ortho isomer as the one that yields three triderivatives when a third substituent is added; the meta is the one that yields 3, and the para is the one that yields 1.

6.3 ALTERNATIVE HYPOTHESES

But by this time other chemists had proposed alternative hypotheses, two of which made all the (neighboring) carbon-carbon bonds as well as all the carbon atoms equivalent and could therefore claim to be equally well supported by the success of predictions 1 and 4. Moreover, since they did not involve any double bonds, they could claim to explain the chemical stability of the benzene ring better than Kekulé's oscillation hypothesis K-Osc. Adolf Claus proposed

> Hypothesis CC (Claus "centric" or "diagonal" hypothesis): the fourth valence of each carbon atom is directed toward the one on the opposite side of the ring (C^1-C^4, etc.).

While this formula (Figure 6.3) is more symmetric than Kekulé's alternating bond hypothesis (K-2), it seems to allow only two diderivatives; ortho and para would be equivalent, since in both isomers the substituents are bonded to carbon atoms that are bonded to each other. There was some disagreement among chemists as to whether CC provides only two diderivatives instead of the observed three, whether para bonds could count as legitimate valence bonds, whether they would risk some degree of unsaturation, and whether they would somehow interfere with each other when they crossed in the center. Albert Ladenburg proposed

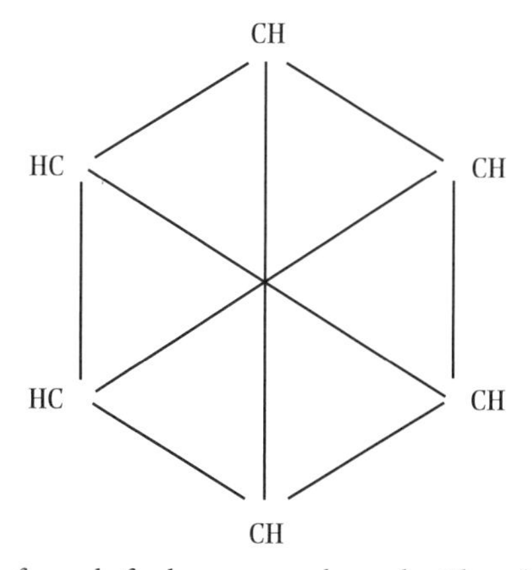

Figure 6.3 Claus centric formula for benzene as shown by Claus (1867).

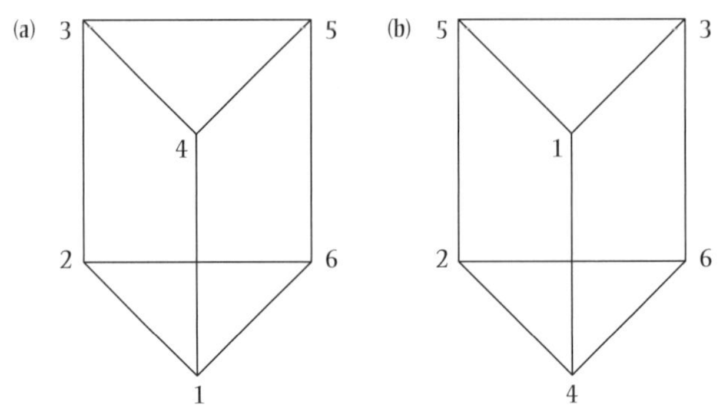

Figure 6.4 Ladenburg prism model for benzene (LP) (a) as shown by Ladenburg (1869b); (b) as shown by Lachman (1899, p. 44). Note the difference in numbering of the carbon atoms; Lachman points out that they should be numbered in accordance with Körner's method, with the ortho position being 1-2, etc., so that, for example, the para position is that which furnishes but one tri-derivative. This has some disadvantages, e.g., the ortho atoms are not adjacent as they are in Kekulé's formula.

> Hypothesis LP (Ladenburg "prism" hypothesis): abandoning the assumption that all carbon atoms are in the same plane, Ladenburg proposed that they are located at the vertices of a triangular prism. All carbon-carbon bonds are equal in length, thereby evading one of the objections to CC, but a vertical bond (forming the edge of two squares) is not quite equivalent to a horizontal bond (forming the edge of a triangle and a square; see Figure 6.4).

Nevertheless, LP was clearly in agreement with the empirical facts that confirmed predictions P-1 and P-4. Because of the redefinition of ortho, meta, and para by means of Körner's procedure (see above), chemists renumbered the carbon atoms in LP so that ortho atoms are no longer adjacent (Figure 6.4(b)).

6.4 RECEPTION OF BENZENE THEORIES 1866–1880

Within a decade after the publication of Kekulé's theory, the hexagonal structure (K-1) was widely accepted by chemists, according to historian Alan J. Rocke. But if we look for a connection between this acceptance and the empirical tests of predictions based on his hypotheses, we find a curious situation. Rocke states that "the number of post-1865 confirmed predictions is legion," and "other theories were relatively quickly excluded by failed predictions . . . one could not ask for a more classic case study of the predictive success of a theory . . . and, more generally, the application of predictive *criteria* for both confirming and falsifying instances, using hypothetic-deductive method."

But which version of Kekulé's theory was actually being tested? On one hand, the empirical equivalence of all carbon atoms, and of all carbon-carbon bonds, in determining the number of isomers of substitution compounds, was often counted in favor of Kekulé's theory—without mentioning that LP could account for the same facts. On the other hand, Kekulé's hypothesis K-2 (with alternating double and single bonds) was presented in textbooks with the assertion that it predicted only three diderivatives, in agreement with observations, because substitution at the 1 and 2 positions is equivalent to substitution at the 1 and 6 positions—without any mention of the oscillation hypothesis (K-Osc) that Kekulé himself had proposed in order to justify that claim, or even any explicit assertion that the theoretical difference between the two hypothetical ortho isomers would be too small to produce a detectable difference in properties. On the basis of this general indifference to the issue, Rocke argues that we might as well proceed as if all the above-mentioned hypotheses (K-2, K-Osc, LP, CC) are "minor variants of a single theory."

An even more serious problem (at least from my viewpoint) noted by Rocke is that "it is difficult in this case to make a general distinction between *predictions* (i.e., of outcomes of experiments not yet performed) and *retrodictions* (i.e., comparing data from experiments predating the formulation of the theory, or from experiments that were performed for reasons other than testing the theory)."

One of the few chemists who gave a full discussion of the reasons for adopting Kekulé's theory was Johannes Wislicenus. In an 1874 textbook he listed

1. the fact that benzene can be synthesized from acetylene (the alternating single/double bond structure is the simplest one resulting from this synthesis);
2. benzene is most likely to combine with 2, 4, or 6 atoms of chlorine or bromine, but no more than 6;
3. many derivatives such as $C_6H_4(CO@OH)_2$ can be formed and then one can regenerate the original compound;
4. the chemical properties of benzene show the complete equivalence of the six hydrogen atoms; also the monoderivative has only one isomer.

Ladenburg's prism (Hypothesis LP) was mentioned by only a few writers in this period. It was ignored not because of any inferior predictive capacity but perhaps rather because its three-dimensional structure was difficult to represent on a page, especially when discussing compounds with two or more joined benzene rings.

During this period the structures proposed by Claus and others were almost completely ignored by writers of textbooks, reference works, and journal articles.

6.5 NEW EXPERIMENTS, NEW THEORIES 1881–1900

The introduction and widespread acceptance of ideas about the three-dimensional structure of molecules, starting with the publication of the famous articles by J. H. van't Hoff and J. A. Le Bel in 1874, transformed chemical theory in the late nineteenth century. In particular, the new stereochemistry encouraged chemists to consider three-dimensional structures for benzene based on the tetrahedral bonds of the carbon atom, and made them wonder how any merely two-dimensional structure such as Kekulé's ring could represent reality.

While stereochemical models of benzene enjoyed some popularity in the late nineteenth century, none of them proved to be a serious competitor. More important for benzene theory was the investigation of physical properties such as heats of formation and refractivity. While these measurements did not ultimately count as evidence against Kekulé's theory, they forced chemists to reassess the status of that theory, and to revive *old* negative evidence that had been disregarded.

The first stage in that reassessment was:

Prediction 5. Hypotheses K-2 and K-Osc imply that benzene should have the physical properties corresponding to a molecule with three single carbon-carbon bonds and three double bonds, whereas CC and LP imply the physical properties corresponding to nine single bonds.

In 1880, Julius Thomsen concluded from his thermochemical researches that the heat produced or absorbed in chemical reactions is related to the number of single and double bonds in a molecule. Applying his formulas to benzene, he argued that its heat of combustion shows it to contain nine single carbon-carbon bonds rather than three single and three double bonds. This result would support benzene models such as CC and LP in which each carbon atom is bonded to three others.

Thomsen's conclusions contradicted those of Julius Wilhelm Brühl, who developed a relation between the index of refraction of molecules, their atomic refractions, and their bonds. Brühl concluded that benzene does have three carbon-carbon double bonds, in agreement with Kekulé's formula.

Thomsen's thermochemical refutation of Kekulé's formula was widely cited in the chemical literature for a few years. Thomsen's interpretation of his data was later judged to be fallacious by Walther Nernst and other physical chemists. But it served to undermine the complacent acceptance of Kekulé's formula, which though not rejected had to share the stage with competitors such as the centric and prism formulas. The fact that the alternating single/double bond structure K-2 actually predicted two ortho isomers could no longer be easily overlooked but had to be explicitly mentioned as a disadvantage. A few chemists were thus led to recall the oscillation hypothesis K-2a, though they did not seem eager to adopt it. Others favored Ladenburg's prism—especially if it could be represented in a more convenient two-dimensional projection—and Adolphe Wurtz asserted in 1884 that this structure was now preferred by most chemists.

The popularity of Ladenburg's prism did not last very long, because of

Prediction 6. In chemical reactions, the carbon atoms at ortho positions behave as if they are adjacent, while carbon atoms at para positions behave as if they are not adjacent. This is true for plane hexagonal models such as K-2, K-Osc, and CC, but not for LP (see Figure 6.4(b)).

Adolf von Baeyer, in an extensive series of experiments, produced strong chemical evidence against LP. The essence of his rather complicated argument was that the outcome of a particular chemical reaction involving the breaking of two para-bonds in benzene could not reasonably be explained by hypothesis LP, but was just what one would expect according to hypothesis K-2. I have summarized this result, rather vaguely, as a test of P-6. (The chemical literature suggests that no one else had a precise understanding of Baeyer's argument.)

Although Baeyer's benzene research was occasionally derided as a futile attempt to solve an insoluble or ill-defined problem, it did have a significant impact on chemical theory by knocking out a major competitor of Kekulé's formula. Chemical writers during the next two decades agreed that Baeyer had empirically refuted hypothesis LP, even though they did not try to explain exactly how he had done so.

Baeyer also tested the centric hypothesis, revived and extended by Claus in 1882. Claus assumed that the para-bond was different from the other carbon-carbon bonds, so the ortho and para diderivatives would be distinguishable isomers. His hypothesis would thereby imply three diderivatives, as observed. But, Baeyer argued, in what I will call Prediction 7, "if this assumption is correct, it is to be expected that after rupturing one bond the presence of the other two could be shown." Experiment proved, however, that only double linkings remained, and so it was concluded that benzene contained no para-bonds. To this Claus objected that the para -bonds might rearrange themselves to form double linkings. In fact this did happen (in the reduction of terephthalic acid) but the same result could also be obtained from Kekulé's formula. Baeyer concluded that "the addition of hydrogen in the para position by no means requires the presence of a para linking in the ring."

So even though Claus (according to Baeyer) made a successful novel prediction, he got no credit for it because the same prediction *could have been made* from the established (Kekulé) theory. On the positive side, Baeyer made three important contributions to the benzene problem.

First, in 1888 he proposed a different version of the centric hypothesis, similar to one suggested a year earlier by Henry Edward Armstrong:

> **Hypothesis ABC** (Armstrong-Baeyer centric hypothesis). The fourth valence of each carbon atom is directed toward the center; instead of forming a bond that extends all the way to the opposite carbon atom (Claus's para-bond), it stops short of the center, being somehow neutralized or paralyzed by the fourth valences of the other carbon atoms (Figure 6.5). Armstrong described this as a "residual affinity."

Baeyer admitted that ABC could not be tested: "in its nature it is incapable of proof, for the premise that the valences are held in position by forces that admit of no disturbance except by complete alteration, precludes experimental study. The formula is merely the best picture we have of the complete symmetry of the ring, of its stability toward reagents, and of its difference from the other, better-known forms of union between carbon atoms. *By definition* it avoids these difficulties." In particular, it is hard to see how it could be empirically distinguished from hypothesis K-Osc. Nevertheless it gained some popularity in the 1890s and early 1900s. A few chemists even regarded its indefiniteness as a point in its favor.

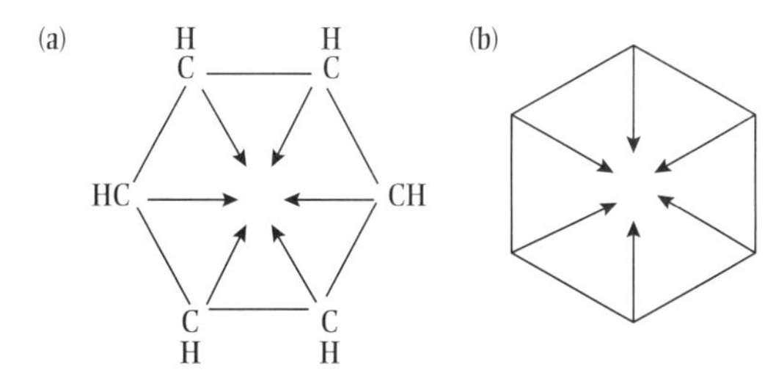

Figure 6.5 The Armstrong-Baeyer centric model (ABC), (a) as shown by Armstrong (1887); (b) as shown by Baeyer (1890).

Second, he proposed that the stability of molecular structures depends on the amount of strain experienced by the bond angles. Baeyer's strain theory was qualitatively confirmed by quantum chemistry.

Baeyer's third contribution, this one specific to benzene theory, was his suggestion that one should abandon the assumption, underlying all earlier work, that there is a single structure of the benzene molecule that governs its chemical and physical properties. Perhaps instead its structure depends on the particular reaction it undergoes. If taken seriously, this suggestion would cast doubt on the reality of the valence bond itself and thus undermine the credibility of chemical structure theory in general. Along with the recognition of tautomerism in other cases, it may also have encouraged a more favorable attitude toward Kekulé's oscillation hypothesis, which could be interpreted as compatible with this viewpoint.

The influence of the oscillation hypothesis can be seen in a more elaborate three-dimensional benzene structure proposed in 1897 by J. Norman Collie. By incorporating vibrations, Collie could make his structure compatible with both the Kekulé and centric hypotheses. Although too complicated to use in practice or to subject to conclusive empirical tests, Collie's model survived as one that was sometimes considered the most likely to represent the true structure of benzene.

Other spatial formulas, which assumed a fixed array of atoms, could be tested by

Prediction 8: any three-dimensional molecule not superimposable on its own mirror-image must have distinguishable optical isomers (enantiomorphs), according to stereochemical theory.

This would apply to derivatives of benzene; such isomers were not observed. So the absence of optical isomers of benzene derivatives was taken as a refutation of LP and other three-dimensional formulas. The existence of such alternatives was sometimes acknowledged in textbooks and reviews, but only to point out that they had been empirically disproved.

At the end of the nineteenth century Kekulé's theory was no longer as dominant as it had been in the 1870s, yet it had survived the challenge of numerous competitors and stringent experimental tests. Francis Japp was exaggerating only slightly when he proclaimed in his Kekulé Memorial Lecture on December 15, 1897:

> Kekulé's memoir on the benzene theory is the most brilliant piece of scientific prediction to be found in the whole range of organic chemistry. What Kekulé wrote in 1865 has since been verified in every essential particular. Not only have the various substitution derivatives been discovered, in the number and with the properties required by the theory, but various observations which appeared to contradict this theory have been proved erroneous.

Even the skeptic Marcellin Berthelot had to admit that despite all its faults, Kekulé's formula was the one most widely used by chemists. This statement is easily confirmed by a glance at books and journals published at the end of the century. But a few of those chemists continued to be frustrated by their failure to establish *the* structure of this "devious hydrocarbon" and would have agreed with A. Lachman's conclusion that "the final answer has yet to be given to the tantalizing question: What is the constitution of C_6H_6? When will it come, and whence?"

6.6 THE FAILURE OF AROMATIC EMPIRICISM 1901–1930

Baeyer's benzene research comprised the high point of hypothetico-deductive method (his refutation of LP) and the beginning of its decline (his proposal of the untestable ABC). Textbooks published in the early 1900s continued to present the hypotheses of Kekulé, Ladenburg, Claus, Armstrong, and Baeyer, but seem to have abandoned any hope of determining which one was really correct: each had advantages and disadvantages in explaining the chemical reactions of aromatic compounds, and the evidence from physical properties such as heat of formation and refractivity (Prediction 5) was inconclusive. As Keane wrote in 1909, despite much research "the experimental solution of the problem has nevertheless baffled the best efforts of chemists for forty years."

One indication of frustration is that scientists who can't think of anything better to do start to make lists of the conditions that must be satisfied by any acceptable theory. Thus Victor Meyer and Paul Jacobson asserted that a good benzene formula must (1) agree with the correct number of isomers of substitution products; (2) allow an explanation of condensed benzene nuclei; (3) explain relations of benzene derivatives to their hydrides and addition products; (4) give a cause for the peculiarity of aromatic character; and (5) give an account of why *six* carbon atoms are necessary and sufficient to form an "aromatic nucleus." (Note that making a *novel* prediction is not one of the requirements.) As the history of other long-standing unsolved problems suggests, the evidence that turns out to be crucial in theory-change may not be related to any criterion that appears on such lists, which instead merely serve to indicate the domain of the problem.

Several new structures were proposed around the turn of the century but most of them failed to attract any significant support.

During this period there was only one clear-cut test of a prediction:

> **Prediction 9.** According to K-Osc, cyclic hydrocarbons such as C_4H_4 and C_8C_8 should have chemical properties similar to those of benzene, the properties we now call aromaticity, since the same oscillation mechanism should work if there are an even number of carbon atoms and if the molecule is planar. An even more general prediction about all cyclic hydrocarbons C_nH_n seems to follow from ABC and CC.

In 1911 Richard Willstätter synthesized cyclo-octatetraene, C_8H_8 (COT). From its formula one might think COT would be analogous to benzene, with a symmetrical eight-carbon ring. As Meyer and Jacobson had pointed out, a satisfactory theory must explain why six carbon atoms are necessary and sufficient to form a nucleus with the aromatic properties of benzene; but all four hypotheses implied that a nucleus with eight carbon atoms should also be aromatic. COT was not; it behaved just like an olefin with unsaturated double bonds. Although only a few chemists explicitly recognized that the properties of COT refuted one or all of the four accepted benzene structure hypotheses, this event must have aggravated a widespread suspicion that none of those hypotheses is really correct.

To escape the refutation of Prediction 9, chemists could again resort to an ad hoc hypothesis, this time one that smacked of numerology:

> **Hypothesis AS**: the special properties of benzene (and molecules like naphthalene formed by fusing benzene rings) are due to the aromatic sextet. This was defined by J. W. Armit and Robert Robinson in 1925 as a group of six electrons that form a group that resists disruption; the concept was endorsed by F. R. Goss and C. K. Ingold, who traced the germ of the idea back to Eugen Bamberger's papers in the 1890s.

Toward the end of the 1920s, new physical methods such as X-ray diffraction made it possible to determine not only the spatial arrangement of atoms in a molecule but the actual distances between them. One could now test

> **Prediction 10**: According to K-2, the carbon-carbon bonds in benzene should have unequal length since (in other hydrocarbons) single bonds are longer than double bonds.

Crystallographer Kathleen Lonsdale inferred from her X-ray results on $C_6(CH_3)_6$ that benzene (which is not crystalline at room temperature) is planar, and that its bonds must be arranged to give the ring as a whole a center of symmetry; "this last condition quite eliminates the Kekulé static model [K-2], with its three double bonds."

Physics could help to refute chemical hypotheses, but could it generate satisfactory new ones? Reynold C. Fuson, reviewing the state of theoretical organic chemistry at a symposium of the American Chemical Society on January 1, 1930, had some doubts:

> Theoretical organic chemistry appears . . . to be in a transitional phase . . . It is becoming increasingly difficult to reconcile experimental developments with the classical theories, and as yet relatively few of the host of new theoretical ideas seem to have achieved any considerable success. The present state of disorder is due chiefly to the influence of modern physics . . . The results [of attempts to apply electronic conceptions] . . . have frequently brought confusion rather than clarity.

NOTES FOR CHAPTER 6

176 *The Benzene Problem 1865–1930* This chapter is based on my article "Dynamics of Theory Change in Chemistry: Part 1. The Benzene Problem 1865–1945," published in *Studies in History and Philosophy of Science* 30 (1999): 21–79, by permission of Elsevier Science Ltd.; this article should be consulted for further details and references to publications before 1999. For a more comprehensive account of Kekulé's life and work see Alan J. Rocke, *Image and Reality: Kekulé, Kopp, and the scientific Imagination* (Chicago: University of Chicago Press, 2010).

176 *"The benzene problem offers an 'excellent example of the scientific method'"* C. K. Ingold, *Structure and Mechanism in Organic Chemistry* (Ithaca, NY: Cornell University Press, 1953), pp. 156–157.

Section 6.1

176 *had only recently begun to embrace the hypothetico-deductive method* See Brush, "Benzene 1865–1945," note 3.

177 *had been proposed only a few years before Kekulé published his benzene theory* Although most chemists had by then accepted P-I, which implied that the composition of benzene is C_6H_6, one can find French textbooks published as late as 1869 in which the older atomic weight C = 6 is still used, leading to the formula $C_{12}H_6$ for benzene. See the books by L. Troost, É. Bouant, R. Jagnaux, J.-P.-L. Giardin cited by Brush, "Benzene 1865–1945," note 5. According to M. J. Nye, "it was not until the 1890s that Berthelot, L. J. Troost and other French equivalentists finally adopted in their publications and teaching the atomic notation then used universally elsewhere"; Nye, "The Nineteenth-Century Atomic Debates and the Dilemma of an 'Indifferent Hypothesis,'" *Studies in History and Philosophy of Science* 7 (1976): 245–268, on p. 262. A. Kekulé, "Sur la Constitution des Substances Aromatiques," *Bulletin de la Société Chimique de Paris*, 3 (1865): 98–110; "Note sur quelques Produits de Substitution de la Benzine," *Bulletin de l'Academie Royale de Belgique* [series 2] 19 (1865): 551–563; "Untersuchungen über aromatische Verbindungen," [Justus Liebig's] *Annalen der Chemie und Pharmacie* 137 (1865): 129–196.

177 *In 1872 K-2 was revised to read* Kekulé, "Über einige Condensationsproducte des Aldehyds," [Justus Liebig's] *Annalen der Chemie und der Pharmacie* 162 (1872): 309–320.

Section 6.2

178 *"the best possible evidence of the reality of chemical atoms"* Hinrichs, *The Principles of Chemistry and Molecular Mechanics* (Davenport, Iowa: Day, Egbert & Fidlar, 1874), p. 60 (emphasis in original).

179 *As was shown by Wilhelm Körner* Körner, "Studi sull' isomeria della cosi dette sostanze aromatiche a sei atomi di carbonio," *Gazzeta Chimica Italiana* 4 (1874): 305–446; long abstract by H. Armstrong in *Journal of the Chemical Society* 29 (1876): 204–224; Rocke, *Image*, pp. 216–218.

181 *This prediction was confirmed by Beilstein and Kurbatow* F. Beilstein and A. Kurbatow, "Ueber die Chlorderivate des Benzols," *Annalen der Chemie und Pharmacie* 192 (1878): 228–240.

Section 6.3

181 *Adolf Claus proposed Hypothesis CC* "Theoretische Betrachtungen und deren Anwendung zur Systematik der organischen Chemie," *Berichte über Verhandlungen der Naturforschenden Gesellschaft zu Frieburg I. B.*, 4 (1867): 116–381, on p. 367. See Brush, "Benzene 1865–1945," Fig. 3 (p. 28) and note 23.

181 *Albert Ladenburg proposed Hypothesis LP* "Bemerkungen zur aromatische Theorie," *Berichte, Deutsche Chemische Gesellschaft*, 2 (1869): 140–142. See Brush, "Benzene 1865–1945," Fig. 4 (p. 29) and note 24.

Section 6.4

182 *(K-1) was widely accepted by chemists* Alan J. Rocke, "Kekulé's Benzene Theory and the Appraisal of Scientific Theories," in *Scrutinizing Science: Empirical Studies of Scientific Change*, edited by Arthur Donovan, Larry Laudan, and Rachel Laudan (Doredrecht: Kluwer, 1988), pp. 145–161.

182 *"the number of post-1865 confirmed predictions is legion"* Rocke, "Kekulé's Theory," p. 152.

183 *"minor variants of a single theory"* Rocke, ibid.

183 *"It is difficult in this case to make a general distinction"* Rocke, "Kekulé's Theory," p. 153.

Section 6.5

185 *Claus objected that the para-bonds might rearrange themselves* Baeyer, Address at Celebration of 25th anniversary of Kekulé's Theory, *Berichte, Deutsche Chemische Gesellschaft* 23 (1890): 1272–1287, quoted from the translation in A. Lachmann, *The Spirit of Organic Chemistry* (New York: Macmillan, 1899, reprinted 1909); see also T. A. Koeppel, *Benzene-Structure Controversies 1865–1920*, PhD dissertation, University of Pennsylvania, 1973, pp. 285f.

185 *"in its nature it is incapable of proof"* Baeyer, quoted by Lachmann, *Spirit*; W. A. Tilden, "The Constitution of the Terpenes and of Benzene," *Journal of the Chemical Society* [London] 53 (1888): 879–888.

186 *"Kekulé's memoir on the benzene theory"* Japp, "Kekulé Memorial Lecture," *Journal of the Chemical Society* [London] 73 (1898): 97–138, on p. 135.

187 *"the final answer has yet to be given"* Lachman, *Spirit*, p. 59.

Section 6.6

187 *"the experimental solution of the problem"* C. A. Keane, *Modern Organic Chemistry* (London: Scott, 1909), p. 76.

187 *the evidence that turns out to be crucial* This was what happened to theories of the origin of the solar system in the 1960s and 1970s: see S. G. Brush, *A History of Modern Planetary Physics*, vol. 3, *Fruitful Encounters* (New York: Cambridge University Press, 1996), pp. 95–97.

188 *"this last condition quite eliminates the Kekulé static model"* Kathleen Lonsdale, "The Structure of the Benzene Ring in $C_6(CH_3)_6$," *Proceedings of the Royal Society of London* A123 (1929): 494–515.

188 *"have frequently brought confusion rather than clarity"* Fuson, "Some recent Advances in Theoretical Organic Chemistry," *Chemical Reviews* 7 (1930): 347–368.

7

The Light Quantum Hypothesis

7.1 BLACK-BODY RADIATION

The quantum hypothesis originated in physicist Max Planck's search for a mathematical formula to describe the distribution of energy over frequencies in black-body radiation (BBR). Black-body radiation is the continuous spectrum of electromagnetic radiation emitted by any substance in thermal equilibrium, apart from the line spectrum composed of a discrete set of frequencies characteristic of a particular substance. (By the mid-nineteenth century the distinction between three modes of heat transfer—conduction, convection, and radiation—had been established.)

In 1879, physicist Josef Stefan showed that according to available experimental data, the total energy emitted by of a hot body is proportional to the fourth power of its absolute temperature *T*. In 1884, physicist Ludwig Boltzmann derived the T4 law by a combination of thermodynamic and electromagnetic reasoning, showing that there must be a mechanical pressure as well as an energy density associated with electromagnetic radiation in order to satisfy the second law of thermodynamics, and using the relation between pressure and energy given by Maxwell's electromagnetic theory.

In 1896, Wilhelm Wien proposed a formula for the energy density as a function of frequency (ν) and absolute temperature (T):

$$\rho(v,T) = \alpha v^3 exp\ (-\beta v / T)$$

(where α and β are constants), and showed that this function fitted the experimental data available at that time.

7.2 PLANCK'S THEORY

During the years 1897–1899, Max Planck tried to construct a thermodynamic-electromagnetic theory of black-body radiation. Presumably his interest was because of (or at least stimulated by) the very active experimental group working on BBR in Berlin. Planck had been appointed professor of physics at the University of Berlin, and took on the job of editing for publication the job of editing the lecture notes of Gustav Robert Kirchhoff, his predecessor, for publication. Planck was a critic of atomistic theories, but as it happened Kirchhoff, though also skeptical of atomism, included the kinetic theory of gases in his lectures. Planck got involved in a minor polemic with Ludwig Boltzmann (the leading advocate of kinetic theory after Maxwell's death in 1879) about the derivation of Boltzmann's *H*-theorem in Kirchhoff's lectures, so it appears that he had to become an expert on the subject in self-defense. This expertise allowed him to develop a statistical interpretation of BBR and started him on the road to the discovery of the quantum theory (though he didn't quite get there).

At this point, we have to dispose of two myths about the origin of the quantum theory.

The first myth is the ultraviolet catastrophe. Lord Rayleigh had supposedly suggested that the BBR distribution corresponds to equal distribution of energy among all the modes of vibration of the ether, considered at that time to be an elastic solid. This would mean an infinite amount of energy goes into the high frequency vibrations—that is, ultraviolet and X-ray vibrations. This purely hypothetical result was called the ultraviolet catastrophe. Planck, according to the myth, proposed the quantum hypothesis in order to avoid the ultraviolet catastrophe.

This myth is easily demolished by simply reading the relevant 1900 papers of Rayleigh and Planck. Rayleigh clearly states that his formula applies only to low frequencies; he proposed an exponential factor that would reduce the energy at high frequencies and make the total energy finite. Planck does not even mention Rayleigh's paper, much less use its result as a reason to propose his own formula. Yet one still occasionally sees the myth in print, not just in popular works on history of science but also in the title of a children's book.

The second myth is more plausible, and seems to be still believed by some physicists, although it has been abandoned by most historians who have examined the original papers of Planck. To explain it we must look at what Planck did in 1900.

Planck postulated a system of atomic oscillators that could absorb and emit electromagnetic radiation. He then calculated the relation between the entropy S and energy U of an oscillator of any frequency, assuming Wien's formula is valid. In order to study the thermodynamic stability of his system he calculated the second derivative of *S* with respect to *U*, and found that it is inversely proportional to *U*; that is, its reciprocal is proportional to *U*. But then more accurate data showed that Wien's formula is not quite valid after all, so he proposed an empirical correction to the mathematical formula for the reciprocal of the second derivative, making it now proportional to *U* plus another term proportional to U^2. The result for the energy of an oscillator as a function of frequency and temperature was now:

$$\rho(\nu,T) = \left[8\pi\nu^2 / c^3\right] h\nu / (e^{h\nu/kT} - 1)$$

(where *h* is a new physical constant, now called Planck's constant). This is Planck's distribution law for black-body radiation. It turned out to give an excellent fit to the new experimental data, so he now had to figure out how to interpret his theoretical formula.

Planck's new formula corresponded to the following formula for the entropy of an oscillator with frequency <:

$$S = k\{(1 + U/h<) \log (1 + U/h<) - (U/h<) \log U/h<\}$$

where "log" means the natural logarithm. He now tried to interpret his formula using Boltzmann's formula $S = k \log W$, where W is the probability of a thermodynamic state; that is the number of microstates (defined by atomic configurations and speeds) corresponding to that macrostate (defined by variables like temperature, pressure, and density).

Planck assumed that the energy of each oscillator is an integral multiple of a basic energy element $h\nu$. That assumption is needed in order that the quantity W be finite, so that one can use the standard techniques of combinatorial analysis. If one considers a collection of N oscillators among which is distributed a total amount of energy $E = Ph\nu$, in such a way that each oscillator has an integral number of energy elements, then one can derive a formula for W that satisfies the new formula for S. The number of arrangements is

$$W=(N+P-1)!/P!(N-1)!.$$

where "!" means "factorial": $N! = N(N-1)(N-2)\ldots 1$. When N is large, this agrees with the above formula for the entropy S.

Did Planck intend to propose a *physical* quantization of energy at this point, or did he merely introduce energy elements for mathematical convenience in doing combinatorial calculations? Thomas Kuhn has argued that the latter is the case, and that the quantum discontinuity was first seriously proposed not by Planck but by Einstein and Ehrenfest, and only retrospectively associated with Planck's 1900 hypothesis. Kuhn's interpretation is now generally accepted by those historians of physics who have read Planck's 1900 papers.

There is one sentence in the second 1900 paper that seems to settle the matter: "If the ratio [of the total energy E to the energy element,] thus calculated is not an integer, we take for P an integer in the neighborhood." (See the notes to this chapter for context of this quotation.)

One additional piece of evidence is in Planck's Nobel lecture. Discussing the introduction of the quantum of action, he said that it could be interpreted in two ways. It might be just a

> fictional quantity, then the whole deduction of the radiation law was in the main illusory and represented nothing more than an empty nonsignificant play on formulae, or the derivation of the radiation law was based on a sound physical conception. . . . Experiment had decided for the second alternative. That the decision should be made so soon and so definitely was due not to the proving of the energy distribution law of heat radiation, still less to the special derivation of that law devised by me, but rather should it be attributed to the restless forward-thrusting work of those research workers who used the quantum of action to help them in their own investigations. The first impact in this field was made by A. Einstein.

The fact that Planck, as a result of a mathematical calculation, was led to a revolutionary conclusion that he himself did not intend or even accept (until much later) seems to be a striking example of what Eugene Wigner called "the unreasonable effectiveness of mathematics in the natural sciences." We have already noted that the Copernican and Newtonian theories of planetary motion were first accepted because they led to successful calculations, based on physical ideas (the Earth's motion, action at a distance) that were rejected at the time.

7.3 FORMULATION OF THE LIGHT-QUANTUM HYPOTHESIS

Planck's 1900 papers on the frequency distribution of black-body radiation are generally regarded as the beginning of the development of the old quantum theory. As noted in the last part of Section 3.5, this theory produced several novel predictions, which brought Nobel Prizes to the theorists who made the predictions and experiments who confirmed them. Here I focus on just one aspect of that theory: the hypothesis that light (and electromagnetic radiation in general) has a particle nature in addition to its well-established wave nature. This light quantum hypothesis (LQH) was first proposed by Albert Einstein in 1905, and, as shown by Thomas Kuhn, it marked the introduction of the quantum as a physical entity rather than just a mathematical device.

In the rest of this chapter I describe the reception of Albert Einstein's light quantum hypothesis by the European-American physics community. How and when did it become generally agreed that light (and electromagnetic radiation in general) has a particle character along with its well-established wave character?

This section introduces Einstein's hypothesis and relates it to the physics of its time. Section 7.4 reviews the reasons why the wave theory of light was accepted at the beginning of the twentieth century. Sections 7.5, 7.6, and 7.7 present the theoretical proposals of Einstein and Compton and their experimental testing by Millikan, Compton, and others; these sections are relatively brief because several other historians have already written excellent accounts. Section 7.8 takes up at greater length the reception of the LQH and other particulate theories *before* the publication of Compton's definite work in 1923, in an attempt to resolve an old question that is central to this chapter: what consensus, if any, had the physics community reached? This is still controversial among historians, and more evidence is needed. Section 7.9 reviews the acceptance of the LQH *after* 1923, and calls attention to a new question: if physicists agreed that the Compton effect, along with the refutation of the Bohr-Kramers-Slater alternative hypothesis, provided the strongest evidence, why did physics textbooks continue to give equal or greater weight to the photoelectric effect? Section 7.10 gives an example of the phenomenon mentioned in Chapter 1: too much faith in predictions may encourage fraudulent attempts to confirm them. Conclusions are presented in Section 7.11.

Albert Einstein wrote in 1905 that "the wave theory of light . . . has been excellently justified for the representation of purely optical phenomena and it is unlikely ever to be replaced by another theory." But "optical observations refer to time averages and not to instantaneous values," and may not apply to "phenomena of the creation and conversion of light." Instead, "black-body radiation, photoluminescence, the production of cathode rays by ultraviolet light, and other phenomena . . . can be better understood on the assumption that the energy of light is distributed discontinuously in space." After reviewing the theory of black-body radiation and comparing it with the theory of systems of molecules in gases and in dilute solutions, using Boltzmann's statistical theory of entropy, Einstein discussed photoluminescence: "Consider monochromatic light which is changed . . . to light of a different frequency." If we assume that "both the original and the changed light consist of energy quanta" whose magnitude is proportional to their frequency, it is clear that "the energy of a final light quantum can, according to the energy conservation law, not be larger than that of an initial light quantum," and hence its frequency must be lower; "this is the well-known Stokes's rule." Then Einstein applied his quantum hypothesis to photo-electric phenomena and derived an equation stating that the maximum kinetic energy of the electron ejected from a solid by light must be equal to the energy of one quantum of light minus a constant characteristic of the solid (which may be related to the work needed to free a single electron from a molecule). From this equation he predicted that the maximum kinetic energy of a photoelectron should vary linearly with the frequency of the light striking the metal, since the photoelectron is ejected as a result of being hit by just one quantum of light.

Einstein later showed that Planck's law could be derived from the assumption that the radiation is in thermal equilibrium with molecules whose internal energy is distributed over a discrete set of states, as postulated in Bohr's 1913 theory of the hydrogen atom. In this important paper, first published in 1916 and reprinted in a more widely circulated journal in 1917, he introduced the concept of stimulated emission of radiation, which provided (many years later) the theoretical basis for the laser. He also showed (as had been stated earlier by Johannes Stark) that the light quanta carry not only energy $h\nu$, but also momentum $h\nu/c$ in a definite direction.

According to Christa Jungnickel and Russell McCormmach, "[T]he reasoning behind Einstein's proposal of light quanta in 1905 did not convince his colleagues at the time: Planck, Laue, Wien, Sommerfeld, and other early supporters of Einstein's relativity theory all rejected his hypothesis of light quanta. Their principal argument was that interference phenomena—and also diffraction, refraction, and other phenomena of physical optics—demanded a wave interpretation of light." Lorentz "was also unpersuaded . . . German physicists did not see the need for so 'radical' a step, as Planck put it. . . . Only one other leading German physicist advocated light quanta at this time, the experimentalist Johannes Stark." Niels Bohr became a strong opponent of the LQH, although his atomic theory assumed that electrons absorb or emit radiation in quantized amounts in going from one state to another. But Henri Poincaré in France and Paul Ehrenfest in Russia argued that the existence of quanta could be deduced directly from Planck's law.

Another reason for giving some consideration to particle theories was the well-known fact that X-rays seemed to behave like particles in some respects; yet they also behaved like light, as was shown by the discovery of X-ray diffraction. By 1920 it was generally agreed that infrared, light, ultraviolet, X-rays, and gamma ways are all part of a single electromagnetic spectrum: they are all the same kind of entity, differing from each other only in the value of quantitative parameter, wavelength, or frequency. Hence if X-rays can behave like particles as well as like waves, so can light.

7.4 THE WAVE THEORY OF LIGHT

As is well known, the Newtonian particle theory of light was replaced in the early nineteenth century by a wave theory, as a result of the work of physicist and physician Thomas Young and physicist Augustin Fresnel. To understand the reception of the light quantum hypothesis, we need to consider which properties of light were believed, a century later, to prove its wave nature. One of them, as indicated in the last few words of the song quoted at the beginning of Section 7.3, was *interference*: the fact that the intensity of a pattern formed by combining two beams of light could in some places be *less* than the intensity of either one observed separately, as well as *more* in other places. This can easily be explained by the wave theory, if the waves are vibratory motions of a medium and those motions can either reinforce or cancel each other. The particle theory seemed to offer no explanation at all for this phenomenon.

Another property is the *speed of light in substances of different density.* According to the Newtonian particle theory, light should travel faster in a denser medium (water or glass) than in a rarer one (air or vacuum), while according to the wave theory as developed by physicist and astronomer Christiaan Huygens, the opposite should be the case. This experiment could not be done until the middle of the nineteenth century, because of the difficulty of measuring the speed of light in a terrestrial laboratory. When it was done in 1850 by physicists A.H.L. Fizeau and J.B.L. Foucault, the result was unequivocally in favor of the wave theory. But that was too late to have any impact on the debate, since by that time almost all physicists had already accepted the wave theory for other reasons. Nevertheless, it was used in early twentieth-century textbooks as a justification for the wave theory, since it seemed clearly to refute the particle theory and was perhaps considered easier to explain to students than the evidence from interference.

James Clerk Maxwell's electromagnetic theory of light predicted the existence of waves longer and shorter than visible light. Physicist Heinrich Hertz's experimental confirmation of this prediction was regarded as support for the existence of an ether, but not for any comprehensible physical mechanism; Hertz could not understand Maxwell's theory and simply declared that Maxwell's theory *is* Maxwell's equations.

Another successful prediction was radiation pressure, confirmed by physicist P. N. Lebedev in Russia and by physicists E. F. Nichols and G. F. Hull in the United States around 1900. (This would probably have been regarded as evidence for a particle theory, if Maxwell's theory had not predicted it.)

At the beginning of the twentieth century the wave theory was still generally accepted by physicists. The only disagreement was on the physical nature of the waves and of the medium in which they travel. The Fresnel-Young model of vibrations in a space-filling ether was somewhat implausible, because the ether would have to be an elastic solid in order to support the transverse waves needed to account for polarization. Maxwell's electromagnetic wave theory, while somewhat more abstract, seemed to avoid this difficulty and thus was usually employed in more advanced works. A few authors asserted that there was no satisfactory theory of light since neither the elastic solids nor electromagnetic fields really *explain* light; they just give *equations* from which observable results can be computed. Surprisingly, several texts presented the Newtonian particle theory as an alternative because it provided a simpler explanation of rectilinear propagation and shadows, even though it had to be (reluctantly) abandoned because the more complex wave theory was required in order to explain more complex optical phenomena.

One of the first indications that the wave theory might need to be amended came from studies of the ionization of gases by X-rays. Only a very small number of the gas molecules emit electrons. If the X-rays are electromagnetic waves, continuously filling the space occupied by the gas, one would expect that all (or none) of the molecules would be ionized. This kind of phenomenon led J. J. Thomson, before Einstein proposed the LQH, to suggest a particulate theory. His particles were not tiny spheres or parallelepipeds, as in Newton's theory, but tubes of electric force as conceived by Michael Faraday. He did not reject Maxwell's electromagnetic wave theory, but suggested that the wave front, rather than being continuous, might be speckled by regions of high intensity.

7.5 EINSTEIN'S HEURISTIC VIEWPOINT

In the 1905 publication summarized in Section 7.3, sometimes misleadingly called his photoelectric effect paper, Einstein proposed the following equation for the maximum kinetic energy of an electron ejected from a solid by a single quantum of energy hv, assuming that the energy needed to bring it from inside to the surface of the solid is p:

$$\tfrac{1}{2}Mv^2 = hv - p$$

(this is the notation used by Millikan and others in the 1910s and later). I will call this Einstein's photoelectric equation; it amounts to a *prediction* to be tested by experiment. According to historian Roger Stuewer, the empirical data available in 1905 were not adequate to confirm it, so it should be regarded as a *novel* prediction.

Einstein proposed, according to the title of his paper, a heuristic viewpoint—an exploratory approach intended to uncover new facts without any commitment to a definite theory. He clearly did not want to *replace* the wave theory by a particle theory, although he implied that the wave theory might be derivable from a particle theory. He did not, in 1905, propose what was later called wave-particle duality or complementarity, which requires that waves and particles coexist on the same ontological level. But he did, in a famous letter to Konrad Habicht, refer to his light quantum hypothesis as "very revolutionary"—a term he did not use to describe his relativity theory.

Here is Karl K. Darrow's insightful definition of the word "heuristic" as used by Einstein:

> [It seems to] describe a theory which achieves successes though its author feels at heart that it really is too absurd to be presentable. The implication is, that the experimenters should proceed to verify the predictions based upon the idea quite as if it were acceptable, while remembering always that it is absurd. If the successes continue to mount up, the absurdity may be confidently accepted to fade gradually out of the public mind.

This definition could apply to the quantum theory as a whole: it started out as a set of formulae obtained by juggling equations without taking too seriously the physical principles implied by those equations. One simply follows the "philosophy of as if" described by philosopher Hans Vaihinger. When the equations gave results in agreement with empirical observations, the physics had to be revised in order to agree with the equations.

7.6 WHAT DID MILLIKAN PROVE?

During the decade after Einstein proposed his Light Quantum Hypothesis, and the photoelectric equation derived from it, empirical evidence gradually accumulated for the equation but the hypothesis was not generally accepted. Physicists argued that Einstein's equation could be just as well explained by other hypotheses that were compatible with the wave theory of light.

Robert A. Millikan provided in 1916 the most definitive experimental proof of the equation; he also found that the value of Planck's constant *h* obtained from his experiments was the same as that deduced from Planck's law of black-body radiation. This could be seen as a link between the LQH and the more general quantum theory, which had been successfully applied to the specific heats of solids (Einstein, Debye) and the spectrum of hydrogen (Bohr). Yet he also expressed an uncompromising rejection of the hypothesis from which the photoelectric equation had been derived. He wrote:

> We are confronted . . . by the astonishing situation that [the facts of photoelectric phenomena] were corrected and accurately predicted nine years ago by a form of quantum theory which has now been pretty generally abandoned. . . . the semi-corpuscular theory by which Einstein arrived at this equation seems at present to be wholly untenable.

Millikan has been criticized for refusing to accept the light-quantum hypothesis that his own experiments confirmed, and some historians have tried to explain his puzzling behavior. But was it really so puzzling? From a logical point of view an experiment cannot *confirm* a hypothesis, unless you can prove that no other hypothesis could lead to the same empirical result (recall the fallacy of affirming the consequent, Section 1.2). In this case there were indeed several other hypotheses that could explain Einstein's equation. More importantly, it does not make sense to abandon a hypothesis like the wave theory of light, which made several confirmed predictions and explained most of the observable properties of light, in favor of a hypothesis that is credited with only one confirmed prediction, along with plausible explanations of a few other phenomena, but *fails* (as of 1916) to explain wave properties like interference, or the fact that light travels faster in a less dense medium. Isn't it unreasonable to fault Millikan for refusing to accept what we *now* consider to be the right answer, even though some of the best evidence for the LQH had not yet been uncovered in 1916?

Einstein received the Nobel Prize in 1921. Physicists are often surprised to learn that he did not get it for relativity, but for quantum theory. With the recent recognition by historians that Einstein rather than Planck was the originator of the *physical* quantum hypothesis (see above, Section 7.2), the decision of the Nobel Prize electors makes a little more sense. But why would they give him the prize for a theory that was not yet accepted?

The answer is, they didn't. Although the original draft citation mentioned Einstein's *theory* of the photoelectric effect, it was changed to Einstein's *equation*. To have discovered the quantitative nature of the phenomenon was important enough to deserve the prize even if the discovery was made with the help of what Bohr and others still considered a dubious theory.

7.7 THE COMPTON EFFECT

Between 1905 and 1922, other empirical evidence both for and against the light quantum hypothesis emerged. In addition to the large amount of research on the photoelectric effect, Eddington's announcement that he had confirmed Einstein's light-bending prediction gave more credibility to the latter's assertion that the luminiferous ether is unnecessary (and its putative properties are even more implausible than they seemed in the nineteenth century). Hence, proponents of light waves could not rely on a material medium to propagate those waves, a difficulty not faced by corpuscularists. Moreover, the fact that gravity can act on light is easy to understand if light consists of particles that have mass, but mystifying if light is simply a wave motion. (The light quantum does not have mass in the ordinary sense, but does have an effective mass determined by its frequency.)

As historian Roger Stuewer described in detail in 1971, the establishment of the corpuscular nature of electromagnetic radiation in the 1920s was not simply the outcome of research on black-body radiation and the photoelectric effect; instead, it owed much to the study of X and gamma rays. Many of the X-ray phenomena were most easily explained, by G. G Stokes and others, by assuming that the rays consisted of localized pulses rather than continuously extended waves. It was observed that when X-rays were scattered from matter, some of the secondary scattered rays were softer (less penetrating) than the primary rays.

In May 1923, Compton published a 20-page paper in the *Physical Review* laying out his theory of the scattering of X-rays by electrons, with supporting experimental data. The essence of the Compton effect, as this kind of scattering quickly came to be called, is that one can calculate both the change in wavelength of the scattered X-ray and the momentum of the recoil electron by treating both as particles with specified energy and momentum, each of these two quantities (summed over all particles) being conserved in the collision. The increase in wavelength of the X-ray is a simple function of the angle θ between the incident and scattered ray:)$\Delta\lambda = (2h/mc)\sin^2(\frac{1}{2}\theta)$, independent of the wavelength. Moreover,

> any particular quantum of X-rays is not scattered by all the electrons in the radiator, but spends all of its energy upon some particular electron. This electron will in turn scatter the ray in some definite direction, at an angle with the incident beam. This bending of the path of the quantum of radiation results in a change in its momentum. As a consequence, the scattering electron will recoil with a momentum equal to the change in momentum of the X-ray.

According to Compton, "the electrons which recoil in the process of the scattering of ordinary X-rays have not been observed." Within two months of the publication of Compton's paper, physicist C.T.R. Wilson reported the observation of these recoil electrons, using his new cloud chamber method. Compton immediately pointed out that Wilson's observation confirmed his prediction.

Compton now began a campaign to prove that his effect provided stronger evidence for the light-quantum hypothesis than did the photoelectric effect. In addition to the fact that the Compton effect confirms the conservation of momentum as well as energy while the photoelectric effect involves only energy, he argued that his theory produced a confirmed *novel prediction*, the existence and properties of *recoil electrons*, whereas Einstein's photoelectric hypothesis did not:

> In view of the fact that these recoil electrons were unknown at the time this theory was presented, their existence and the close agreement with the predictions as to their number, direction and velocity supplies strong evidence in favor of the fundamental hypotheses of the quantum theory of scattering.

His experiments show

> that X-rays, and so also light, consist of discrete units, proceeding in definite directions, each unit possessing the energy *hL* and the corresponding momentum *h/8*. So in a recent letter to me Sommerfeld has expressed his opinion that the discovery of the change of wavelength of radiation, due to scattering, sounds the death knell of the wave theory of radiation.

In a popular article, Compton reviewed the evidence that the wave theory of light should be revised. Einstein was credited with reviving "the old Newtonian idea of light corpuscles in the form of quanta," but

> since the idea of light quanta was invented primarily to explain the photoelectric effect, the fact that it does so very well is no great evidence in its favor. The wave theory explains so satisfactorily such things as the reflection, refraction and interference of light that the rival quantum theory could not be given much credence unless it was found to account for some new theory for which it had not been especially designed. This is just what the quantum theory has recently accomplished in connection with the scattering of X-rays.

The wave theory, Compton pointed out, predicts that scattered X-rays will have the same wavelength as the primary (incident) rays. Quantum theory explains why some of them have longer wavelengths and predicts the existence of recoil electrons, later discovered by Wilson and confirmed by Compton's group, which also found that the number of cloud chamber tracks, their direction, and range agree with the predictions of quantum theory.

Compton wrote

> "Since their very existence was unknown before they were predicted by the quantum theory, these recoil electrons must be taken as a strong support of the theory of radiation quanta."

In a paper in the *Physical Review*, Compton stated his claim in a different way: there are now several phenomena most simply explained by Einstein's LQH, but none that necessarily demand it. Thus the photoelectric effect can be explained by wave theory if you postulate a mechanism inside the atom to store energy until a quantum is received. But noncorpuscular explanations of the Compton effect, while possible, are not plausible.

This argument was apparently not very convincing, so Compton went back, in three later publications, to his previous assertion that the discovery of recoil electrons confirmed a novel prediction whereas Einstein's LQH merely explained known facts, and thus was not as strong evidence as that from the Compton effect. The denigration of the photoelectric effect was somewhat unfair, since according to Stuewer, Einstein's light quantum hypothesis was "a necessary consequence of very fundamental assumptions: in no sense did he propose it in an ad hoc fashion to 'explain' certain experiments. . . . [his] prediction that the maximum photoelectron energy depends linearly on the frequency of the incident radiation" was bold, since "the experimental situation was highly uncertain" and other (nonlinear) relations were being proposed. McCormmach argues that Einstein's theory was more successful in winning support than Thomson's largely because the former made quantitative predictions while the latter did not. Moreover, Millikan also found that the constant *h* in the photoelectric equation has the *same numerical value* as that deduced from other phenomena such as black-body radiation.

But Compton might have been correct in thinking that a prediction of a qualitatively new phenomenon (recoil electrons) would count as better evidence than the quantitative refinement of a qualitatively known phenomenon. From my viewpoint the relevant question (especially for those who want to know whether novel predictions are better evidence than retrodictions) is: did other physicists accept Compton's claim about recoil electrons?

Two physicists who clearly did not accept that claim were Niels Bohr and H. A. Kramers. They were so desperate to rescue the wave theory of light that they were willing to give up the absolute validity of the laws of conservation of energy and momentum in interactions between X-rays and electrons. Following a suggestion of C. G. Darwin and with the somewhat reluctant assistance of John C. Slater, they developed a theory that reduced those laws to statistical averages, denying a direct causal connection between the incident X-rays and the scattered X-rays and electrons. Their theory disgusted Einstein so much that he exclaimed, in a famous statement whose context is often forgotten, that if it were true he would rather be a cobbler than a physicist. This is a precursor of his even more famous assertion, "God does not play dice."

This new attack on the LQH, called the "Copenhagen putsch" by Wolfgang Pauli, yielded a new novel prediction that could be directly tested. The test was conducted by Walther Bothe and Hans Geiger, and by Compton and A. W. Simon; both groups concluded that the Bohr-Kramers-Slater theory was wrong, and reconfirmed the reality of light quanta. (But the theory did have some influence on the later development of quantum mechanics.) Bohr now abandoned his opposition to the LQH and invented a new concept, complementarity, to explain how (or rather, assert that) pairs of apparently incompatible concepts such as waves and particles can both be valid at the same time.

Compton and Wilson shared the 1927 Nobel Prize in physics. One might think that this award completed the triumph of the LQH. But the award to Compton was not for establishing the particle character of electromagnetic radiation, but simply for his discovery of the effect named after him. And Compton was, at least for a few minutes, robbed of his triumph by another putsch, this one engineered by Erwin Schrödinger. In the presentation speech, Manne Siegbahn stated:

> Compton deduced a new kind of corpuscular theory, with which all experimental results showed perfect agreement within the limit of experimental error . . . this theory predicts recoil electrons . . . It was a triumph for both parties [Compton and Wilson] when these recoil electrons were discovered by Wilson's experimental method both by Wilson himself and, independently, by another investigator. Hereby the second chief phenomenon of the Compton effect was experimentally verified [the first was the change in wavelength of the scattered X-rays], and all observations proved to agree with what had been predicted in Compton's

> theory. . . . The Compton effect has, through the latest evolutions of the atomic theory, *got rid of the original explanation based upon a corpuscular theory. The new wave mechanics, in fact, lead as a logical consequence to the mathematical basis of Compton's theory*. Thus the effect has gained an acceptable connection with other observations in the sphere of radiation (emphasis added).

Of course Compton, in his acceptance speech, made it quite clear that his discovery was, indeed, that "all electromagnetic radiation is constituted of discrete quanta proceeding in definite directions." For most physicists, this became an established fact, and the only remaining question was how much credit Compton should receive for establishing it. But for a few, the apparent particle behavior of radiation could be reduced to its true wave nature in accordance with Schrödinger's theory or, for more dogmatic antirealists, the wave-particle controversy itself could be declared a nonissue.

7.8 RECEPTION OF NEO-NEWTONIAN OPTICS BEFORE 1923

In 1918 Einstein wrote to Michele Besso: "I no longer have doubts about the *reality* of quanta in radiation, though I'm still quite alone in this conviction." More recently, seven well-known historians of physics—Martin Klein, Helge Kragh, Thomas Kuhn, Jagdish Mehra, Abraham Pais, Helmut Rechenberg, and Roger Stuewer—stated that most or nearly all physicists rejected the light quantum hypothesis before the discovery of the Compton effect. This is also the view of the Historic Sites Committee of the American Physical Society's Forum on the history of physics, as engraved on a plaque at Washington University in St. Louis.

Einstein and the historians may well be correct, but I wish they had provided a little more evidence to support this assertion. Max Planck's opinion that Einstein had "overshot the target" in some of his speculations, in particular his LQH, is often quoted, but does not necessarily represent the consensus of the physics community. By contrast, historians Jungnickel and McCormmach made a more limited claim (quoted in Section 7.3): that "Planck, Laue, Wien, Sommerfeld and other early supporters of Einstein's relativity theory" rejected the hypothesis; and, more importantly, these historians backed up their claim with specific references to the writings of the four named physicists. Also, John Stachel has provided ample evidence that Bohr rejected the photon concept until the mid-1920s, and even after the discovery of the Compton effect he insisted on the primacy of the wave nature of light.

Why do we want to know who the opponents as well as the supporters of the LQH were? First, in order to judge the claim that the LQH was not accepted until after the discovery of the Compton effect; second, in order to confirm that those physicists who did accept it after 1923 did so because of the evidence from the Compton effect, and to determine how much the confirmation of Einstein's photoelectric equation also counted. I recognize that questions of the type "was X accepted at time T? Why?" cannot be answered simply by counting votes on each side, since some votes are obviously more important than others (and will have a greater influence on other voters). I am not going to propose a definitive answer based on my own estimate of the importance of early twentieth-century physicists, but will leave that to the judgment of readers. However, I do think it is possible to confirm or refute the statement that "nearly all physicists did *not* accept X during a time interval from T_a

to T_b" by examining a reasonably large sample of publications during that interval. In my experience it is more effective to focus on monographs, review articles, and textbooks than on research articles, because the former are more likely to make statements about the nature of light and give reasons for those statements.

Of course we also want to know about the early support for Einstein's theory, even if it came from a minority of physicists. One of the first challenges to the historiographic consensus came from a Russian philosopher, Rinat H. Nugayev. Nugayev disputed the views of Klein and Pais expressed at an Einstein centennial meeting, pointing out that they had been challenged by two physicists who described their own experiences in the 1910s, H. D. Smyth and Walther Gerlach. Smyth recalled that the particle nature of light was accepted at Princeton in 1918–1919, while Gerlach remembered that the discovery of X-ray diffraction "enlivened the discussion about Einstein's light quantum theory."

A recollection that reinforces Smyth's but was published much closer to the time was that of Karl K. Darrow, a physicist who was active in popularizing the quantum theory in the 1920s and 1930s. Darrow wrote in 1937:

> For anyone who studied physics in the years just before the war . . . [the photoelectric effect] was the *pièce de conviction*, the grand piece of evidence which undeniably spoke for the corpuscular nature of light . . . How enthusiastically our teachers used to speak of it! How strongly they used to stress those of its features which harmonized with the corpuscular theory of light, but apparently not with the undulatory.

Historian Alexei Kozhevnikov argues that there was

> a general change of attitude in physics following the end of World War I . . . by 1920 light quanta grew out of oblivion into an extremely popular concept and began to be widely understood as particles or corpuscles. Traditional historiography saw the explanation of this change in the discovery of the Compton effect in 1923, but the development had already been in place for several years before that and was crowned by, rather than caused by, Compton's landmark achievement. . . . Rather than being caused by new experimental or theoretical developments, the revival of light quanta appears more like a shift in the prevailing fashion among physicists. . . . Most of the authors who started using this concept soon after the end of the war actually belonged to a younger generation who also favored different approaches to physical problems.

Kozhevnikov points to Henry Small's analysis of citations in 16 major physics journals in the 1920s, showing that the annual rate of citations of Einstein's 1917 paper on the quantum theory of radiation was rising in the early 1920s; it was one of the most frequently cited papers in the decade, second only to Compton's 1923 paper.

Going back a decade, we learn from McCormmach that "in Europe at this time [1907–1910], the interpretation of the quantum theory as one of light quanta was held only by a very small minority, while in Britain the situation was reverse and nearly everyone who had any point of view at all considered the theory to be based on an atomic constitution of radiation, or of energy in general" (he mentions Larmor, Schuster, and Jeans.)

From Roger Stuewer's comprehensive book on the Compton effect, I infer that the *major* physicists were almost evenly divided. On the procorpuscular side were (in addition to Einstein) Louis de Broglie, William H. Bragg, James Jeans, Henri Poincaré, Erwin Schrödinger, and Johannes Stark; opponents included Niels Bohr, H. A. Lorentz, O. W. Richardson, Arnold Sommerfeld, and J. J. Thomson. But Thomson, included on that list as an opponent of Einstein's LQH, held views that were (or were often seen as) corpuscularian. Sommerfeld by 1922 was on the verge of accepting the LQH. Richardson was struggling to understand how radiation "behaves as though it possessed at the same time the opposite properties of extension and localisation." H. A. Kramers and J. C. Slater might be considered opponents on the basis of their coauthorship of the Bohr-Kramers-Slater paper (see next section), but both were more favorable before they came under Bohr's influence.

While Lorentz in his public statements was critical of the LQH, he described its advantages as well as its disadvantages in a long 1909 letter to Einstein. He concluded:

> It is a real pity that the light quantum hypothesis encounters such serious difficulties, because otherwise the hypothesis is very pretty, and many of its applications that you and Stark have made of it are very enticing. But the doubts that have been raised carry so much weight with me that I want to confine myself to the statement: "If we have a ponderable body in a space enclosed by reflecting walls and filled with ether, then the distribution of the energy between the body and the ether proceeds *as if* each degree of freedom of the ether could take up or give off energy only in portions of the magnitude $h\nu$." As you see, not much is gained thereby, the "*as if*" would have to be elucidated through further analysis.

It is probably not fruitful to focus narrowly on the reception of the LQH while ignoring the other corpuscular theories of radiation discussed in the early twentieth century. Some of those theories were first proposed to explain X-rays, which were not definitely known to be adequately described by Maxwell's electromagnetic wave theory until the discovery of X-ray diffraction in 1912. Before that it was reasonable to suppose that X-rays are corpuscular in nature, on the basis of their known properties. But after 1912 it was more reasonable to suppose that if X-rays and light are essentially the same phenomenon, differing only by having different values of a numerical parameter (wavelength or frequency), then all the arguments for the corpuscular nature of X-rays would also imply a corpuscular nature for light. Otherwise, if X-ray diffraction (or another wave property) had not been discovered until after 1923, the Compton effect would not initially have been considered a proof that *visible* light is corpuscular.

The relevant question is then: which scientists were supporters or opponents of the corpuscular nature of light (not necessarily limited to the Einstein LQH) before 1923? Opponents would generally insist on the absolute validity of the wave theory of light, while supporters would argue that the wave theory, while adequate to account for many aspects of light such as interference, diffraction, and polarization, failed to explain several newly discovered properties of electromagnetic radiation, and therefore had to be modified in some way.

I have surveyed supporters and opponents of a corpuscular aspect of electromagnetic radiation based on publications in the period 1916–1922. There were 28 supporters, including well-known physicists such as W. H. and W. C. Bragg, A. S. Eddington,

James Jeans, H. A. Kramers, Walther Nernst, Erwin Schrödinger, Johannes Stark, and J. J. Thomson—three of them (Eddington, Kramers, and Schrödinger), who became opponents, at least temporarily, after 1922.

On the other side were 18 opponents, including Niels Bohr, Max Born, A. H. Compton, K. T. Compton, Peter Debye, Max Laue, H. A. Lorentz, R. A. Millikan, Max Planck, and Arnold Sommerfeld. Two of them (including A. H. Compton) were converts who became supporters after 1922.

Is this a case of Planck's principle—are the supporters of the new idea younger that the supporters of the old? Yes, on the average: mid-forties compared to mid-fifties—but the opponents are not ready to retire.

Perhaps the best example (other than Einstein) of a physicist who strongly supported the LQH before 1923 is Fritz Reiche. Reiche showed explicitly what was vaguely alluded to in much of the literature I have examined: the evidence for the corpuscular nature of light, and for the LQH in particular, did not come from just one phenomenon like the photoelectric effect; it came from many experiments and theoretical calculations, all pointing in the same direction. Even if the corpuscular theory could not yet explain as many phenomena as the wave theory, it was moving ahead rapidly and would soon take the lead.

In the terminology of philosopher Imre Lakatos, it was a "progressive research programme," or as philosopher Gonzalo Munevar expressed it, it offered "promise more than performance." Millikan recognized this fact in his 1924 Faraday lecture: he asserted that although he could not accept the LQH even after the discovery of the Compton effect, "the times are, however, pregnant with new ideas, and atomic conceptions in the field of ether waves seem to hold at the moment the master-key to progress."

7.9 THE IMPACT OF COMPTON'S DISCOVERY

Niels Bohr and a few other leading physicists were persuaded to abandon their opposition to the light quantum hypothesis by the Compton effect, after the confirmation of Compton's predictions and the refutation of the Bohr-Kramers-Slater theory. Here I want to go beyond the leaders to see how other physicists reacted to this new evidence, and whether they gave it more or less weight than the confirmation of Einstein's theory of the photoelectric effect.

One measure of the impact of Compton's paper is the number of citations it received. As noted above, according to Henry Small it was the most frequently cited paper in 16 major physics journals in the decade 1920–1929. There were 78 citations in seven years, but this number underestimates its impact; by 1926 it was not even necessary to give a citation to Compton's original paper when discussing his effect. Yet the texts of these papers tell us little or nothing about the relative importance of the photoelectric effect and the Compton effect in persuading the author to accept the light quantum hypothesis.

Two of the earliest books to educate physicists about the Compton effect were Millikan's second edition of *The Electron* and Sommerfeld's fourth edition of *Atombau und Spektrallinien*, both published in 1924. Millikan called it "the best evidence yet found in favor of Einstein's hypothesis of localized light quanta"—a hypothesis that "is having new and remarkable successes" despite the difficulty of reconciling it with the wave properties of light. Sommerfeld wrote that the Compton effect was the most important discovery that could have been made in the present state of physics, one that had changed his own views in the direction of the extreme light quantum hypothesis. The great prestige of these physicists and their previous skepticism about the reality of light quanta must have made a deep impression on many readers.

In 1926 another important physicist, Walther Gerlach, published a graduate text in German, followed by an English translation in 1928. He asserted that "the Compton effect more than all other quantum phenomena necessitates the assumption of light quanta and their directed emission."

I did not find any discussion of the Compton effect in undergraduate textbooks published before 1928. In that year two books (one in German, one in English) recognized it as important evidence for the LQH, more or less comparable to the photoelectric effect. Two others mentioned the Compton effect, but seemed to consider it somewhat weaker evidence. In 1929 two more authors featured the Compton effect as being perhaps the best evidence for the particle nature of light: one was an American physicist writing on the history of physics, the other a German physicist giving guest lectures at Ohio State University. A Dutch Jesuit was one of the first to state explicitly that the Compton effect "shows even more clearly [than the photoelectric effect] that light has an atomic structure," in a German book published in 1929 and translated into English in 1930.

I have attempted to examine all the books and review articles by physicists in English, German, or French, in the 30 years following Einstein's publication of his light quantum hypothesis, to see whether the authors accepted or rejected it, and what reasons they gave. A few publications by chemists and astronomers were included. Publications by Einstein and A. H. Compton are not included, since they were promoters rather than receivers of the theory. This summary is based on more than 250 books and articles that appeared in the two decades beginning with Millikan's confirmation of Einstein's photoelectric equation: 1916–1935. They have been divided into two major categories: (1) monographs and review articles, directed to an audience of physicists; (2) textbooks and popular articles, directed to students and the public. The viewpoint of each publication was assigned to one of the following categories: strongly supports LQH (or other corpuscular theory), leans toward LQH, neutral, leans against LQH (or doesn't mention it but supports the wave theory of light), and strongly rejects the light quantum hypothesis. The distinction between LQH and other corpuscular theories, which is significant before 1921 (see previous section), is mostly ignored by writers in this later period; I noted only a handful who rejected the LQH but supported another corpuscular theory.

Besides the photoelectric effect and the Compton effect, there were several other possible reasons to accept the LQH.

My survey shows that for 1916–1920, monographs (including technical reviews) were about evenly split between supporters of a particulate character of light (along with its wave properties) and those who rejected it (or did not mention it while endorsing the wave theory of light). A slim majority (about 57%) of the monographs published in the period 1921–1925 favored the LQH or a similar corpuscular theory, and this is true even though only 5% mentioned the Compton effect as evidence for it; most gave the photoelectric effect as the only evidence.

The majority in favor of the LQH grew to about 84% in 1926–1930; about 58% of all authors specified the Compton effect as being either stronger evidence than the photoelectric effect, or at least as strong. 35% mentioned the photoelectric effect as being better evidence.

In the final half-decade, 1931–1935, the balance shifted even more strongly toward the LQH, favored by about 92%. Now a slightly higher proportion (62%) of all authors supported the Compton effect, while 54% mentioned the photoelectric effect.

Among textbooks and popular articles published in 1916–1920, only one (by Comstock and Troland) out of 18 favored the light quantum hypothesis. In 1921–1925 it was supported by almost one-third (31%). This increased to about 70% in 1926–1930, and to 84% in 1931–1935. But now we see the beginning of a split between the monographs and the textbooks regarding the reason for adopting the hypothesis. In the half-decade 1926–1930, only 28% of textbook authors mentioned the Compton effect as evidence (stronger than or as strong as the photoelectric effect) for the LQH, compared with 62% of monograph authors. In the same period 52% of textbook authors mentioned the photoelectric effect compared with 35% of monograph authors.

In the half-decade 1931–1935, 84% of textbook authors favored the LQH. But only 49% mentioned the Compton effect, while 55% cited the photoelectric effect as a reason for supporting the LQH. This was about the same proportion as among the monograph authors (54%). While the gap between monographs and textbooks might seem to be narrowing with time, it should be noted that I found only a few (13) monographs published in 1931–1935, compared with the much larger number of textbooks (55), so the data for monographs may not be representative.

My provisional conclusion is that starting around 1926, when the Compton effect was probably fairly well known to most physicists active in research or teaching, authors of books and reviews directed to physicists were more likely to call it the most important evidence for the LQH than were the authors of textbooks and popular articles, who tended to cite the photoelectric effect more often.

Why the difference? My guess is that the Compton effect was considered more elegant physics. It is a direct application of the beloved conservation laws for energy and momentum, and it involves no adjustable parameters. From a minimum of assumptions it gives you maximum results (a simple formula for the change of wavelength of the X-ray, and relations between the scattered X-ray and the recoil electron). It combines the best features of classical physics with the one formula of quantum theory that is familiar (though not necessarily comprehensible) to all physicists, $E = h\nu$ along with its relativistic corollary, p [momentum] = $h\nu/c$. Moreover, it survived a dramatic challenge from one of the most authoritative physicists in the world (Bohr).

The photoelectric effect, on the other hand, gives you only one result, the maximum energy of the ejected electron, at the cost of introducing a variable parameter (the energy needed to bring the electron to the surface of the metal), and it took lots of tedious work to nail down that result.

But it's much easier to explain the photoelectric effect to students. You don't have to do any algebra or trigonometry. Moreover, it has an interesting practical application that should be familiar to almost all students: the electric eye that automatically opens a door when you approach it, or prevents an elevator door from hitting you.

It is not obvious that physicists would have accepted the LQH on the basis of either the photoelectric effect or the Compton effect alone, or that acceptance would have come earlier if the chronological order of the discovery of the two effects (and their theoretical explanation) had been reversed. One exception to the wave properties of light, no matter how elegant or well-documented, probably would not have been enough. The photoelectric effect, along with corpuscular theories based on X-ray research, created doubts about the absolute validity of the wave theory of light but not enough to overthrow it. It took at least two discoveries, both of which could be explained by the same hypothesis, to tip the balance.

There is one other reason why the Compton effect might have carried more weight, pointed out by Compton himself: it involved a confirmed novel prediction. Compton's assertion that the photoelectric effect did *not* have this virtue was somewhat misleading; Einstein predicted a linear relation between maximum kinetic energy and frequency at a time when such a relation not been established. But Compton also predicted a *qualitatively new* phenomenon: the recoil electron, and one could argue that this should count more than a quantitative prediction about a qualitatively known phenomenon. However, only a few authors even mentioned the fact that Compton had predicted recoil electrons, and none of them except Compton himself stated that his theory was more likely to be valid *because* he predicted them *before* they were discovered.

Popperians may find some consolation in the fact that while physicists did not give Compton extra credit for the novelty of his prediction, some of them did praise the Bohr-Kramers-Slater theory for being so precise in its predictions as to be immediately and clearly falsifiable—and falsified.

A valid theory should make testable predictions whenever possible, but that does not mean it must make confirmed novel predictions *before* it is accepted as valid. Quantum mechanics is a counterexample: a theory that was widely accepted before its novel predictions were tested (see next chapter). If one must have an explicit criterion for acceptance (a requirement I find rather dubious), here's a better one from the 1920s discussion of the nature of light: "A theory must pass a very strict test nowadays; it must not only be accurate, it must be a convenient and powerful instrument of thought." As Einstein himself stated, the light quantum hypothesis was not intended to end the debate about the nature of light, but rather to open a new and more fruitful period of research. Compton's own theory of his effect was certainly not the final answer either, but it provoked others like Schrödinger, Dirac, Klein, and Nishina to work out a more accurate and comprehensive quantum-mechanical theory. It was not until the 1960s that our present understanding of quantum optics began to emerge, with the work of Roy Glauber, John Hall, and Theodor Hänsch, recently recognized by the award of the 2005 Nobel Prize in physics.

7.10 RUPP'S FRAUDULENT EXPERIMENTS

Although Compton's claim that his evidence for the LQG was better because it achieved confirmed novel predictions was not persuasive to most physicists, such predictions do have some rhetorical value. In particular, some experimenters seem to believe that they can enhance their own reputations by confirming a prediction made by a highly respected scientist. As noted in Chapter 1, this belief occasionally leads an experimenter to commit fraud, or perhaps to select the data that confirm the theory while ignoring or rejecting those that disprove it. Some critics have accused Arthur Eddington of doing this in his test of the gravitational light bending predicted by Einstein's general theory of relativity, although the fact that later observations roughly confirmed his results seems to have preserved his reputation.

A notorious example is the physicist Emil Rupp. Rupp published in 1926 a study of light emitted by canal rays, including their interference; this attracted considerable attention. Einstein suggested that Rupp's work be extended to provide a test of the hypothesis that light is emitted instantaneously (as implied by the LQH) rather than over a period of time (as expected from the classical wave theory of light). Einstein proposed a collaboration with Rupp, which the latter eagerly accepted. Physicist Robert d'Escourt Atkinson argued that Einstein's proposed experiment would not work because Rupp's results were invalid, due to thermal motion and Doppler shifts (effects that Rupp had recognized but underestimated).

One of Einstein's proposals to Rupp involved rotating a mirror. According to historian Jeroen van Dongen,

> It seems unavoidable to conclude that Rupp never observed the mirror rotation bringing about interferences and that he just reported confirming what he believed to be Einstein's correct prediction . . . Einstein . . . had wanted confirmation of his analysis which may have made him insufficiently critical . . . Einstein was convinced . . . that his theoretical analysis was correct, and expected Rupp to find results that were in complete agreement with his analysis. He therefore pressed him long enough until he got the results he expected. Once Einstein believed that Rupp had found such confirmation, he apparently felt no further need to scrutinize the latter's work, or, for that matter, to attend to the Atkinson publication.

Later it was shown that Rupp's earlier canal ray experiments had been forged (along with several other experiments), and the supposed confirmation of the LQH had no legitimate basis. Rupp retreated in disgrace, and the episode was largely forgotten or suppressed by physicists. It did, however, play a part in the subsequent history of quantum mechanics and its interpretation.

7.11 CONCLUSIONS

The establishment of the particle nature of light—without denying its wave nature—was a revolutionary event in physics, as Einstein suspected in 1905. It was not accomplished by any single discovery such as the photoelectric effect or the Compton effect. It was not (as some physicists initially thought) a reversion to the Newtonian corpuscular theory, since that theory, like the wave theory of light, presupposed a mechanistic view of nature. Instead, it was the result of an accumulation of the theoretical and experimental efforts of many physicists, trying to explore and understand what might be called anomalies in the behavior of electromagnetic radiation. Einstein was the driving force in this effort, expending enormous energy in "hatching this favorite egg of mine," as he put it, yet he was never quite successful in finding a solution that satisfied his own criteria. Bohr, who resisted the light quantum hypothesis to the bitter end (even though the success of his 1913 atomic model was one of the factors that helped persuade other physicists to accept that hypothesis), was perhaps the first to realize that the mechanistic view would have to be abandoned in order to accept the quantum view of nature. In this paradigm switch, one had to give up not a particular theory such as the wave theory of light, but the *criteria for judging theories*, such as the requirement for a visualizable mechanistic explanation. Only then could the wave and particle theories peacefully coexist.

As we know from twentieth-century political history, fanatical proponents of one extreme doctrine may, when they finally abandon it, become equally fanatical proponents of the extreme opposite view. A remarkable example from the history of physical optics is Robert Alexander Houstoun, author of several commercially successful textbooks. He resisted the LQH into the 1930s, then apparently decided that his beloved wave theory was completely wrong and resurfaced in the 1960s as a fervent advocate of—*not* the light quantum hypothesis—but of Newton's original corpuscular theory.

Of course one cannot convert the physics community by simply proclaiming that a new paradigm must be accepted. One needs empirical evidence. I suggest three major facts, to which each physicist might give a different weight, but all of which were needed to explain the conversion of (almost) the entire community: (1) the Compton effect; (2) the photoelectric effect; and (3) all the other phenomena, especially those involving X-rays, specific heats of solids at low temperatures, and atomic spectra, which could not plausibly be explained by a wave theory but could (more or less accurately) be explained by some kind of quantum theory. The establishment of the light quantum hypothesis was a major step toward the victory of the quantum worldview, but it was not the first or the last.

NOTES FOR CHAPTER 7

191 *The Light Quantum Hypothesis* The first two sections of this chapter are based on my book *Statistical Physics and the Atomic Theory of Matter, from Boyle and Newton to Landau and Onsager* (Princeton, NJ: Princeton University Press, 1984). For further details and references see Hans Kangro, *Early History of Planck's Radiation Law* (New York: Crane, Russak, 1976); Thomas S. Kuhn, *Black-Body Theory and the Quantum Discontinuity* (New York: Oxford University Press, 1978).

SECTION 7.1

191 *black-body radiation* The ideal black body is a cavity within which thermal equilibrium is maintained, equipped with a pin hole through which the radiation emerges to be sampled by an outside observer. The frequency-distribution of this radiation is independent of the nature of the cavity and depends only on the absolute temperature *T*.

191 *In 1884, physicist Ludwig Boltzmann derived the T^4 law* For details and references see S. G. Brush, *The Kind of Motion We Call Heat: A History of the Kinetic Theory of Gases in the 19th Century* (Amsterdam: North-Holland, 1976), pp. 514–519.

191 *In 1896, Wilhelm Wien proposed a formula* Wien, "Ueber die Energievertheilung im Emissionsspectrum eines schwarzes Körpers," *Annalen der Physik* [series 3] 58: 662–669; English translation in *Philosophical Magazine* [series 5] 43 (1897): 214–220.

SECTION 7.2

191 *Planck was a critic of atomistic theories* Brush, *Kind of Motion*, pp. 641–642.

191 *Planck got involved in a minor polemic with Ludwig Boltzmann* G. R. Kirchhoff, *Vorlesungen über die Theorie der Wärme* (Leipzig: Teubner, 1894); Brush, *Kind of Motion*, pp. 643–644.

192 *Rayleigh had supposedly suggested* Lord [Third Baron] Rayleigh, "Remarks Upon the Law of Complete Radiation," *Philosophical Magazine* [series 5] 49 (1900): 539–540.

192 *Planck does not even mention Rayleigh's paper* Planck, "Ueber eine Verbesserung der Wien'schen Spectralgleichung," *Verhandlungen der Deutschen Physikalischen Gesellschaft* 2 (1900): 202–204; "Zur Theorie des Gesetzes der Energieverteilung im Normalspectrum," ibid. 237–245. English translations of both papers were published by D. ter Haar, *The Old Quantum Theory* (Oxford: Pergamon Press, 1967), reprinted with extensive notes by Hans Kangro, *Planck's Original Papers in Quantum Physics, German and English edition* (London: Taylor and Francis, 1972).

192 *one still occasionally sees the myth in print* Margaret Mahy, *Ultra-Violet Catastrophe! Or the Unexpected Walk with Great-Uncle Magnus Pringle* (New York: Parents' Magazine Press, 1975); other references in Brush, *Statistical Physics*, p. 263. Mahy was a prolific author of children's books; see Paul Vitello, "Margaret Mahy, Author of Children's Books, Dies at 76," *New York Times*, July 2012, p. 22.

192 *Planck postulated a system of atomic oscillators* Planck showed that the distribution law for BBR is related to the energy-distribution law for the oscillators by the equation

$$\rho(v,T) = 8\pi v^2 / c^3 U(v,\ T).$$

where U is the energy of an oscillator with frequency v and c is the speed of light.

192 *calculated the relation between the entropy S and energy U of an oscillator* He analyzed the relation between the energy (U) and entropy (S) of an oscillator, for various possible BBR distribution laws, and came to the conclusion that Wien's formula, which seemed to be the most satisfactory on experimental grounds (up to 1899), corresponded to a very simple and natural formula

$$S = -(U/av)\ log\ (U/ea'v),$$

where a, e and a' are constants.

192 *In order to study the thermodynamic stability* Now suppose we want to apply the second law of thermodynamics in the form: entropy tends toward a maximum value. To find that maximum, according to the methods of calculus we must find the rate of change (slope, or derivative) of S as a function of U, and set it equal to zero (dS/dU = 0); then, to be sure that we have a maximum rather than a minimum, we must calculate the rate of change of the rate of change, i.e., the second derivative. If it's positive we have a minimum, and if its negative we have a maximum. (Think of a hill where S is the height and U is the horizontal distance; then as you climb up the hill from the left, the slope dS/dU decreases and becomes zero at the top of the hill; the rate of change of the slope is always negative).

Planck's formula for S has the property that the second derivative of entropy with respect to energy is

$$d^2S / dU^2 = -(1/av)(1/U),$$

or, taking the reciprocal of each side,

$$1/(d^2S / dU^2) = -(av)\ U$$

which is always negative, since u, v, and U are always positive. This means that if the energy of an individual oscillator happens to deviate from its equilibrium value, it will return to equilibrium with an increase in entropy; equilibrium corresponds to an absolute maximum of entropy. Planck thought that he could make his theory perfectly consistent with the second law of thermodynamics by adopting Wien's formula for black-body radiation.

192 *he proposed an empirical correction to* But more accurate experiments showed that Wien's law breaks down at high temperatures and low frequencies, so the above formulae for S and its derivatives need to be corrected. In physical science (physical chemistry in particular) it sometimes happens that you have a linear relation between two variables when both are small, but when the independent variable becomes larger the dependent variable increases more rapidly than does the independent variable. In such cases, in the absence of any further knowledge, one tries to fit the data by adding a nonlinear term, e.g., the square of the independent variable. Planck writes that he had already started to "construct completely arbitrary expressions for the entropy which still seem to satisfy just as completely all requirements of the thermodynamic and electromagnetic theory." In this case the independent variable is U and the dependent variable is the reciprocal of the second derivative of S with respect to U; so that one could write

$$1/\left(d^2S/dU^2\right) = -[(a\nu)\ U + \beta U^2]$$

One can then integrate twice with respect to U and get a formula for the entropy,

$$S = k\{\left(1 + U/h\nu\right)\ ln\ \left(1 + U/h\nu\right) - U/h\nu lnU/h\nu\}.$$

192 *The result for the energy of an oscillator* From the above formula for S one can calculate the distribution law,

$$\rho\left(\nu,T\right) = \left[8\pi\nu^2/c^3\right]h\nu/(e^{h\nu/kT} - 1)$$

Kangro, *Planck's Papers*, pp. 5 (German), 37 (English).

193 *He now tried to interpret his formula using Boltzmann's formula* Although this equation was engraved on Boltzmann's tombstone, it did not appear in this form in any of his publications—and for a good reason, which is also crucial to the interpretation of Planck's derivation. In Newtonian mechanics, the microstates of a system of particles are defined by specifying the positions and momenta of all the particles. Since there are continuous variables, W is infinite. Boltzmann always worked with *differences* in entropy in which case, by taking limits carefully, one can make the infinities cancel out. The entropy of a single state (including the state at absolute zero temperature) can be finite only in quantum theory.

193 *The number of arrangements is* $W = (N+P-1)!/P!(N-1)!$ One way to prove this is to represent each arrangement by a sequence of P dots with $N-1$ lines for the boundaries between the oscillators: (..././/.../.../../.). The numerator gives the total number of ways of arranging the dots and lines, regarded simply as $N + P - 1$ permutable elements. The denominator represents permutations of the dots among themselves, which should not count as different arrangements, since each quantum of energy is

identical and not labeled; it also represents permutations of the lines among themselves (implying that the oscillators are identical and unlabeled).

193 *one sentence . . . seems to settle the matter* "The question is how in a stationary state this energy is distributed over the vibrations of the resonators and over the various colors of the radiation present in the medium, and what will be the temperature of the total system."

"To answer this question we first of all consider the vibrations of the resonators and try to assign to them certain arbitrary energies, for instance, an energy E to the N resonators v, E' to the N' resonators v', . . . The sum

$$E + E' + E'' + \ldots = E_0$$

must, of course, be less than E_t [total energy]. The remainder $E_t - E_0$ pertains then to the radiation present in the medium. We must now give the distribution of the energy E over the N resonators of frequency v. If E is considered to be a continuously divisible quantity, this distribution is possible in infinitely many ways. We consider, however—this is the most essential point of the whole calculation—E to be composed of a well-defined number of equal parts and use thereto the constant of nature $h = 6.55 \text{ x } 10{-}^{27}$ erg sec. This constant multiplied by the common frequency v of the resonators gives us the energy element, in erg, and dividing E by, we get the number P of energy elements which must be divided over the N resonators. *If the ratio thus calculated is not an integer, we take for P an integer in the neighborhood*" (emphasis added). Kangro, *Planck's Papers* pp. 8 (German), 40 (English).

193 *It might be just a "fictional quantity"* Planck, *Die Entstehung und bisherige Entwicklung der Quantentheorie* (Nobel Prize Lecture) Leipzig: Barth, 1920. Translation by R. Jones and D. H. Williams, in Planck, *A Survey of Physical Theory* (New York: Dover, 1960), pp. 102–114, on p. 109.

193 *what Eugene Wigner called "the unreasonable effectiveness" Communications on Pure and Applied Mathematics* 13 (1960): 601–614.

Section 7.3

194 Sections 7.3 through 7.9 and 7.11 of this chapter are based on my article "How Ideas Became Knowledge: The Light Quantum Hypothesis 1905–1935," *Historical Studies in the Physical and Biological Sciences* 37 (2007): 205–246, by permission of the Regents of the University of California; this paper should be consulted for further details and references. I thank J. L. Heilbron, Alexei Kozhevnikov, Gonzalo Munevar, and Roger Stuewer for many valuable suggestions and corrections. The research was greatly facilitated by the magnificent collections at the Niels Bohr Library, American Center for Physics in College Park, Maryland. I also made extensive use of the Max Born Collection at the Engineering and Physical Sciences Library and the Interlibrary Loan Office at McKeldin Library, both at the University of Maryland, and the *Physics Citation Index 1920–1929* (Philadelphia: Institute for Scientific Information, 1981). I thank the staff of all these institutions for their assistance.

194 *Albert Einstein wrote in 1905* Einstein, "Ueber einen der Erzeugung und Verwandlung des Lichtes betreffenden heuristischen Gesichtspunkt," *Annalen der Physik* [series 4] 17 (1905): 132–148; reprinted in *The Collected Papers of Albert Einstein*, vol. 2 (Princeton, NJ: Princeton University Press, 1989): 134–166, with an extensive editorial note. Quotations are from the English translation in Dirk ter Haar, *The Old Quantum Theory* (Oxford: Pergamon Press, 1967), pp. 91–107.

195 *According to Christa Jungnickel and Russell McCormmach* Jungnickel and McCormmach, *Intellectual Mastery of Nature: Theoretical Physics from Ohm to Einstein*, vol. 2, *The Now Mighty Theoretical Physics 1870–1925* (Chicago: University of Chicago Press, 1986), p. 305.

Section 7.4

195 *the Newtonian particle theory of light was replaced* Jed Z. Buchwald, *The Rise of the Wave Theory of Light: Optical Theory and Experiment in the early Nineteenth Century* (Chicago: University of Chicago Press, 1983).

Section 7.5

197 *in a famous letter to Konrad Habicht* Einstein to Habicht, May 18 or 25, 1905, in Einstein, *Collected Papers*, vol. 5, edited by Martin J. Klein, A. J. Kox and Robert Schulmann (1993), pp. 31–32, on p. 31; English translation by Anna Beck, vol. 5, pp. 19–20, on p. 20.

197 *Karl K. Darrow's insightful definition of the word "heuristic"* Darrow, *Introduction to Contemporary Physics* (New York: D. Van Nostrand Co., 1926), pp. 116–117.

197 *the "philosophy of* as if" Vaihinger, *The Philosophy of As if"* (1911 English translation-London: Routledge and Kegan Paul, 1924).

Section 7.6

198 *"We are confronted . . . by the astonishing situation"* Millikan, "A direct Photoelectric Determination of Planck's '*h*', *Physical Review* [series 2] 7 (1916): 355–388, on pp. 355–383.

198 *To have discovered the quantitative nature of the phenomenon* "The predictions of Einstein's theory [of light quanta] have received such exact experimental confirmation in recent years, that perhaps the most exact determination of Planck's constant is afforded by measurements on the photoelectric effect. In spite of its heuristic value, however, the hypothesis of light-quanta, which is quite irreconcilable with so-called interference phenomena, is not able to throw light on the nature of radiation." Niels Bohr, "The Structure of the Atom. Nobel Lecture, December 11, 1922," in *Nobel Lectures, Physics 1922–1941* (Singapore: World Scientific, 1998) pp. 7–43, on p. 14.

Section 7.7

199 *As historian Roger Stuewer described in detail in 1971* Stuewer, "William H. Bragg's corpuscular Theory of X-Rays and γ-Rays," *The British Journal for the History of Science* 5 (1971): 258–281. See also Bruce R. Wheaton, *The Tiger and the Shark: Empirical Roots of Wave-Particle Dualism* (Cambridge: Cambridge University Press, 1983).

199 *some of the secondary scattered rays were softer* In 1921, before he had adopted the corpuscular hypothesis himself, Compton mentioned this explanation as proposed by the Canadian physicist Joseph Alexander Gray:

"Prof. J. A. Gray (*Franklin Institute Journal*, November, 1920) . . . showed that if the primary rays came in thin pulses, as suggested by Stokes's theory of X-rays, and if these rays are scattered by atoms of electrons of dimensions com-parable with the thickness of the pulse, the thickness of the scattered pulse will be greater than that of the incident pulse. He accordingly suggests that the observed softening of the secondary rays may be due to the process of scattering." Compton, "The Softening of secondary X-rays," *Nature* 108 (1921): 3667–367, on 366. Stuewer, *Compton Effect.*

Compton believed that his own data refuted Gray's hypothesis; at that time he was trying several other ways to explain the data, without yet having settled on any particular theory. But during the following year he decided that a somewhat different explanation, the light-quantum hypothesis, might be valid. He credited O. W. Richardson for the idea "that as the electron absorbs a quantum h*ν* of energy, the momentum of the absorbed radiation is also transferred to the electron." But Compton then explicitly rejected that idea. See A. H. Compton, "Secondary Radiation Produced by X-rays, and Some of Their Applications to Physical Problems," *Bulletin of the National Research Council*, 4, part 2, no. 20 (1922): 24; *Scientific Papers of Arthur Holly Compton: X-Rays and other Studies*, edited by Robert S. Shankland (Chicago: University of Chicago Press, 1973), pp. 382–401. On the origin of Compton's formula $p = E.c$ for the momentum of a light quantum, see R. S. Shankland, appendix 1 in ibid., pp 756–758.

199 *any particular quantum of X-rays is not scattered* A. H. Compton, "A Quantum Theory of the Scattering of X-rays by light Elements," *Physical Review* [series 2] 21 (1923): 483–502.

199 *"the electrons which recoil in the process of the scattering"* Compton, ibid., p. 496.

199 *"In view of the fact that these recoil electrons were unknown"* A. H. Compton, "The Scattering of X-rays," *Journal of the Franklin Institute* 198 (1924): 57–72, on p. 68.

200 *"that X-rays, and so also light, consist of discrete units"* Ibid., p. 70.

200 *"the idea of light quanta was invented primarily to explain"* A. H. Compton, "Light Waves or Light Bullets," *Scientific American* 133 (October 1925): 246–247. According to Roger Stuewer (private communication), Compton's statement that Einstein invented the LQH "primarily to explain the photoelectric effect" is clear proof that he never read Einstein's 1905 paper.

200 *"Since their very existence was unknown before they were predicted"* Ibid.

200 *Compton stated his claim in a different way* A. H. Compton and A. W. Simon, "Directed Quanta of Scattered X-rays," *Physical Review* [series 2] 26 (1925): 289–299.

200 *Einstein's light quantum hypothesis was "a necessary consequence"* Stuewer, "Non-Einsteinian Interpretations of the Photoelectric Effect," in Stuewer, ed., *Historical and Philosophical Perspectives of Science* (Minneapolis: University of Minnesota Press, 1970), pp. 246–273, on pp. 247–248.

201 *Two physicists who clearly did not accept that claim* Bohr, Kramers and Slater, "The Quantum Theory of Radiation," *Philosophical Magazine* 47 (1924): 785–802. On Bohr's opposition to the particle nature of light see John Stachel, "Bohr and the Photon," in *Quantum Reality, Relativistic Causality, and Closing the Epistemic Circle: Essays in Honour of Abner Shimony*, edited by Wayne C. Myrvold and Joy Christian (n.p.: Springer, 2009), pp. 69–83. Paradoxically, Bohr's collaborator H. A. Kramers had derived the Compton effect before Compton published it, but was apparently discouraged from publishing it by Bohr. See M. Dresden, *H. A. Kramers* (New York: Springer, 1987), cited by Abraham Pais, *Niels Bohr's Times, In Physics, Philosophy, and Polity* (Oxford: Clarendon Press, 1991), p. 238, who says "the matter never came up during the many discussions I had with Kramers in the 1940s and 1950s." According to the way Louisa Gilder imagined it, Slater heard about this from Kramers' wife Anna 'Storm' Petersen: see *The Age of Entanglement: When Quantum Physics was Reborn* (New York: Random House/Vintage Books, 2008), p. 63–64.

201 *that if it were true he would rather be a cobbler than a physicist* Letter from Einstein to Hedwig and Max Born, April 29, 1924, in Max Born, editor, *The Born-Einstein Letters* (New York: Walker, 1971), p. 82. For an account of Einstein's objections to the Bohr-Kramers-Slater theory see Wolfgang Pauli, letter to Bohr,

October 2, 1924, in *Niels Bohr Collected Works*, vol. 5, edited by K. Stolzenburg (Amsterdam: North-Holland, 1984), pp. 414–418.

201 *"God does not play dice"* Letter from Einstein to Max Born, December 4, 1926, in *Born-Einstein Letters*, p. 91.

201 *This new attack on the LQH, called the "Copenhagen putsch"* Wolfgang Pauli, letter to Kramers, July 27, 1925, in *Niels Bohr Collected Works*, vol. 5, pp. 439–442.

201 *The test was conducted by Walther Bothe and Hans Geiger* W. Bothe and H. Geiger, "Experimentelles zur Theorie von Bohr, Kramers und Slater," *Die Naturwissenschaften* 13 (1925): 440–441; "Über das Wesen des Compton-effekts; ein experimenteller Beitrag zur Theorie der Strahlung," *Zeitschrift für Physics* 32 (1925): 639–663. A. H. Compton and A. W. Simon, "Directed Quanta of Scattered X-Rays," *Physical Review* 26 (1925): 289–299; Compton, "On the Mechanism of X-Ray Scattering," *Proceedings of the National Academy of Sciences* 11 (1925): 303–306.

201 *But the theory did have some influence on the later development* According to Max Jammer, BKS introduced "a radically new approach. By interpreting Einstein's spontaneous emission as a process 'induced by the virtual field of radiation' and Einstein's induced transitions as occurring 'in consequence of the virtual radiation in the surrounding space due to other atoms,' it paved the way for the subsequent quantum-mechanical conception of probability as something endowed with physical reality and not merely a mathematical category of reasoning." Jammer, *The Conceptual Development of Quantum Mechanics* (New York: Tomash Publishers/ American Institute of Physics, 2nd ed., 1989), p. 191 and notes 136 and 137.

202 *"Compton deduced a new kind of corpuscular theory" Nobel Lectures in Physics, 1922–1941* (Singapore: World Scientific, 1988), pp. 170–171.

202 *"all electromagnetic radiation is constituted of discrete quanta"* "X-Rays as a Branch of Optics" (Nobel Lecture, December 12, 1927), in *Papers of Arthur Holly Compton*, p. 541.

Section 7.8

202 *"I no longer have doubts"* Letter, July 29, 1918, in *Collected Papers of Albert Einstein*, vol. 8, edited by Robert Schulmann et al. (Princeton, NJ: Princeton University Press, 1998), Doc. 591, p. 836; English translation, p. 613.

202 *seven well-known historians of physics* See Brush, "How Ideas Became Knowledge: The Light Quantum Hypothesis 1905–1935," *Historical Studies in the Physical and Biological Sciences*, 37 (2007): 205–246, on p. 227, note 67.

202 *This is also the view of the Historic Sites Committee* Ben Bederson, "Report on Historic Cites Committee (HSC) Activities," *History of Physics Newsletter* 11, no. 3 (Fall 2010): 14–17, on p. 14.

202 *Planck's opinion that Einstein had "overshot the target"* Proposal for Einstein's membership in the Prussian Academy of Science by Planck, Nernst, Rubens and Warburg, June 12, 1913, in Einstein, *Collected Papers*, volume 5, Doc. 445, pp. 526–528, on p. 527; English translation, pp 336–338, on p. 337.

202 *Stachel has provided ample evidence* Stachel, "Bohr and Photon."

203 *the discovery of X-ray diffraction "enlivened the discussion"* Gerlach, "Reminiscences of Albert Einstein from 1908 to 1930), in *Albert Einstein: His Influence on Physics, Philosophy, Politics*, edited by P. C. Aichelburg and R. U. Sexl, (Wiesbaden, 1979) pp. 189–200, on p. 191.

203 *"For anyone who studied physics in the years just before the war" The Renaissance of Physics* (New York: Macmillan, 1937), pp. 177–178.

203 *"a general change of attitude in physics"* Kozhevnikov, "Einstein's Fluctuation Formula and the Wave-Particle Duality," in *Einstein Studies in Russia*, edited by Y. Balashov and V. Vizgin (Boston: Birkhäuser, 2002, pp. 181–228.

203 *we learn from McCormmach that "in Europe at this time"* Russell McCormmach, "J. J. Thomson and the Structure of Light," *British Journal for the History of Science* 3 (1967): 362–387, on p. 375.

204 *"as though it possessed at the same time the opposite properties"* O. W. Richardson, *The Electron Theory of Matter* (Cambridge: Cambridge University Press, 2nd ed. 1916), p. 507.

204 *in a long 1909 letter to Einstein* Lorentz to Einstein, May 6, 1909, in Einstein, *Collected Papers*, vol. 5, Doc. 152, pp. 170–180, on p. 176; English translation, p. 112.

205 *a "progressive research programme"* Lakatos, "Falsification and the Methodology of Scientific Research Programmes," in *Criticism and the Growth of Knowledge*, edited by I. Lakatos and A. Musgrave, (New York: Cambridge University Press, 1970), pp. 91–196.

205 *it offered "promise more than performance"* Munevar, private communication to SGB; see also his "Reflections on Hull's Remarks," in *Prematurity in Scientific Discovery*, edited by E. B. Hook (Berkeley: University of California Press, 2002), pp. 342–345.

205 *"the times are, however, pregnant with new ideas"* R. A. Millikan, "Atomism in modern Physics," *Journal of the Chemical Society* 125 (1924): 1405–1417, on p. 1417.

Section 7.9

205 *"the best evidence yet found"* R. A. Millikan, *The Electron: Its Isolation and Measurement and the Determination of some of its Properties* (Chicago:University of Chicago Press, Second edition 1924), p. 256.

206 *"the Compton effect more than all other quantum phenomena" Matter, Electricity, Energy: The Principles of Modern atomistics and Experimental Results of Atomic Investigation*, translated from the 2nd German ed. of 1926 (New York: Van Nostrand, 1928), on p. 262.

206 *the Compton effect "shows even more clearly"* Theodor Wulf, S. J., *Modern Physics: A general Survey of its Principles*, translated from the 2nd German ed. of 1929 (New York: Dutton, 1929; London: Methuen, 1930).

208 *"A theory must pass a very strict test nowadays"* T. L. Eckersley, "The Compton Scattering and the Structure of Radiation," *Philosophical Magazine* 2 (1926): 267f, on p. 286.

Section 7.10

208 *leads an experimenter to commit fraud* See Brush, "Prediction and Theory Evaluation: The Case of Light Bending," *Science* 246 (1989): 1124–1129, on p. 1127 and publications cited in note 86 therein; S. Weinberg, *Dreams of a Final Theory* (New York: Pantheon, 1992), pp. 95–96.

209 *Rupp published in 1926 a study of light emitted by canal rays* "Interferenzuntersuchungen an Kanalstrahlen," *Annalen der Physik* 79 (1926): 1–34.

209 *Einstein suggested that Rupp's work be extended to provide a test* "Vorschlag zu einem die Natur des elementaren Strahlungs-Emissionsprocesses betreffenden Experiment," *Naturwissenschaften* 14 (1926): 300–301.

209 *Atkinson argued that Einstein's proposed experiment would not work* "Ueber Interferenz von Kanalstrahlenlicht," *Naturwissenschaften* 14 (1926): 599–600.

209 *"It seems unavoidable to conclude that Rupp never observed"* "Emil Rupp, Albert Einstein, and the Canal Ray Experiments on Wave-Particle Duality," *Historical*

Studies in the Physical and Biological Sciences 37, Supplement (2007): 73–119, on pp. 98, 103, 116. See also Dongen, *Einstein's Unification* (Cambridge: Cambridge University Press, 2010), pp. 81–86.

209 *Rupp's earlier canal ray experiments had been forged* Anthony D. French, "The Strange Case of Emil Rupp," *Physics in Perspective* 1 (1999): 3–21; Dongen, "Emil Rupp," pp. 102–111.

209 *the subsequent history of quantum mechanics* Joeren van Dongen, "The Interpretation of the Einstein-Rupp Experiments and their Influence on the History of Quantum Mechanics," *Historical Studies in the Physical and Biological Sciences* 37, Supplement (2007): 121–131.

Section 7.11

209 *"hatching this favorite egg of mine"* The "egg" metaphor comes from his letter to Jakob Laub, December 31, 1909, in Einstein, *Collected Papers*, vol. 5, Doc. 196, on p. 227; English translation, p. 146.

8

Quantum Mechanics

By 1925, the old quantum theory was approaching its final measure: the new physics was a series of clever cadenzas isolated one from the other, and it lacked a theme to tie them together.

—*physicist and historian* John Rigden *(1987)*

This afternoon I spent two hours at Einstein's. . . . Einstein summed up his relations to [the theory of quanta] in the following way: the situation is desperate, it's impossible to understand anything!

—*physicist* V. Ya. Frenkel, *November 1925*

8.1 THE BOHR MODEL

Quantum mechanics grew out of the interaction of two major subjects of nineteenth-century physical science: atoms and light. More precisely, it combined research on the properties of atoms and molecules with the study of electromagnetic radiation, drawing not only from physics and chemistry, but also the new science of astrophysics.

In Section 2.5, I summarized the attempt of physicist Niels Bohr to explain the spectrum and other properties of atoms by assuming that negatively charged electrons revolve around a positively charged nucleus, by analogy with the motion of planets around the sun. He assumed that electrons could absorb or emit electromagnetic waves only when jumping from one to another of a discrete set of orbits. These waves have frequencies corresponding to the energy-difference between the orbits, in accordance with Planck's quantum hypothesis. Bohr was thereby able to explain the Balmer and other series of spectral lines, including the value of the Rydberg constant. He could also predict that a certain spectral line previously attributed to hydrogen was actually produced by singly ionized helium (with a nucleus about four times as massive as the proton orbited by a single electron). Bohr's model was widely accepted by 1916.

Further confirmation of Bohr's theory came from James Franck and Gustav Hertz's experiment in 1914. They bombarded a gas (mercury) with electrons of increasing energy; there were only elastic collisions (no radiation emitted) until they reached a certain critical energy, then they suddenly observed inelastic collisions with emitted radiation at a particular frequency. The energy loss of the electron corresponded to the energy difference between the ground state and an excited state of the atom. (This experiment was *not* originally done for the purpose of testing Bohr's theory, which they did not understand; it was only interpreted this way afterward.) In 1915–1916, Arnold Sommerfeld generalized Bohr's model to include elliptical orbits in three dimensions. He treated the problem relativistically (using Einstein's formula for the increase of mass with velocity), and found that the orbit of an electron would no longer be fixed in space. The orbit itself would move (precess) so that the point nearest the nucleus (corresponding to the perihelion in the solar system) would slowly move. This would produce a correction to the energy levels that was proportional to the square of the atomic number Z. It would also be proportional to the square of (what became known as) the fine-structure constant $\alpha = 2\pi e^2/hc$. The energy level for an orbit with radial quantum number n' and azimuthal quantum number k is given by the formula

$$E_{nk} = -Z^2\ Rhc\,[1/n^2 + \left(\alpha^2 Z^2 / n^4\right)(n/k - 3/4)]$$

where n = n' + k. Friedrich Paschen confirmed the result predicted from this formula for ionized helium (Z = 2).

According to historian Max Jammer, this success of Sommerfeld's fine-structure formula "served also as an indirect confirmation of Einstein's relativistic formula for the velocity dependence of inertial mass." It was thus one of the many empirical facts supporting relativity theory (Chapter 11). But I remind the reader that it is a logical fallacy to assume that the truth of a prediction proves the truth of the hypothesis from which the prediction was derived (since some other hypothesis could have generated the same prediction).

The Bohr model of the atom provided a research program difficult and exciting enough to challenge some of the world's brightest theoretical physicists during the decade after 1913, and stimulated much new experimental work in atomic physics. Yet despite its early triumphs, there were certain mysterious aspects of Bohr's theory that made it unsatisfactory, especially if one tried to think of the electron actually moving around in space in one orbit and then jumping to another. One obvious objection was presented even before Bohr's paper was published: Rutherford wrote to Bohr in 1913, on receiving the manuscript of his paper, as follows:

> There appears to be one grave difficulty in your hypothesis, which I have no doubt you fully realize, namely, how does an electron decide what frequency it is going to vibrate at when it passes from one stationary state to another? It seems to me that you would have to assume that the electron knows beforehand where it is going to stop.

The struggles of physicists to use or modify Bohr's model in the period before 1925 have been comprehensively described by historians John Heilbron, Paul Forman, Helge Kragh, and others. While by 1919 that model, according to Kragh, "enjoyed general acceptance among experts in atomic and quantum theory," it was not considered really satisfactory. I turn to one especially vivid account. In a comprehensive survey of atomic theory published in 1926, just after the introduction of quantum mechanics, the British physicist E. N. Da C. Andrade wrote that in the old quantum theory, the use of half-integer quantum numbers is necessary (to explain the anomalous Zeeman effect), but "repulsive":

> The [old] quantum theory has no justification but its success, and actually cannot be reconciled with certain wave phenomena, such as interference. It represents a postulate beyond which, at present, we cannot see; and all we can do is to be profoundly thankful that so far the elephantine philosophers have not offered to help us hatch out our quantic egg. . . . the willingness of the theoretical physicists to sacrifice dynamical sanity for the temporary description of an empirical rule has been occasionally bewildering. The quantum theory has become, in its higher branches, a collection of numerical receipts, to which a superficial resemblance to a coherent mathematical theory has been given to hide their ad hoc character . . . Some of the rules recall irresistibly the teaching of the alchemists, or the witches' kitchen in Faust.

Andrade's curious expression "hatch out our quantic egg" (echoed in Einstein's September 1925 letter to Ehrenfest, Section 8.7), suggests that what was needed was not a revolutionary change to an incommensurable new paradigm, but a new improved theory growing organically out of the old one. Historian Suman Seth argues that Thomas Kuhn, in his 1960s interviews with quantum physicists, was looking for a 1920s crisis as described in his own theory of scientific revolutions, and found it by asking leading questions. On the contrary, Seth writes, "Sommerfeld and the members of his School did not register a sense of crisis." Instead, Sommerfeld in 1929 said that "the new development does not signify a revolution but a joyful advancement of what was already in existence, with many fundamental clarifications and sharpenings."

Einstein, who believed in what Seth calls a "physics of principles" (rather than Sommerfeld's "physics of problems"), perceived a crisis in November 1925 (see quote at beginning of this chapter). Later he strongly attacked quantum mechanics despite its success in solving many problems—he found no joy in its basic principles.

8.2 THE WAVE NATURE OF MATTER

To set the stage for the introduction of quantum mechanics, let us recall the situation in physics around 1920. First, Planck's quantum hypothesis had been generally accepted as a basis for atomic theory, and the photon concept had been given strong support by Robert A. Millikan's experimental confirmation of Einstein's theory of the photoelectric effect. Even more definitive proof of the particle nature of electromagnetic radiation came from A. H. Compton's demonstration in 1923 that X-rays are scattered by electrons as if they were classical particles with a definite momentum.

Yet despite the early success of Bohr's theory of the hydrogen atom, no one had found a satisfactory way to give quantitative accounts of the spectra of atoms and molecules (including the effect of electric and magnetic fields without introducing arbitrary additional hypotheses tailored to fit each particular case). In some cases these hypotheses led to formulae that turned out to be not just approximately, but exactly, correct. The discovery that these formulae could be derived much more easily and consistently from quantum mechanics was a convincing reason to accept the new theory, so the immense effort needed to derive them from the old quantum theory was not completely wasted.

Although the photoelectric effect, as interpreted by Einstein, suggested that light energy comes in discrete units, this did not by any means persuade physicists that the wave theory of light should be abandoned in favor of a particle theory. The experiments on interference and diffraction of light, and indeed all the properties of electromagnetic radiation discovered in the nineteenth century, seemed to require wave properties. Even the quantum theory, which assigned each particle of light an energy proportional to its frequency ($E = h\nu$), depended on a wave property, frequency. Thus one was led to the paradoxical conclusion that light behaves in some respects like particles, and in other respects like waves. This conclusion was reluctantly accepted by many physicists before 1923, whether or not they adopted the specific light quantum hypothesis as developed by Einstein (see previous chapter). The wave/particle puzzle was one of several anomalies that undermined confidence in the old quantum theory.

Another puzzling aspect of the physical theories developed in the first two decades of the twentieth century was their failure to prescribe definite *mechanisms* on the atomic level. Thus Bohr's theory specified the initial and final states of an atomic radiation process, but did not describe just how the electron got from one orbit to another. The theory of radioactive decay was even less definite: it simply predicted the *probability* that a nucleus would emit radiation, but did not state when or how any particular nucleus would decay. It appeared that the *determinism* or *causality* characteristic of the Newtonian world machine—the predictability, at least in principle, of the detailed course of events of any atomic process, given the initial masses, velocities, and forces—was slipping away. Just as James Clerk Maxwell and Ludwig Boltzmann had shown that the second law of thermodynamics was only statistically valid as a description of observable processes, so some physicists began to suspect that Newton's laws of motion and the laws of conservation of energy and momentum might, in the atomic world, be true only on the average, but not necessarily at every instant of time. Yet an uncritical application of that idea could also be misleading, as was shown by the failure of the Bohr-Kramers-Slater theory (Section 7.7).

But if neither the wave theory nor the particle theory of light could be proved exclusively correct, and if the classical laws of mechanics and electromagnetism could no longer be relied on to describe atomic processes, clearly something new was needed, some fundamental law from which all these partially or statistically valid theories could be derived. And perhaps even the basic Newtonian presuppositions about the nature and purpose of theories would have to be modified.

It is a curious fact that radical conceptual changes in a science are sometimes initiated by people who did not receive their initial professional training in that science. A person who first approaches the problems of a discipline with the mature perspective of an outsider, rather than being indoctrinated previously with the established methods and attitudes of the discipline, can sometimes point to unorthodox—though when seen in retrospect—remarkably simple solutions to those problems. Thus, as is well known to historians of science, the generalized law of conservation of energy was introduced into physics from engineering, physiology, and philosophy, though its value was accepted readily enough by physicists. Of course the outsider character of the innovator may also create greater resistance to the innovation, and a derogatory attitude to the innovator despite (or perhaps because of) his eventually being proved right. Hannes Alfvén (see Chapter 1) as well as Louis de Broglie suffered from this treatment.

Louis de Broglie received his first degree in history, intending to go into the civil service. But he became intrigued by scientific problems as a result of discussions with his brother, physicist Maurice de Broglie, and by reading the popular scientific expositions of mathematician Henri Poincaré. Although he managed to acquire sufficient competence in using the standard tools of theoretical physics to satisfy his professors, his real interest was in the more fundamental problems of the nature of space, time, matter, and energy.

In his doctoral dissertation of 1924 (based in part on papers published a year earlier), de Broglie proposed a sweeping symmetry for physics: just as photons behave like particles as well as like waves, so electrons should behave like waves as well as like particles. In particular, an electron of mass m moving with velocity v will have a wavelength λ given by the simple formula

$$\lambda = h / mv \tag{1}$$

where h is Planck's constant. He pointed out that the quantization of allowed orbits in the Bohr model of the hydrogen atom can be deduced from this formula, if one makes the natural assumption that the circumference of the orbit must be just large enough to contain an integral number of wavelengths, so that a continuous waveform results (see Figure 8.1).

De Broglie's hypothesis suggests that the reason why atomic energy levels are quantized is the same as the reason why the frequencies of overtones of a vibrating string are quantized: the waves must exactly fill up a certain space. Hence a theory of atomic properties might be constructed by analogy with the theory of vibrations of mechanical systems reminiscent of the old program of Pythagoras and Kepler, who based many of the laws of nature on the laws that govern the operation of musical instruments.

Although de Broglie's hypothesis was based originally on theoretical arguments, it was soon realized that it could be checked rather directly by experiments of a kind that had already been performed by physicist C. J. Davisson, with C. H. Kunsman. In these experiments, electrons emitted from an electrode in a vacuum tube were allowed to strike a metal surface, and the scattering of the electron beam at various angles was measured. Walter Elsasser pointed out that Davisson's results could be

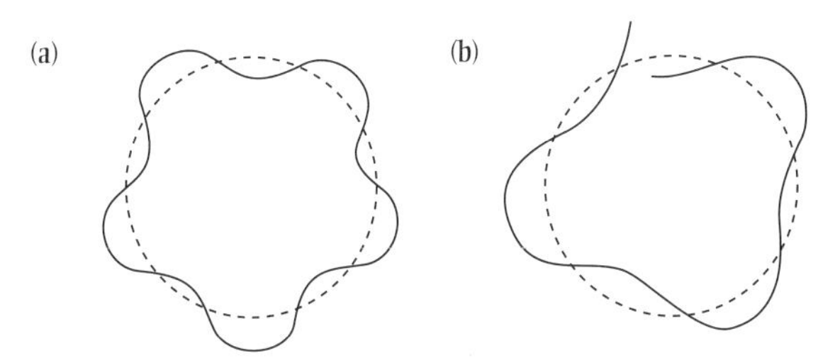

Figure 8.1 Only certain wavelengths will fit around a circle, e.g., in (a) but not in (b)

interpreted as showing a *diffraction* of the beam of electrons as they passed through rows of atoms in the metal crystal, and that other experiments on electron scattering by C. Ramsauer could be attributed to *interference* of waves that somehow determined the path of the electrons. In both cases the apparent wavelength associated with the electrons agreed with the values calculated from de Broglie's equation.

One of the first attempts to confirm de Broglie's prediction was made by physicist E. G. Dymond in 1926. His report on electron scattering in helium claimed to give qualitative support to wave mechanics, although he recognized that his results did not agree quantitatively with Elsasser's conclusions about the nature of the interference pattern that should be observed. Although Dymond's experiment was cited for a short time as a confirmation of de Broglie's hypothesis, it was eventually found that the effects he observed could not be attributed to diffraction, and this confirmation was withdrawn. I mention this case to illustrate one of the hazards of stressing prediction as the essence of the scientific method: it occasionally leads experimenters to misinterpret their own results in a way favorable to the theory.

In the summer of 1926 Davisson went on vacation to England, where he attended a meeting of the British Association for the Advancement of Science at Oxford. There he was surprised to hear physicist Max Born, in a lecture, cite Davisson and Kunsman's 1923 experiment as confirmatory evidence for de Broglie waves. Since other scientists seemed so excited by his "relatively feeble peaks" in unpublished data he had brought along, and because of the connection with the new wave mechanics, Davisson made an effort to learn about the new theory. As a result he was able to design a new experiment, with L. H. Germer. Their report, published in 1927, was quickly acclaimed as the definitive confirmation of de Broglie's hypothesis. At the same time, J. J. Thomson's son G. P. Thomson demonstrated the phenomenon of electron diffraction by still another method. Later experiments showed conclusively that not only electrons but protons, neutrons, heavier nuclei, and indeed all material particles share this wavelike behavior.

8.3 SCHRÖDINGER'S WAVE MECHANICS

> Schrödinger . . . was guided by the mathematical apparatus, whose inner logic, so to speak, showed the way for the new physics to travel.
>
> —*philosopher* HANS REICHENBACH, *1933*

Before showing how this cluc was followed up, I must acknowledge that (as is well known to historians of physics) the solution to the problem of atomic structure was first discovered by Werner Heisenberg; his theory, subsequently known as matrix mechanics, was published in 1925, a few months before Schrödinger's wave mechanics. It was developed in more detail by Heisenberg, together with physicists Max Born and Pascual Jordan; physicist Wolfgang Pauli was able to deduce the energy levels of the hydrogen atom, but most physicists found it rather difficult to use.

Although the two theories are generally regarded as being mathematically equivalent, Schrödinger's is much more widely used and seems to convey a more definite (though perhaps more misleading) physical picture than Heisenberg's, and therefore will be given a more prominent place here.

According to the report by Born and Heisenberg presented at the 1927 Solvay Conference:

> The most noticeable defect of the original matrix mechanics consists in the fact that at first it appears to give information not about actual phenomena, but rather only about possible states and processes. It allows one to calculate the possible stationary states of a system; further it makes a statement about the nature of the harmonic oscillation that can manifest itself as a light wave in a quantum jump. But it says nothing about when a given state is present, or when a change is to be expected. The reason for this is clear: matrix mechanics deals only with closed periodic systems, and in these there are indeed no changes. In order to have true processes . . . one must direct one's attention to a *part* of the system; this is no longer closed and enters into interaction with the rest of the system.

Heisenberg did play an important role in the reception of quantum mechanics, even though his matrix theory lost out in the competition with wave mechanics. His analysis of the helium spectrum led to the first successful quantitative calculation of the energy levels and other properties of that atom, and thus showed that quantum mechanics could be applied to more complex atoms. His discovery of ortho and para helium inspired a similar discovery of two kinds of hydrogen, which was one of the achievements cited in the award of his Nobel Prize. The near-simultaneous discovery of quantum mechanics, by two completely different routes, helped to convince physicists that the theory must be valid, while showing at the same time that one did not have to accept the physical reality of waves in order to accept the general theory. More famous in the long run was his indeterminacy principle (often misleadingly called the uncertainty principle), which influenced philosophical interpretations of quantum mechanics and made the theory known to a wider public.

Erwin Schrödinger's early publications dealt with the acoustics of the atmosphere, the thermodynamics of lattice vibrations, and statistical mechanics. His discovery of the fundamental differential equation in quantum mechanics was directly related to his interest in the statistical problems of gas theory, especially Einstein's attempt in 1925 to apply the Bose counting procedure to a system of particles with finite mass. As a result of this work on gas theory he became acquainted with Louis de Broglie's dissertation on the wave nature of matter; wave mechanics is the outcome of his attempt to reconcile the apparently inconsistent ideas of Boltzmann, Bose, Einstein, and de Broglie.

Schrödinger explained his theory by using the analogy that mathematician and physicist William Rowan Hamilton, at the beginning of the nineteenth century, had pointed out between mechanics and geometrical optics. He asked: what is the generalization of Newtonian particle mechanics analogous to the well-known generalization of geometrical optics to the wave theory of light?

The wave theory of light was based on an equation (originally derived by mathematician Jean le Rond d'Alembert in the eighteenth century) that looks something like this:

$$\nabla^2 \varphi - \left(1/u^2\right) d^2\varphi/dt^2 = 0.$$

Here the symbol φ (Greek letter phi) represents a small motion of a small piece of the invisible ether that supposedly fills space. At any time (t) the motion varies from one place to another; it is a mathematical function that depends on the variables x, y, z (space coordinates) as well as on t. The symbol ∇^2 (pronounced "del-squared") followed by any such function means: compute the rate of change with time of the rate of change with time of the value of that function as you move a short distance in the x direction, do the same for motions in the y and z directions, and add those three quantities together. The term d^2/dt^2 followed by any function means: compute the rate of change of the rate of change of the value of the function, as time goes on.

If the function happened to be the position of your car on a straight road, measured in miles from a fixed point, then the rate of change with time would be the *speed* and the rate of change of the rate of change with time would be the *acceleration*. So whenever you see d^2/dt^2, think of the accelerator pedal in your car.

Finally, the letter u in d'Alembert's equation means the speed at which a wave-like disturbance travels in the ether. (For further information about this equation, see the notes to this chapter.)

Back to the twentieth century and Schrödinger's question. According to Louis de Broglie's hypothesis, the wavelength of a particle with mass m, speed v, moving in a force field with potential energy V and total energy E, is

$$\lambda = h / \sqrt{(2m[E - V])} \quad (7)$$

Now comes the tricky part. In theoretical physics or electrical engineering, it is convenient to represent a periodic wave motion by using the imaginary number i defined as the square root of minus one [$i = \sqrt{(-1)}$]. You don't need to know what i really is, since it's not a real number anyway; all you have to remember is that $i^2 = -1$. Also, some numbers are complex: they are made up of a real number and an imaginary number (e.g., $3 + 4i$). As Leonhard Euler and other mathematicians proved in the eighteenth century, the complex-valued function $e^{-i\omega t}$ can represent a quantity oscillating with a frequency ω, which for a wave of length λ, moving at a speed u, is equal to $2\pi u/\lambda$.

Schrödinger now *postulated* that φ, which depends on the variables x, y, z, t, can be written as the product of a function that depends only on t, multiplied by another function that depends only on x, y, and z. Why? Because he knew that many problems in theoretical physics can be more easily solved if that is true, especially if we know or assume that a periodic wave motion is involved. So he wrote

$$\varphi = e^{-i\omega t} \psi(x, y, z) \quad (8)$$

where ψ (Greek letter psi) stands for the (time-independent) *wave function*.

Substituting Eq. (8) into Eq. (7) and using de Broglie's hypothesis, he obtained

$$\nabla^2 \psi + \left(8\pi^2 m / h^2\right)(E - V)\psi = 0. \quad (9)$$

This is the simplest version of the *Schrödinger equation*, the most important equation in twentieth-century physical science since, suitably generalized, it is now believed to determine (apart from relativistic corrections) the behavior of all material systems. If you followed the last few paragraphs and read the corresponding notes, you now understand the basic idea of quantum mechanics about as well as it is possible to understand it.

Now comes the first great stunning achievement of wave mechanics: Schrödinger applied his equation to the hydrogen atom, assuming that the nucleus (a proton) is infinitely more massive than the electron. He found that the quantization of energy levels did not have to be arbitrarily introduced, but followed directly from the mathematical process of solving the equation. When E is negative, the equation has solutions only for a discrete set of values E_n, n = 1 to ∞, exactly the infinite set derived from the Bohr model. These are called the *eigenvalues,* a bastard word half-translated from the German word *Eigenwert,* meaning proper or characteristic values. The function ψ_n corresponding to E_n is called the *eigenfunction* for that energy. When E is positive there is a continuum of eigenvalues from 0 to ∞, just as in the Bohr model. The two other quantum numbers that had to be postulated in the Bohr model also come out naturally from the process of solving the equation, including the plausible condition that the function ψ must be single-valued at any point in space. So Schrödinger succeeded immediately in reproducing all the *observable* properties of the Bohr atom, from an equation that could be easily generalized to more complicated systems (though the equation could not easily be *solved* for those systems).

The most important generalization does what Born and Heisenberg, in the passage quoted above, admitted that their original matrix theory did not do. Schrödinger postulated how the wave function changes with time:

$$\left(h / 2\pi i\right)\left(\partial\varphi / \partial t\right) = -\left(h^2 / 8\pi^2 m\right)\nabla^2\varphi + V\varphi. \tag{10}$$

(This equation is valid only when relativistic effects can be ignored; see the next chapter.)

What is the physical meaning of the wave function? That was a subject of intense controversy in the 1920s, and I will not go into the subject here except to point out that since the numerical value of ψ is in general a complex number, it can't have any physical meaning itself. The quantity that does have a physical meaning is the square of the wave function, or more precisely the product of ψ multiplied by its complex conjugate ψ*. If ψ is decomposed into its real and imaginary parts, ψ = A + iB (where A and B are both real), then ψ* is defined as A − iB. Thus ψψ* = (A + iB)(A − iB) = $A^2 + B^2$ (remembering that by definition $I^2 = -1$, and noting that the cross terms cancel out).

Max Born proposed that ψψ * measures the probabilityf that a measurement on the particle will find it to be in the corresponding state: "The motion of particles conforms to the laws of probability but the probability itself is propagated in accordance with the law of causality."

Imagine a weather map of the United States that shows for a single day the probability of rain in each square mile area, using the colors of the spectrum (red for zero probability, e.g., up to violet for 100% probability). Whether it actually rains that day in a specified area is a matter of chance, but meteorologists can calculate approximately, using the laws of physics, how the *pattern* of probabilities will change from one day to the next. That's what Schrödinger's equation (with a term giving the rate of change of ψ with time instead of zero on the right hand side) does on the atomic level.

At this point you may suspect that wave mechanics is an abstract mathematical game that can't possibly have anything to do with the structure of atoms. What would it take to persuade you that in some way, wave mechanics does indeed govern the atomic world? Some physicists argued that the Davisson-Germer experiment provides the desired grounding in empirical fact. Schrödinger himself apparently declined to resort to this argument, stating in a letter to Wilhelm Wien that the experiment was a vindication of de Broglie's theory, but not necessarily for his own.

8.4 THE EXCLUSION PRINCIPLE, SPIN, AND THE ELECTRONIC STRUCTURE OF ATOMS

Early in the 1920s it had already become clear that the Bohr model of the hydrogen atom might be extended to a more general theory of atomic structure. An atom of atomic number Z (nuclear charge $Q = Ze$) would contain Z electrons in the neutral state; these electrons would occupy the various energy levels available to the single electron in the hydrogen atom. However, it was necessary to adopt some apparently arbitrary assumptions about how many electrons could occupy each energy level. The lowest level, whose quantum number is $n = 1$, could hold 2 electrons; the next one, $n = 2$, could hold 8; the next, $n = 3$, could hold 18, and so forth (except that there was not yet a general formula by which one could predict these numbers for higher n). It was suggested that the different electrons occupying the same energy level might be distinguished by assigning them other quantum numbers. For example, a quantum number might determine the eccentricity of an elliptical orbit. Ordinarily the elliptical orbit might have the same energy as a circular orbit, but it might be affected differently by a magnetic field; this idea was suggested by the splitting of spectral lines in magnetic fields (the Zeeman effect).

In 1925, Wolfgang Pauli suggested that all possible electron states could be accounted for by assigning each electron a set of four quantum numbers:

$$\begin{aligned} n &= 1,2,3,4,5,\ldots \\ l &= 0,1,\ldots,n-1 \\ m &= -l,-l+1,\ldots,l-1,l \\ m_s &= -\tfrac{1}{2},+\tfrac{1}{2} \end{aligned}$$

Note that I am using the modern definitions of these quantum numbers, not the definitions originally proposed by Pauli—but the basic idea is the same. The first three quantum numbers could be related to some conceivable physical property of the electron orbit-energy, angular momentum, and so on. Moreover, it was found that they could be *derived* from the Schrödinger equation, so they did not have to be postulated ad hoc. However, the fourth quantum number, m_s, could only be attributed to some mysterious two-valuedness of the electron.

Using this set of quantum numbers, Pauli could then postulate his exclusion principle: no two electrons in an atom may occupy the same state defined by values of these quantum numbers. (As we will see in Section 8.6, Enrico Fermi and Dirac successfully applied this principle to the free electron gas, and their theory was subsequently developed into the modern electron theory of metals.)

Shortly after Pauli announced his new quantum number, physicists George Uhlenbeck and Samuel Goudsmit proposed that this number could reasonably refer to a rotation or spin of the electron itself. The two values +½ and −½ correspond to clockwise or counterclockwise rotation with respect to some direction in space—defined, for example, by a magnetic field. In 1928, P.A.M. Dirac in England found that spin comes naturally out of a relativistic form of quantum mechanics (see next chapter).

The fact that only two electrons can occupy the lowest level, $n = 1$, now follows immediately, since l and m must both be zero, and m_s can have the two values $m. = +½$ or $-½$. For $n = 2$, 1 may be either 0 or 1, and for $1 = 0$ you can have only $m = 0$, whereas for $l = 1$ you can have three values of m: -1, 0, or + 1. This gives four possible combinations of (l, m) values, and for each of these there are two possible values of m_s; so the total number of states is eight.

From this set of rules physicists could immediately derive the electron-shell theory of the atom. A shell consists of all the electrons that have a given value of n. Thus the first shell is filled by 2 electrons, the second by 8, the third by 18, and so forth. A closed shell corresponds to a rare-gas atom—thus helium has 2 electrons, and neon has $2 + 8 = 10$. Things get a little more complicated after that, because the energies of the electron states of a many-electron atom are not quite the same as those for hydrogen. In fact it is surprising that they are similar at all, since the potential energy of interaction between electrons is of the same order of magnitude as that between an electron and the nucleus. However, it was possible to work out a fairly good approximate theory of the electronic structure of all the elements, and to predict correctly the order in which the various states would be filled as more electrons were added.

The chemical similarity between elements in the same column of the periodic table now had an obvious explanation. Lithium and sodium, for example, both contain one electron outside a core of closed shells; only this outer electron is involved in chemical reactions. Since the outside electron has an orbit almost entirely outside the core, it is not acted on simply by the entire attractive force of the positive nucleus (charge $Q = Ze$); instead, most of this force is canceled by the core of $Z - 1$ electrons, which repel it. So the outer electron is fairly likely to be removed from the atom under favorable circumstances. Conversely, atoms like fluorine ($Z = 9$) are one electron short of a closed shell, and tend to want to pick up that electron.

There is a widespread belief among physicists that quantum mechanics, which incorporates the exclusion principle, can in principle explain all of chemistry. However, it would be historically more accurate to say that the explanatory arrow goes to some extent in the opposite direction: the principle was introduced, at least in part, to explain the periodic table, and thus reveals an influence of chemistry on physics.

The introduction of electron spin offered a new approach to the fine structure of the spectra of hydrogen and hydrogen-like atoms, which had apparently been adequately explained by Sommerfeld's relativistic theory in 1916; in fact, the detailed spectrum of the ion He^+ was *predicted* by Sommerfeld (Equation 4, section 8.1) and later confirmed experimentally by Paschen. But in 1926 Heisenberg and Pascual Jordan showed that Sommerfeld's formula could be derived from the Uhlenbeck-Goudsmit hypothesis, and that "the relativistic explanation of the doublets collapses." This is a good example of the logical fallacy of affirming the consequent (Section 1.2): assuming that a theory is correct just because it leads to a confirmed novel prediction.

8.5 BOSE-EINSTEIN STATISTICS

In June 1924 the Indian physicist Satyendranath Bose sent Einstein a short paper in English on the derivation of Planck's radiation law, asking him to arrange for its publication in the *Zeitschrift fur Physik.* Einstein translated it into German himself and added a note stating that he considered the new derivation an important contribution; it was published a few weeks later. Bose derived Planck's distribution law without making any use of classical mechanics and electrodynamics or of Bohr's correspondence principle. Instead he used a purely combinatorial method, assuming that radiation consists of quanta of various frequencies, there being N quanta of frequency v. Bose assumed, in effect, that the quanta themselves are indistinguishable, so that permutations of quanta do not count as separate arrangements. This is the key to the new derivation, though Bose did not discuss it explicitly. He simply gave a formula for the number of ways of distributing the N quanta in these cells such that there are a specific number of vacant cells, a specified number with one quantum, and so on. He was then able to derive Planck's distribution law for black-body radiation without relying on assumptions from classical electromagnetic theory.

Bose's work supported the light quantum hypothesis (Chapter 7) by suggesting that electromagnetic radiation could be treated as a gas of particles with *zero* (rest) mass, as long as one recognized their indistinguishability, and this was presumably one reason why Einstein was attracted to Bose's approach. He immediately applied it to a monatomic ideal gas of particles with *finite* mass. He found that for a specified particle mass, volume, and temperature, the total number of particles with finite energy has a maximum value. What happens if you try to put more particles into the system? He suggested that in this case some atoms will have to go into the lowest quantum state with zero kinetic energy; the phenomenon would be analogous in some ways to the condensation of a saturated vapor. Yet there are no attractive forces to hold the atoms together in any particular region of space.

Einstein's prediction of a quantum condensation of gases, published in 1925, was not taken seriously by other physicists for more than a decade. One reason for this was that Uhlenbeck, in his 1927 dissertation, criticized Einstein's mathematical treatment of the problem; he argued that the result was obtained only by replacing the sum over energy levels by an integral, and that if the sum was computed accurately, no condensation would occur. Although Uhlenbeck's dissertation was published only in Dutch, his research director was Ehrenfest, and presumably his criticism became known to the physics community through informal communications. Uhlenbeck retracted his criticism in 1938, and Fritz London made Bose-Einstein statistics the basis of a theory of the superfluidity of liquid helium.

8.6 FERMI-DIRAC STATISTICS

In the late 1920s, theoretical physicists were starting to discuss quantum mechanics in terms of two new kinds of quantum statistics, Bose-Einstein and Fermi-Dirac, to be contrasted with the classical version now called Boltzmann statistics. The alternative Fermi-Dirac statistics is based on the exclusion principle, proposed in 1925 by Wolfgang Pauli in order to explain the anomalous Zeeman effect in atomic spectra, and more generally to provide a systematic method of accounting for the filling of allowed orbitals by electrons (Section 8.4). Despite Pauli's initial reluctance to accept it, electron spin became an important feature of Fermi-Dirac statistics.

Enrico Fermi presented his quantum theory of the ideal gas, based on the Pauli exclusion principle, to the Accademia dei Lincei in February 1926, and submitted a comprehensive paper on this theory to the *Zeitschrift fur Physik* on March 26; Dirac's paper was presented to the Royal Society of London on August 26, 1926. Both Fermi and Dirac presented essentially the same derivation for their distribution law, but emphasized different applications.

Dirac pointed out that the difference between Bose-Einstein and Fermi-Dirac statistics corresponds to the difference between wave functions that are symmetric and antisymmetric with respect to interchanges of particles. Suppose we have a wave function that satisfies the Schrödinger equation for two electrons. It will be a function of six space variables, plus two variables to specify the electron spins: $\psi(x_1, y_1, z_1, s_1, x_2, y_2, z_2, s_2)$, which for the present purpose we abbreviate as ψ (1, 2). "Symmetric" in this case means that the function remains the same, $\psi (2, 1) = \psi (1, 2)$, whereas "antisymmetric" means that it changes sign, $\psi (2, 1) = -\psi (1, 2)$. In both cases the square of the wave function (or, if it is complex-valued, the wave function multiplied by its conjugate, $\psi\psi^*$) remains the same, this being the quantity that could have a direct physical significance.

If particles 1 and 2 are in the same quantum state and the wave function is antisymmetric, it must be zero (since zero is the only number that is equal to minus itself). If the wave function of a particle is zero everywhere, it doesn't exist. Thus the Pauli principle is automatically satisfied for antisymmetric wave functions.

Whereas the ideal Bose-Einstein gas (one in which there are no forces between the particles) can undergo a condensation to a state in which a finite fraction of the particles are in the ground state, the ideal Fermi-Dirac gas behaves very differently because at most only one particle can be in the ground state. At absolute zero temperature the states with energies up to a fixed energy E_F, called the Fermi energy, will be filled; states with higher energies will be empty. The states up to the Fermi energy comprise the Fermi sea. At higher temperatures a few particles close to the top of the Fermi sea may acquire enough energy to jump to states with energies higher than E_F, but a particle further down in the Fermi sea cannot absorb a small amount of energy. As a result, the specific heat of the gas is very small at low temperatures.

Pauli and Sommerfeld quickly used Fermi-Dirac statistics to construct a theory of electrons in metals. It had already been realized before 1926 that many properties of metals could be explained by assuming the electrons are free to move throughout the system, but this assumption seemed inconsistent with the fact that electrons made hardly any contribution to the specific heat of the metal. It turned out that the ideal Fermi-Dirac gas gave rather good results by ignoring the electrostatic forces between electrons. J. E. Lennard-Jones, an expert on statistical mechanics who is famous for his formula for interatomic forces, wrote that the quantum-mechanical explanation for the small contribution of electrons to the specific heat "disposes of a well-known difficulty in the electron theory of metals, and constitutes one of the most striking successes of the new statistics."

Electrostatic forces have essentially no effect on electrons inside the Fermi sea; interactions between a pair of electrons cannot scatter them into different states (as one expects in a classical theory) because there are no unoccupied states available. The probability of a scattering process involving enough energy to kick one electron up above the surface of the Fermi sea is very small unless the electrons are already close to the surface.

8.7 INITIAL RECEPTION OF *QUANTUM MECHANICS*

In October 1927, Karl K. Darrow wrote that wave mechanics "has captivated the world of physics in a few brief months," not because of its successful predictions or its superior agreement with experience but "because it seems natural or sensible or reasonable or beautiful."

What two things do the following scientists have in common: Hans Bethe, Peter Debye, P. P. Ewald, Herbert Fröhlich, Werner Heisenberg, Walter Heitler, Alfred Landé, Fritz London, Wolfgang Pauli, Linus Pauling, Albrecht Unsöld, and Gregor Wentzel?

First, all made significant contributions to the development or applications of quantum mechanics; second, all studied with Arnold Sommerfeld. (We might also add S. Chandrasekhar, who was influenced to become an astrophysicist by a brief encounter with Sommerfeld). The Sommerfeld school at the University of Munich, according to historian Suman Seth, was "the most successful educational institution for theoretical physics in the twentieth century." Seth writes that Bethe, in a 1964 interview with Thomas Kuhn, "remembered the enthusiasm with which members of the Munich School adopted what they saw as an almost miraculous new means of solving previously intractable problems." Bethe exclaimed that "you could now do all this . . . one could *do* the Stark effect and *do* the intensities . . . I think this first seminar [on quantum mechanics] was simply fascinated by the fact that it now worked."

Just as Sommerfeld was the doctor-father of many twentieth-century theoretical physicists, he himself had a distinguished academic genealogy, which included German mathematicians going back to the seventeenth century, as traced by a contemporary physicist, Roman Jackiw, who studied with Bethe.

It is difficult to disentangle the initial reception of quantum mechanics from the development of the theory itself. If we consider quantum mechanics to be Heisenberg's matrix mechanics, then the first published response was the paper "Zur Quantenmechanik" by Max Born and Pascual Jordan, which appeared in *Zeitschrift für Physik* on November 28, 1925. This was quickly followed by Paul Dirac's "The Fundamental Equations of Quantum Mechanics," in *Proceedings of the Royal Society of London*, December 1, 1925. The next month Wolfgang Pauli completed a paper deducing the hydrogen spectrum from matrix mechanics, and it was published on March 27, 1926. Why did these four physicists consider Heisenberg's theory important enough to cause them to drop other work and devote their efforts to it?

Born and Jordan began by stating that Heisenberg's "theoretical approach . . . appears to us of considerable potential significance. It represents an attempt to render justice to the new facts by setting up a new and really suitable conceptual system instead of adapting the customary conceptions in a more or less artificial and forced manner." But more remains to be done: "His hypotheses have been applied only to simple examples without being fully carried through to a generalized theory." Going beyond Heisenberg's paper they introduced the techniques of matrix analysis from mathematics, and derived the law of conservation of energy and the Bohr frequency relation.

Dirac's motivation for working on Heisenberg's quantum mechanics, as far as can be detected from his published paper, is that it "suggests that it is not the equations of classical mechanics that are in any way at fault" for their failure to describe transitions from one state to another, "but that the mathematical operations by which physical results are deduced from them require modification. *All* the information supplied by the classical theory can thus be made use of in the new theory." This gave him the opportunity to work out an interesting new mathematical theory, which he called quantum algebra. He did not yet consider it necessary to predict any specific experimental facts from his version of the theory.

Pauli, on the other hand, was very much concerned with the details of atomic spectra that could not be correctly explained by the old quantum theory. He called Heisenberg's theory a "considerable advance over the previous theory of multiply periodic systems," and, as further extended by Born, Jordan, and Dirac, "a consistent mathematical system." So he was willing to carry out elaborate matrix calculations that Heisenberg, Born, Jordan, and Dirac had not yet successfully undertaken. He calculated the energy levels for an atom with a single electron, including the case when external electric and magnetic fields are present. He was able to reproduce the known levels for the Balmer series of hydrogen and the corresponding Stark effect; "difficulties disappear which had arisen in the old theory through the additional exclusion of singular motions in which the electron comes arbitrarily near the nucleus, and which became particularly evident for the case of crossed electric and magnetic fields." In other words, the singular states in which the electron's orbit intersected the nucleus had to be arbitrarily excluded in the old quantum theory, but disappeared as a logical consequence of the basic postulates of the new theory. However, Pauli concluded that quantum mechanics could not yet explain the anomalous Zeeman effect; he was not satisfied with the Goudsmit-Uhlenbeck hypothesis, which attempted to account for the effect by using electron spin.

One of the reasons why Pauli resisted the electron-spin explanation of the anomalous Zeeman effect was a missing factor of ½, which was needed to achieve agreement with experiment. This factor was supplied by L. H. Thomas, who derived it from an improved relativistic treatment, but Pauli could not understand or accept this calculation until after heated arguments with Bohr and Goudsmit. After Pauli withdrew his objections, Heisenberg and Jordan published the solution in June 1926. It agreed with a formula previously derived from the old quantum theory by Sommerfeld in 1922 (without postulating electron spin), which was known to agree with experiment, so it was not a novel prediction. It was nevertheless regarded as a major achievement of quantum mechanics.

Niels Bohr, perhaps the most influential quantum physicist at this time, gave matrix mechanics his blessing in a section added to the end of his lecture "Atomic Theory and Mechanics," published in *Nature* as a supplement to the issue of December 5, 1925. He reported that Heisenberg "has taken a step probably of fundamental importance by formulating the problems of the quantum theory in a novel way by which the difficulties attached to the use of mechanical pictures may, it is hoped, be avoided." He noted the fact that Born and Jordan, in a paper "to appear shortly," could show that the theory contains "a conservation theorem analogous to the energy law of classical mechanics" as well as being consistent with the postulates of the old quantum theory and the correspondence principle. In a note added just prior to publication, he mentioned Pauli's unpublished deduction of "the Balmer formula for the hydrogen spectrum as well as the influence on this spectrum of electric and magnetic fields. This is a very important result, since Pauli's analysis shows how the new theory avoids the difficulty involved in the older account of the spectral evidence in the necessary exclusion of the stationary states corresponding to the singular solution of the electronic motion." In treating the radiative properties of atoms, "Heisenberg's theory represents a real advance. In particular it allows us, in the phenomena of scattering, to recognise the presence of electrons bound in atoms in a way completely analogous to the classical theories, which . . . enabled the number of electrons in an atom to be counted from measurements of X-rays."

In April 1925 Bohr, preparing to give a lecture at Cambridge, had written to Rutherford about "our present theoretical troubles which are of an alarming character indeed." In a letter to Rutherford dated January 27 of the next year, his mood had greatly improved thanks to the work of Heisenberg: "prospects have with a stroke been realized, which although only vague[ly] grasped have for a long time been the centre of our wishes. . . . we now see the possibility of developing a quantitative structure." Moreover, the electron spin idea proposed by Goudsmit and Uhlenbeck "allows in a remarkable way to overcome the apparently grave difficulties in the spectral theory which have puzzled us so much in later years. . . . I am at present just as optimistic . . . as I was when I gave my first lectures in Cambridge 5 years ago." Two days later (January 29, 1926), Bohr wrote to physicist C. W. Oseen: "the atomic theory has passed through a serious crisis in these last years, and at times we have been thoroughly bewildered. Now we again see brighter days ahead, first of all because of the development of the new quantum mechanics which, resting upon ideas of Kramers and especially of Heisenberg, has been shaped into a wonderful theory by Born. In fact, this theory actually promises to satisfy so many vague wishes that gradually had come more and more to the fore . . . At the same time as we apparently thus finally attain a basis for the construction of a quantitative atomic theory, the very foundation of our views on atomic structure has received an exceedingly welcome supplement in the assumption, introduced by Goudsmit and Uhlenbeck, according to which the electron, in addition to its electric charge, possesses a magnetic moment caused by a rotation about its own axis . . . [when I heard about this] I was seized by great enthusiasm. In fact, all the difficulties which in the last few years have accumulated in the analysis of the fine structure of spectra, the Zeeman effect and related phenomena, seem to disappear completely."

A discordant note was sounded (but only in private) by Einstein, who wrote to Paul Ehrenfest in September 1925: "Heisenberg has laid a big quantum egg."

The rapid transition from optimism about matrix mechanics to delight in wave mechanics within a few months is illustrated by three letters of Einstein in the glorious spring of 1926. On March 7, having tasted Heisenberg's egg as cooked by Max Born and Pascual Jordan, he wrote to Born's wife Hedwig:

> The Heisenberg-Born concepts leave us all breathless, and have made a deep impression on all theoretically oriented people. Instead of dull resignation, there is now a singular tension in us sluggish people.

On April 16 he wrote to Erwin Schrödinger:

> The idea of your work springs from pure genius.

Ten days later, to Schrödinger:

> I am convinced that you have made a decisive advance with your formulation of the quantum condition, just as I am convinced that the Heisenberg-Born method is misleading.

And then, in cold December, in a letter to Max Born, appears Einstein's most famous statement:

> Quantum mechanics is certainly imposing. But an inner voice tells me that it is not yet the real thing. The theory says a lot, but does not really bring us any closer to the secret of the "old one." I, at any rate, am convinced that He is not playing at dice.

Max Planck, now the grand old man of quantum theory, was tempted to violate his own sociological principle (that new ideas win out, not by converting the established leaders of science but by waiting for them to retire or die); on April 2, 1926, just before his sixty-eighth birthday, he wrote to Schrödinger,

> I read your article the way an inquisitive child listens in suspense to the solution of a puzzle that he has been bothered about for a long time, and I am delighted with the beauties that are evident to the eye, but I have to study it much more closely and in detail to be able to grasp it completely.

Arnold Sommerfeld, still very active at age 57, immediately became a strong and effective advocate for wave mechanics. On February 3, 1926, he wrote to Schrödinger:

> This is really terribly interesting . . . I was just preparing an outline for lectures in London that whistled the same old tune. Then, like a thunderstroke, your manuscript dropped in. My impression is this: Your method is a replacement of the new quantum mechanics of Heisenberg, Born, Dirac, and certainly a simpler one, so to speak an analytic solution of their algebraic problem.

More important than his kind words to Schrödinger were his actions at his own institute in Munich. Historian Michael Eckert surmised: "Sommerfeld admired Schrödinger's wave mechanics because of its mathematical beauty, but he could not bring himself to respect his own pupil Heisenberg's cumbersome matrix theory." He immediately made wave mechanics the subject of his seminar, attended by several stars of the next generation, including Hans Bethe, Walter Heitler, Fritz London, and Linus Pauling. Each student was given a topic from Schrödinger's theory to present to the group; many of them continued on to do outstanding research in wave mechanics.

In 1928, on a lecture tour in India, Sommerfeld met a young student, Subrahmanyan Chandrasekhar, who was fascinated by atomic physics, and gave him the galley proofs of his forthcoming paper on Fermi-Dirac statistics. Chandra, as he is now known to all astrophysicists, later called that encounter the "single most important event" in his life; it led to the research for which he received the 1983 Nobel Prize in physics.

With Albrecht Unsöld, who wrote "probably the first doctoral dissertation in the area of Schrödinger's wave mechanics" and later became a prominent astrophysicist, Sommerfeld recalculated the fine structure of the hydrogen spectrum. They found several previously unknown lines, and their predictions were (mostly) confirmed by physicists Norton A. Kent, Lucien B. Taylor, and Hazel Pearson at the California Institute of Technology (Caltech). These were among the first published confirmations of novel predictions from quantum mechanics, but I found no evidence that they persuaded anyone to accept the new theory.

Let's look now at the way a major physicist who was *not* directly involved in the development of quantum theory but contributed to both theoretical and experimental physics reacted to quantum mechanics. Wilhelm Wien, now in his sixties but still active, closely followed the new ideas. On first reading Schrödinger's papers he wrote to the author (March 22, 1926):

> It is in the highest degree amazing and delightful how in such quick steps you have advanced, and your new result, which at least opens a bridge to matrix theory, leaves no doubt that you have chosen the right path.

Two months later (May 13, 1926), he praised wave mechanics because it was not a radical departure from traditional ideas:

> There can hardly be any more doubt that you have solved the problem of atomic vibrations, and indeed in a delightful way in the closest harmony with the classical theory. The entire quantum theory is no longer so strange as it was up to now. Also Planck, with whom I have spoken on this subject, was completely enthusiastic and completely agrees with my opinion.

But the problem of radiation—that is, the coupling between individual vibrations and the electromagnetic field—is still not solved..

On June 19, 1926, Wien gave a public lecture in Munich on the past, present, and future of physics. He proposed to describe "the struggles that have now lasted three decades and whose end is not yet in sight." After reviewing the historical background of quantum mechanics he stated that wave mechanics is reminiscent of Pythagorean number mysticism, but if it works it would *replace* number mysticism by the cool logic of physical thought. Wien pointed out that the theory is only half a year old and it's too soon to make a final judgment on its validity.

As a tentative conclusion, he noted that we know that the laws of mechanics do not apply in realm of the atom, likewise the theory of light. Near the smallest parts of matter there is a wave system that spreads over the entire space of the atom. This wave mechanics is a development that should not surprise us too much, in the light of the contentious past history of physics; it is part of an ongoing battle between different ideas.

By August he was urging Schrödinger to visit for a few days, to thrash out some difficulties he now perceived to afflict the new theory (August 20, 1926):

> I'd like to discuss some theoretical questions with you. One of the most serious difficulties for wave mechanics is the photoelectric effect. I've spoken to Bohr and

> Born in Oxford about this, and both have the conviction that it is not compatible with wave mechanics. The former holds a space-time representation to be impossible, the latter wants to give up the law of causality. I don't agree with either one. It's not clear to me how the photoelectric effect will be explained. . . . It seems to me that one can't get at it without statistics.

A year later Wien was still dissatisfied with the existing interpretations of quantum mechanics (letter to Schrödinger, September 6, 1927):

> Against the view of de Broglie and Born I have nothing to substitute, if one considers it as a preliminary means of representation. We also cannot demand that in such a short time all questions of quantum theory must be completely clarified but can we also not say that we know at present nothing better? However I am completely convinced that one should never be satisfied that in the elementary processes "probabilities" occur, therefore causality will be given up. It is an error that the younger generation easily falls into, that it should be content to find the philosopher's stone. There will be enough for the younger generation to do.

When he heard about the Davisson-Germer experiment he told Schrödinger that it was "a great triumph for your theory," to which Schrödinger replied: "Not a triumph for my theory, rather for that of de Broglie." Indeed, at that time, de Broglie's pilot wave theory was still a serious rival to Schrödinger's version of wave mechanics, and de Broglie pointed to this result as a confirmation of his theory.

Wien has been remembered as a physicist who rejected quantum mechanics in 1926 but accepted it later because of the Davisson-Germer experiment. I don't think this is accurate in view of the evidence quoted above, but additional documents may be found to confirm it or tell a more complicated story.

8.8 THE COMMUNITY IS CONVERTED

> We of the younger generation accepted almost unquestioningly both the formalism of quantum mechanics and its interpretation, and under that banner we went forth into the wonderful world of atomic physics. Those were the days when ['] every morning brought a noble chance,[/] And every chance brought out a noble knight ['] Days when even lesser knights could find rewarding tasks.
>
> —H.B.G. CASIMIR *(1983)*

> Heisenberg and Schrödinger had knifed a sack of gemstones, and the race was on to pick out the diamonds.
>
> *—physicist* GRAHAM FARMELO *(2009)*

There are a number of published studies of the reception of quantum mechanics. But perhaps the best indication of the initial impact on research is given in the paper by historians A. Kozhevnikov and C. Novik. They cite 203 papers, completed and submitted for publication from July 1925 (when Heisenberg's first paper was written) through February 1927. There were 80 authors from 14 countries. Although

Kozhevnikov and Novik do not estimate the total size of the physics community at this time, it is certainly remarkable that so many physicists had already started to use or at least mention the new theory less than two years after its publication. Ten of these papers were reviews, translations, or preliminary communications; putting those aside, they are left with 166 papers reporting original research. Although I have not examined all these papers, my impression is that they rarely give a specific reason for adopting quantum mechanics but make it clear that they think it works: it is the best if not the only way to solve the problem in which they happen to be interested. This view was strengthened by the fact that matrix mechanics and wave mechanics were soon shown to be mathematically equivalent, or at least they

Table 8.1 Reasons for Accepting Quantum Mechanics (QM) According to 10 Authors of Monographs and Review Articles, 1927 and 1928

Half-quantum numbers (8): These had to be arbitrarily postulated to explain spectral lines and zero-point energy (postulated by Nernst in his theory of specific heats at low temperatures); now they are a natural consequence of the QM postulates.
Stark effect (8): The old theory had to arbitrarily exclude electron orbits that would pass through the nucleus (the azimuthal quantum number can't be zero); this follows directly from QM, which also gives better agreement with experiment.
Compton effect (6): Calculation of details (distribution of scattering angles, etc.) gives better agreement with experiment and resolves wave/particle dilemma by deriving both aspects from one theory.
Wave nature of matter (5): De Broglie's hypothesis is supported by electron diffraction experiments of Davisson and Germer, G. P. Thomson, and others.
Zeeman effect (4): QM explains anomalous Zeeman effect with help of electron-spin postulate. That postulate is not itself arbitrary, since Dirac (1928) showed it to be a consequence of special relativity theory. QM also explains the Paschen-Back effect (the anomalous Zeeman effect becomes normal as field strength increases).
Helium atom (4): Old quantum theory completely failed to explain this and other systems with two or more electrons. Heisenberg introduced an approximation scheme based on the resonance concept, which eventually led to good values of spectral frequencies and ground-state energy.
Electronic properties of metals (4): According to Fermi-Dirac statistics, electrons behave like a degenerate gas at low temperatures, with all states filled up to Fermi level; electrons below Fermi level make almost no contribution to specific heat because they can't absorb small amounts of energy. Quantum theory of metals can explain paramagnetism (Pauli), Wiedemann-Franz law, etc.
Scattering and absorption of radiation by atoms (4): Explains why resonant frequency corresponds to spectroscopic frequencies (Kramers dispersion formula) rather than to frequencies of motions of electrons in atoms.
Hydrogen spectrum (4): QM gives frequencies of both discrete and continuous states from single equation, without invoking arbitrary quantum conditions; it also allows calculation of intensities of lines (previously estimated only roughly from correspondence principle).
Intensities of spectral lines (3): Can be calculated directly; in the old quantum theory they could only be approximated using the correspondence principle.

yielded identical results for almost all practical purposes. Thus each was enhanced by the success of the other, and those who found matrix mechanics too complicated and abstract could work more easily and confidently with wave mechanics. The disagreement between Schrödinger and the supporters of matrix mechanics about the physical meaning of quantum mechanics did not stop the gold rush to extract as many results as possible from this fertile theory.

According to Kozhevnikov and Novik, more than half of the 80 authors were under age 30 at the beginning of 1926. But Planck's principle does not seem to apply here, since "the majority of the authors worked or were educated before in the old quantum theory tradition," yet there was little resistance to the replacement of the old theory by the new. This quick and easy acceptance of quantum mechanics was "probably because the old quantum theory was already in a crisis and the appearance of something like quantum mechanics was expected by a majority of scientists." So it was not necessary for the old quantum theory generation to retire before the new quantum mechanics generation could take over; on the contrary, many of the former found it quite easy to apply wave mechanics, especially if they had worked in areas of theoretical physics involving partial differential equations.

A quick survey of the titles listed by Kozhevnikov and Novik suggests that the most popular topic is the interpretation of molecular spectra by applying quantum mechanics to rotating or vibrating models (15 papers). Several theoretical treatments of scattering (collisions, dispersion) were also included (10), as were calculations of the dielectric constants of dipole gases (9) and treatments of the hydrogen atom (8). There were a few papers on magnetism and the Stark effect (5 each), the Compton effect and quantum statistics (4 each), and the Zeeman effect (3). These numbers suggest that chemical applications of quantum mechanics were already becoming a major factor in its widespread acceptance.

Review articles, monographs, and textbooks are a better source of information about the reasons for accepting quantum mechanics. They are directed in part at readers who are not yet experts on the new theory but want to understand its significance, scope, and methods. I found six books and four articles, published in 1927 and 1928. Four authors are British (Allen, Birtwistle, Flint, Richardson); two are German (Jordan, Westphal); one Austrian (Thirring); and two Americans (Darrow, Van Vleck).

Table 8.2 lists the achievements of quantum mechanics described by two or more authors. There is a fairly clear consensus that the primary advantage of quantum mechanics is its greatly superior ability to explain, in a consistent and accurate way, details of the hydrogen spectrum, involving half-quantum numbers and the effect of electric fields (the Stark effect). In addition, it resolves the puzzling inconsistency between wave and particle aspects of electromagnetic radiation by deducing both from one theory. But see also Table 8.3 for reasons for accepting Quantum Mechanics (QM) claimed by authors of textbooks and popularizations between 1929–1932.

8.9 NOVEL PREDICTIONS OF QUANTUM MECHANICS

Given the instrumentalist flavor of the writings of its advocates, especially Heisenberg and Bohr, one might have expected that quantum mechanics would have been put forward with strong claims about its confirmed novel predictions—but that was not the case. Instead, the advocates argued that quantum mechanics accounted at least as well for the facts explained by the old theory, explained several anomalies that its predecessor had failed to resolve, and gave a single consistent method for doing calculations in place of a collection of ad hoc rules. With the possible exception of the Davisson- Germer experiment, none of the reasons mentioned in the 1927–1928 surveys (Table 8.1) could be considered a confirmed novel prediction of quantum mechanics. Even after 1928, when there were some confirmed novel predictions, most physicists did *not* ascribe any extra significance to them because of their novelty. (A few exceptions will be noted in the following pages.) So although quantum mechanics is often praised because all of its predictions have been confirmed, as far as I can determine from a survey of books and articles, the confirmation of its *novel* predictions played essentially no role in the rapid acceptance of quantum mechanics by researchers in atomic physics before 1929, and only a secondary role in its reception by the rest of the physics community from 1929 to 1935.

It appears that the earliest novel predictions of quantum mechanics were:

1 hydrogen molecules should exist in two forms, ortho and para, detectable by spectroscopic and specific heat measurements, and capable of being physically separated;
2. new lines in the fine structure of hydrogen;
3. the intensities of Stark components in the spectra of hydrogen and helium (i.e., lines shifted by an external electric field) have numerical values different from those previously measured by experimentalists, including Stark himself;
4. the scattering of pairs of identical particles, known as Mott scattering, is significantly different from the scattering of nonidentical but similar pairs of particles;
5. the electron distribution in an atom is a continuous function of distance from the nucleus (and perhaps of angle measured a direction corresponding to an external field), rather than a discontinuous function as assumed by the Bohr model. I have looked into ***some other alleged novel predictions*** and decided that they should not be counted.

Ortho and Para Hydrogen

Table 8.1 does not include the discovery that diatomic molecules like H_2 can have two forms, known as ortho and para, because the spins of their two nuclei can be aligned parallel or antiparallel. Transitions from one form to the other are so slow that for practical purposes they can be regarded as distinct substances, with similar but slightly different spectra. The prediction was made by Friedrich Hund in 1927, and Heisenberg (announced at the Solvay Congress in 1927, published 1928). It was used to explain the anomalous specific heats of hydrogen by David M. Dennison in 1927. The two forms were physically separated in 1929 by K. F. Bonhoeffer and P. Harteck.

The theoretical discovery of the two forms of hydrogen was one of the two achievements for which Heisenberg received the Nobel Prize (the other was matrix mechanics), though his part in the discovery was indirect and he did not even mention it in his Nobel lecture. I found no publications that mentioned this discovery as a reason for accepting quantum mechanics until 1929, when Rutherford noted that Hund and Dennison had proposed that the hydrogen molecule has ortho and para forms, by analogy with Heisenberg's theory of the helium spectrum.

Fine Structure of Balmer Lines of Hydrogen

In 1926 Sommerfeld and Unsöld used wave mechanics to calculate frequencies and intensities of lines in the spectrum of hydrogen. They obtained more accurate values for most of the lines and predicted 5 lines not previously detected. The following year, physicists Norton A. Kent, Lucien B. Taylor, and Hazel Pearson at Caltech confirmed this prediction.

Stark Effect Intensities

In 1929, physicist J. Stuart Foster and his student Laura Chalk published their experimental results on the intensities of Stark components of hydrogen. They concentrated on those components for which Schrödinger's wave-mechanical calculation disagreed most strongly with Stark's original experimental values, and they found that Schrödinger's values were accurately confirmed. They had published a preliminary announcement of some of their results in October 1926, showing rough agreement with Schrödinger's predictions but giving so few details that this experiment could hardly be regarded as a crucial confirmation. This paper was not cited in any major physics journal before 1930 except by Foster himself, and there was only one mention in a book.

Foster's observations on the Stark effect in helium also confirmed quantum mechanics, but it is not clear from his published report whether he made the observations before or after he had done the calculations. He implied that the calculations were done first and that additional observations at high fields were then made to test them, but the ambiguity of his exposition is itself an indication of how little importance he ascribed to novelty.

Foster himself, though an important figure in the development of Canadian physics, is rarely mentioned in histories of quantum mechanics; if he helped Heisenberg and Schrödinger win their Nobel prizes, they left no acknowledgement on the public record. Chalk, later known by her married name Laura Chalk Rowles, made only a token effort to claim credit for her contribution to the establishment of quantum mechanics; she is likewise overlooked by most historians, even those with a special interest in the contributions of women.

Mott Scattering

Another early novel prediction from ***quantum mechanics*** was physicist N. F. Mott's (1930) calculation of the scattering of identical particles, following the methods proposed by Max Born and Robert Oppenheimer. His formula indicated that the scattering at a 45° angle should be twice as great as for nonidentical particles, showing the effect of Pauli's exclusion principle and the indistinguishability of identical particles. Mott's prediction was confirmed experimentally for α particles by James Chadwick (1930) and by P.M.S. Blackett and F. C. Champion (1931), for electrons by E. J. Williams (1930), and for protons by C. Gerthsen (1931). Again, there is a significant ambiguity about the novelty of one of those predictions: Williams said that he had made the observations in 1927, and Mott acknowledged that he had known of the results before publication, but didn't say if he knew them before he did his calculation.

Mott later recalled that "this was one of the few totally unexpected things predicted by quantum mechanics before any sign of it was found experimentally, and impressed Rutherford a good deal." Nevertheless there is no evidence that this confirmed novel prediction converted anyone, who had not yet accepted quantum mechanics, although some who were already converted cited it as additional evidence.

Electron Distribution in Atoms

In 1927, chemist Linus Pauling calculated the electron distribution in the helium atom from wave mechanics. A. H. Compton used C. S. Barrett's data on the scattering of X-rays by helium to test Pauling's result, and found "close agreement." The wave mechanical distribution "differs by more than the probable experimental error from the electron orbits given by Bohr's theory . . . The striking similarity between this distribution predicted by the [new] quantum theory and that coming from our interpretation of the scattering experiments is the more convincing when it is noted that there are no arbitrary constants to make the two curves correspond. This agreement is a strong argument in favor of a continuous electron distribution, as predicted by the wave mechanics, as opposed to the Bohr quantum theory of definite orbits."

As noted in the previous chapter, Compton was one of the few physicists who believed in the special evidential value of novel predictions, at least in the case of his own confirmation of the light quantum hypothesis. So he is consistent in claiming this evidential value for predictions made and tested by other physicists. But I have found only one other physicist who cited electron distribution as a confirmed prediction of quantum mechanics before 1936.

8.10 THE HELIUM ATOM

If quantum mechanics was going to provide an accurate account of the properties of any system other than the hydrogen, its proponents would have to develop practical methods for solving its basic equations for atoms and molecules with more than one electron. The simplest such system is the helium atom, consisting of a positively charged nucleus and two negatively charged electrons. The simplest property of this system is the ionization potential or ground-state energy: the energy required to remove one electron from the lowest bound state and boost it to the continuum of unbound states. According to the old Bohr quantum theory this energy is

$$I_{He} = 28.9 \text{ electron volts } (\text{eV}).$$

In 1923, H. A. Kramers developed another model, still based on electron orbits, from which he derived a value of 20.63 eV. The observed value, reported by Theodore Lyman in 1924, was 24.46 eV.

After the introduction of quantum mechanics, an attempt to calculate I_{he}, using a simple perturbation treatment of the Schrödinger equation, was made by A. Unsöld. He obtained a result similar to Kramers's, 20.3 eV. Georg Kellner , with a more elaborate calculation, obtained the result 23.75.

In 1926 a young Norwegian physicist, Egil Hylleraas, was granted a fellowship by the American International Education Board to study abroad. Having been inspired by Max Born's book *Dynamik der Kristallgitter* (1915), he had started his own research on the atomic theory of crystals, and eagerly pursued the opportunity to work with the great expert on this subject. But when he arrived in Göttingen he was "heartily disappointed" to learn that Born had already abandoned the subject and was deeply involved in the development of quantum mechanics. In the library of Born's institute he found an American physicist, E. H. Kennard, who was reading about Schrödinger's wave equation and asked Hylleraas why he was not doing the same thing, since everyone else was working on the new theories. Hylleraas, with some reluctance, abandoned his beloved crystal theory and took up quantum mechanics, accepting Born's suggestion to work on the vexed problem of the helium atom.

Heisenberg had shown in principle how the problem might be attacked, using the postulate that the two electrons must be not only identical but indistinguishable (the basis of Mott scattering). As noted in Section 8.6, this means that the wave function may be symmetric or antisymmetric under an interchange of the coordinates of the two electrons. Heisenberg proposed to approximate the actual wave function by a sum of symmetric and antisymmetric functions, using the term "resonance" to describe the idea that the two electrons (mathematically) change places in such a way as to minimize the energy of the system.

Hylleraas wrote that a recent paper by J. C. Slater suggested to him the idea of a mutual polarizing effect between the distributions of the two electrons. The use of a parametric effective nuclear charge, to be found by a variational method, was suggested by Kellner. And so Hylleraas embarked on an arduous task, the calculation that Born argued was "the simplest crucial test of the correctness of wave mechanics in general . . . Its application to the helium atom—in particular to the ground state." He was assisted by some of the best mathematical physicists in the world, who happened to be in Göttingen at that time: in addition to Born, mathematicians Richard Courant and David Hilbert had just published their classic textbook on mathematical methods and were lecturing on the subject, and the young physicist Eugene Wigner, apparently the first physicist who appreciated the usefulness of group theory, was also there.

Hylleraas submitted the report of his first calculation in *Zeitschrift für Physik* in March 1928, with a note added in proof in April. He used the well-known variational procedure of Walter Ritz, and carrying the calculations through to the eleventh approximation, he found a result for the ionization potential of helium corresponding to 24.35 eV. This differed by only 0.15% from Lyman's experimental value, 24.46 eV. He recalled in 1963 that his result "was greatly admired and thought of as almost a proof of the validity of wave mechanics, also, in the numerical sense. The truth about it, however, was, in fact, that its deviation from the experimental value by an amount of one-tenth of an electron volt was on the spectroscopic scale quite a substantial quantity and might as well have been taken to be a disproof." After worrying about this discrepancy for "a year or so," he realized that it was probably due to the incompleteness of the set of functions he had used in his calculation, and that this defect could be remedied by a change of coordinates. This had, "to my astonishment and to my great satisfaction as well, almost the effect of a miracle," and reduced the discrepancy to an amount so small even on the spectroscopic scale that it was within the limits of observational error.

Hylleraas recalled: "The result of the new method was published in the first half of 1929, but I was unaware of its recognition until I had presented it myself before Det Skandinaviske Naturforskermöto in Copenhagen in September of the same year." This meeting, held at Bohr's Institute, was his first opportunity to meet many of the leading Nordic physicists: "I was somewhat struck by the spontaneous acclamation that followed my report, and in a happy mood I returned to my seat at a peaceful place far back." Lost in his own daydreams, he paid no attention to the report of the next speaker ("a very young bright-haired Swede," the experimentalist Bengt Edlén), after which the great Niels Bohr suddenly called upon him to give his opinion on "whether my method might possibly be applied" to Edlén's problem, the positive lithium ion. He avoided embarrassing himself "by keeping a bit silent and looking as wise as I possibly could," and later developed "a most exciting collaboration" with Edlén in research on other atomic systems.

Although it seemed in the 1930s that Hylleraas had completely solved the problem of calculating the ground state or ionization potential of the helium atom, later research turned up small errors in his work that disturbed the agreement between theory and experiment. Nevertheless, the way toward ever-more-accurate calculations was clear; reviewing recent work in 1964, Hylleraas declared that Heisenberg's explanation of the helium atom was "one of the greatest achievements of quantum mechanics."

8.11 REASONS FOR ACCEPTING *QUANTUM MECHANICS* AFTER 1928

I chose the four-year period 1929–1932 in order to see whether any of the early novel predictions and explanations mentioned above had joined the list of reasons for accepting quantum mechanics, assuming (correctly as it turned out) that Anderson's discovery of the positron would not be mentioned in books and review articles before 1933. The authors represent a broader segment of the physics community than those included in section 8.8. I found 21 books and 5 articles that surveyed the achievements of quantum mechanics, excluding those that discussed only its application to a special topic such as the electron theory of metals.

No single achievement was cited in more than half of these publications. As indicated in Table 8.2, the most frequently mentioned achievement was the determination of properties of the helium atom (ionization potential and spectrum). This was important, because the old quantum theory had completely failed to attain even a rough estimate of these properties. By treating successfully a system with three strongly interacting particles, quantum mechanics promised to provide a method for studying more complex atoms and molecules, thus opening the road to an explanation of chemistry. Next came the deduction of the hydrogen spectrum. The significance of these two achievements in comparison with the deduction of half-quantum numbers and the Stark effect, which topped the earlier list, was more evident to the larger group of physicists. But only one of the early novel predictions—the existence of ortho and para forms of the hydrogen molecule—attracted much attention. At same time, for many authors who were not active participants in the development and application of quantum mechanics, a blanket statement to the effect that quantum mechanics could explain everything, including properties for which the old quantum theory failed to give a satisfactory account, was a convenient way to characterize the success of the new theory.

Table 8.2 Reasons for Accepting Quantum Mechanics (QM) Mentioned at Least Twice By Authors of Monographs and Review Articles, 1929–1932

Helium atom (12): Explained by resonance concept.
Hydrogen spectrum (10): Explained from a single equation without invoking arbitrary "quantum conditions."
General statement (8): QM explains all the phenomena that were explained correctly (i.e., in agreement with experiment) by the old quantum theory, and also explains correctly all the phenomena that the old theory did not explain correctly.
Ortho and para hydrogen (8): QM successfully predicted a new fact: there are two kinds of H_2 molecules. In the ortho version, the nuclear spins are in the same direction; in the para versions, they are in opposite directions.
Half-quantum numbers (8): Follow from Quantum Mechanics postulates rather than being arbitrarily postulated.
Zeeman effect (8): Explained with electron-spin postulated.
Intensities of spectral lines (8): Calculated directly, not approximately from Correspondence Principle.
Electronic properties of metals (8): Calculated from Fermi-Dirac statistics.
Scattering and absorption of radiation by atoms (8). Explained in terms of spectroscopic frequencies.
Wave nature of matter (8): De Broglie's hypothesis is supported by electron diffraction experiments.
Stark effect (7): Exclusion of electron orbits passing through nucleus is derived from Quantum Mechanics rather than arbitrarily postulated.
Compton effect (7): Quantum Mechanics resolves wave/particle dilemma by deriving both aspects from one theory.
Electron spin (6): This can now be deduced from Dirac's relativistic equation rather than arbitrarily postulated.
Interatomic forces and chemical valence (6): These can be calculated in simple cases (e.g., interaction of two hydrogen atoms) with the hope of accurate approximations for more complex systems.
Mott scattering (2): When two particles are identical their scattering is significantly different from that of two nonidentical particles with the same masses and charges.
Periodic Table (2): QM offers a semiquantitative explanation based on Pauli principle, concept of electron shells with finite capacity.

Table 8.3 Reasons for Accepting Quantum Mechanics (QM) According to Authors of Textbooks and Popularizations, 1929–1932

Wave nature of matter (11): De Broglie's hypothesis is supported by electron diffraction experiments.
General statement (8): Quantum Mechanics explains all the phenomena that were explained by old quantum theory and also all phenomena that old theory did not explain.
Electronic properties of metals (2): Calculated from Fermi-Dirac statistics.
Radioactive decay (2)

Another reason to accept wave mechanics, especially after the Davisson-Germer experiment, was that it could be seen as a natural development of classical wave theory, one that retained a certain amount of visualizability. This was the view of physicist Charles Galton Darwin, as summarized by historian Jaume Navarro: "the central physical entity was neither the particle nor the wave but the wave function . . . The corpuscular aspect appeared as a consequence of the process of measurement." As Darwin wrote in 1929,

> We are led to the conception of a sub-world [in which] ψ is an unthinkably complicated function of all the variables associated with all the particles of the world; nothing definite is happening at all, but it expresses simultaneously everything that could possibly happen.

Two decades later, physicist Richard P. Feynman gave a mathematical expression of a similar idea in his famous integral-over-all-paths formulation of quantum mechanics.

One question remains: if quantum mechanics was such a spectacularly successful theory, and was generally accepted by most physicists by 1928, why did it take so long (seven years) for its creators to be honored by Nobel prizes? Heisenberg won the prize in 1932, and Schrödinger in 1933 (shared with Dirac). The answer remained a mystery (except to a few insiders) until after 1976, when the Nobel Foundation decided that the part of its archives 50 or more years in the past could be opened to researchers. As we will see in the next chapter, the answer is quite relevant to one of the basic questions discussed in this book.

NOTES FOR CHAPTER 8

219 *Quantum Mechanics* This case study has not previously been published in full, although its conclusion—that quantum mechanics was accepted by atomic physicists and others before any of its novel predictions had been confirmed—has been mentioned in two places: "Dynamics of Theory Change: The Role of Prediction," *PSA 1994 (Proceedings of the 1994 Biennial Meeting of the Philosophy of Science Association*, vol. 2 (East Lansing, MI: Philosophy of Science Association, 1995), pp. 133–145; "Remembering Rabi: A Challenge . . . ," *Physics Today* 60, no. 6 (June 2007): 10.

219 *By 1925, the old quantum theory was approaching its final measure* John S. Rigden, *Rabi: Scientist and Citizen* (New York: Basic Books, 1987), p. 39.

219 *This afternoon I spent two hours* Letter from Yakov Frenkel to his parents from Berlin, 20 November 1925, in V. Ya. Frenkel, *Yakov Ilich Frenkel: His Work, Life, and Letters*, translated from Russian by A. S. Silbergleit (Basel: Birkhäuser Verlag, 1996), p. 75.

Section 8.1

219 *it combined research on the properties of atoms* General works on the history of quantum mechanics: Max Jammer, *The Conceptual Development of Quantum Mechanics* (New York: McGraw-Hill, 1966; 2nd ed., New York: Tomash/AIP Press, 1989); Jagdish Mehra and Helmut Rechenberg, *The Historical Development of Quantum Theory*, 6 vols. (New York: Springer, 1982–2001); J. L. Heilbron, "Quantum Historiography and the Archive for History of Quantum Physics," *History of Science* 7 (1968): 90–111; J. L. Heilbron, Bruce Wheaton, *et al.*, *Literature on the History of Physics in the 20th Century* (Berkeley: Office for History of Science and Technology, University of California, 1981), pp. 334–337; Stephen G. Brush and Lanfranco Belloni, *The History of Modern Physics: An International Bibliography* (New York: Garland, 1983), pp. 181–185. I also recommend the book by Louisa Gilder, *The Age of Entanglement: When Quantum Physics was Reborn* (New York: Knopf, 2008; Vintage paperback reprint, 2009), which is a semifictionalized, mostly nonmathematical account of the history of quantum mechanics leading to experiments inspired by John Bell's ideas.

219 *James Franck and Gustav Hertz's experiment* Frank and Hertz, "Über Zusammenstösse zwischen Elektronen und den Molekülen des Quecksilberdampfes und die Ionisierungsspannung desselben," *Verhandlungen der Deutschen Physikalischen Gesellschaft* 16 (1914): 457–467 (1914); "Über Kinetik von Elektronen und Ionen in Gasen," *Physikalische Zeitschrift* 17 (1916); 409–416; "Die Bestätigung der Bohrschen Atomtheorie im optischen Spektrum durch Untersuchungen der unelastischen Zusammenstösse langsamer Elektronen mit Gasmolekülen," *Physikalische Zeitschrift* 20 (1919): 132–143, as cited by Jammer, *Conceptual Development*, p. 92, note 105.

220 *a correction to the energy levels that was proportional* A. Sommerfeld, "Zur Quantentheorie der Spektrallinien, *Annalen der Physik* 51 (1916): 1–94, 125–167; Jammer, *Conceptual Development*, p. 97; John L. Heilbron, "The Kossel-Sommerfeld theory and the Ring Atom," *Isis* 58 (1967): 451–485; Helge Kragh, "The Fine Structure of Hydrogen and the Gross Structure of the Physics Community, 1916–26," *Historical Studies in the Physical Sciences* 15, part 2 (1985): 67–125.

220 *Rutherford wrote to Bohr in 1913* Rutherford to Bohr, 20 March 1913, quoted by Abraham Pais, *Niels Bohr's Times* (Oxford: Clarendon Press, 1991), pp. 152–153, from Niels Bohr, *Collected Works*, Amsterdam: North-Holland, 1981, Vol. 2, p. 583.

220 *The struggles of physicists to use or modify Bohr's model* John L. Heilbron, *Historical Studies in the Theory of Atomic Structure* (New York: Arno Press, 1981); "Lectures on the History of Atomic Physics 1900–1922," in *History of Twentieth Century Physics*, ed. C. Weiner, 40–108 (New York: Academic Press, 1977); Paul Forman, "The Doublet Riddle and Atomic Physics circa 1924," *Isis* 59 (1968): 156–174; Forman, "Alfred Landé and the Anomalous Zeeman Effect," *Historical Studies in the Physical Sciences* 2 (1970): 153–261.Daniel Serwer, "Unmechanischer Zwang: Pauli, Heisenberg, and the Rejection of the Mechanical Atom, 1923–1925," *Historical Studies in the*

Physical Sciences 8 (1977): 189–256. Carsten Jensen, “Two one-electron Anomalies in the Old Quantum Theory,” *Historical Studies in the Physical Sciences* 15, part 1 (1984): 81–106; Kragh, “Fine Structure.”

220 *that model, according to Kragh, “enjoyed general acceptance”* H. Kragh, “Resisting the Bohr Atom: The early British Opposition,” *Physics in Perspective* 13 (2011): 4–35, on p. 11. For further references see notes to Section 2.5.

221 *the use of half-integer quantum numbers is necessary* E. N. Da C. Andrade, *The Structure of the Atom* (London: Bell, 3rd ed. 1926), p. 510.

221 *“hatch out our quantic egg”* Andrade, *Structure,* p. 703.

221 *“the witches’ kitchen in* Faust” Andrade, *Structure,* p. 708.

221 *was looking for a 1920s crisis* Seth, *Crafting the Quantum: Arnold Sommerfeld and the Practice of Theory, 1890–1926* (Cambridge, MA: MIT Press, 2010), p. 266.

221 *“the new development does not signify a revolution”* Seth, *Crafting the Quantum,* p. 266–268; see also Seth, “Crisis and the Construction of Modern Theoretical Physics,” *British Journal for the History of Science* 40 (2007): 25–51.

Section 8.2

222 *the immense effort needed to derive them* Aside from the original Bohr model for the hydrogen atom, a good example is the Landé g-factor. See Forman, “Doublet Riddle” and “Landé.”

223 *an electron of mass* m *moving with velocity v* L. de Broglie, “Ondes et Quanta,” *Comptes Rendu hebdomadaires des Séances de l’Academie des Sciences, Paris* 177 (1923): 507–510; “Recherches sur la Théorie des Quanta,” *Annales de Physique* 3 (1925): 22–128. A partial English translation of the second paper is in Gunther Ludwig, *Wave Mechanics* (New York: Pergamon Press, 1968), pp. 73–93. See also Louis de Broglie and Léon Brillouin, *Selected Papers on Wave Mechanics* (London: Blackie, 1928); Jared W. Haslett, “Phase Waves of Louis de Broglie,” *American Journal of Physics* 40 (1972): 1315–1320; Jammer, Conceptual Development, pp. 242–259; Dominique Pestre *et al., Le Découverte des Ondes de Matière* (Paris: Technique & Documentation-Lavoisier, 1994). On the anticipation of wave-particle duality by his brother Maurice de Broglie in 1922 see Mary Jo Nye, “Aristocratic Culture and the Pursuit of Science: The de Broglies in modern France,” *Isis* 88 (1997): 397–421; M. De Broglie, *Les Rayons X* (Paris: Blanchard, 1922), p. 22.

For photons, equating Planck’s quantum formula for energy to Einstein’s relation between energy and mass gives $h\nu= mc^2$. Since $\nu\lambda= c$, we have $m = h\nu/c^2 = h/c\lambda$, or $\lambda= h/mc$. De Broglie postulates that a similar formula holds for all particles, replacing c by v. But the phase wave travels at a speed c^2/v, which is greater than c if $v < c$. Hence it can’t carry energy (according to relativity theory) but only guides the particle.

224 *the apparent wavelength associated with the electrons* C. Davisson and C. H. Kunsman, “The Scattering of Electrons by Nickel,” *Science* 54 (1921): 522–524; “The Scattering of low speed Electrons by Platinum and Magnesium,” *Physical Review* 22 (1923): 242–258; W. Elsasser, “Bemerkungen zur Quantenmechanik freier Elektronen,” *Naturwissenschaften* 13 (1925): 711; Jammer, *Conceptual Development*, pp. 252–253; Mehra and Rechenberg *Historical*, vol. 6, part 1, pp. 373–380. According to Jeremy Bernstein, physicists in the 1920s did not know that Davisson’s early (1921) experiments had “in effect, confirmed de Broglie’s hypothesis.” This was “brought to light by the physicist Richard Gehrenbeck, writing in the January 1978 issue of the monthly *Physics Today*.” *Hans Bethe, Prophet of Energy* (New York: Basic Books, 1979), 18. See next note for Gehrenbeck’s work. For

counterexamples to Bernstein's statement see W. C. Dampier-Whetham, *A History of Science and its Relations with Philosophy and Religion* (Cambridge: Cambridge University Press, 1929), p. 414; N. F. Mott, *An Outline of Wave Mechanics* (Cambridge: Cambridge University Press, 1930), p. 7; J. A. Eldridge, *The Physical Basis of Things* (New York: McGraw-Hill, 1934), p. 365; A. Haas, *Materiewellen und Quantenmechanik. Eine Einführung auf Grund der Theorien von De Broglie, Schrödinger, Heisenberg und Dirac.* (Leipzig: Akademische verlagsgesellschaft m.b.h., 1934), p. 43; and the 1966 edition of Jammer's book cited above.

Friedrich Hund proposes a somewhat more hypothetical precursor, a 1923 paper by William Duane; see Hund, *History of Quantum Theory* (New York: Harper & Row/Barnes & Noble, 1974), pp. 142–143 (translated by G. Reece from the German edition of 1967).

224 *One of the first attempts to confirm de Broglie's prediction* E. G. Dymond, "Scattering of Electrons in Helium," *Nature* 118 (1926): 336–337; "On Electron Scattering in Helium," *Physical Review* [series 2] 29 (1927): 433–441. Dymond and E. E. Watson, "Electron Scattering in Helium," *Proceedings of the Royal Society of London* A122 (1929): 571–582.

224 *Their report . . . was quickly acclaimed as the definitive confirmation* C. Davisson and L. H. Germer, "The Scattering of Electrons by a single Crystal of Nickel," *Nature* 119 (1927): 558–560; "Diffraction of Electrons by a Crystal of Nickel," *Physical Review* [series 2], 30 (1927): 705–740; Davisson, "Are Electrons Waves?" *Journal of the Franklin Institute* 205 (1928): 597–623. Richard K. Gehrenbeck, *C. J. Davisson, L. H. Germer, and the Discovery of Electron Diffraction* (PhD dissertation, University of Minnesota, 1973); "Electron Diffraction: Fifty Years Ago," *Physics Today* 31, no. 1 (Jan. 1978): 34–41; Richard Schlegel, "Who Discovered Matter Waves?" *Physics Today* 31, no. 7 (July 1978): 9, ll; reply by Gehrenbeck, ibid., 11, 13; P. Goodman, ed., *Fifty Years of Electron Diffraction* (Boston: Reidel, 1981).

224 *G. P. Thomson demonstrated the phenomenon* G. P. Thomson and A. Reid, "Diffraction of Cathode Rays by a thin Film," *Nature* 119 (1927), 890; Thomson, "The Diffraction of Cathode Rays by thin Films of Platinum," *Nature* 120 (1927): 802; Thomson, "Experiments on the Diffraction of Cathode Rays," *Proceedings of the Royal Society of London* A117 (1928): 600–609; Thomson, "Early Work in Electron Diffraction," *American Journal of Physics* 29 (1961), 821–825. Thomson was the son of J. J. Thomson, who had received the Nobel Prize in Physics (1906) for his part in the discovery of the election and proof that it is a particle. In 1937, Davisson and G. P. Thomson were jointly awarded the Nobel Prize in Physics for demonstrating the wave nature of the electron.

Section 8.3

224 *"Schrödinger . . . was guided by the mathematical apparatus"* Hans Reichenbach, *Atom and Cosmos* (New York: Macmillan, 1933, translated from 1930 German book), p. 253.

224 *The solution . . . was first discovered by Werner Heisenberg* Werner Heisenberg, "Über die quantentheoretische Umdeutung kinematischer und mechanischer Beziehungen," *Zeitschrift für Physik* 33 (1925): 879–893; Max Born and Pascual Jordan, "Zur Quantenmechanik," *Zeitschrift für Physik* 34 (1925): 858–888; Born, Heisenberg and Jordan, "Zur Quantenmechanik II," *Zeitschrift für Physik* 35 (1926): 557–615. Wolfgang Pauli, "Über das Wasserstoffspektrum vom Standpunkte der neuen Quantenmechanik," *Zeitschrift für Physik* 36 (1926): 336–363. English

translations (abridged) of these four papers are in *Sources of Quantum Mechanics*, edited by B. L. Van der Waerden (Amsterdam: North-Holland, 1967). Heisenberg and Pascual Jordan, "Anwendung der Quantenmechanik auf das Problem der anomalen Zeemaneffekte," *Zeitschrift für Physik* 37 (1926): 263–277.

Useful accounts of matrix mechanics and its history may be found in Jammer, *Conceptual Development*, Chapter 5; Edward MacKinnon, "Heisenberg, Models, and the Rise of Matrix Mechanics," *Historical Studies in the Physical Sciences* 8 (1977): 137–188; Mehra and Rechenberg, *Historical* vol. 3; David C. Cassidy, *Uncertainty: The Life and Science of Werner Heisenberg* (New York: Freeman, 1992), Chapters 10 and 11; Mara Beller, "Matrix Theory before Schrödinger: Philosophy, Problems, Consequences," *Isis* 74 (1983): 469–491. On the equivalence proof by Pauli and others see B. L. Van der Waerden, "From Matrix Mechanics and Wave Mechanics to Unified Quantum Mechanics," in *The Physicist's Conception of Nature*, edited by J. Mehra, pp. 276–293 (Dordrecht: Reidel, 1973).

225 *"The most noticeable defect of the original matrix mechanics"* "Quantum Mechanics" (translation of paper presented at 1927 Solvay Conference), in Guido Bacciagaluppi and Antony Valentini, *Quantum Theory at the Crossroads: Reconsidering the 1927 Solvay Conference* (Cambridge: Cambridge University Press, 2009), pp. 372–401, on p. 383.

225 *(often misleadingly called the uncertainty principle)* Werner Heisenberg, "Über den anschaulichen Inhalt der Quantenmechanik," *Zeitschrift für Physik* 43 (1927): 172–198. English translation in J. A. Wheeler and Zurek, *Quantum Theory and Measurement* (Princeton, NJ: Princeton University Press, 1983). Cassidy, *Uncertainty.*

225 *what is the generalization of Newtonian particle mechanics* Erwin Schrödinger, "Quantisierung als Eigenwertproblem," *Annalen der Physik* 79 (1926): 361–376, 489–527; 80 (1926): 437–490; 81 (1926): 109–139. *Collected Papers on Wave Mechanics* (London: Blackie, 1928; New York: Chelsea, 1982). English translations (abridged) of the first, second and fourth papers are in *Wave Mechanics*, ed. Gunther Ludwig (Oxford: Pergamon Press, 1968). "An Undulatory Theory of the Mechanics of Atoms and Molecules," *Physical Review* [series 2] 28 (1926): 1049–1070; Walter Moore, *Schrödinger: Life and Thought* (Cambridge: Cambridge University Press, 1989); Linda Wessels, "Schrödinger's Route to Wave Mechanics," *Studies in History and Philosophy of Science* 10 (1979): 311–340. Helge Kragh, "Erwin Schrödinger and the Wave Equation: The Crucial Phase," *Centaurus* 26 (1982): 154–197.

226 *the letter u in d'Alembert's equation means the speed* The basic equation for wave propagation was derived by the French mathematician Jean Ie Rond d' Alembert in the eighteenth century. The equation determines a function N, which represents a small displacement of a small part of a vibrating substance. In the simplest case, a vibrating string whose ends are fixed at two points on a horizontal line (x-axis), the displacement in the vertical direction, φ, at a time t, would vary with x, being equal to zero at the two fixed points and oscillating between positive and negative values in between. Moreover, if one looks at a particular point x one would see that the displacement at that point varies with t. So we are looking for a function that represents the displacement as a function of the two variables x and t.

To find the equation, d'Alembert started from two laws of classical physics. The first is Hooke's law, which states that for an elastic substance like a string, the force acting to pull it back to its original position ($y = 0$) is proportional to the displacement φ. The second is Newton's law, usually written $F = ma$, but more conveniently written as (dividing both sides by m) $a = F/m$. The greater the force acting on an object with mass *m*, the greater will be its acceleration *a*; the greater the mass, the smaller its acceleration. As Galileo proved, the most suitable definition of acceleration is the rate at which the speed changes with time.

Now we use the mathematical calculus invented independently by Newton and Leibniz, but we follow Leibniz's more convenient notation, using the letter *d* to represent a very small change in whatever quantity is written after it. We define the speed *s* as the ratio of a very small change in distance, *dx*, to the corresponding very small interval in time, *dt*, taking the limit as *dx* and *dt* go to zero. This limit operation is called "taking the derivative of *x* with respect to *t*." Similarly the acceleration will be the ratio of a very small change in speed, *ds*, to a very small time interval *t*; this is written $d(dx/dt)/dt$ or d^2x/dt^2. This is called "taking the second derivative of *x* with respect to *t*" or "finding how fast the change changes." It was a rather abstract concept in the seventeenth century, but nowadays every teenager knows how to use the accelerator of a car to change its speed. The result of d'Alembert's analysis of the vibrating string is that the second derivative of the displacement (φ) with respect to *t* is proportional to the second derivative of the displacement with respect to *x*:

$$d^2 \varphi/dt^2 = u^2 d^2 \varphi/dx^2$$

where *u* is a constant which turns out to be the speed at which the wave travels. He could then generalize this equation to three spatial dimensions, so that the displacement φ is a function of *x, y,* and *z*:

$$d^2\varphi / dt^2 = u^2[d^2\varphi / dx^2 + d^2\varphi / dy^2 + d^2\varphi / dy^2]$$

The sum of three second derivatives appears so often in theoretical physics that it has been assigned a special symbol, ^{2} (pronounced "del squared"); it is also called the Laplacian operator after the French mathematical astronomer Laplace, who used it to analyze many physical phenomena. Thus d'Alembert's equation is usually written

$$\nabla^2\varphi - (1/u^2)d^2\varphi / dt^2 = 0.$$

226 *According to Louis de Broglie's hypothesis, the wavelength of a particle* Start with the equation from mechanics, stating that the total energy of a particle is the sum of its kinetic and potential energies:

$$E = \tfrac{1}{2}mv^2 + V \rightarrow (N1)$$

Its momentum, defined as $p = mv$, is found by solving Eq. (N1) for *v* and then multiplying that by *m*; the result is

$$v^2 = (E - V)/(\tfrac{1}{2}m);\ v = [(E - V)/(\tfrac{1}{2}m)];\ p = mv = [2m(E - V)]$$

Therefore the de Broglie wavelength is

$$\lambda = h/mv = h/[2m(E - V)]$$

the complex-valued function $e^{-i\omega t}$ can represent a quantity oscillating with a frequency ω: Wave motion is most accurately described by using the trigonometric functions, cosine and sine of an angle θ (measured in radians), abbreviated cos θ and sin θ. As θ goes from 0 to 2π, cos θ goes from 1 to 0 and back to 1; sin θ goes from 0 to 1 and back to 0. This cycle corresponds to moving a distance of one wavelength (say, from one positive peak to the next), so if the speed of the wave is u, a cycle corresponds to an angle $\theta = 2\pi u/\lambda$. The combination $2\pi u/\lambda$ is sometimes called the circular frequency and is designated by the Greek letter ω (omega). The Swiss mathematical Leonhard Euler showed in 1740 that

$$e^{i\omega r} = \cos \omega t + i \sin \omega t.$$

226 *using de Broglie's hypothesis, he obtained* The operator ∇^2 affects only the space coordinates (x, y, z) but not the time coordinate t, so the first term in Eq. (3) becomes $e^{-i\omega t}\nabla^2\psi$. An interesting feature of the exponential function e^{at} is that if you take its time derivative (d/dt) you get the same exponential function multiplied by the constant a. If you do that twice, you get the factor a^2. So:

$$d^2\varphi/dt^2 = (-i\omega)^2 e^{-i\omega r}\psi = -\omega^2 e^{-i\omega r}\psi = -(2\pi u/\lambda)^2 e^{-i\omega r}\psi$$
$$(1/u^2)d^2\varphi/dt^2 = -(2\pi/\lambda)^2 d^2\varphi/dt^2 e^{-i\omega r}\psi$$

Now, remembering that $\lambda = h/\sqrt{(2m[E - V])}$, and dividing through by $e-^{i\omega r}$ we get Eq. (5) in the text.

227 *Now comes the first great stunning achievement* Schrödinger, *Collected Papers*. The following summary is taken from S. G. Brush, *Statistical Physics and the Atomic Theory of Matter, from Boyle and Newton to Landau and Onsager* (Princeton, NJ: Princeton University Press, 1983), pp. 135–137.

227 *So Schrödinger succeeded immediately* Brush, *Statistical Physics*, pp. 138–139. Pauli could do that from matrix mechanics but with much more effort: Pauli, "Über das Wasserstoffspektrum vom Standpunkt der neuen Quantenmechanik," *Zeitschrift für Physik* 36 (1926): 336–363. Reprinted in Pauli's *Collected Scientific Papers*, vol. 2, pp. 252–279. English translation in van der Waerden, *Sources*. Pauli managed to calculate the wavelengths of lines in the hydrogen spectrum, but not their intensities.

227 *the experiment was a vindication of de Broglie's theory* See the Wien-Schrödinger correspondence in November 1927, quoted by Mehra and Rechenberg, *Historical* vol. 5, part 2, pp. 866–867. They also assert that Schrödinger "certainly was happy about the experimental confirmation of matter waves. Anyway, his colleagues in Berlin considered wave mechanics as the real theoretical fulfilment of the idea of matter waves" (ibid.).

227 *Born proposed that $\psi\psi^*$ measures the probability* Born, "Quantenmechanik der Stossvorgängen," *Zeitschrift für Physik* 38 (1926): 803–827, on p. 804, as translated by Jammer, *Conceptual*, p. 302.

227 *"The motion of particles conforms to the laws of probability"* Ibid.

Section 8.4

228 *Pauli suggested that all possible electron states could be accounted for* Wolfgang Pauli, "Über den Zusammenhang des Abschlusses der Elektronengruppen im Atom mit der Komplexstruktur der Spektren," *Zeitschrift für Physik* 31 (1925): 765–783;

English translation in D. ter Haar, *Old Quantum Theory* (Oxford: Pergamon Press, 1967). Pauli, "Remarks on the History of the Exclusion Principle," *Science* 103 (1946): 213–215; see also Hermann Weyl's "Encomium," ibid., 216–218. Pauli gave further details (but omitting some items) in his Nobel Lecture (1946), in his *Collected Scientific Papers,* vol. 2, pp. 1080–1096. John C. Slater, *Solid State and Molecular Theory: A Scientific Biography* (New York: Wiley, 1975), Chapter 4; Karl von Meyenn, "Paulis Weg zum Ausschliessungsprinzip," *Physikalische Blätter* 36 (1980): 293–298; 37 (1981): 13–19. Brush, *Statistical Physics*, pp. 217–219; J. L. Heilbron, "The Origins of the Exclusion Principle," *Historical Studies in the Physical Sciences* 13 (1983): 261–310. For a comprehensive historical and philosophical analysis see Michela Massimi, *Paulu's Exclusion Principle: The Origin and Validation of a Scientific Principle* (Cambridge: Cambridge University Press, 2005).

228 *proposed that this number could reasonably refer to a rotation or* spin G. E. Uhlenbeck and S. Goudsmit, "Ersetzung der Hypothese vom unmechanismen Zwang durch eine Forderung bezüglich des inneren Verhaltens jedes einzelnen Electrons," *Naturwissenschaften* 13 (1925): 953–954; "Spinning Electrons and the Structure of Spectra," *Nature* 117 (1925), 264. It was later pointed out that A. H. Compton and others had previously made similar proposals, which were ignored—a case of premature discovery? See Compton, "The Magnetic Electron," *Journal of the Franklin Institute* 192 (1921): 145–155. For historical accounts and recollections of participants see Markus Fierz and V. F. Weisskopf (eds.), *Theoretical Physics in the Twentieth Century: A Memorial Volume to Wolfgang Pauli* (New York: Interscience, 1960); articles by R. Kronig and B. L. van der Waerden; Goudsmit, "Entdeckung des Elektronenspins," *Physikalische Blätter* 21 (1965): 445–453; "Guess Work: The Discovery of the Electron Spin," *Delta* [Amsterdam] 15, no. 2 (1972): 77–91; Goudsmit, "Fifty Years of Spin: It Might As Well Be Spin," *Physics Today* 29, no. 6 (June 1976): 40–43; Uhlenbeck, "Fifty Years of Spin: Personal Reminiscences," ibid. 43–48; A. Pais, "George Uhlenbeck and the Discovery of Electron Spin," *Physics Today* 42, no. 12 (December 1989): 34–40; Dwight E. Neuenschwander, "Sam Goudsmit: Physics, Editor, and More. FHP Session at the APS 2010 Meeting," *History of Physics Newsletter* 11, no. 3 (Fall 2010): 8–9, 11, 20.

228 *can in principle explain all of chemistry* "The general theory of quantum mechanics is now almost complete . . . The underlying physical laws necessary for the mathematical theory of a large part of physics and the whole of chemistry are thus completely known." P.A.M. Dirac, "Quantum Mechanics of Many-Electron Systems," *Proceedings of the Royal Society of London* A123 (1929): 714–733, on p. 714.

228 *and thus reveals an influence of chemistry on physics* Mansel Davies, "C. R. Bury, L. Vegard, and the Electronic Interpretation of the Periodic Table: A Note," *Archive for History of Exact Sciences* 41 (1990): 185–187; W. Heitler, Interview with J. L. Heilbron, 18 III 63, *Archive for History of Quantum Physics*, p. 7; E. Fermi, "Über die Anwendungen der statistische Methode auf die Problem des Atombaues," in *Quantentheorie und Chemie*,.ed. H. Falkenhagen, 95–111 (Leipzig: Hirzel, 1928); M. A. El'iashevich, "The Mendeleev Periodic Law, Atomic Spectra, and Atomic Structure: On the History of the Physical Interpretation of the Periodic Table of the Elements," *Soviet Physics Uspekhi* 13 (1970): 1–23; Helge Kragh, "The First Subatomic Explanations of the Periodic System," *Foundations of Chemistry* 3 (2001): 129–143; E. R. Scerri, *The Periodic Table and its Significance* (New York: Oxford University Press, 2007).

229 *"The relativistic explanation of the doublets collapses"* Heisenberg and Jordan, "Quantenmechanik II." The quotation is from Heisenberg's letter to Goudsmit, 19 February 1926, translated by Kragh, "Fine Structure," who explains the new calculation.

Section 8.5

230 *a short paper in English on the derivation of Planck's radiation law* S. N. Bose, "Plancks Gesetz und Lichtquantenhypothese," *Zeitschrift für Physik* 26 (1924): 178–181. English translation in H. A. Boorse and L. Motz, *World of the Atom* (New York: Basic Books, 1966); also in O. Theimer and B. Ram, "The Beginning of Quantum Statistics," *American Journal of Physics* 44 (1976): 1056–1057. William A. Blanpied, "Satyendranath Bose: Co-Founder of Quantum Statistics," *American Journal of Physics* 40 (1972): 1212–1220; Kameshwar Wali, "The Man Behind Bose Statistics," *Physics Today* 59, no. 10 (Oct. 2006): 46–52; Brush, *Statistical Physics*, p. 158, note 1.

230 *He was then able to derive Planck's distribution law* See Brush, *Statistical Physics*, pp. 158–159.

230 *some atoms will have to go into the lowest quantum state* Albert Einstein, "Quantentheorie des einatomigen idealen Gases," *Sitzungsberichte, Akademie der Wissenschaften, Berlin (1924)*: 261–267; (1925): 3–14. English translation in a forthcoming volume of *The Collected Papers of Albert Einstein*. Agostino Desalvo, "From the Chemical Constant to Quantum Statistics: A Thermodynamic Route to Quantum Mechanics," *Physis* 29 (1992): 465–537.

230 *criticized Einstein's mathematical treatment of the problem* G. E. Uhlenbeck, *Over statistische methoden in de theorie der quanta* ('s-Gravenhage: M. Nijhoff, 1927).

230 *presumably his criticism became known* See for example Ehrenfest's letter to Einstein, quoted by Uhlenbeck, "Some reminiscences about Einstein's visits to Leiden," in *Some Strangeness in the Proportion, A Centennial Symposium to Celebrate the Achievements of Albert Einstein*, edited by H. Woolf, 524–525 (Reading, MA: Addison-Wesley, 1980). For the views of Schrödinger and others on this point see Paul Hanle, *Erwin Schrödinger's Statistical Mechanics, 1912–1925* (New Haven, CT: PhD dissertation, Yale University, 1975), pp. 232–250.

230 *Uhlenbeck retracted his criticism in 1938* B. Kahn and G. E. Uhlenbeck, "On the Theory of Condensation," *Physica* 5 (1938): 399–416; Fritz London, "The λ-Phenomenon of Liquid Helium and the Bose-Einstein Degeneracy," *Nature* 141 (1938): 643–644; "On the Bose-Einstein Condensation," *Physical Review Series* 2, 54 (1938): 947–954; Brush, *Statistical Physics*, pp. 172–203; Kostas Gavroglu, "From Defiant Youth to Conformist Adulthood: The Sad Story of Liquid Helium," *Physics in Perspective* 3 (2001): 165–188.

230 *Fermi and Dirac presented essentially the same derivation* Enrico Fermi, "Sulla quantizzazione del gas perfetto monoatomico," *Rendiconti dell' Accademia dei Lincei* 3, no. 3 (1926): 145–149; "Zur Quantelung des idealen einatomigen Gases," *Z. Physik* 36 (1926): 902–912; *Collected Papers*, edited by Emilio Segrè *et al.* (Chicago: University of Chicago Press, 1962–1965), vol. 1, pp. 178f, 186f; P.A.M. Dirac, "On the Theory of Quantum Mechanics," *Proceedings of the Royal Society of London* A112 (1926): 661–677; Fermi, letter to Dirac, 25 October 1926, M/f 59, Sect. 2, Archive for History of Quantum Physics (cited hereafter as AHQP).

For historical accounts see Emilio Segré, *Enrico Fermi, Physicist* (Chicago: University of Chicago Press, 1970); Lanfranco Belloni, "A Note on Fermi's

Route to Fermi-Dirac statistics," *Scientia*, Annus LXXII, 113 (1978): 421–430; Brush, *Statistical Physics*, 159–162; Mehra and Rechenberg, *Historical* vol. 5, part 2 (1987): 746–771; Dan Cooper, *Enrico Fermi and the Revolutions of Modern Physics* (New York: Oxford University Press, 1999).

Fermi in his 1926 paper showed that at low temperatures (T) the pressure of a Fermi-Dirac gas containing *n* particles in a fixed volume (say, 1 liter) is

$$p = ah^2n^{5/3}/m + b\,mn^{1/3}k^2T^2/h^2 + \ldots$$

where *a* and *b* are numerical constants. This may be compared with the ideal gas law,

$$p = knT$$

The striking difference is that for an ideal gas the pressure goes to zero when the absolute temperature T goes to zero, regardless of the density *n*. The fact that the Fermi-Dirac gas has a *finite* pressure at absolute zero,

$$p = ah^2n^{5/3}/m$$

is a consequence of the quantum "zero-point energy."

230 *Pauli and Sommerfeld quickly used Fermi-Dirac statistics* W. Pauli, "Über Gasentartung und Paramagnetismus," *Zeitschrift für Physik* 41 (1927): 81–102; A. Sommerfeld, "Zur Elektronentheorie der Metalle," *Naturwissenschaften* 15 (1927): 825–832; 16 (1927): 374–381; "Zur Elektronentheorie der Metalle auf Grun der Fermischen Statistik," *Zeitschrift für Physik* 47 (1928): 1–32, 43–60.

For historical accounts see Lillian Hoddeson and Gordon Baym, "The Development of the Quantum Mechanical Electron Theory of Metals: 1900–1928," *Proceedings of the Royal Society of London* A371 (1980): 8–23; Hoddeson, Baym, and Michael Eckert, "The Development of the Quantum-Mechanical Electron Theory of Metals, 1928–1933," *Reviews of Modern Physics* 59 (1987): 287–327; Hoddeson, Ernest Braun, Jürgen Teichmann, and Spencer Weart, eds., *Out of the Crystal Maze: Chapters from the History of Solid State Physics*: (New York: Oxford University Press, 1992); Eckert, "Propaganda in Science: Sommerfeld and the Spread of the Electron Theory of Metals," *Historical Studies in the Physical Sciences* 17, part 2 (1987): 191–233; "Sommerfeld und die Anfäng der Festkörperphysik," *Deutsches Museum Wissenschaftliches Jahrbuch* (1990): 33–71; S. G. Brush, "Francis Bitter and 'Landau Diamagnetism,'" *Journal of Statistical Physics* 2 (1970): 195–197; Mehra and Rechenberg, *Historical* vol. 6, part 1 (2000), pp. 572–632.

230 *that many properties of metals could be explained* E. Riecke, "Zur Theorie des Galvanismus und der Wärme," *Annalen der Physik* [series 3], 66 (1898): 353–389, 545–581, 1199–1200; H. A. Lorentz, *The Theory of Electrons and its Applications to the Phenomena of Light and Radiant Heat* (Leipzig: Teubner/New York: Stechert, 1909; 2nd ed. 1915).

230 *"disposes of a well-known difficulty"* J. E. Lennard-Jones, "Some recent Developments of Statistical Mechanics," *Proceedings of the Royal Society of London* 40 (1928): 320–337.

Section 8.7

232 *Wave mechanics "has captivated the world"* Karl K. Darrow, "Contemporary Advances in Physics. XIV. Introduction to Wave-Mechanics," *Bell System Technical Journal* 6 (1927): 653–701, on p. 653.

232 *all studied with Arnold Sommerfeld* The list of physicists influenced by Sommerfeld includes also Paul Epstein, Max von Laue, Wilhelm Lenz, and Otto Laporte. See S. S. Schweber, "Weimar Physics: Sommerfeld's Seminar and the Causality Principle," *Physics in Perspective* 11 (2009): 261–301. Physicist Edward U. Condon (1902–1974) recalled: "everyone of my generation grew up in atomic physics by way of [Sommerfeld's] great *Atombau und Spektrallinien*, a large group have profited by the stimulation of his lectures on his American visits, a fortunate few have derived boundless stimulation from the opportunity of working in his Institut für Theoretischen Physik in the former brighter days," quoted by L. Brown and H. Rechenberg, *The Origin of the Concept of Nuclear Forces* (Bristol, UK: Institute of Physics Publishing, 1996), p. 217.

232 *"the most successful educational institution for theoretical physics"* Seth, *Crafting the Quantum*, p. 3. See also Paul Forman and Armin Hermann, "Sommerfeld, Arnold (Johannes Wilhelm)," *Dictionary of Scientific Biography*, edited by C. C. Gillispie, vol. 12 (1975), pp. 525–532, on pp. 529–530.

232 *Bethe exclaimed that "you could now* do *all this"* Seth, *Crafting the Quantum*, p. 264; Kuhn, "Interview with Hans Bethe, 01/17/1964," p. 9 (Archive for History of Quantum Physics).

232 *he himself had a distinguished academic genealogy* Roman Jackiw, "Hans Bethe, My Teacher," *Physics in Perspective* 11 (2009): 98–103, on p. 100.

232 *The first* published *response was the paper "Zur Quantenmechanik"* M. Born and P. Jordan, "Zur Quantenmechanik," *Zeitschrift für Physik* 34 (1925): 858–888. English translation (abridged) in B. L. Van der Waerden, *Sources*, pp. 277–306. Born, letter to Einstein, 15 July 1925, English translation in *The Born-Einstein Letters: The Correspondence between Albert Einstein and Max and Hedwig Born, 1916–1955, with commentaries by Max Born*, translated by Irene Born (New York: Walker, 1971), pp. 83–88; P. Jordan, "Early Years of Quantum Mechanics: Some Reminiscences," in *The Physicist's Conception of Nature*, edited by Jagdish Mehra (Dordrecht: Reidel, 1973), pp. 294–299.

232 *Dirac's "The Fundamental Equations"* P.A.M. Dirac, "The Fundamental Equations of Quantum Mechanics," *Proceedings of the Royal Society of London* A109 (1926): 642–653. Reprinted in van der Waerden, *Sources*, pp. 307–320. Dirac, interview with T. S. Kuhn and E. P. Wigner, 1 April 1962, in AHQP); Dirac, "Recollections of an Exciting Era," in *History of Twentieth Century Physics*, edited by C.Weiner, 109–146 (New York: Academic Press, 1977). Here Dirac recalls that when R. H. Fowler asked him to look at the proofs of Heisenberg's paper and give his opinion, he was struck by the non-commutative property of matrix multiplication and had the idea of relating it to the Poisson bracket formula of classical mechanics. After his paper was published he sent a copy to Heisenberg, whose "very kind" reply is quoted on pp. 124–125; further correspondence is discussed on pp. 126–131. His own letters to Heisenberg were confiscated by American military authorities at the end of World War II, in 1956, and "presumably . . . are now lying somewhere among the secret files of the American Atomic Energy Commission" (p. 126).

On Dirac's life and work see Joan Bromberg, "Dirac's Quantum Electrodynamics and the Wave-Particle Equivalence," Weiner, *History,* pp. 147–157; Mehra and

Rechenberg, *Historical* vol. 4, part 1 (1982); Michelangelo de Maria and Francesco La Teana, "Dirac's 'Unorthodox' Contribution to Orthodox Quantum Mechanics (1925–1927)," *Scientia* 118 (1983): 595–611; A. Pais, "Playing with Equations, the Dirac Way," in *Reminiscences About a Great Physicist: Paul Adrien Maurice Dirac*, edited by B. N. Kursunoglu and E. P. Wigner (New York: Cambridge University Press, 1987), pp. 93–116; Helge Kragh, *Dirac: A Scientific Biography* (Cambridge: Cambridge University Press, 1990); Olivier Darrigol, *From C-numbers to Q-numbers: The Classical Analogy in the History of Quantum Theory* (Berkeley: University of California Press, 1992); Peter Galison, "The Suppressed Drawing: Paul Dirac's hidden Geometry," *Representations* 72 (2000): 145–166; Andrew Warwick, *Masters of Theory: Cambridge and the Rise of Mathematical Physics* (Chicago: University of Chicago Press, 2003), pp. 490–511; Helge Kragh, *Dirac: A Scientific Biography* (Cambridge: Cambridge University Press, 1990); Graham Farmelo, *The Strangest Man: The Hidden Life of Paul Dirac, Mystic of the Atom* (New York: Basic Books, 2009).

232 *The next month Wolfgang Pauli completed a paper* W. Pauli Jr., "Über das Wasserstoffspektrum vom Standpunkt der neuen Quantenmechanik," *Zeitschrift für Physik* 36 (1926): 336–363. English translation in van der Waerden, *Sources*, pp. 387–415.

232 *Born and Jordan began by stating that Heisenberg's "theoretical approach"* Born and Jordan, in van der Waerden, *Sources*, p. 277. See Born, Heisenberg, and Jordan, "Quantenmechanik II"; Jordan, "Early Years"; Born, interview with T. S. Kuhn et al., October 17, 1962, AHQP.

233 *Dirac's motivation for working on Heisenberg's quantum mechanics* Dirac, quoted from van der Waerden, *Sources*, p. 307. "I just now have a letter from [Edward U.] Condon, saying that Dirac is at Göttingen, and is the real master of the situation, when he talks Born just sits and listens to him open-mouthed . . . Condon asked him if he would like to visit America, and he replied, 'There are no physicists in America.'" Letter from R. T. Birge to J. H. Van Vleck, 10 III 1927 (AHQP, 49). The letter continues: "That is worse than Pauli, whom [sic] I understand credits America with only two, whom [sic] I presume are Epstein and yourself."

233 *was very much concerned with the details of atomic spectra* Pauli, in van der Waerden, *Sources*, p. 391, 396, 413–415. Correspondence with Bohr in *Niels Bohr Collected Works*, vol. 5, pp. 445–461. See his letter to Bohr, 17 XI 1925, in which he explains the significance and limitations of his results more clearly than in the published paper (Bohr's *Collected Works*, vol. 5, pp. 449–453, with Bohr's reply, ibid. P. 454–456). For a review of these and other difficulties, see Mara Beller, "Matrix Theory before Schrödinger: Philosophy, Problems, Consequences," *Isis* 74 (1983): 469–491; Mehra and Rechenberg, *Historical* vol. 3 (1982), Chapter V.5.

233 *Heisenberg and Jordan published the solution in June 1926* W. Heisenberg and P. Jordan, "Anwendung"; Mehra and Rechenberg, *Historical* vol. 3, pp. 266–278; Edward M. MacKinnon, *Scientific Explanation and Atomic Physics* (Chicago: University of Chicago Press, 1982); Jammer, *Conceptual Development*, p. 146; K. Von Meyenn, "Between Scylla and Charybdis: Wolfgang Pauli and the transition from the old to the new Quantum Theory," *Archives des Sciences* 54 (2001): 117–128.

233 *regarded as a major achievement of quantum mechanics* Helmut Rechenberg, "Quanta and Quantum Mechanics," in *Twentieth Century Physics*, edited by L. M. Brown, A. Pais, and B. Pippard (Bristol: Institute of Physics Publishing/New York: American Institute of Physics Press, 1995), pp. 143–248, on p. 211.

234 *Bohr . . . gave matrix mechanics his blessing* Bohr, "Atomic Theory and Mechanics," *Nature* 116 (Dec. 5, 1925): 845–852 (quotes from p. 852). See Bohr's *Collected Works*, vol. 5, pp. 273–280 and Introduction by E. Rüdinger and K. Stolzenberg, pp. 219–222.

234 *Bohr, preparing to give a lecture at Cambridge, had written to Rutherford* Cambridge University Library, Rutherford Add. 7653/B147 and B148. Bohr, *Collected Works*, vol. 5, p. 488; vol. 6, p. 457.

234 *"the atomic theory has passed through a serious crisis"* Bohr to Oseen, 29 I 1926, in Bohr's *Collected Works*, vol. 5, pp. 405–408; English translation, pp. 238–240. In a letter to Bohr, 15 XII 25, Oseen regretted Bohr's decision to abandon his own theory in favor of the new one, calling it "a devastating defeat of the atomic theory" (letter from Grandin to SGB, 24 VII 95).

234 *"Heisenberg has laid a big quantum egg"* "Heisenberg hat einen grosse Quantenei geleft" quoted by Mehra and Rechenberg, *Historical* vol. 5, part 2 (1987), p. 637.

234 *"The Heisenberg-Born concepts leave us all breathless" Albert-Einstein-Hedwig und Max Born: Briefwechsel 1916–1955* (Munich: Nymphenburger Verlagshandlung, 1969), commentaries by Max Born; English translation by Irene Born, *The Born-Einstein Letters* (New York: Walker, 1971), pp. 88–89.

235 *"The idea of your work springs from pure genius" Schrödinger-Planck-Einstein-Lorentz, Briefe zur Wellenmechanik*, edited by K Przibram (Vienna: Springer-Verlag, 1963), pp. 21–22. Translation quoted from Walter Moore, *Schrödinger: Life and Thought* (Cambridge: Cambridge University Press, 1989), p. 209. See also *Letters on Wave Mechanics: Schrödinger, Planck, Einstein, Lorentz*, edited by M. J. Klein (New York: Philosophical Library, 1967).

235 *"I am convinced that you have made a decisive advance"* Prizibram, *Briefe*, p. 26; Moore, *Schrödinger*, p. 209.

235 "He *is not playing at dice"* Born, *Born-Einstein Letters*, p. 91.

235 *he wrote to Schrödinger* Przibram, *Briefe*, p. 3. Klein, *Letters*, p. 23.

235 *"like a thunderstroke"* Sommerfeld-Nachlass at Deutsches Museum, Munich, quoted from Michael Eckert, "Propaganda," p. 205.

235 *"Sommerfeld admired Schrödinger's wave mechanics"* Eckert, "Propaganda," pp. 205, 207. See also Hans Bethe, "Sommerfeld's Seminar," *Physics in Perspective* 2 (2000): 3–5. In Bethe's interview with T. S. Kuhn, January 17, 1964, AHQP, he said: "Sommerfeld found wave mechanics tremendously appealing; this was something he could do. He understood the mathematics, he could deal with the mathematics" (p. 9).

235 *Sommerfeld met a young student* Quoted by K. C. Wali, "Chandra: A Biographical Portrait," *Physics Today* 63, no. 12 (December 2010): 38–43, on p. 39.

236 *Sommerfeld recalculated the fine structure of the hydrogen spectrum* Eckert, "Propaganda," p. 207. Sommerfeld and A. Unsold, "Über das Spektrum des Wasserstoffs," *Zeitschrift für Physik* 38 (1926): 237–241; Norton A. Kent, Lucien B. Taylor, and Hazel Pearson, "Doublet-Separation and Fine Structure of the Balmer Lines of Hydrogen," *Physical Review* 30 (1927): 266–283; Kragh, "Fine Structure," p. 122. The *Physics Citation Index (1920–1929)* lists only one citation of the Kent et al. paper, by their Caltech colleague William Houston.

236 *he wrote to the author (March 22, 1926)* "Es ist höchst erstaunlich und erfreulich, in welch schnellen Schritten Sie vorwärtseilen und Ihr neues Ergebnis, das wenigstens die mathematische Brücke zur Matrizentheorie schlägt, lasst wohl keinen Zweifel mehr bestehen, das Sie einen richtigen Weg eingeschlagen haven." *Wilhelm Wien: Aus dem Leben und Wirken eines Physikers, mit persönlich Erinnerungen von*

E. V. Drygalski / C. Duisberg / M. V. Frey / H. Oncken / F. Paschen / M. Planck / E. Rüchardt / E. Rutherford und einem Nachruf von M. V. Laue und E. Rüchardt (Leipzig: J. A. Barth, 1930), p. 72.

236 *he praised wave mechanics because it was not a radical departure* "Es kann kaum mehr einem Zweifel unterliegen, dass Sie das Problem der Atomschwingungen gelöst haben, und zwar in erfreulich Weise in engsten Anschluss an die Klassische Theorie. Die ganze Quantentheorie ist nicht mehre so fremdartig wie bisher. Auch Planck, mit dem ich über die Säche sprach, war ganz begeistert und ist ganz der gleichen Meinung." Wien, *Aus dem Leben*, p. 72.

236 *it is part of an ongoing battle between different ideas* "Kämpfe, die nun schon drei Jahrzehnte dauern und deren Ende noch absusehen ist." Wien, "Vergangenheit, Gegenwart und Zukunft der Physik. Rede, gehalten beim Stiftungsfest der Universität München am 19. Juni 1926," in Wien, *Aus dem Leben*, pp. 120–135, on 120; see also 132–134.

236 *By August he was urging Schrödinger to visit* "Es wäre mir sehr lieb, wenn ich einige theoretische Fragen eingehender mit Ihnen besprechen könnte. Eine der ernstesten Schwierigkeiten für die Wellenmechanik scheint mir doch im lichtelektrischen Effekt zu liegen. Ich habe über diese Fragen eingehend mit Bohr und Born in Oxford gesprochen, die beide der Überzeugung sind, dass der lichtelektrische Effekt mit der Wellenmechanik nicht vereinbar ist. Ersterer hält die Darstellung in Raum und Zeit für unmöglich, letzterer will das Kausalgesetz preisgeben. Ich vermag den beiden auf diesen Wegen nicht zu folgen, um so weniger, als sie noch nicht gezeigt haben, dass nun das Preisgeben jener Grundlagen wirklich zu einem befriedigendem System führen kann. *Wie* der lichtelektrische Effekt erklärt werden soll, ist mire allerdings auch nicht klar. . . . Mir scheint daher, dass man dem lichtelektrischen Effekt ohne Statistik nicht beikommen kann." Wien, *Aus dem Leben*, pp. 73–74.

237 *A year later Wien was still dissatisfied* "Gegen die Auffassung de Broglies und Borns habe ich nichts einzuwenden, wenn man sie als eine vorläufige Darstellungsweise betrachtet. Wir können ja such verlangen, dass in so kurzer Zeit alle Fragen der Quantentheorie restlos geklärt werden, warum also nicht sagen, wir wissen es vorläufig nicht besser? Aber ich bin vollkommen davon überzeugt, dass man-sich niemals damit zufrieden geben wird, dass in den elementare Vorgängen 'Wahrscheinlichkeiten' vorkommen, also die Kausalität ausgeschaltet wird. Es ist ein Irrtum, dem die jüngere Generation verfällt, dass ihr bescheiden sei, den Stein der Weisen zu finden. Es wird für die nächsten Generationen noch genug zu tun geben." *Wilhelm Wien* (ref. 43i), 75.

237 *"Not a triumph for* my *theory"* Letter from Wien to Schrödinger, 22 XI 27; letter from Schrödinger to Wien, 25 XII 27. Mehra and Rechenberg, *Historical* vol. 5, part 2, p. 867.

237 *de Broglie's pilot wave theory was still a serious rival* Bacciagaluppi and Valentini, *Crossroads*.

237 *de Broglie pointed to this result as a confirmation* "The New Dynamics of Quanta," in Bacciagaluppi and Valentini, *Crossroads*, pp. 341–364, on p. 360. Schrödinger refers to de Broglie's theory as "four-dimensional wave mechanics." His own, which he asserts is the only one that can describe the motion of more than one electron, he calls "multidimensional wave mechanics." Schrödinger, "Wave Mechanics," in Bacciagaluppi and Valentini, pp. 406–424, on p. 406.

237 *Wien has been remembered as a physicist who rejected quantum mechanics* The "Sommerfeld group was convinced by the Schrödinger equation. But there was Wien and a few other experimentalists who were not convinced and they were

then convinced by the Davisson-Germer experiment." Hans Bethe, Interview with T. S. Kuhn, 17 I 64, AHQP, p. 10. "Wilhelm Wien [in Sommerfeld's seminar, autumn 1926] did not believe any part of the quantum theory." Rudolf Peierls, *Bird of Passage: Reflections of a Physicist* (Princeton, NJ: Princeton University Press, 1985), 27.

Section 8.8

237 *"We of the younger generation accepted almost unquestioningly"* H.B.G. Casimir, *Haphazard Reality: Half a Century of Science* (New York: Harper & Row, 1983), pp. 55–56.

237 *"Heisenberg and Schrödinger had knifed a sack of gemstones"* Farmelo, *Strangest Man*, p. 102.

237 *a number of published studies* The most comprehensive survey of the early reception of quantum mechanics is in the books by Jagdish Mehra and Helmut Rechenberg, *Historical* vol. 4, part 2, *The Reception of the new Quantum Mechanics 1925–1926* (New York: Springer-Verlag, 1982), and vol. 5, *Erwin Schrödinger and the Rise of Wave Mechanics*, Part 2, *The Creation of Wave Mechanics: Early Response and Applications* (New York: Springer-Verlag, 1987). Other studies are limited to single countries. For Britain: Jeff Hughes, " 'Modernists with a Vengeance': Changing Cultures of Theory in Nuclear Science, 1920–1940," *Studies in History and Philosophy of Modern Physics* 29 (1998): 339–367.

For Canada: Y. Gingras, "La Physique à McGill entre 1920 et 1940: La Réception de la Mécanique Quantique par une Communauté périphérique," in *Science, Technology and Medicine in Canada's Past*, edited by R. A. Jarrell and James P. Hull, 105–128 (Thornhill, Ontario: Scientia Press, 1991); reprinted from *Journal of the History of Canadian Science, Technology and Medicine, HSTC Bulletin* 5, no. 1 (Jan. 1981) :15–39.

For France: Chieko Kojima, "Acceptance of Quantum Theory in France," *Historia Scientiarum* 10, no. 2 (2000): 156–162.

For Italy: Vincenzo Fano, "How Italian Philosophy Reacted to the Advent of Quantum Mechanics in the Thirties," In *The Nature of Quantum Paradoxes*, edited by G. Tarozzi and A. Van der Merwe, 385–401 (Boston: Kluwer, 1988).

For Poland: Wieslaw A. Kaminski and Stanislaw Szpikowski, "Recepcja mechaniki kwantowej w Polskich osrodkach naukowych w latach dwudziestych," *Studia I Materialy z Dziejów Nauki Polskiej, Seria C: Historia Nauk Matematycznych, Fizyko-chemicznych I Geoloiczno-geograficznych* 23 (1979): 105–109 (not seen; citation from *Isis Cumulative Bibliography* 71, No. 5, (1980): 202).

For USSR: V. P. Vizgin, "At the Intersection of Traditions and Innovations: Perceptions of Schrödinger's Wave Mechanics in the USSR," in *Revolution in Sciences, Sciences in Revolution*, ed. J. Janko, 89–113 (Prague: Institute of Czechoslovak and General History, Czechoslovak Academy of Sciences, 1989).

For the United States: Stanley Coben, "The Scientific Establishment and the Transmission of Quantum Mechanics to the United States, 1919–1932, "*American Historical Review 76* (1971): 442–466; this article was strongly criticized by Van Vleck in a lecture titled "American Physics Before the War: Fact and Fiction," at the Washington, DC meeting of the American Physical Society, April 22, 1974. Nancy Cartwright, "Philosophic Problems of Quantum Theory: The Response of American Physicists," in *The Probabilistic Revolution*, ed. L. Kruger *et al.*, vol. 2, pp.: 417–435 (Cambridge, MA: MIT Press, 1987); Elisabetta Donini, "Aspetti scientifici e di contestro storico nel passaggio della meccanica quantistica negli Stati Uniti," in *Fisica e Società negli Anni '20* (Milano: CLUP/CLUED, 1980), and article by Robert Seidel in same book; Katherine Sopka, *Quantum Physics in America: The Years through 1935* (New York: American Institute of Physics/Tomash Publishers, 1988);

237 *the best indication of the initial impact* A. Kozhevnikov and C. Novik, "Analysis of Informational Ties Dynamics in Early Quantum Mechanics (1925–1927)," *Acta Historiae Rerum Naturalium necnon Technicarum*, special issue 20 (1989): 115–159; also published in *Revolution in Sciences, Sciences in Revolution*, ed. J. Janko (Prague: Institute of Czechoslovak and General History Czechoslovak Academy of Sciences, 1989), 115–159. Some of this research is described by Jagdish Mehra, "Erwin Schrödinger and the Rise of Wave Mechanics. III. Early Response and Applications," *Foundations of Physics* 18 (1988): 107–184; see also Mehra and Rechenberg, *Historical* vol. 5, part 2 (1987)

238 *were soon shown to be mathematically equivalent* According to Mehra and Rechenberg, *Historical* vol. 5, part 2, p. 578, "One of the most enthusiastic responses [to wave mechanics] came from . . . Pauli, having already been informed about the new forthcoming theory (prior to publication) in early February 1926 by Sommerfeld . . . A few weeks later Pauli demonstrated in a letter to Pascual Jordan in Göttingen, that the wave mechanical method . . . yielded the same energy values for atomic systems (Pauli to Jordan, 12 IV 1926). Schrödinger himself has arrived at exactly the same conclusion nearly four weeks earlier . . . Schrödinger's and Pauli's result . . . stimulated many theoreticians in Europe and America to attack old and new quantum theoretical problems with the help of Schrödinger's wave mechanics." For more on the equivalence proof and its impact see Mehra and Rechenberg, *Historical* vol. 5, part 2, p. 636–684; F. A. Muller, "The Equivalence Myth of Quantum Mechanics," *Studies in History and Philosophy of Modern Physics* 28 (1997), 35–61.

239 *the old quantum theory was already in a crisis* Kozhevnikov and Novik, "Analysis," p. 128. Suman Seth disputes the view that there was a generally perceived crisis in the 1920s, arguing that the crisis was seen mainly by Bohr and his followers but not by the Sommerfeld group, which was more interested in solving specific problems than in discussing fundamental principles. Seth, "Crisis and the Construction of Modern Theoretical Physics," *British Journal for the History of Science* 40 (2007): 25–51.

239 *many of the former found it quite easy to apply wave mechanics* Sommerfeld "accepts completely the wave mechanics, as a good mathematician would, for all quantum theory problems are thus reduced to the boundary conditions of the old mathematical physics." L. Pauling, letter to A. A. Noyes, 17 XII 1926, at Oregon State University; quoted by T. Hager, *Force of Nature* (New York: Simon & Schuster, 1995), p. 635.

239 *I found six books and four articles* Herbert Stanley Allen, *The Quantum and its Interpretation* (London: Methuen, 1928); George Birtwistle, *The new Quantum Mechanics* (Cambridge: Cambridge University Press, 1928); Karl K. Darrow, "Contemporary Advances in Physics. XIV. Introduction to Wave-Mechanics," *Bell System Technical Journal* 6 (1927): 653–701; Henry Thomas Flint, *Wave Mechanics, Being one Aspect of the New Quantum Theory* (New York: Dutton, 1928); Arthur Haas, *The World of Atoms: Ten Non-Mathematical Lectures*, translated from German (New York: Van Nostrand, 1928); *Materiewellen und Quantenmechanik: Eine elementare Einführung auf Grund der Theorien de Broglies Schrödingers und Heisenbergs* (Leipzig: Akademische Verlagsgesellschaft mbH, 1928), translated as *Wave Mechanics and the new Quantum Theory* (London: Constable, 1928); Pascual Jordan, "Die Entwicklung der neuen Quantenmechanik," *Naturwissenschaften*

15 (1927): 614–623, 636–649; O. W. Richardson, "On the Present State of Atomic Physics" (presidential address), *Proceedings of the Physical Society of London* 39 (1927): 171–186; J. H. Van Vleck, "The new Quantum Mechanics," *Chemical Reviews* 51 (1928): 467–507; Wilhelm Westphal, *Physik: Ein Lehrbuch für Studierende an den Universitäten und technischen Hochschulen* (Berlin: Springer, 1928). See also the letter from Van Vleck to Dana, 20 XI 28 on the literature of quantum mechanics (Archives for History of Quantum Physics, 49-1).

Section 8.9

240 *With the possible exception of the Davisson-Germer experiment* See notes for Section 8.2 As Schrödinger himself pointed out, this was a test of de Broglie's hypothesis, not of his own theory (letter to Wien, quoted above). Moreover, as a prediction of Schrödinger's theory it was not completely novel since Walter Elsasser had already suggested in 1925 that Davisson's earlier experiments supported the wave nature of the electron.

Another phenomenon sometimes mentioned as early evidence for wave mechanics is the Ramsauer effect; see Mehra and Rechenberg, *Historical* vol. 1, part 2 (1982), pp. 620–627 and vol. 6, part 1, pp. 371–372, 380–383. Although Mehra and Rechenberg claim that it "played a role in establishing the fundamental matter-wave concept" (ibid., p. 383), their own survey of the literature shows that the quantum mechanical explanation of this effect was still not generally accepted as late as 1928.

240 *could be considered a confirmed novel prediction* Max Jammer writes (*Conceptual Development*, p. 277): "Just as Mulliken's [1924, 1925] observations of the shift in the band spectrum of boron monoxide was the earliest evidence for the superiority of Heisenberg's matrix mechanics over the old quantum theory, so was Kiuti's observation of the shift in the line pattern of the hydrogen Stark effect [1925] the earliest proof of the superiority of Schrödinger's wave mechanics. Both formalisms found support in evidence gathered prior to their establishment."

240 *all of its predictions have been confirmed* "Most of the accurate detail which will not fit into the Sommerfeld orbit models can be predicted and, in a sense, explained [by quantum mechanics]," C. B. Bazzoni, *Kernels of the Universe: Modern Discoveries in Regard to the Fundamental Constitution of Things* (New York: Doran, 1927), p. 152; "Predictions made by the wave mechanics are in accord with experiment," N. F. Mott, *An Outline of Wave Mechanics* (Cambridge: Cambridge University Press, 1930), p. 54; "In general where there is a difference between the predictions of Schrödinger's theory and those of the older quantum theory, the experimental results are in favour of the former theory," H. S. Allen, *Electrons and Waves* (London: Macmillan, 1932), p. 263; quantum mechanics "forms a wonderfully complete and coherent system enabling one to codify existing knowledge and predict, within limits, phenomena not as yet observed," F. A. Lindemann, *The Physical Significance of the Quantum Theory* (Oxford: Clarendon Press, 1932), p. 9; "Many new phenomena have been predicted [by wave mechanics] and subsequently verified experimentally," F. K. Richtmyer, *Introduction to Modern Physics* (New York: McGraw-Hill, 1934), p. 701 (then cites Compton test of Pauling prediction of electron distribution; has previously [p. 678] cited discovery of electron diffraction following prediction by de Broglie); "Errors and omissions excepted, wave mechanics enables us to predict sensory experience so far as we have any reason to suppose that sensory experience is predictable," Arthur Eddington, *New Pathways in Science* (Cambridge: Cambridge University Press, 1935), p. 44; "To

date, all experiments magnificently confirm all quantum predictions with impressive precision"—A. Zeilinger, "The Quantum Centennial," *Nature* 408 (2000): 639–641, on p. 639.

240 *some other alleged novel predictions* One promising candidate is Dirac's calculation of the intensity of scattering at different angles in Compton scattering (I thank Helge Kragh for this suggestion). Similar results were derived by Walter Gordon and Ivar Waller. Dirac claimed that "this is the first physical result obtained from the new mechanics that had not been previously known" (P.A.M. Dirac, "Relativity Quantum Mechanics with an Application to Compton Scattering," *Proceedings of the Royal Society of London* A111 (1926), 405–23, on p. 421), but in a footnote he admits that his result for unpolarised incident radiation had been obtained from the correspondence principle by Breit (1926). Thus it is arguably not really a prediction of quantum mechanics as distinguished from earlier theories. But it is an example in which a theoretician makes a prediction that *disagrees* with a previous experimental result (Compton's), and is vindicated by later experiments. The prediction was confirmed by C. S. Barrett and J. A. Bearden, "The Polarizing Angle for X-Rays," *Physical Review* [series 2] 29 (1927), 352–53; reported to Dirac in a letter from Compton, August 21, 1926, quoted by Helge Kragh, *Dirac: A Scientific Biography* (New York: Cambridge University Press, 1990, p. 29; but not published in full until July 1928, by Barrett, "The Scattering of X-Rays from Gases," *Physical Review* [series 2] 32 (1928), 22–29, and by Fritz Kirchner, "Experimentelle Untersuchungen über die Richtungsverteilung von Röntgenstrahlen ausgelösten Elektronen," *Physikalische Zeitschrift* 27 (1926), 799–801, announced at a meeting on 24 September 1926. But although Dirac's paper was widely cited (Kragh, *Dirac*, p. 28), the experimental tests of the predictions attracted little notice in the physics journals of the 1920s: Henry Small, *Physics Citation Index 1920–1929* (Philadelphia: Institute for Scientific Information, 1981; A. H. Compton and Samuel K. Allison give a detailed survey in *X-Rays in Theory and in Experiment* (New York: Van Nostrand, 1935).

240 *One of the two achievements for which Heisenberg received the Nobel Prize* The idea came from Heisenberg's paper on the application of the "resonance" concept to the approximate treatment of two-electron atoms such as helium: "Mehrkörperproblem und Resonanz in der Quantenmechanik," *Zeitschrift für Physik* 38 (1926): 411–426; 41 (1927): 239–267. With the realization that protons, like electrons, may have spin ½ and obey Fermi-Dirac statistics, Heisenberg and Friedrich Hund proposed that H_2 molecules could also have two forms: Hund, "Zur Deutung der Molekelspektren II," *Zeitschrift für Physik* 42 (1927): 93–120; Heisenberg, discussion remark following Bohr's paper, *Atti del Congresso Internazionale dei Fisici, 11–20 Settembre 1927, Como-Pavia-Roma*, vol. 2 (Bologna: Zanichelli, 1928) on p. 297.

According to E. A. Hylleraas, "Heisenberg's theory of the helium atom was conceived in Norway. . . . walking in the mountains for several weeks entirely alone . . . One day, when trying to pass a stream, he fell into the water and had a very narrow escape. Back in some hotel he wrote his famous paper. How much the cold bath may have contributed to clearing his mind, I cannot tell." Hylleraas, "Reminiscences from early Quantum Mechanics of Two-Electron Atoms," *Reviews of Modern Physics* 35 (1963): 421–431, on pp. 424–425.

David M. Dennison worked out the thermodynamic consequences of the hypothesis and found that it could explain anomalous specific heat data: "A note on the specific heat of the Hydrogen molecule," *Proceedings of the Royal Society of London* A115 (1927): 483–486; Heisenberg, discussion remark in *Electrons and Photons, Rapports et Discussions du Cinquième Conseil de Physique tenu à Bruxelles du 24 au 29 Octobre 1927* (Paris: Gauthier-Villars, 1928), pp. 271–272. Dennison, interview with Thomas S. Kuhn, 30 I 64, transcript in *Archive for History of Quantum Physics*, pp. 14–16; "Recollections of Physics and of Physicists during the 1920s," *American Journal of Physics* 42 (1974): 1051–1056.

Later it was shown that the two forms could be physically separated: K. F. Bonhoeffer and P. Harteck, "Über Para-und Orthowasserstoff," *Zeitschrift über Physikalische Chemie* B4 (1929): 113–141. For a contemporary review of this subject see A. Farkas, *Orthohydrogen, Para-ydrogen, and Heavy Hydrogen* (Cambridge: Cambridge University Press, 1935).

241 *had proposed that the hydrogen molecule has ortho and para forms* E. Rutherford, "Recent Reactions between Theory and Experiment," *Nature* 124 (1929): 878–880, 892, on p. 879. There is a rather vague mention in the paper by N. M. Bligh, "The Evolution of the new Quantum Mechanics," *Science Progress* 23 (1929): 619–632, on p. 630. For other earlier references see also Arthur Haas, *Quantum Chemistry* (London: Constable, 1930, translated from 1929 German edition), p. 62; Niels Bohr, "Chemistry and the Quantum Theory of Atomic Constitution" (Faraday Lecture, 8 May 1930), *Nature* 125 (1930), 788–789; *Journal of the Chemical Society* (1932): 349–384.

In connection with the ortho/para phenomenon, quantum mechanics predicts that certain lines in the spectra of diatomic molecules will be absent if the two atoms are identical but present if they are different. This prediction led to the discovery of two previously-undetected isotopes of oxygen, ^{17}O and ^{18}O, by W. F. Giauque and H. L. Johnston: "An Isotope of Oxygen, Mass 18," *Nature* 123 (1929); "An Isotope of Oxygen of Mass 17 in the Earth's Atmosphere," *Nature* 124 (1929), 1–2; *Low Temperature, Chemical, and Magneto Thermodynamics: The Scientific Papers of William F. Giauque*, Volume I (1923–1949) (New York: Dover Publications, 1969), pp. 86–92, 102–103, 126–127, 149–155, 609, 618–619. "Thus the 18-16 molecule has every rotation state where the 16-16 molecule has only alternate levels. Such an excellent confirmation of the predictions of wave mechanics in this regard has not heretofore been possible since the presence of nuclear spin usually permits all states to exist although not in equal amount" (*Low Temperature*, p. 102).

241 Norton A. *Kent, Lucien B. Taylor, and Hazel Pearson at Caltech confirmed* Sommerfeld and Unsöld, "Spektrum"; Kent et al. "Doublet-Separation"; Kragh, "Fine Structure," pp. 115–122; Sommerfeld, "Zur Feinstruktur der Wasserstofflinien: Geschichte und gegenwärtige Stand der Theorie," *Naturwissenschaften* 28 (1940): 417–423, reprinted in his *Gesammelte Schriften*, edited by Fritz Sauter, Band III (Braunschweig: F. Vieweg, 1968), pp. 857–863.

241 *Foster and his student Laura Chalk published their experimental results* Erwin Schrödinger, "Quantisierung als Eigenwertproblem. Dritte Mitteilung: Störungstheorie mit Anwendung auf den Starkeffekt der Balmerlinien," *Annalen der Physik* [series 4], 80 (1926): 437–490. For English translation see his *Collected Papers on Wave Mechanics* (New York: Chelsea Pub. Co., 2nd ed. 1978), pp. 62–101. Mehra and Rechenberg, *Historical* vol. 5, part 2 (1987), pp. 700–701 provide a summary of his calculation. See also P. S. Epstein, "The Stark Effect from the Point of View of Schroedinger's Quantum Theory," *Physical Review* [series 2], 28 (1926): 695–710; Mehra and Rechenberg, *Historical* vol. 5, part 2, pp. 717–723; Ivar Waller, "Der Starkeffekt zweiter Ordnung bei Wasserstoff und die Rydberg Korrection

der Spektra von He und Li+," *Zeitschrift für Physik* 38 (1926): 635–646; Gregor Wentzel, "Eine Verallgemeinerung der Quantenbedingungen für die Zwecke der Wellenmechanik," *Zeitschrift für Physik* 38 (1926): 518–529; J. Stuart Foster and Laura M. Chalk, "Observed Relative Intensities of Stark Components in Hydrogen," *Nature* 118 (1926): 592; Foster and Chalk, "Relative Intensities of Stark Components in Hydrogen," *Proceedings of the Royal Society of London* 123 (1929): 108–118.

241 *not cited in any major physics journal before 1930* Henry Small, *Physics Citation Index 1920–1929* (Philadelphia: Institute for Scientific Information, 1981); G. Birtwistle, *The New Quantum Mechanics* (Cambridge: Cambridge University Press, 1928), p. 178.

241 *Foster's observations on the Stark effect in helium* Foster, "Stark Patterns Observed in Helium," *Proceedings of the Royal Society of London* A114 (1927): 47–65; Foster, "A Quarter-Century of Research in Physics" (presidential address), *Transactions of the Royal Society of Canada* 4 3, section III (1949): 1–13.

241 *Foster himself, though an important figure* Jerry Thomas, "John Stuart Foster, McGill University and the Renascence of Nuclear Physics in Montreal, 1935–1950," *Historical Studies in the Physical Sciences* 14 (1984): 357–377. Robert E. Bell. "John Stuart Foster, 1890–1964," *Biographical Memoirs of the Royal Society of London* 12 (1966) 147–161 (personal disclosure: his son, John S. Foster, Jr., was director of Lawrence Livermore National Laboratory when I was employed there in 1959–1965).

241 *they left no acknowledgement on the public record* Heisenberg recalled, but only in an unpublished interview (1963), that Foster's work on the empirical testing of quantum mechanics was "very exciting." Interview with T. S. Kuhn, 19 February 1963, *AHQP* Tape 50b, p. 15. In 1949 Foster reviewed his research on the Stark effect but did not claim to have established quantum mechanics by his experiments (Foster, "Quarter-Century"). Thomas ("Foster") discusses Foster's research on the Stark effect but does not mention his confirmation of a prediction from quantum mechanics. Bell, "John Stuart Foster," discusses Foster's 1927 paper on helium and notes that it includes some new experimental results that are compared with theory, but does not mention his papers with Laura Chalk on hydrogen.

241 *made only a token effort to claim credit* "My results were the first published data to check the validity of the new Schrödinger's wave mechanics." See the chapter "Long Experience and a Happy Existence" by Laura [Chalk] Rowles in *Our Own Agenda, Autobiographical Essays by Women Associated with McGill University*, edited by Margaret Gillett and Ann Beer, (Montreal: McGill-Queen's University Press, 1995), pp. 33–46.

241 *Mott's (1930) calculation of the scattering of identical particles* N. F. Mott, "The Collision between two Electrons," *Proceedings of the Royal Society*, A126 (1930): 259–267. See also N. F. Mott and H. S. W. Massey, *Theory of Atomic Collisions* (Oxford: Clarendon Press, 1933), pp. 73–74.

241 *Mott's prediction was confirmed* J. Chadwick, "The Scattering of α-Particles in Helium," *Proceedings of the Royal Society of London* A128 (1930): 114–122; P.M.S. Blackett and F. C. Champion, "The Scattering of Slow Alpha Particles by Helium," *Proceedings of the Royal Society of London* A130 (1931): 380–388; E. J. Williams, "Passage of slow β-Particles through Matter—Production of Branches," *Proceedings of the Royal Society of London* A128 (1930): 459–468; Chr. Gerthsen, "Streuungsmessungen von H-Strahlen in Wasserstoff als Beitrag zur Klärung ihrer Wellennatur," *Annalen der Physik* [series 5], 9 (1931): 769–786.

241 *Williams said that he had made the observations in 1927* "These observations were carried out in 1927, before the new quantum theory calculations were made, and the results were first mentioned in a Ph.D. thesis (Cambridge) presented in September, 1929"—Williams, "Passage" (cited in note 63), p. 459. Mott, "Collision," p. 204.

242 *"one of the few totally unexpected things* predicted *by quantum mechanics"* N. F. Mott, "Notes by Prof. Mott on his Personal Experiences of the Development of Quantum Physics" (March 18, 1962), *Archive for History of Quantum Physics*, tape 66, p. 1–3. He recalls that Rutherford said: "Well, Mott if you think of anything else like that, come and tell me."

242 *no evidence that this confirmed novel prediction converted anyone* E. Rutherford, J. Chadwick and C. D. Ellis, *Radiations from Radioactive Substances* (Cambridge: Cambridge University Press, 1930), p. 234; C. G. Darwin, *The New Conceptions of Matter* (London: Bell & Sons, 1931), p. 186.

242 *"a strong argument in favor of a continuous electron distribution"* A. H. Compton, "The Determination of Electron Distributions from Measurements of scattered X-rays," *Physical Review* [series 2], 35 (1930): 925–938. L. Pauling, "The Theoretical Prediction of the Physical Properties of Many-Electron Atoms and Ions: Diamagnetic Susceptibility and Extension in Space," *Proceedings of the Royal Society of London A* 114, no. 767 (1 March 1927): 181–211; Barrett, "The Scattering of X-rays from Gases," *Physical Review* [2], 32 (1928): 22–29.

242 *only one other physicist who cited electron distribution* F. K. Richtmyer, *Introduction to Modern Physics* (New York: McGraw-Hill, 2nd edition, 1934), p. 701.

Section 8.10

242 *The observed value, reported by Theodore Lyman* Lyman, "The Spectrum of Helium in the extreme Ultraviolet," *Astrophysical Journal* 60 (1924): 1–14; Kramers, "Über das Modell des Heliumatoms," *Zeitschrift für Physik* 13 (1923): 312–341; Darrigol, *From c-Numbers*, pp. 159–181. Darrigol describes in detail "the catastrophe of helium" in the old quantum theory (p. 331), but mentions only briefly Heisenberg's success in finding the route to the solution.

243 *He obtained a result similar to Kramers's* A. Unsöld, "Beiträge zur Quantenmechanik der Atome," *Annalen der Physik* 387, Issue 3 (1927): 355–393. [check]

243 *Kellner, with a more elaborate calculation, obtained the result 23.75* See Heisenberg's 1927 paper "Mehrkörper II"); Georg W. Kellner, "Die Ionisierungsspannung des Heliums nach der Schrödingerischen Theorie," *Zeitschrift für Physik* 44 (1927): 91–109; Mehra and Rechenberg, *Historical* vol. 3, chapter V.6 (pp. 282–301) and vol. 5, pp. 735–746.

243 *Heisenberg proposed to approximate the actual wave function* See J. C. Slater, *Solid-State and Molecular Theory: A Scientific Biography* (New York: Wiley, 1975), Chapter 6 for a concise explanation of this calculation. Slater explains his own attempt to solve the helium atom problem in Chapter 19; he abandoned it because "By the time I wrote up this helium work for publication, in 1928, Hartree's papers about the self-consistent field were coming out . . . I felt that the added refinements I was trying to introduce with my helium work might not be necessary . . . another reason for dropping the helium problem was that this was when Hylleraas started his very important series of papers" (p. 153). He notes (p. 155) that in 1933, H. M. James and A. S. Coolidge at Harvard applied methods similar to those of Hylleraas to the H_2 molecule, with equally good results.

243 *a recent paper by J. C. Slater suggested to him the idea* Hylleraas, "Reminiscences" p. 425; reprinted in *Selected Scientific Papers of Egil A. Hylleraas*, edited by John

Midtdal, Knut Thalberg, and Harald Wergeland (Trondheim: Norges tekniske høgskole Press, 1968), vol. 2, pp 419–429. The problem was originally assigned by Born to another young physicist, Biemüller, but he became ill and could not complete the work so Hylleraas agreed to work on it.

243 *He was assisted by some of the best mathematical physicists* Hylleraas, "Reminiscences . . ." pp. 425–426. Note the propitious timing of the publication of Courant and Hilbert's book, *Methoden der mathematischen Physik* (Berlin: Springer, 1924), According to Jammer, *Conceptual Development*, "it contained precisely those parts of algebra and analysis on which the later development of quantum mechanics had to be based; its merits for the subsequent rapid growth of our theory can hardly be exaggerated" (p. 217). But Laszlo Tisza (born 1907) doesn't agree: in a 2004–2006 interview he said that after the publication of Courant and Hilbert's *Methods of Mathematical Physics* (1924), which treated eigenvalue problems, some mathematicians thought the methods might be applied to Bohr's theory of the hydrogen atom. But "the hydrogen spectrum consists of a discrete part that converges to the ionization limit where it is joined by the continuous spectrum . . . when Erwin Schrödinger derived the hydrogen spectrum by solving a wave-mechanical eigenvalue problem . . . Hilbert commented . . . 'who would have believed that the singularity in the differential equation would bring the convergence of the spectrum into the finite?' Of course . . . Schrödinger came to the Coulomb potential [which supplied the singularity] by physical reasons; the mathematical singularity was a byproduct." Tisza, "Adventures of a Theoretical Physicist. Part I: Europe," *Physics in Perspective* 11 (2009): 46–97, on pp. 60–61.

243 *apparently the first physicist who appreciated the usefulness of group theory* E. P. Wigner, "Über nicht kombinierende Terme in der neuen Quantentheorie II," *Zeitschrift für Physik* 40 (1927): 883–892; *Gruppentheorie und ihre Anwendung auf die Quantenmechanik der Atomspektren* (Braunschweig: Vieweg, 1931); *Symmetries and Reflections* (Woodbridge, CT: Ox Bow Press, 1979), Part I.

243 *the well-known variational procedure of Walter Ritz* Paul Forman, "Ritz, Walter," *Dictionary of Scientific Biography* 11 (1975): 475–481; Hylleraas, "Über den Grundzustand des Heliumatoms," *Z. Physik* 48 (1928): 469–494; reprinted in his *Selected Scientific Papers*, vol. I, pp. 20–45.

243 *"might as well have been taken to be a disproof"* Hylleraas, "Reminiscences . . . " p. 427.

244 *"almost the effect of a miracle"* Ibid.; he suggests that he may have gotten a hint toward this trick from a paper by J. C. Slater. Hylleraas published his improved result in "Neue Berechnung der Energie des Heliums im Grundzustande, sowie des tiefsten Terms von Ortho-Helium," *Zeitschrift für Physik* 54 (1929): 347–366; reprinted in his *Selected Scientific Papers*, vol. I, pp. 57–76.

244 *later developed "a most exciting collaboration" with Edlén* Hylleraas, "Reminiscences," pp. 427–428.

244 *later research turned up small errors* He reported that the "excellent agreement with the experimental value" was not quite so excellent; his theoretical value was "false, even though at the end I believe it will prove to be the right one." He had made an arithmetic error (transposing two successive digits), first detected 20 years later when Chandrasekhar, Herzberg, and Elbert obtained a different result by independent calculations. E. A. Hylleraas, "Recent Calculations of the Energy Values of Two-Electron Atoms," lecture at the Symposium on the Quantum Theory of Molecules, Stockholm-Uppsala, March 20–25, 1955, published in *Festskrift til Prof. Bjørn Helland-Hansen* (Bergen, 1956), reprinted in *Selected Scientific Papers*, Vol. 2, pp. 226–236, on p. 229.

244 *"one of the greatest achievements of quantum mechanics"* Hylleraas, "The Schrödinger Two-electron Atomic Problem," *Advances in Quantum Chemistry* 1 (1964): 1–33, on p. 13.

Section 8.11

244 *I found 21 books and 5 articles* Herbert Stanley Allen, *Electrons and Waves: An Introduction to Atomic Physics* (London: Macmillan, 1932); Charles B. Bazzoni, *Energy and Matter: Building Blocks of the Universe* (New York: University Society, 1932); N. M. Bligh, "The Evolution of the New Quantum Mechanics," *Science Progress* 23 (1929): 619–632; Eugène Bloch, *L'ancienne et la nouvelle Théorie des Quanta* (Paris: Hermann, 1930); William Lawrence Bragg, *Recent Advances in Physics* (London: Broadway Press, 1930); Niels Bohr, "Chemistry and the Quantum Theory of Atomic Constitution" (Faraday Lecture), *Nature* 125 (1930): 788–789, also in *Journal of the Chemical Society* 131 (1932): 349–384; Florian Cajori, *A History of Physics* (New York: Macmillan, rev. ed. 1929); Gaetano Castelfranchi, *Recent Advances in Atomic Physics* (London: Churchill, 1932); Edward U. Condon and Philip M. Morse, *Quantum Mechanics* (New York: McGraw-Hill, 1929); C. G. Darwin, *The New Conceptions of Matter* (London: Bell, 1931); J. Frenkel, *Einführung in die Wellenmechanik* (Berlin: Springer, 1929); Arthur Haas, *Quantum Chemistry* (London: Constable, 1930) and *The New Physics* (London: Methuen, 3rd ed. 1930) (these two books are counted as one publication); Otto Halpern and Hans Thirring, *The Elements of the new Quantum Mechanics* (London: Methuen, 1932); Paul R. Heyl, *New Frontiers of Physics* (New York: Appleton, 1930); Alfred Landé, *Vorlesungen über Wellenmechanik* (Leipzig: Akademische Verlagsgesellschaft, 1930); Paul Langevin, "L'Orientation actuelle de la Physique," in *L'Orientation actuelle des Sciences* by L. Brunschvicg et al., pp. 29–62 (Paris: Alcan, 1930); F. A. Lindemann, *The Physical Significance of the Quantum Theory* (Oxford: Clarendon Press, 1932); Arthur March, *Die Grundlagen der Quantenmechanik* (Leipzig: Barth, 1931); N. F. Mott, *An Outline of Wave Mechanics* (Cambridge: Cambridge University Press, 1930); W. Pauli, Jr., "Allgemeine Grundlagen der Quantentheorie des Atombaues," in *Müller-Pouillets Lehrbuch*, Bd. 2, Teil 2, 11. Aufl., pp. 1709–1842 (Braunschweig: Vieweg, 1929); Arthur Edward Ruark and Harold Clayton Urey, *Atoms, Molecules, and Quanta* (New York: McGraw-Hill, 1930); E. Rutherford, "Recent Reactions Between Theory and Experiment: The Raman Effect: The Constitution of Hydrogen Gas," *Nature* 124 (1929): 878–880, 892; Rutherford, James Chadwick and C. D. Ellis, *Radiations from Radioactive Substances* (Cambridge: Cambridge University Press, 1930) (the article and the coauthored book are counted as one publication); Arnold Sommerfeld, *Wave-Mechanics* (New York: Dutton, 1930); G. Temple, *An Introduction to Quantum Theory* (London: Williams & Norgate, 1931).

246 *the view of physicist George Galton Darwin* Navarro, "A 'Dedicated Missionary': Charles Galton Darwin and the new Quantum Mechanics in Britain," *Studies in History and Philosophy of Modern Physics* 40 (2009): 316–326, on p. 324.

246 *As Darwin wrote in 1929, "We are led to the conception"* Darwin, "A Collision Problem in the Wave Mechanics," *Proceedings of the Royal Society of London* 124 (1929): 375–394, on p. 393 as quoted by Navarro, "Darwin," p. 324.

246 *Feynman gave a mathematical expression of a similar idea* Feynman, "Space-Time Approach to Non-Relativistic Quantum Mechanics," *Reviews of Modern Physics* 20 (1948): 367–387. See also Farmelo, *Strangest Man*, p. 333.

9

New Particles

This chapter examines two episodes in the history of modern physics. In each case, a theoretical hypothesis led to the prediction of the existence of a previously unobserved particle with specified properties, and a particle with those properties was subsequently discovered. By examining the writings of physicists in the years following the discovery, I try to determine whether the success of the prediction significantly increased the acceptance of the theory by scientists. The two cases to be examined are P.A.M. Dirac's prediction of the positron from his relativistic quantum theory of the electron (supplemented by the "hole hypothesis"), and physicist Hideki Yukawa's prediction of the meson from his theory of forces between nucleons. They were selected because they have been widely celebrated as examples of successful novel prediction and a substantial technical literature is available for each case.

As in other cases, I am especially interested in determining the role of *novelty*: what difference did it make to the evaluation of the theory that the prediction came *before* the discovery? Here are three views on this point expressed by modern particle theorists.

Physicist Sheldon Glashow asserts that "to be valid, a theory must not simply explain the results of experiments that have already been done, but must point the way to new ideas and new observations." Thus Einstein's prediction of gravitational light bending was "more impressive" because it was an "absolutely new and unexpected effect."

Physicist Frank Wilczek states that "in most fields of science, theories get judged by their ability to predict the results of new experiments and observations." But he also admits that in fields like cosmology we may have to be satisfied with "postdiction" of earlier events, based on the time-reversal invariance of the laws of nature.

Physicist Yuval Ne'eman says:

> The importance attached to a successful prediction is associated with human psychology rather than with the scientific methodology. It would not have detracted at all from the effectiveness of the eightfold way if the Σ− [omega-minus particle] had been discovered *before* the theory was proposed. But human nature stands in great awe when a prophecy comes true, and regards the realization of a theoretical prediction as an irrefutable proof of the validity of the theory.

According to Glashow "you cannot predict something that has already been measured," but unfortunately many other scientists do use the word "prediction" ambiguously, to mean either the deduction of a known fact or the forecast of a new fact (Section 1.3). This ambiguity in itself suggests that scientists do not attach as much importance to temporal novelty as Glashow, Wilczek, and Ne'eman imply.

One reason for demanding that theorists make *novel* predictions is the suspicion that a theory can be manipulated in order to obtain a known result. Less often, it is recognized that the opposite can happen: an empirical result may be influenced by the desire to confirm (or refute) a theoretical prediction. Thus, physicist John Ziman writes: if a scientist "did not already know the answer to the hypothetical experiment, then he could not have 'cooked' his theory to agree with it; therefore it is more convincing. The element of predictability in good science is just what makes us believe in it so strongly." But, he writes, it is fortunate that Anderson was unaware of Dirac's theory, just as Dirac was unaware of Anderson's observation of the positron, so "neither party could have cooked his argument or his results. It is not surprising that the identity of the experimental particle with its theoretical model was immediately acceptable to physicists." But just because the new particle's behavior is consistent with Dirac's previously published theory, does that mean that physicists *therefore* accepted Dirac's theory? And what difference does it make if we accept physicist Graham Farmelo's statement that at the time of his discovery of the position? He said:

> Anderson had earlier spent several evenings a week struggling through Oppenheimer's evening lectures on Dirac's hole theory, so it is practically certain that he knew about the anti-electron within it. Yet he did not make the connection. . . . It beggars belief that Oppenheimer never pointed out the connection between Dirac's theory and Anderson's experiment to Dirac, to Anderson or to anyone else. Yet that appears to be what happened.

9.1 DIRAC'S PREDICTION AND ANDERSON'S DISCOVERY OF THE POSITRON

As in the discovery of the old quantum theory and the new quantum mechanics, mathematics played an essential role in the development of the first relativistic wave equation. Dirac postulated that this equation should satisfy the condition of Lorentz invariance, meaning that a transformation of the space-time coordinate system should leave the equation unchanged. Here the time variable t must appear in the combination ict, where $i = \sqrt{-1}$, c = speed of light. Moreover, he started with a *first-order* equation, meaning that it contains *first* derivatives of the wave function with respect to x, y, z, and ict, whereas the original Schrödinger equation contains *second* derivatives with respect to the variables. Even worse, the so-called *time-dependent Schrödinger equation* contains these second space derivatives along with a first derivative with respect to time, so it is obviously not Lorentz-invariant.

Again for purely mathematical reasons, Dirac argued that the wave function P must have four components, two of which correspond to positive energy with spin $+\frac{1}{2}$ and $-\frac{1}{2}$, and the other two correspond to negative energy with spin $+\frac{1}{2}$ and $-\frac{1}{2}$.

Thus Dirac's equation *explained* why the electron has two possible spin states, without have to arbitrarily postulate those states as Uhlenbeck and Goudsmit had done. (Remember that in physics one can explain something by deriving it from a basic postulate, but that does not necessarily mean that one understands it.)

At this point the mathematics collides with the physics: the theory predicts an infinite number of negative energy states as well as an infinite number of positive energy states. The latter is not disturbing—Newtonian physics and nonrelativistic Quantum mechanics also predict that you can keep giving more energy to a particle, making it do faster and faster. But what does negative energy mean?

One possibility is that the particle has negative mass, hence negative kinetic energy ($\frac{1}{2}mv^2$). Physicist George Gamow called it a "donkey electron": according to Newtonian mechanics, if you push a negative-mass particle in one direction, it moves in the opposite direction.

But Dirac proposed something even more crazy: there is an infinite sea of negative-energy states filling all space. Normally all the states are filled with electrons, no more than one per state according to the Pauli Exclusion principle. We can move freely through this sea without feeling any resistance; an electron inside the sea cannot resist us because it cannot absorb energy, since all the states above it (closer to zero) are filled. But if an electron happens to receive enough energy all at once (from a γ ray, for example) to kick it up to an unoccupied positive energy state, then it will appear as a normal positive-energy electron with negative change. At the same time, it will leave behind an empty state—a hole—in the sea of negative-energy electrons. This hole, according to Dirac, will behave just like an electron with positive electric charge. So, Dirac wrote in 1931,

> an encounter between two hard γ -rays, (of energy at least half a million volts) could lead to the creation simultaneously of an electron and anti-electron.

The discovery of the positron (as we would now say) occurred in August 1932, when physicist Carl D. Anderson decided that some of his cosmic-ray tracks whose curvature in a magnetic field indicated a positive electric charge must have been made by particles with masses comparable to that of the electron. He later denied that he made the discovery as a consequence of Dirac's prediction. He wrote in 1961 that

> the discovery of the positron was wholly accidental . . . despite the fact that the existence of [Dirac's] theory was well known to nearly all physicists, including myself, it played no part whatsoever in the discovery of the positron.

Anderson argued that before the discovery of the positron,

> the Dirac theory, in spite of its successes, carried with it so many novel and seemingly unphysical ideas, such as negative mass, negative energy, infinite charge density, etc. Its highly esoteric character was apparently not in tune with most of the scientific thinking of that day. . . . This kind of thinking prevented most experimenters from accepting the Dirac theory wholeheartedly and relating it to the real physical world until after the existence of the positron was established on an experimental basis.

This seems to imply that experimenters *did* accept Dirac's theory because of the discovery of the positron, yet Anderson himself did not mention the theory in five papers on the positron he published in 1933. In December of that year he did mention it in an address to the American Physical Society, but there and in a major paper in 1934 he noted discrepancies between the theory's predictions and observations of the numbers of positive and negative electrons produced in showers.

Anderson's opinion of Dirac's theory was more favorable in an address at the International Conference on physics in 1934, and in his Nobel Lecture in 1936. Other physicists were not so reluctant to believe that the positron was the particle predicted by Dirac's theory. The first public statement to this effect was by physicist R. N. Langer, Anderson's colleague at Caltech.

Physicists P.M.S. Blackett and G.P.S. Occhialini, Dirac's colleagues at Cambridge University, confirmed Anderson's discovery and credited Dirac's theory with the quantitative prediction of the positron's annihilation by combination with the electron. In 1933 Blackett stated more explicitly that the Dirac theory had predicted the existence of particles having the same properties as the positron and that the experimental results thus supported the essence of the theory. In later publications Blackett reiterated that the discovery of the positron confirmed the validity of Dirac's hole theory.

Physicist J. R. Oppenheimer was a strong advocate of Dirac's theory in the American physics community. With physicist Milton Plesset, he used the theory to calculate the probability of (e^+, $e-$) pair creation by radiation impinging on nuclei. Their 1933 paper begins:

> The experimental discovery of the positive electron gives us a striking confirmation of Dirac's theory of the electron.

But what some physicists considered the most striking confirmation of the theory was not the *existence* of the positron as a new particle—rather it was the fact that it was *created out of radiation*. Conversely, when a positron meets an electron, they annihilate each other, leaving only radiation—a rather direct illustration of Einstein's famous equation $E = mc^2$.

9.2 THE RECEPTION OF DIRAC'S THEORY

Physicist Gregor Wentzel recalled in 1960 that "the hole theory was given much credit for the prediction of the positron." I have tried to collect all the contemporary evidence for this statement (which is found in much of the secondary literature).

In examining the reaction of other physicists to the confirmation of Dirac's theory by the discovery of the positron, we must note that the theory had already been favorably received by some physicists before 1932, according to physicist Friedrich Hund and historians Karl Grandin, Helge Kragh, and D. F. Moyer. It had been successfully used to calculate more accurate wavelengths for the spectral lines of hydrogen, and to compute the Compton scattering of photons by electrons. In particular, it had been shown that one gets the correct low-energy limit for Compton scattering from the Dirac equation only if the negative-energy states are included. So we need to find evidence that specifically links the acceptance of the Dirac theory with the discovery and properties of the positron.

For convenience I have divided my survey of the initial reception of hole theory and the positron into two parts: first, the reactions of a few physicists who had established their reputations by 1932 and whose views would therefore be likely to influence others; second, a more systematic analysis of a large number of references to the Dirac theory and positrons in selected journals, before and after Anderson announced his discovery.

The decision to award the 1933 Nobel Prize in physics to Dirac reflects the judgment of the most influential physicists, including one whose name may not be familiar to you. C. W. Oseen chaired the committee in the Swedish Academy that screened nominations for the physics prize. According to historian Robert Marc Friedman, who has studied the records of this committee, Oseen was responsible for the curious fact noted at the end of the last chapter: the founders of quantum mechanics did not receive the Nobel Prize until several years after that theory was recognized as one of the most important achievements in twentieth-century science. Before 1932, despite many nominations and private communications from leading physicists, Oseen argued that quantum mechanics did not deserve the prize since it had not made any successful novel predictions, and therefore did not represent new knowledge.

Oseen finally changed his mind in 1932 because of Anderson's discovery of the positron, predicted by Dirac: a *successful* novel prediction. Thanks to Dirac and Anderson, Heisenberg received the Nobel Prize in 1932, while Dirac and Schrödinger shared the 1933 prize (Anderson had to wait until 1936, Pauli until 1945, and Born until 1954).

Several leaders of the quantum revolution, including Bohr, de Broglie, Einstein, Heisenberg, Kramers, Pauli, and Schrödinger, actively studied and discussed Dirac's equation in the years 1930–1932. Their initial reaction to the positron discovery was a mixture of delight in the success of Dirac's theory and insistence that the search for a more satisfactory theory must continue. There were three serious problems, as articulated especially by Pauli. First, negative energy states did not make sense, especially since the theory requires an infinite number of them. Second, holes as entities that could freely move from not being in one negative energy state to not being in another negative state made even less sense. Third, even if one could swallow negative energy states and holes, the theory was not *symmetrical* with respect to charge: why should negative charges be more real than positive ones?

Though Pauli did not accept Dirac's hole theory, he realized it could not be ignored, because of its success in predicting the positron. The solution was to have an open discussion at which Dirac could defend his theory and other physicists could advance better alternatives. So Pauli urged Paul Langevin to invite Dirac to speak at the upcoming 1933 Solvay Congress.

This was done; Dirac did speak at the congress. His Cambridge colleague P.M.S. Blackett presented a report on the positron and a long survey of research on cosmic rays, both of which stated that Dirac's theory was supported by empirical facts. Pauli, having put Dirac's theory on the agenda for the meeting, could now point out that it suffered from defects: not only the above-mentioned three, but also infinite self-energy and vacuum polarization. Bohr, previously skeptical, now hailed the confirmation of Dirac's theory, but urged his colleagues to explain the behavior of positrons *without* using that theory.

The most striking opinion voiced at the 1933 Solvay Congress was Ernest Rutherford's:

> It seems to a certain degree regrettable that we had a theory of the positive electron before the beginning of experiments. . . . I would be more pleased if the theory had appeared after the establishment of the experimental facts.

Perhaps he was aware of the tendency of experimentalists to observe what theorists said they should observe (see Sections 1.8 and 7.10).

Pauli and other physicists tried to replace Dirac's theory by an alternative that could yield the same predictions without the defects of the hole theory. These efforts were not successful until the 1940s (see Section 9.3).

To discover the response of the physics community as a whole, I have examined 276 papers on the Dirac equation and the positron, published in scientific journals in the year before Anderson announced his discovery and the three years following. The number of papers on electron theory that discussed Dirac's relativistic wave equation during the 12 months before October 1932 gives an indication of the acceptance of Dirac's theory, which can serve as a base line for comparison with its standing after the discovery.

Of the papers published in the three years after Anderson's discovery of the positron, only 27 (or 24 if we don't count papers by Dirac himself) explicitly stated that Dirac's theory was confirmed or supported by that discovery. *None* of these papers stated that Dirac's theory should receive more credit because of the novelty of its prediction of the positron.

In the three-year period following the positron discovery, Dirac's relativistic wave equation was discussed in 73 papers that did not mention the positron, and the positron was discussed in 64 papers that did not mention Dirac's equation. There were 77 papers that mention ed both Dirac's equation and the positron, but did not state that the latter confirmed the former. The results are summarized in Table 9.1.

Thus by the second year after the discovery of the positron, a clear majority of those physicists who discussed the Dirac equation associated it with the new particle, even though they did not choose to make a clear statement that the theory had gained empirical support. At the same time, many experimental papers on the positron that ignored the Dirac theory were appearing. It was not until the third year,

Table 9.1 Opinions of Physicists on the Relation between Dirac's Theory and the Discovery of the Positron

Year	Previous	First	Second	Third	Total
Positron confirms Dirac	–	7	19	1	27
Mention both positron & Dirac	–	15	38	24	77
Dirac but not positron	33	26	23	23	105
Positron but not Dirac	–	23	34	6	63
Total	33	73	114	54	272

after the initial phase of excitement about the discovery had passed, that a majority of papers on the positron mentioned Dirac. By the third year, however, some theorists had gone back to arguing about the difficulties of Dirac's hole theory as if they had never heard of the positron.

9.3 THE TRANSFORMATION OF DIRAC'S THEORY

Despite or perhaps because of the discovery of the positron, it was clear to the leading theorists that the hole theory had to be revised. One could not simply say that the discovery confirmed Dirac's *equation* without giving support to the objectionable hole concept, because the prediction had been directly based on that concept. Aside from the paradoxes and infinities associated with the sea of negative-energy states, it was now clear, both from the basic postulates of the Dirac theory itself and from the observed properties of positrons and electrons, that positive and negative charges had to be treated symmetrically. As Dirac himself pointed out in the 1935 edition of his monograph on quantum mechanics, there was no more reason to postulate a sea of negative-energy *electrons* than a sea of negative-energy *positrons*, so that negatively charged electrons could just as well be interpreted as holes in the sea of positrons.

It was also obvious to the experts that as soon as you introduced a sea of negative-energy particles in your model, you no longer had a one-particle system but a many-particle system. It was well known that in such a system, even if there are no interparticle forces, a single particle could not be accurately described by the solution of a single-particle wave equation.

From the viewpoint of the quantum field theorists of the next generation, the natural way to resolve those difficulties was second quantization, as proposed by physicists Vladimir Fock, Wendell Furry, Oppenheimer, Heisenberg, and Wolfgang Pauli and Victor Weisskopf. The Dirac wave function becomes (as explained by Julian Schwinger) an "operator field unifying the electron and positron as two alternative states of a single particle . . . With this formalism the vacuum becomes again a physically reasonable state with no particles in evidence. The picture of an infinite sea of negative energy electrons is now best regarded as an historical curiosity, and forgotten." From this point of view, Furry and Oppenheimer and Pauli and Weisskopf "showed how quantum field theory naturally incorporates the idea of antimatter, without introducing unobserved particles of negative energy" and thus "settled the matter."

Indeed, most modern textbooks do accept the second quantization approach as the solution to the problem. Dirac's hole theory is still praised as an important step toward the modern theory, but nevertheless as a hypothesis that ultimately had to be discarded.

The folk history of the field theorists fails to acknowledge, much less explain, one curious fact: Dirac's hole theory survived in the physics literature, long after it was supposedly replaced by second quantization, and despite its failure to account for other data from scattering experiments and spectroscopy. Pauli, Heisenberg, Oppenheimer, and their associates did pursue the second-quantization approach in the period 1935–1949. In some of these calculations the hole concept survived, often with holes in a sea of negative-energy positrons being included on an equal basis with holes in a sea of negative-energy electrons.

Full acceptance of the second-quantized field theory as a *replacement* for the hole hypothesis had to wait for midcentury, when physicists Julian Schwinger, Sin-ItiroTomonaga, and Richard P. Feynman showed that a refined, renormalized quantum electrodynamics could give a more accurate description than Dirac's theory of empirical facts, such as the Lamb shift in the hydrogen spectrum. Even then there was some resistance to completely abandoning the hole theory. Perhaps this was because the new quantum electrodynamics did not seem to *predict* antiparticles in the sense that Dirac's hole theory did; it simply postulated their existence. This perception is not entirely accurate, since some theorists did deduce the existence of antiparticles from general principles such as relativistic invariance.

In addition to the disturbing feature that one must subtract infinite quantities in order to get a finite answer (renormalization), the new quantum electrodynamics did not provide an intuitively understandable mechanism for the creation and annihilation of antiparticles in the sense that Dirac's hole theory did. The major exception is John Wheeler and Richard Feynman's proposal to regard positrons as electrons moving backward in time, an elaboration of suggestions made earlier by De Donder and Stueckelberg. Though widely publicized in popular books on modern science, the Wheeler-Feynman proposal was not taken seriously by other physicists except as a convenient way to label terms in a calculation. Similarly, the abstract proof of the existence of antiparticles from relativistic invariance, while tacitly accepted by physicists, is not given much prominence in modern treatments of the subject.

In this case the ability to predict a new particle was not the most important criterion in choosing a theory; it could not override the strong theoretical objections to Dirac's hypothesis. (According to historian Jagdish Mehra, Dirac considered the fact that his equation gave a natural explanation of spin more important than the fact that it predicted a new particle.) Yet the successful prediction did force theorists to seek, and eventually to find, satisfactory ways to overcome those objections. The improved theory was not completely developed and accepted by the community until new empirical data (the Lamb shift) showed that Dirac's theory was not quantitatively accurate. Physicists reluctantly gave up a theory that had made a sensationally successful prediction of a completely new phenomenon (antiparticles) in favor of a more consistent (though still imperfect) theory that could provide a more accurate non-novel prediction of the details of the hydrogen spectrum and other aspects of the interaction of charged particles with radiation.

9.4 YUKAWA'S THEORY OF NUCLEAR FORCES

Physicist Hideki Yukawa's 1935 proposal of a new intermediate-mass particle to carry the force between protons and neutrons, and the 1937 discovery of the μ meson, are generally considered milestones in the history of nuclear and elementary-particle physics. To what extent did the discovery of the predicted particle lead physicists to accept the theory on which the prediction was based? In this section and the next, I focus on the reception of the Yukawa theory during the three years after the discovery of what is now called the μ meson. Discussion of nuclear forces in physics journals ceased shortly after the beginning of World War II, as did open communication among scientists in Japan, Germany, France, Great Britain, and the United States, so it is reasonable to terminate the historical period at that point.

To review briefly the origin of the prediction: following the discovery of the neutron and the establishment of the proton-neutron model for the nucleus, physicists wanted to understand the nature of the forces between protons and neutrons. Presumably there must be some kind of short-range attraction, strong enough to overcome the electrostatic repulsion of protons within a nucleus, but not so strong as to make the nucleus collapse to a point.

Quantum mechanics offered the example of the exchange force based on Heisenberg's concept of resonance of an atomic system between two states (Section 10.1). If the system contains two nuclei or more nuclei, resonance may be visualized as a rapid shuttling of an electron between them, although in quantum mechanics this just means that the wave function for the system has to be a combination of terms corresponding to the electron being associated with each nucleus.

In his theory of the nucleus, Heisenberg treated the neutron as a compound particle (p^+ + e−) and proposed to describe the *force* between a proton and a neutron in terms of the exchange of an electron between two protons. According to historian Arthur I. Miller, the exchange concept was now beginning to change into a migration concept involving the actual motion of the light particle from one nucleon to the other.

In 1933, Enrico Fermi proposed a theory of ß (beta) decay, incorporating the hypothetical particle suggested by Pauli and named the neutrino by Fermi. According to Fermi a neutron can transform into a proton while emitting an electron and a neutrino (ν, Greek letter nu):

$$\mathrm{n} \rightarrow \mathrm{p} + \mathrm{e}^- + \nu$$

Heisenberg suggested in 1934 that such a transformation might also account for the force that holds neutrons and protons together in the nucleus. This became known as the Fermi-field theory of nuclear forces. According to historians Laurie Brown and Helmut Rechenberg, this theory would be the major rival to Yukawa's theory.

Yukawa was a careful student of Heisenberg's work. In October 1934, after thinking about the Heisenberg theory, he decided to propose a new particle to carry the proton-neutron force. Using the standard equation for the electromagnetic potential with an additional term $1/R^2$ depending on the range *R* of the force, Yukawa obtained a neutron-proton potential in which the classical Coulomb potential e^2/r is multiplied by an exponential function $e^{-r/R}$. The new factor would go rapidly to zero for distances *r* much greater than *R*. The force constant *e* (for electric charge) is replaced by a new parameter *g*, measuring the strength of the nuclear force. When interpreted in terms of quantum field theory, the potential corresponds to a quantum whose mass is inversely proportional to *R*. Assuming *R* is approximately 2×10^{-13} cm, he found that the mass of the new particle should be about 200 times the mass of the electron.

Yukawa noted that since such a particle has never been observed, the theory might seem to be wrong; however, he argued that "in the ordinary nuclear transformation, such a quantum can not be emitted into outer space" because the energy difference between proton and neutron is less than the rest-mass energy of the particle. But, he suggested, "the massive quanta may also have some bearing on the shower produced by cosmic rays."

Yukawa's theory initially attracted little attention and less support, even in Japan. I looked at 41 papers on nuclear forces published in the 12 months preceding the publication of the Neddermeyer-Anderson discovery; none of them mention Yukawa.

9.5 DISCOVERY OF THE MUON AND RECEPTION OF YUKAWA'S THEORY

There is some dispute as to when the muon was discovered, but physicists in the late 1930s most often cited the paper published by physicists Seth H. Neddermeyer and Carl D. Anderson in the May 15, 1937 issue of *Physical Review*. In a note added in proofs, they stated that evidence for a similar particle had been presented by physicists J. C. Street and E. C. Stevenson on April 29 at a meeting of the American Physical Society (after the Neddermeyer-Anderson paper was submitted).

None of these initial reports mentioned Yukawa's theory. Anderson wrote in 1961 that the discovery of muons "was based on purely experimental measurements and procedures, with no guide from any theoretical considerations." Yukawa's theory was "unknown to the workers engaged in the experiments on the muon until after the muon's existence was established. . . . he published it in a Japanese journal which did not have a general circulation in this country."

Anderson stated that Yukawa's theory would have had a significant influence if it had been known to experimenters, because

> for a period of almost two years there was strong and accumulating evidence for the muon's existence, and it was only the caution of the experimental workers that prevented an earlier announcement of its existence. I believe that a theoretical idea like Yukawa's would have appealed to the people carrying out the experiments and would have provided them with a belief that maybe after all there is some need for a particle as strange as a muon, especially if it could help explain something as interesting as the enigmatic nuclear forces.

As Rutherford and others have noted, experimenters sometimes are motivated to confirm a theoretical prediction in cases where the data are dubious.

As it happened, Yukawa's theory did become widely known to physicists almost immediately after the publication of the Neddermeyer-Anderson discovery. According to Brown and Rechenberg, "the new particles were hailed as being identical to Yukawa's 'heavy quanta' and they aroused world-wide interest in his theory." His hypothetical particle was sometimes called the yukon, even in papers that did not mention Yukawa by name. The question was: is the yukon identical to the observed mesotron or meson (the name generally adopted by 1941)?

As with Dirac's theory of the positron, it was Oppenheimer who first played the role of publicist, though not that of advocate. According to physicist Robert Serber, who was working with him at that time, Oppenheimer received a reprint of Yukawa's 1935 paper directly from the author. His attitude was generally hostile, and he did not take the Neddermeyer-Anderson observation seriously until he learned that Street had obtained a similar result.

On June 1, 1937 (just after the Neddermeyer-Anderson paper was published), Oppenheimer and Serber sent a letter to *Physical Review* pointing out the connection between the range of nuclear forces and the mass of the recently discovered particle. "In fact," they wrote, "it has been suggested by Yukawa that the possibility of exchanging such particles of intermediate mass would offer a more natural explanation of the range and magnitude of the exchange forces between proton and neutron than the Fermi theory of the electron-neutrino field."

But then, rather than arguing that the discovery of the new particle provided evidence in favor of Yukawa's theory, Oppenheimer and Serber criticized the theory on the grounds that it did not reconcile the "saturation character of nuclear forces with the apparent equality of like and unlike particle forces and with the magnetic moments of neutron and proton." Yukawa's ideas "cannot be regarded as the elements of a correct theory, nor serve as any argument whatever for the existence of the particles," although they might ultimately "prove relevant to an understanding of nuclear forces."

Physicists in Zürich (Pauli, G. Wentzel, and E. C. G. Stueckelberg) apparently learned of Yukawa's theory independently of Oppenheimer. The fact that the Pauli reprint collection includes Yukawa's 1935 paper suggests that Yukawa sent reprints to several physicists. The second published reference to Yukawa's paper was by Stueckelberg, who thought it highly probable that the new particle was the one predicted by Yukawa's theory.

Yukawa himself called attention to his own prediction and its confirmation by the discovery of the new particle, in a short note published in Japan. But a letter by Yukawa and his collaborators responding to the Oppenheimer-Serber criticism, elaborating the Yukawa theory and showing how it was supported by evidence, was rejected for publication in Physical Review.

Physicist Nicholas Kemmer was one of the first European physicists to develop Yukawa's theory. "Although it is premature to draw definite theoretical conclusions from the present experimental knowledge," he wrote in December 1937, "it is certainly suggestive that a Yukawa particle with a mass of the observed order of magnitude ($100m_{el}$) does indeed give nuclear forces of the correct range." But he immediately rejected the specific theory proposed by Yukawa because it could not correctly account for the lowest energy level of the deuteron; instead he proposed to use a more complicated vector wave function.

Two months later Kemmer used a phrase that was to become standard: the discovery of the meson had "aroused considerable interest in Yukawa's suggestion," but no claim was made that the confirmation of Yukawa's prediction was to be considered evidence in favor of his theory. The most he would say, in a paper with Herbert Fröhlich and Walter Heitler, was that the Yukawa meson exchange process "can account for the nuclear forces in a reasonable way."

Physicist Homi Bhabha, who probably learned about Yukawa's theory from Heitler, was also an early enthusiast for meson theory. He was the first to point out that the lifetime of the new particle should be longer when in motion than when at rest because of the Lorentz time-dilation effect; observations on the meson thus offered another experimental test of the special theory of relativity. This new support for relativity was welcome, especially in Nazi Germany where Einstein's theory was still under attack. Heisenberg, who used this evidence to defend relativity, "felt always grateful to the muons."

In January 1938 Bhabha and Kemmer visited Zürich, where they tried to sell Pauli on the virtues of Yukawa's theory. But Pauli, as usual, was skeptical; he complained to Weisskopf and Heisenberg that the new theory still didn't solve the fundamental difficulties of the older theories of quantum electrodynamics and ß-decay because it still suffered from divergences (one would have to eliminate an infinite term to get a finite result). But meson theory nevertheless fascinated him and he eventually devoted considerable time to it, even though he was never satisfied that it agreed with experimental results.

Heisenberg (not surprisingly in view of the origin of Yukawa's ideas) quickly became a public advocate of the meson theory of nuclear forces, even though he admitted its shortcomings in correspondence with Pauli. He mentioned it favorably in eight papers published in 1938 and 1939, including a long review article with Hans Euler that was frequently cited during the next two years. In their review, Euler and Heisenberg stated that while Yukawa had predicted the existence of a particle like the meson, it would be premature to speak of a definitive confirmation of his theory; nevertheless it seemed natural to analyze the experimental results in terms of the Yukawa theory.

Heisenberg's statements grew gradually stronger. He told the Physikalisches Colloquium at Hamburg, on December 1, 1938, that it is reasonable to identify the meson with Yukawa's hypothetical particle. In mid-1939 his paper in *Zeitschrift für Physik* asserted that the discovery of the meson and the proof of its decay have led to a far-reaching confirmation of Yukawa's theory, so that this theory can hardly be doubted as far as its general principles are concerned. The impact of Heisenberg's endorsement should not be underestimated; recall that other physicists regarded him so highly that many of them devoted several years to developing the atomic bomb, in large part because they feared Heisenberg would develop it for Hitler.

Early in 1938 Heisenberg asked Bohr what he thought of the "now so fashionable" Yukawa theory. I didn't find a reply from Bohr or any definite statements in Bohr's publications, but he did write to Robert A. Millikan later that year: "I do not know whether one shall admire most the ingenuity and foresight of Yukawa or the tenaciousness with which the group in your Institute kept on in tracing the indications of the new effects."

Was the discovery of the meson considered to be a confirmation of Yukawa's theory of nuclear forces? As in the case of Dirac's prediction of the positron, one finds relatively few explicit statements of this kind, but many criticisms of the theory itself. Here are the results of my survey of the physics literature. In the three years following the discovery of the meson, there are only nine statements that it confirms Yukawa's prediction and thus supports his theory. None states that Yukawa's prediction is more important because it came before rather than after the discovery of the muon. Only one, a review by Heitler, makes a point of reminding the reader that the prediction came before the discovery.

Neddermeyer and Anderson did not mention Yukawa in their publications during the first two years after their discovery report, even though they were aware that the name "yukon" had been proposed for the new particle. At the Chicago Cosmic-Ray Symposium in June 1939 their reference to Yukawa was somewhat unfavorable: the mean life of the particle is 15 to 20 times as great as that estimated from his theory, and more observations are needed "to find out whether the mesotrons can be identified with the particles postulated by Yukawa to account for nuclear forces."

As in the case of Dirac's prediction of the positron, we can measure the extent to which scientists voted with their feet without explicitly stating that the meson's discovery provided evidence for Yukawa's theory. There are two indicators that I have used in analyzing a total of 298 papers during a four-year period (one before and three after the Neddermeyer-Anderson announcement in May 1937). Nuclear forces were discussed, without mentioning Yukawa's theory, in 78 papers, as compared to 30 papers in which that connection was made. But 41 of those 78 were in the year before the meson discovery, leaving only 38 in the three years after—still a majority for those years. There is a third alternative: 30 papers that discussed meson theory but without mentioning Yukawa by name. (These papers often cited papers by Euler and Heisenberg, Bhabha, Bethe, Heitler, or Kemmer.)

Second, 75 papers on mesons did not mention its possible connection with the particle postulated by Yukawa, while 57 papers did (but only eight of those actually stated that this connection confirmed Yukawa's theory). In principle there is also a third alternative here, since some experimentalists cited papers by other meson-theorists in analyzing their data, but I did not find it possible in most cases to establish that they acknowledged a connection between the cosmic-ray particle and the quantum proposed to carry nuclear forces, so these papers have all been included in the "not Yukawa" category.

To summarize: by the third year after the discovery of the meson, most papers on nuclear forces (37 out of 44) associated them with mesons even if they did not explicitly credit Yukawa's theory. But while a majority of papers on mesons mentioned Yukawa in the second year after the discovery, the proportion slipped below half in the third year. While there is not an exact correspondence between papers on nuclear forces and theoretical papers, or between papers on mesons and experimental papers, the above statistics do suggest that theorists were significantly more enthusiastic about Yukawa's theory than experimentalists.

What was the status of Yukawa's theory around 1940? Some theorists considered it natural to identify mesons with a particle that carries nuclear forces, even though it was generally admitted that Yukawa's 1935 theory would have to be substantially modified in order to give an accurate description of those forces. Others accepted the qualitative idea that nuclear forces are carried by an intermediate-mass particle, but concluded that the meson could not be that particle because it does not have the quantitative properties predicted by theory; in particular, its lifetime is too long. The apparent absence of an *uncharged* meson meant that the theory could not explain proton-proton or neutron-neutron attraction.

Among experimenters, the most favorable assessment came from physicists E. J. Williams and G. E. Roberts, in connection with their observations of the decay of mesons, which they regarded as additional evidence for "the remarkable parallel between the meson and the Yukawa particle." But other experimentalists, and those theorists primarily interested in fitting scattering data with nuclear force laws, suggested that there may be a more massive particle responsible for nuclear forces, since the meson's mass is significantly smaller than that estimated for the yukon.

Looking back from the perspective of later decades, one physicist asserted that the discovery of the muon "appeared to be a remarkable verification of the correctness of Yukawa's speculations"; others recalled that the identity of the meson with the yukon was adopted for the next decade despite the quantitative discrepancies between their properties. This was later contradicted by physicist R. E. Marshak, who said in 1989 that despite the apparent spectacular confirmation of Yukawa's hypothesis by the meson discovery, "the pre-World War II period ended with great frustration on the meson-theoretic front." Yukawa recalled in 1949 that "already in 1941, the identification of the cosmic-ray mesons with the meson, which was supposed to be responsible for nuclear forces, became doubtful," and lamented that "the intervention of the mu meson completely destroyed the picture which I had had in mind."

Perhaps I can best describe the situation of Yukawa's theory by comparing it with that of Dirac's hole theory in the mid-1930s. A major difference is that Dirac's theory was well known to theorists before the discovery of the positron, so we can compare their attitudes before and after the discovery, whereas Yukawa's theory was known to very few theorists until after the discovery of the muon. Both theories encountered serious mathematical difficulties because infinite quantities frequently arose in calculations involving perturbation expansions in powers of the coupling constant. These problems were more serious for meson theory because the force constant g, when expressed in appropriate dimensionless units, is greater than 1, whereas the corresponding electromagnetic constant is less than 1. Neither Yukawa's original 1935 model nor any of its modifications achieved quantitatively accurate predictions, whereas Dirac's theory did give fairly accurate results if one ignored or subtracted away the infinite terms. On the other hand, the qualitative idea that nuclear forces are transmitted by the exchange of an intermediate-mass particle was intuitively plausible to physicists in the 1930s, whereas the idea of negative-energy particles was not.

While physicists often do not state explicitly that a successful prediction confirms a theory (but simply use or don't use the theory), they were relatively more likely to say that about Dirac's theory than about Yukawa's. Finally, in each case the number of papers linking the new particle with the theorist who predicted it reached a maximum in the second year and declined thereafter.

In 1940, Sin-itiro Tomonaga (who was later to share a Nobel Prize with Julian Schwinger and Richard Feynman for quantum electrodynamics) and G. Araki calculated from Yukawa's theory the probabilities that a yukon would be captured by an atomic nucleus or would spontaneously decay to an electron plus a neutrino. They found that, taking into account the Coulomb electrostatic force between the nucleus and the yukon, "the probability for negative mesons being captured is always larger than the probability of disintegration," but the opposite is true for positive mesons. The experimental test of this theoretical prediction would later show that the mesotron (now called the μ meson) is *not* the yukon.

During World War II, when communications were difficult between physicists in Japan and other countries, an informal group of Japanese physicists called the Meson Club met twice a year. At a meeting in June 1942 the suggestion emerged that there are two mesons, the nuclear force meson (Yukon) and the cosmic-ray meson (mesotron)—different particles, but related. According to Brown and Rechenberg, "Yasutaka Tanikawa was the first to suggest that [the hypothesis of] two different mesons might explain the anomalies." Another two-meson theory was independently proposed by physicists Robert E. Marshak and Hans Bethe in 1947.

Two modern experts on elementary particles, Val L. Fitch and Jonathan L. Rosen, wrote in 1995:

> The experiment which definitely showed that the meson discovered in 1937 was not Yukawa's particle was performed by Conversi, Pancini and Piccioni (1945). Using magnetized iron as a charge selector, they showed that both positive and negative mesons appeared to decay when coming to rest in carbon, while if the particles came to rest in iron only the positive ones decayed. Since negative mesons with Yukawa-like properties would be expected to interact even in elements as light as carbon, the evidence was unmistakable that the mesons could not be the strongly interacting Yukawa type.

This conclusion was not explicitly drawn by Conversi's group but by Fermi, Edward Teller, and V. F. Weisskopf in 1947.

9.6 THE TRANSFORMATION OF THE YUKON

In 1945, G. P. S. Occhialini (who, with Blackett, had confirmed Anderson's discovery of the positron) joined the group at Bristol (UK) led by physicist Cecil Frank Powell. They were using photographic plates to study nuclear reactions and cosmic rays. They persuaded the Ilford Company to make special photographic emulsions which were much more sensitive than those normally used. Another new member of the group, Cesar M. G. Lattes, made a further improvement to the emulsion: adding boron to prevent the particle traces from fading. They were soon flooded with a large amount of high-quality data, too much for the physicists to examine in detail. They employed a team of scanners, mostly women, to inspect the many photographic plates and look for the few events that might be of interest.

As recalled by Lattes, "after a few days of scanning, a young lady, Marietta Kurz, found an unusual event: one stopping meson and, emerging from [the end of its track] a new meson." This was interpreted as the decay of a heavier meson to a lighter one. After a month the group had found 40 examples of decaying heavy mesons, including 11 in which the secondary meson's track ended on the same plate so its range could be determined. In the 1947 publication by Lattes, Occhialini, and Powell, reporting this discovery, the primary meson was called π and the secondary μ.

Positive and negative π mesons were found to have the appropriate mass and other properties for a particle carrying the nuclear force between protons and neutrons. All that remained was to find a neutral π meson. This was done in 1950 at Berkeley by Jack Steinberger, W.K.H. Panofsky, and J. Stellar, using the electron synchrotron. According to Brown and Rechenberg, it was the first time that a new particle had been discovered using an accelerator.

Physicists had already concluded that the pion is in fact the yukon. Yukawa was awarded the 1949 Nobel Prize "for his prediction of the existence of mesons on the basis of theoretical work on nuclear forces"; the pion has about the right mass and lifetime, and its discovery was therefore considered "a brilliant vindication of Yukawa's fundamental ideas" despite the fact that "it has not yet been possible to give a theory for the nuclear forces, which yields results that are in good quantitative agreement with the experiments."

Strenuous efforts by theoretical physicists to develop a satisfactory meson field theory of nuclear forces, comparable to quantum electrodynamics, did not succeed. Yukawa's theory remains valid as a semiempirical approximation that is still useful in describing nuclear forces. According to L. M. Brown, "after its acceptance in 1937, Yukawa's theory became the dominant fundamental theory of nuclear forces until the present standard model of quarks and gluons became established in the 1970s, and it is still the main way of interpreting intermediate nuclear physics."

9.7 CONCLUSIONS

In both cases studied here, the discovery of the predicted particle had a major impact on theoretical and experimental research. There was a significant increase in publications on the theory that led to the prediction, whether that theory had previously been well known (Dirac's relativistic wave equation for the electron) or almost completely unknown (Yukawa's meson theory of nuclear forces). Thus our results support the claims of Karl Popper and others that empirical confirmation of a prediction provides corroboration of the hypothesis that yielded the prediction—provided one does not confuse corroboration with verification. As I interpret Popper's use of the term, corroboration does not increase the probability that a hypothesis is true; it merely makes it more reasonable to pursue that hypothesis than one that has not been corroborated. In this minimalist sense, both theories were certainly corroborated by the discovery of the particles they predicted. Conversely, the hypothesis that the mesotron (μ meson) is the yukon could be falsified by testing its predictions, and the falsification advanced science by forcing physicists to keep looking for, and eventually find, a particle that did have the right properties (B meson).

Despite his careful restriction of the meaning of corroboration, Popper (like some other philosophers) still argued that successful predictions are essential for the progress of science, and specifically mentioned Dirac's prediction of antiparticles and Yukawa's meson theory as examples.

But these two cases provide little or no evidence for the claim of Popper and others that *novelty* increases the importance of a prediction. Because of the ambiguous use of the term "prediction" by scientists (Section 1.3), it is impossible to determine how much weight they intended to give to novelty in saying, for example, that Dirac's electron theory was confirmed by his successful prediction of the positron, unless the novelty was explicitly mentioned.

It is clear that the confirmation of a prediction, whether novel or not, is only one factor governing the response to a theory. The case of the positron (and to a lesser extent the meson; see Table 9.2) shows that theoretical objections to a hypothesis can prevent its full acceptance despite the strongest empirical support. The idea that positrons are holes in a sea of negative-energy electrons was undermined not only by severe problems of inconsistency with principles accepted before the idea was proposed, but also by the new idea of particle-antiparticle symmetry that entered physics as a result of the discovery itself.

In each of these cases, the hypothesis that postulated a new particle was either abandoned or substantially revised within two decades of its apparent confirmation by the discovery of that particle. Physicists preferred a theory that gave a satisfactory *explanation* of the new particle to one that predicted its existence and

Table 9.2 Opinions of Physicists on The relation between Yukawa's Theory and the Discovery of Mesons

	Previous	First	Second	Third	Total
Meson confirms Yukawa	2	4	2	8	
Nuclear forces, not Yukawa	41	21	10	7	79
Nuclear forces and Yukawa	0	28	34	16	78
Meson theory, not Yukawa	0	2	7	21	30
Mesons, not Yukawa	2	29	23	21	75
Mesons and Yukawa	0	18	26	13	57
Total	43	85	94	76	298

Columns do not sum to total number since several papers mention both nuclear forces and mesons.

properties in advance. Dirac's hole theory of the electron and Yukawa's meson theory of nuclear forces are both regarded as major steps forward in our understanding of elementary particles; neither is now considered an established truth about the world.

NOTES FOR CHAPTER 9

269 For details and references see my paper "Prediction and Theory Evaluation: Subatomic Particles," published in *Rivista di Storia della Scienza*, Series II, Vol. 1, no. 2 (dicembre 1993), pp. 47–152; also available on my website, http://punsterproductions.com/~sciencehistory/pdf/PTESP.pdf.

269 *New Particles* The chapter is based on research supported by the National Science Foundation, the National Endowment for the Humanities, and the General Research Board of the University of Maryland; the paper was completed while the author was a member of the School of Social Science, Institute for Advanced Study, Princeton, with support from the Andrew W. Mellon Foundation. Fior valuable comments and suggestions I thank S. Adler, L. M. Brown, O. W. Greenberg, A. E. Levin, S. S. Schweber, D. Shapere, and G. Snow.

269 *Einstein's prediction of gravitational light bending was "more impressive"* S. L. Glashow and B. Bova, *Interactions: A Journey through the Mind of a Particle Physicist and the Matter of this World* (New York: Warner, 1988).

269 *Wilczek states that "in most fields of science"* F. Wilczek and B. Devine, *Longing for the Harmonies: Themes and Variations from Modern Physics* (New York: Norton, 1988), p. 41.

269 *"The importance attached to a successful prediction"* Y. Ne'eman and Y. Kirsh, *The Particle Hunters* (New York: Cambridge University Press, 1986), p. 207.

270 *"you cannot predict something that has already been measured"* Glashow and Bova, *Interactions*, p. 164.

270 *"neither party could have cooked his argument or his results"* Ziman, J. M. *Public Knowledge: An Essay Concerning the Social Dimension of Science* (Cambridge: Cambridge University Press, 1968), pp. 41, 51.

270 *"Anderson had earlier spent several evenings a week"* Farmelo, *The Strangest Man: The Hidden Life of Paul Dirac, Mystic of the Atom* (New York: Basic Books, 2009), p. 212.

Section 9.1

270 *Gamow called it a "donkey electron"* G. Gamow, *Biography of Physics* (New York: Harper, 1961), pp. 263–265; *Thirty Years that Shook Physics: The Story of Quantum Theory* (Garden City, NY: Doubleday, 1966), pp. 127–129, 206, 217.

270 *Dirac wrote in 1931, "an encounter between two hard γ-rays"* Dirac, "Quantised Singularities in the Electromagnetic Field," *Proceedings of the Royal Society of London* A133 (1931): 60–72, on p. 61. Two γ-rays are needed because transformation of a single photon into an electron and a positron would not conserve both energy and momentum.

270 *"the discovery of the positron was wholly accidental"* Anderson, "Early Work on the Positron and Muon," *American Journal of Physics* 29 (1961): 825–830, on p. 825.

270 *The discovery of the positron (as we would now say)* "To Anderson's annoyance, the editor of *Science News Letter* in which a photograph of one of the renegade tracks was first published, suggested calling the new particle the *positron*, and Anderson left it at that." Per F. Dahl, *From Nuclear Transmutation to Nuclear Fission, 1932–1939* (Bristol and Philadelphia: Institute of Physics Publishing, 2002), p. 107.

270 *"carried with it so many novel and seemingly unphysical ideas"* Anderson, "Early Work."

270 *he noted discrepancies in . . . the numbers of positive and negative electrons* Anderson, "The Positive Electron," *Physical Review* 43 (1933): 491–494; "Positrons from Gamma Rays," *Physical Review* 43 (1933): 1033; "The Positron," *Nature* 133 (1934): 313–316; Anderson et al., "The Mechanism of Cosmic-Ray Counter Action," *Physical Review* 45 (1934): 352–363.

272 *Anderson's opinion of Dirac's theory was more favorable* Anderson and S. D. Neddermeyer, "Fundamental Processes in the Absorption of Cosmic-Ray Electrons and Photons," *International Conference on Physics, London, 1934*, edited by J. H. Awbery, *Papers and Discussions*, Volume I (London: Physical Society, 1935), pp. 171–187; Anderson, "The Production and Properties of Positrons," (Nobel Lecture, December 12, 1936) in *Nobel Lectures—Physics, 1922–1941* (New York: Elsevier, 1965), pp. 365–376.

272 *"The experimental discovery of the positive electron"* Oppenheimer and Plesset, "On the Production of the Positive Electron," *Physical Review* 44 (1933): 53–55, on p. 53.

Section 9.2

272 *Wentzel recalled in 1960 that "the hole theory was given much credit"* "Quantum Theory of Fields (until 1947)," in M. Fierz and V. F. Weisskopf, editors, *Theoretical Physics in the Twentieth Century* (New York: Interscience Publishers, 1960), pp. 48–77, on p. 58.

272 *favorably received by some physicists before 1932* Karl Grandin, "Intermediate Theoretical Physics," in *Aurora Torealis: Studies in the History of Science and Ideas in Honor of Tore Frängsmyr*, edited by Marco Beretta, Karl Grandin, and Svante

Lindqvist (Sagamore Beach, MA: Science History Publications, 2008), pp. 193–214, on Swedish theoretical physicist Ivar Waller (1898–1991), who mediated between theorists (e.g., Dirac) and experimentalists; includes correspondence with Dirac about the hole theory of electrons and positions. Helge Kragh, *Dirac: A Scientific Biography* (New York: Cambridge University Press, 1994), pp. 64–65. D. F. Moyer, "Evaluations of Dirac's Electron, 1928–1932," *American Journal of Physics* 49 (1981): 1055–1062. F. Hund, *The History of Quantum Theory* (New York: Barnes and Noble, 1974), p. 200.

273 *Oseen finally changed his mind in 1932* Robert Marc Friedman, *The Politics of Excellence: Behind the Nobel Prize in Science* (New York: Freeman/Times Books/Holt, 2001), pp. 175, 221, 252–253.

273 *"It seems to a certain degree regrettable that we had a theory"* Discussion remarks, *Structure et Propriétés des Noyaux Atomiques,* Rapports et Discussions du Septième Conseil de Physique tenu a Bruxelles du 22 au 29 Octobre 1933 sous les Auspices de l'Institut International de Physique Solvay (Paris: Gauthier-Villars, 1934), pp. 177–178.

Section 9.3

275 *an "operator field unifying the electron and positron"* J. Schwinger, "A Report on Quantum electrodynamics," in J. Mehra, editor, *The Physicist's Conception of Nature* (Boston: Reidel, 1973), pp. 413–426, on p. 415.

275 *and thus "settled the matter"* S. Weinberg, "The Search for Unity: Notes for a History of Quantum Field Theory," *Daedalus* 106, no. 4 (1977): 17–35, on p. 24. See also Weinberg, *The Discovery of Subatomic Particles* (New York: Freeman, 1984), p. 162.

276 *his equation gave natural explanation of spin more important* Mehra, "Dirac's Contribution to the early Development of Quantum Mechanics," in J. G. Taylor, editor, *Tributes to Paul Dirac* (Philadelphia: Taylor & Francis, 1987), pp. 63–75.

Section 9.4

276 *milestones in the history of nuclear and elementary-particle physics* For a detailed account of this topic for the period 1932–1950 see Laurie M. Brown and Helmut Rechenberg, *The Origin of the Concept of Nuclear Forces* (Bristol, UK: Institute of Physics Publishing, 1996). See also Joan Bromberg, "The Impact of the Neutron: Bohr and Heisenberg," *Historical Studies in the Physical Sciences* 3 (1971): 307–341.

277 *In his theory of the nucleus, Heisenberg treated the neutron* "Über den Bau der Atomkerne," *Zeitschrift für Physik* 77 (1932): 1–11; 78 (1932): 156–164; 80 (1933): 587–596. Translated extracts from the first two papers, with commentary, are in D. M. Brink, editor, *Nuclear Forces* (Oxford: Pergamon Press, 1965); Brown and Rechenberg, *Origin*, p. 44 note 7, write that Brink omits the parts dealing with electrons in the nucleus and thus misrepresents Heisenberg's theory.

277 *the exchange concept was now beginning to change* Miller, "Werner Heisenberg and the Beginning of Nuclear Physics," *Physics Today* 38, no. 11 (November 1986): 60–68, on pp. 65–66. See also L. M. Brown, "Heisenberg and Nuclear Physics," *Physics Today* 39, no. 6 (June 1986): 116–118; Miller, "Heisenberg and Nuclear Physics," *Physics Today* 39, no. 6 (June 1986): 118–122.

277 *This became known as the Fermi-field theory of nuclear forces* Brown and Rechenberg, *Origin*, pp. 50–59, 63–65, 87.

277 *he argued that "in the ordinary nuclear transformation"* "On the Interaction of Elementary Particles," *Proceedings of the Physico-Mathematical Society of Japan* 17 (1935): 48–57, on pp. 53, 57. Reprinted in R. T. Beyer, editor, *Foundations of Nuclear*

Physics (New York: Dover, 1949), pp. 139–148. Beyer's book contains an extensive bibliography on theories of nuclear structure and related topics.

Section 9.5

278 *the paper published by physicists Seth H. Neddermeyer and Carl D. Anderson* Neddermeyer and Anderson, "Note on the Nature of Cosmic-Ray Particles," *Physical Review* 51 (May 15, 1937): 884–886 (received March 30, 1937).

278 *evidence for a similar particle had been presented* Street and Stevenson, "Penetrating Corpuscular Component of the Cosmic Radiation," *Physical Review* 51 (June 1, 1937): 1005, Abstract of paper presented at American Physical Society meeting, Washington, DC, April 29–May 1, 1937. Street and Stevenson, "New Evidence for the Existence of a Particle of Mass intermediate between the Proton and Electron," *Physical Review* 52 (November 1, 1937): 1003–1004.

278 *the discovery of muons "was based on purely experimental measurements"* Anderson, "Early Work on the Positron and Muon," *American Journal of Physics* 29 (1961): 825–830, on p. 829.

278 *"the new particles were hailed as being identical"* Brown and Rechenberg, *Origin*, pp. 96, 151–152, 204.

278 *His hypothetical particle was sometimes called the yukon* E. E. Whitmer and M. A. Pomerantz, "Evidence for the Existence of a new Elementary Particle," *Journal of Applied Physics* 9 (December 1938): 746–753. R. A. Millikan reported that Bohr had suggested the name "yukon" in an address to the British Association, but Millikan seemed to prefer "mesotron." W.G.F. Swann, according to Millikan, objected to the word "meson" because it is too close to "a word that has come in French to be used as a word for a house of ill fame." Brown and Rechenberg, *Origin*, p. 187.

278 *Oppenheimer received a reprint of Yukawa's 1935 paper* A. Pais, *Inward Bound: Of Matter and Forces in the Physical World* (Oxford: Oxford University Press), p. 433.

278 *His attitude was generally hostile* J. F. Carlson and J. R. Oppenheimer, "Multiplicative Showers," *Physical Review* 51 (February 15, 1937): 220–231, on p. 231. Peter Galison, "The Discovery of the Muon and the Failed Revolution Against Quantum Electrodynamics," *Centaurus* 26, no. 4 (1982): 263–316; R. Serber, "The Early Years," in I. I. Rabi et al., *Oppenheimer* (New York: Scribner's Sons, 1969), pp. 9–20, on p. 15. Brown and Rechenberg, *Origin* (Bristol, UK: Institute of Physics Publishing, 1996), p. 131.

279 *Yukawa's ideas "cannot be regarded as the elements of a correct theory"* Oppenheimer and Serber, "Note on the Nature of Cosmic-Ray Particles," *Physical Review* 51 (June 15, 1937): 1113. Brown and Rechenberg, *Origin*, pp. 130–132.

279 *who thought it highly probable that the new particle was the one predicted* "On the Existence of Heavy Electrons," *Physical Review* 52 (July 1, 1937): 41–42.

279 *Yukawa himself called attention to his own prediction* "On a possible Interpretation of the penetrating Component of the Cosmic Ray," *Proceedings of the Physico-Mathematical Society of Japan* 19 (1937): 712–713; Brown and Rechenberg, *Origin*, pp. 130–131.

279 *was rejected for publication in* Physical Review Brown and Rechenberg, *Origin*, pp. 132–135.

279 *one of the first European physicists to develop Yukawa's theory* "Nature of the Nuclear Field," *Nature* 141 (January 15, 1938): 116–117. See D. C. Cassidy, "Cosmic Ray Showers, High Energy Physics, and Quantum Field Theories: Programmatic

Interactions in the 1930s," *Historical Studies in the Physical Sciences* 12 (1981): 1–39, on pp. 26–27.

279 *discovery of the meson had "aroused considerable interest"* Kemmer, "Quantum Theory of Einstein-Bose Particles and Nuclear Interaction," *Proceedings of the Royal Society of London* A160 (May 4, 1938): 127–153, on p. 127.

279 *the Yukawa meson exchange process "can account for the nuclear forces"* H. Fröhlich, W. Heitler and N. Kemmer, "On the Nuclear Forces and the Magnetic Moments of the Neutron and the Proton," *Proceedings of the Royal Society of London* A166 (May 4, 1938): 154–177, on p. 155.

279 *"felt always grateful to the muons"* Heisenberg, "Cosmic Radiation and fundamental Problems in Physics," *Naturwissenschaften* 63 (1976): 63–67, on p. 64.

279 *he complained to Weisskopf and Heisenberg* Letter to Weisskopf, January 13, 1938, and postcard to Heisenberg, February 22, 1938, both in K. Von Meyenn, editor, *Wolfgang Pauli Wissenschaftliche Briefwechsel mit Bohr, Einstein, Heisenberg, u. a.*, Vol. II (Berlin: Springer, 1985), pp. 547–552. See also correspondence with Kemmer and Heisenberg, ibid., pp. 552–574; Cassidy, "Cosmic Ray Showers," p. 30; Pauli, "Relativistic Theories of elementary Particles," *Reviews of Modern Physics* 13 (July 1941): 203–232; *Meson Theory of Nuclear Forces* (New York: Interscience, 1946); *Collected Scientific Papers*, edited by R. Kronig and V. F. Weisskopf (New York: Interscience/Wiley, 1964), vol. 2, pp. 953–993, 1034–1046, 1053–1072.

Heisenberg admitted its shortcomings in correspondence with Pauli. Meyenn, *Briefwechsel*, pp. 558–572. See also his letter to Bohr, July 1937, about the "unstable 'heavy electron' that Anderson claims to have found," a discovery that Heisenberg did not yet know whether to believe, quoted in Rechenberg and Brown, "Origin of the Concept," p. 27.

280 *a long review article with Hans Euler that was frequently cited* Euler and Heisenberg, "Theoretische Gesichtspunkte zur Deutung der kosmische Strahlung," *Ergebnisse der exakten Naturwissenschaften* 17 (1938): 1–69.

280 *reasonable to identify the meson with Yukawa's hypothetical particle* "Das schwere Elektron (Mesotron) und seine Rolle in der Höhenstrahlung," *Angewandte Chemie*, 52 (January 7, 1939): 41–42. See also his lecture in the Netherlands two months later, "De Atoomkern en Harte Samenstellung," *Nederlandsch Tijdschrift voor Natuurkunde* (1939): 89–98.

280 *led to a far-reaching confirmation of Yukawa's theory* "Zur Theorie der explosionsartigen Schauer in der kosmischen Strahlung. II," *Zeitschrift für Physik* 113 (June 16, 1939): 61–86, on p. 61.

280 *Heisenberg asked Bohr what he thought* Letter, February 9, 1938, in *Niels Bohr Collected Works*, vol. 9 (New York: Elsevier, 1986), p. 586; English translation, p. 587. (No reply is included in this volume.)

280 *he did write to Robert A. Millikan later that year* Millikan, *Electrons (+ and –)* (Chicago: University of Chicago Press, 1947), p. 510.

280 *makes a point of reminding the reader that the prediction came before* "Cosmic Rays," *Reports on Progress in Physics* 5 (1939): 361–389.

280 *they were aware that the name "yukon" had been proposed* "Mesotron (Internediate Particle) as a Name for the new Particles of intermediate Mass," *Nature* 142 (November 12, 1938): 878.

280 *their reference to Yukawa was somewhat unfavorable* Neddermeyer and Anderson, "Nature of Cosmic-Ray Particles," *Reviews of Modern Physics* 11 (July–October 1939): 191–207, on p. 207.

281 *"the remarkable parallel between the meson and the Yukawa particle"* Williams and Roberts, "Evidence for Transformation of Mesotrons into Electrons," *Nature* 145 (January 20, 1940): 102–103, on p. 102.

281 *discovery of the muon "appeared to be a remarkable verification"* C. F. Powell, "Mesons," *Reports on Progress in Physics* 13 (1950): 350–424, on p. 353.

282 *"the pre-World War II period ended with great frustration"* Marshak, "Scientific and Sociological Contributions of the First Decade of the 'Rochester' conferences to the Restructuring of Particle Physics (1950 1960)," in M. De Maria et al., editors, *The Restructuring of Physical Science in Europe and the United States, 1945–1960* (Singapore: World Scientific, 1989), pp. 745–786, on p. 747.

282 *"the identification of the cosmic-ray mesons with the meson"* Yukawa, "Meson Theory in its Developments" (Nobel Lecture, December 12, 1949), in *Nobel Lectures—Physics 1942–1962* (Amsterdam: Elsevier, 1964), 128–134, on p. 131. "Introductory Remarks on Meson Theory," *Reviews of Modern Physics* 29 (April 1957): 213–215, on p. 214.

282 *"Yasutaka Tanika was the first to suggest . . . "* Brown and Rechenberg, *Origin*, p. 278. See also L. M. Brown, "A Tale of two Mesons," *Physics Today* 47, no. 9 (September 1994): 129–130 for further details.

283 *"the meson discovered in 1937 was not Yukawa's particle"* Fitch and Rosner, "Elementary Particle Physics in the Second Half of the Twentieth Century," in L. M. Brown, A. Pais and B. Pippard, editors, *Twentieth Century Physics* (Bristol & Philadelphia: Institute of Physics Publishing and New York: American Institute of Physics, 1995), pp. 635–794, on pp. 652–653. M. Conversi, E. Pacini, and O. Piccioni, "On the Decay Process of Positive and Negative Mesons," *Physical Review* 68 (1945): 232. M. Conversi, "From the Discovery of the Mesotron to that of its leptonic Nature," in B. Foster and P. H. Fowler, editors, *40 Years of Particle Physics* (International Conference, University of Bristol, July 1987) (Bristol: Adam Hilger) . . . , 1988, pp. 1–20 (quoted in Brown and Rechenberg, *Origin*, p. 298).

283 *This conclusion was not explicitly drawn by Conversi's group* Brown and Rechenberg, *Origin*, pp. 299–300.

Section 9.6

283 *The transformation of the Yukon* This section closely follows Brown and Rechenberg, *Origin*, pp. 303–312.

283 *the group at Bristol (U.K.) led by physicist Cecil Frank Powell* C. M. G. Lattes, H. Muirhead, G. P. S. Occhialini, and C. F. Powell, "Processes involving charged Mesons," *Nature* 159 (May 24, 1947): 694–697; Lattes, Occhialini and Powell, "Observations on the Tracks of Slow Mesons in Photographic Emulsion," *Nature* 160 (1947): 453–456, 486–492. Check—is Marietta Kurz mentioned??

283 *Yukawa was awarded the 1949 Nobel Prize* I. Waller, "Physics 1949" in *Nobel Lectures—Physics 1963–1970* (Amsterdam: Elsevier, 1964, pp. 125–127, on pp. 126, 127.

284 *"Yukawa's theory became the dominant fundamental theory"* Brown, "Nuclear Forces, Mesons, and Isospin Symmetry," in *Twentieth Century Physics*, edited by L. M. Brown, A. Pais and B. Pippard (Bristol, UK): Institute of Physics Publishing, 1995), pp. 357–419, on p. 378.

10

Benzene and Molecular Orbitals 1931–1980

> It turned out that while Pauling had enormous success in the short run, he was the hare and the MO was the tortoise that caught up and passed him half a century later.
>
> —*physicist* PHILIP ANDERSON *(2008)*

10.1 RESONANCE, MESOMERISM, AND THE MULE 1931–1945

Although there were several attempts to explain chemical bonding by electronic theories based on the old (pre-1925) quantum theory, these apparently became obsolete when Werner Heisenberg and Erwin Schrödinger developed quantum mechanics in 1925–1926. Thus in theories of benzene we now have to adopt a new fundamental

> **Postulate III.** The structure and reactions of atoms and molecules are determined by quantum mechanics. In practice this usually means the nonrelativistic wave mechanics of Schrödinger, although in some cases one may have to use the relativistic wave equation of P.A.M. Dirac.

A key concept in benzene theories after 1930 is resonance. Quantum-mechanical resonance was first described by Heisenberg, using both his own matrix mechanics and Schrödinger's wave mechanics. It is related to the requirement that the wave function of a many-particle system must be symmetric or antisymmetric in the coordinates of identical particles (see the notes to this chapter for details).

Heisenberg's resonance theory was applied to the bond between two hydrogen atoms by physicists Walter Heitler and Fritz London in 1927. As a first approximation, one can construct a wave function for the system from the ground-state wave functions of the hydrogen atom (see notes).

The calculation shows that the symmetric wave function gives a minimum energy at R = 0.80Å, compared to the experimental value 0.740Å. Without resonance, there would be a much weaker minimum, and the antisymmetric function ($c_2 = -c_1$) has no minimum at all.

Despite the apparent success of the Heitler-London explanation of the hydrogen molecule, it was not universally accepted. In particular, Pauli strongly objected to it. When Heitler presented his theory at conference in Copenhagen in April 1929, as physicist Léon Rosenfeld later recalled:

> Pauli rushed to the blackboard in a state of agitation; striding back and forth, he began to give vent to his annoyance while Heitler sat down on a chair at the corner of the podium. "At great distances," Pauli exclaimed, "the theory is certainly wrong, because there we have Van der Waals attraction; at small distances it is, naturally, likewise completely wrong." At this point he had reached the end of the podium directly opposite Heitler's place. He turned around and went up to him pointing a piece of chalk threateningly in his direction. "And now," he thundered, "a postulate is made, appealing to the simple credulity of physicists, that claims that this model, which is wrong at large distances and wrong at small distances, is nonetheless supposed to be right in the intermediary area!" He had now come very close to Heitler. Heitler suddenly leaned back and the back of the chair gave way with a loud snap, causing poor Heitler to fall over backwards (fortunately without hurting himself all that much).

However, other physicists showed that by using different approximations for the wave functions, results in agreement with experimental measurements of interatomic forces could be obtained. For example, physicist Shou Chin Wang, following a suggestion by physicist and chemist Peter Debye, found that the van der Waals attractive force, proportional to the inverse seventh power of the distance between two atoms, could be explained.

A general conclusion from the early work of Heitler and London was that the formation of covalent bonds depends on resonance energy, the contribution to the electronic energy from integrals like $\int \varphi_a(1)\varphi_b(2)\varphi_b(1)\varphi_a(2)/r_{12}$, in which the electrons change places. $\varphi_a(1)$ is the wave function for electron 1 in atom *a*, $\varphi_b(2)$ is the wave function for electron 2 in atom b, and so on. However, the identification of some part of the total energy of a system as resonance energy is artificial, since it depends on our choice of the one-electron atomic wave functions as the first approximation. One could start from some other first approximation and get a similar final result, without making use of the particular integral that was associated with the resonance energy. So resonance energy is not a physical quantity that can be measured by experiment; rather, it is a mathematical quantity used in a particular kind of approximate calculation.

The Heitler-London approach to chemical bonds was rapidly developed by physicist John C. Slater and chemist Linus Pauling. Slater in 1929 proposed a simple general method for constructing many-electron wave functions that would automatically satisfy the Pauli exclusion principle by being antisymmetric under the interchange of any two electrons (see Section 8.6). Two years later he suggested that the electronic structure of benzene could be understood in terms of resonance between the two Kekulé structures: "By wave mechanics we should have a combination of these two possible forms, resulting in a complete equivalence of the carbons of the ring, a shared valence (essentially 1½ bonds between each pair), and an added stability for the structure on account of the resonance effect on the energy." But he soon recognized that the molecular orbital theory (MO) developed by Hückel, Hund, and Mulliken was better.

Hypothesis VB

Pauling was one of the first theorists to apply the Heitler-London method to chemical bonds; this method became known as the valence-bond theory (VB) because it involved picking out one electron in each of the two combining atoms and constructing a wave function representing a paired-electron bond between them. We will refer to this as the VB hypothesis.

As early as 1928, Pauling reported that "it has further been found that as a result of the resonance phenomenon a tetrahedral arrangement of the four bonds of the quadrivalent atom is a stable one." He did not publish the results until 1931, at which time he acknowledged that Slater had independently discovered "the existence, but not the stability, of tetrahedral eigenfunctions" in 1930. (See Section 8.3 for the definition of eigenfunction.) Pauling proposed that the carbon atom, which in its ground state has only two unpaired electrons in the valence shell (this shell is usually described as $2s^2 2p_x 2p_y$), can be quadrivalent if one of the two 2s electrons is excited to the $2p_z$ orbital. Although it takes some energy to do this, that expenditure of energy is more than compensated by the energy gained in forming four bonds. Moreover, Pauling argued that the four bonds will all be equivalent, rather than one of them being different as might be expected from the fact that the carbon atom has one 2s and three 2p electrons available for bonding. The bond wave functions will actually be mixtures of s and p orbitals, and the bonds will be directed toward the corners of a regular tetrahedron.

Hypothesis MO

At about the same time as Pauling was developing VB, chemical physicist Robert S. Mulliken was developing an alternative technique based on what he called molecular orbitals. This term was first used by Mulliken in 1932, but the concept had been introduced a few years earlier. Physicist Friedrich Hund, beginning in 1926, and Mulliken, in 1928, proposed schemes for assigning quantum numbers to electrons in molecules, based on atomic quantum numbers and their interpretation of molecular spectra.

The first person to write an explicit wave function for a molecular orbital was physicist and chemist John Edward Lennard-Jones, in 1929. He suggested a function of the form

$$\varphi(1) = \varphi_A(1) \pm \varphi_B(1)$$

for an electron in a diatomic molecule; $N_A(1)$ refers to an atomic wave function centered on nucleus A, and $\varphi_B(1)$ to one on nucleus B. Lennard-Jones described his wave function as representing a resonance of the electron between two possible states. Note that this is not the same kind of resonance as that discussed by Heisenberg and by Heitler and London, which involves the interchange of *two* electrons between two possible states. (In the later literature of quantum chemistry the Heisenberg form of resonance is usually referred to as exchange.)

The first comprehensive application of both VB and MO methods to the benzene problem was made by physicist Erich Hückel in 1931. In an interview 40 years later, Hückel recalled: "The fundamental thought was that here apparently the molecule must be viewed as a whole and not merely in terms of the effects of neighbouring atoms. That this must be so could be seen from the 'directing' effects of substituents, which transmit, somehow, the perturbation throughout the ring."

Hückel carried out two calculations for benzene, one by the Heitler-London method (VB), the other by MO. He treated separately the five electrons associated with each carbon atom (called σ electrons) and the remaining one that were free to move throughout the entire ring (called π electrons). In the second method these six electrons behaved like a closed shell that provides the chemical stability characteristic of aromatic molecules.

Since we don't know which method is better, Hückel pointed out, we must try both and compare the results with experiment. He found that the second method (MO) was quantitatively somewhat better for benzene, giving lower energies for most of the quantum states; it also allows us to understand why molecules with exactly six B electrons have a special stability. He extended this result by studying also the negative ion of cyclopentadiene, $C_5H_5^-$, assuming that the extra electron acts like a π electron and the positive ion of cycloheptadiene, $C_7H_7^+$, which, having lost one electron, would also have six π electrons.

Although Hückel was familiar with the earlier chemical theories of benzene, he made it clear that quantum mechanics offered a new and much better approach. In particular, he pointed out that Kekulé's oscillation hypothesis fails to explain the properties of cyclic molecules such as C_4H_4 and C_8H_8, and concluded that K-Osc loses its meaning in quantum theory.

Simple cyclic molecules with N carbon atoms are now called [N] annulenes, thus C_8H_8 would be called [8] annulene.

Prediction 11. According to Hückel's MO theory, monocyclic molecules with 4n+2 mobile (π) electrons, where n is an integer, will have aromatic character. This became known as Hückel's rule. In particular, the rule predicts that neither COT (cyclooctatetraene), with eight mobile electrons, nor cyclobutadiene (CBD), C_4H_4, with 4, is aromatic, even though their structures could be described in Kekulé's theory with alternating double and single carbon-carbon bonds. But $C_{10}H_{10}$, [10] annulene, should be aromatic if it is planar.

This prediction has one important qualification: the molecule has to be planar or very nearly so.

At that time it was already known that COT is not aromatic. The specific predictions about other annulenes were confirmed later.

Such a break from the traditional view of benzene was no part of Pauling's research program. He emphasized on many occasions that his VB method was an outgrowth of chemical theory going back to Kekulé, and that his resonance concept was not really a new idea but only a modernized version of earlier theories. In particular, Pauling identified resonance with the mesomerism concept of chemist C. K. Ingold (see Section 10.2). He did not even claim priority for the suggestion that benzene owed its stability to quantum-mechanical resonance, ascribing that to John C. Slater.

In their 1933 paper on benzene, Pauling and George W. Wheland modestly described their theory as "rather closely similar" to Hückel's first method; it "leads to the same result," but "the calculations are simplified to such an extent that the method can be extended to the naphthalene molecule without undue labor." But their *visual* presentation of the VB hypothesis applied to benzene made it much more striking than Hückel's.

Hypothesis RH

Pauling and Wheland wrote the wave function for benzene as a linear combination of five functions, corresponding to the two Kekulé structures and three other structures called, following Ingold, Dewar structures (Figure 10.1):

$$\varphi = c_1\varphi_1 + c_2\varphi_2 + c_3\varphi_3 + c_4\varphi_4 + c_5\varphi_5$$

where the coefficients c_i are to be obtained by minimizing the energy. This became known as the resonance hybrid hypothesis (RH). Since the two Kekulé structures had the largest weight in the hybrid wave function, the Pauling-Wheland theory of benzene could be described as a modification of K-Osc in which the actual state of the molecule is intermediate between those two structures, ignoring the smaller contribution of the Dewar structures.

Some authors took care to distinguish resonance from tautomerism, a term used to describe oscillation between two molecular states; they insisted that the individual

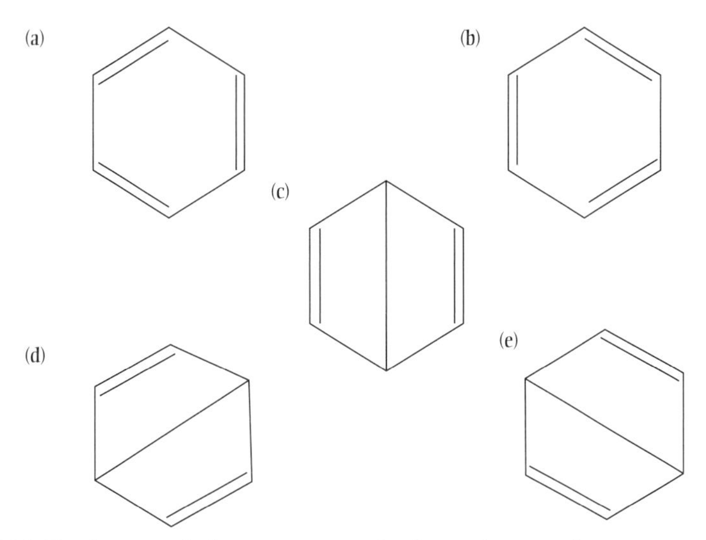

Figure 10.1 The five canonical structures contributing to the normal quantum mechanical state of the benzene molecule, as shown by Pauling and Wheland. The coefficients of the corresponding wave functions (before normalization) are A = B = 1, C = D = E = 0.4341. Approximately 80% of the resonance energy comes from the Kekulé structures (A and B), and about 20% from the three excited structures C, D, E).

structures in the resonance hybrid did not actually exist, and it was certainly not correct to say that a benzene sample consists of a 50% mixture of the two Kekulé forms.

Wheland explained the distinction with a biological analogy:

> A mule is a hybrid between a horse and a donkey. This does not mean that some mules are horses and the rest are donkeys, nor does it mean that a given mule is a horse part of the time and a donkey part of the time. Instead, it means that a mule is a new kind of animal, neither horse nor donkey, but intermediate between the two and partaking to some extent of the character of each. Similarly, the theories of intermediate stages and of mesomerism picture the benzene molecule as having a *hybrid* structure, not identical with either of the Kekulé structures, but intermediate between them.

In one simple but important case, MO yielded a successful (non-novel) prediction (explanation) that could not easily be matched by VB: the magnetic property of diatomic molecules, especially O_2. But early attempts to compare VB and MO for benzene and other molecules generally failed to establish either one as definitely superior to the other. Wheland concluded that VB "seems to give results in somewhat better agreement with experiment" than MO, but the latter "can be extended to a wider variety of problems." Physicists J. H. Van Vleck and Albert Sherman stated that MO is "often the more convenient for purposes of qualitative discussion," whereas VB has been used the more frequently for purposes of quantitative calculation, "partly, but by no means entirely, because of habit." However, it is "meaningless quibbling to argue which of the two methods is the better in refined forms since they ultimately converge." A few authors cautioned that since both were only approximations, a conclusion should not be accepted unless it could be derived by both methods. In the wider context of quantum theories of matter, both VB and MO could be considered pragmatic semiempirical approaches, differing only in calculational details, as contrasted with the more fundamental (but less useful to chemistry) approach of physicists like Heitler and London. Since MO was regarded as being in some way equivalent to VB, it could sneak into discussions of resonance/mesomerism under the guise of delocalization of π-bonds or π-electrons.

In another simple case relevant to benzene theory, the simplest VB treatment and the simplest MO calculation gave radically different results. Cyclobutadiene, C_4H_4, should have resonance energy comparable to that of benzene according to VB, whereas it should have *zero* resonance energy according to MO (in agreement with Hückel's rule, Prediction 11). The fact that this molecule had not yet been observed, apparently because of its instability, could be considered an argument in favor of MO, but not a decisive one.

Much of the research done on the spectrum of benzene and its derivatives in the late 1930s and early 1940s, including an important calculation by physicists Maria Goeppert-Mayer and Alfred L. Sklar, was based on some version of MO, with the implicit assumption or explicit conclusion that this approach was superior to VB for the calculation of spectral frequencies and intensities. Similarly, magnetic properties such as the diamagnetic anisotropy were successfully treated and could be cited as evidence for MO. One could, however, argue that such properties were primarily of interest to physicists or that MO represented a physicist's approach to benzene, while

VB was still more useful to chemists. Mulliken himself admitted that MO departs from "chemical ideology," but insisted that it offers a better conceptual scheme.

10.2 RECEPTION OF QUANTUM THEORIES OF BENZENE 1932–1940

Pauling and Wheland's RH began to be mentioned in the chemical literature with a few months of its publication. An early influential supporter was chemist Nevil Vincent Sidgwick. In earlier writings on the application of the old quantum theory to chemistry, Sidgwick had insisted that the power of theories lies in their predictive character, and had expressed some skepticism about the value of wave mechanics. The Pauling-Wheland paper on benzene and a personal encounter with Pauling apparently converted him. In the Chemical Society's *Annual Report* for 1934, he asserted that the theory of resonance "must now be taken seriously into account by organic as well as by physical chemists." Citing the papers by Pauling and his colleagues on aromatic molecules, he wrote:

> The resonance theory of structure, if, as seems probable, it is to be accepted, is a most important advance in the theory of structural chemistry. It shows that the behaviour of a compound can often not be expressed by a single formula, but only by a combination of two, and this, not in the tautomeric sense that the substance is a mixture of two kinds of molecule, but in the sense that every molecule has, at least to some extent, all the properties represented by the two structures, as well as other properties which are directly due to the resonance, especially an increase of stability, which ensures that the resonance will always occur when it is possible.

The first textbook presentation I have found of the quantum theory of benzene is in *Organic Chemistry*, published in 1935 by chemist Howard J. Lucas, Pauling's colleague at Caltech. Not surprisingly, Lucas gives only RH, ignoring Hückel's molecular orbital theory, but indicating that Pauling's quantum calculations in effect confirm Kekulé's oscillation hypothesis, which is supported by chemical evidence.

Hückel's theory was ignored by most organic chemistry textbooks during the 1930s; it appeared in only a few technical articles and monographs. Surprisingly, it is not even mentioned in his brother's book on theoretical organic chemistry.

During the next few years RH, by itself or as an application of Pauling's VB, gradually began to appear in organic chemistry treatises and textbooks. Following the publication in 1939 of Pauling's classic monograph *The Nature of the Chemical Bond*, quickly followed by a second edition in 1940, RH became the accepted description, though it did not displace K-Osc. In his review of *Nature of the Chemical Bond*, physicist Joseph Mayer (husband of Maria Goeppert-Mayer) worried that because of the excellence of its presentation, VB would eclipse MO even though the latter was needed to explain some phenomena such as the paramagnetism of oxygen. Confirming Mayer's fear 36 years later, Hückel complained in his memoirs that, by delaying the acceptance of MO, Pauling's book retarded the progress of quantum chemistry for 20 years.

Since even the proponents of VB admitted that MO was equally good for calculating many of the properties of benzene and even better for some of them, the question arises: why was RH almost universally adopted in organic chemistry texts before

1950 while MO was rarely even mentioned? What advantages did VB have over MO, or over the earlier theories of benzene?

In these texts the most frequently mentioned justification for the VB theory of benzene was that it explained the resonance energy of this compound as compared to a hypothetical structure with single and double bonds. Pauling's estimate for the resonance energy, 36 to 38 kilocalories per mole, was presented in some texts as if it were a theoretical calculation confirmed by experiment, even though Pauling and Wheland explicitly stated that their estimate was based on a parameter chosen to fit the experimental data. They claimed only that their theory accurately predicted resonance energies for several other aromatic compounds using the same parameter. Protests by other theorists that observed resonance energies have no theoretical significance, and that the Pauling-Wheland calculation ignores the large energy needed to change the bond lengths from their values in the Kekulé structures to the observed lengths, were ignored.

Similarly, the fact that the observed CC bond length in benzene (1.39 Å) is much closer to the double bond length in nonaromatic compounds (1.33 Å) than to the single bond length (1.54 Å) was attributed to the resonance effect, even though Pauling and Wheland had not actually calculated this length from VB theory. Mulliken suggested that the early success of VB was largely due to Pauling's skill as a propagandist rather than to its scientific merits:

> Pauling made a special point of making everything sound as simple as possible and in that way making it very popular with the chemists but delaying their understanding . . . anywhere Pauling went, that became popular.

According to Pauling's biographer Thomas Hager, VB won out in the 1930s because Pauling "was an eloquent teacher and a persuasive writer who knew how to communicate in language chemists could understand . . . When Mulliken talked, people went to sleep . . . He was not much better in print." A Freudian slip conveys the idea succinctly: in a biographical article on Mulliken, chemist H. C. Longuet-Higgins cited Mulliken's works as *Selected Ploddings of Robert S. Mulliken.*

Although personality and rhetoric had some role in the initial popularity of VB, a more important reason is that VB as presented by Pauling really was closely related to accepted theories of the chemical bond. For benzene in particular RH could easily be seen as a modernized version of K-Osc, and by defining it as equivalent to one of the their own concepts, chemists could accept it as a new name for an idea emerging from chemistry rather than an alien import from physics.

It appears that the rapid acceptance of resonance theory by chemists owes something to the opportunism of chemist C. K. Ingold. In July 1933, only a month after the publication of the Pauling-Wheland paper on benzene, he cited their work in support of his claim that intranuclear tautomerism, which he now called mesomerism, arises from wave mechanical degeneracy. The next year he published a long review article on organic reactions in which he again called attention to his own mesomeric effect, calling it "a modernized version of both the oscillation formula of Kekulé and the partial valency formula of Thiele" and noting that it had been studied by Pauling under the name resonance. Ingold evidently regarded himself as a coinventor of the theory of resonance, stating in his Bakerian Lecture to the Royal Society in 1938 that "the physical interpretation of mesomerism was developed independently by

Pauling and me between 1929 and 1933. The central principle employed is . . . the uncertainty principle, or, alternatively, the derivative principle of resonance." Ingold praised his coinventor in carefully chosen words: "the theory of mesomerism has been set out in an elegant, semiquantitative manner by Pauling and Wheland." In later writings Ingold reminded chemists that he had proposed the phenomenon of mesomerism as early as 1926 (i.e., long before Pauling and Wheland's RH) and predicted its effects on dipole moments, quantum mechanics, and then provided an explanation of why the effect occurs.

Ingold's assimilation of resonance to his own mesomerism was self-serving but it also helped Pauling, since the two terms entered the chemical vocabulary hand in hand. For example, in his 1937 presidential address to the Chemical Society of London, N. V. Sidgwick praised the success of Pauling's resonance theory in explaining chemical phenomena, while at the same time advocating the use of Ingold's word mesomerism to describe it. While accepting resonance and mesomerism, organic chemists could maintain that K-Osc, originally proposed only as a Verlegenheitshypothese (hypothesis of embarrassment), had not been abandoned but could still be recognized as a precursor of these modern ideas, demonstrating the prophetic brilliance of the great Kekulé. More generally, as Sidgwick concluded in his presidential address the previous year, "the modern development of the structural theory [i.e., resonance theory], far from destroying the old doctrine, has given it a longer and a fuller life." Sidgwick himself has been credited by Paradowski with helping to "make valence-bond ideas an accepted part of organic chemistry" through his books and articles.

Another reason for the neglect of MO—perhaps decisive for many chemists—was that during the 1930s it did not offer any *visual representation* of benzene to compete

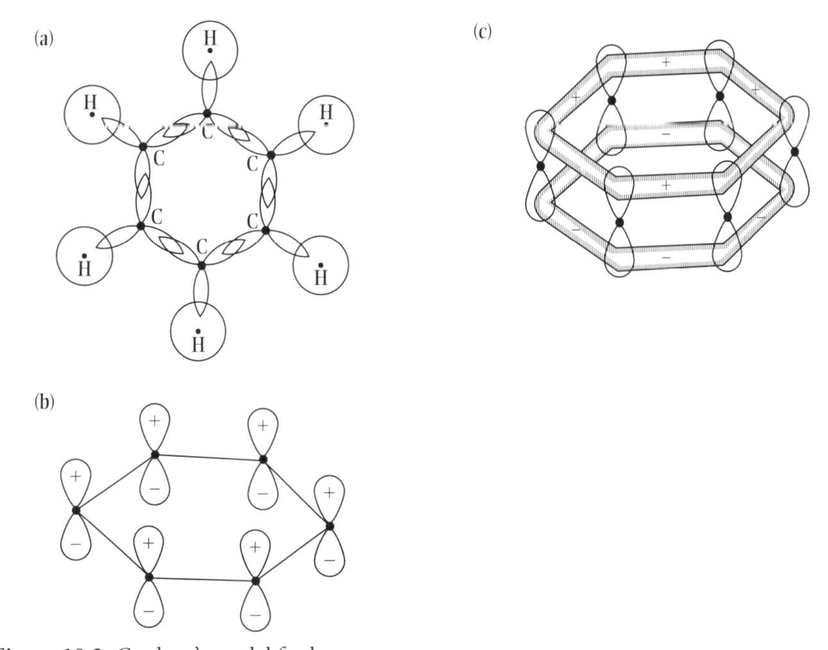

Figure 10.2 Coulson's model for benzene.
(a) σ orbitals, (b) π atomic orbitals, (c) π molecular orbitals.

with the five resonant structures of VB. It was sometimes dismissed as a physicist's theory that offered little help to chemists. But even in modern physics, mathematical theories that fail to offer an *anschaulich* (visualizable) model are often less popular than theories that provide diagrams and pictures. The examples of matrix mechanics versus wave mechanics, and Schwinger's versus Feynman's quantum electrodynamics, are well known to historians of modern physics.

Chemist Charles Coulson, a leader of the MO camp, was very conscious of the need to escape from the "thought-forms of the physicist." But his now-familiar double-donut MO representation of benzene (Figure 10.2) did not begin to emerge until 1941, and did not become widely known until after 1945.

10.3 CHEMICAL PROOF OF KEKULÉ'S THEORY

Does the benzene molecule contain double bonds? At the beginning of the twentieth century, chemist Carl Dietrich Harries developed a method for analyzing organic compounds by using their reactions with ozone. Ozonization became a standard technique for locating double bonds. Harries himself thought that it should be possible to use the method to determine whether benzene has double bonds, but his own results did not appear conclusive. As in other cases, it may be more convenient to work with suitably chosen derivatives of benzene than with benzene itself.

> **Prediction 12**. In accordance with Prediction 2, K-2 (and perhaps also K-Osc, with the tautomeric interpretation) implies that there should be two isomers of an ortho diderivative of benzene; after ozonization one should be able to identify the reaction products of these isomers.

In 1930 chemist Amos G. Cole reinvestigated this problem for his master's thesis at the State College of Washington; his results were presented at a meeting of the American Chemical Society in Cincinnati, and published two years later in a paper coauthored by his mentor, chemist A. A. Levine. Levine and Cole showed that by using *o*-xylene (a benzene derivative in which the hydrogen atoms at two adjacent carbon atoms have been replaced by the methyl group CH_3), it was possible after hydrolysis to obtain three decomposition products: glyoxal, methylglyoxal, and dimethylglyoxal. These are the same products that would be obtained if the two Kekulé structures of o-xylene "merely represented phases, of some complex dynamic structure which may actually exist, since the process of ozonization would tend to stabilize these two phases." One would then obtain two distinct ozonides of benzene: "Since these [three] decomposition products are possible only in the event that two ozonides were previously formed, it follows that there were present in *o*-xylene, the above two structures or their equivalents." Hence "it appears as if *o*-xylene exists in two isomeric configurations."

The Levine-Cole experiment quickly found its way into organic chemistry textbooks as a evidence for Kekulé's formulas K-2 and K-Osc. In a comprehensive review of the ozonization method published in 1940, chemist Louis Long Jr. agreed that this experiment is "a further corroboration of the Kekulé structure"; Levine and Cole "demonstrated the existence of isomeric ortho-disubstitution products of benzene. . . . Since neither form of xylene could have yielded all three oxidation products, the hydrocarbon must have consisted of an equilibrium mixture of the two

Kekulé forms." In the 1930s neither Pauling nor Wheland mentioned this chemical proof of K-Osc, which might seem to undermine their claim that a resonance hybrid is *not* a mixture of actually existing individual structures.

Bolstered by further experiments done by chemists J. P. Wibaut and P. W. Hayman, the ozonization proof of K-Osc continued to be regarded as strong evidence not only for that hypothesis but for the resonance theory of benzene throughout the 1940s and 1950s, and could still be found in organic chemistry textbooks as late as 1970.

10.4 ANTIRESONANCE AND THE RHINOCEROS

At the end of World War II, VB was widely accepted among organic chemists. The resonance/mesomerism theory of benzene, bolstered by the results of the Levine-Cole ozonization experiment, continued to dominate textbooks until the 1960s. Pauling received the 1954 Nobel Prize in chemistry for his research into the nature of the chemical bond.

But around 1950, Pauling's theory encountered assaults from two very different directions: from Russian scientists who argued that the resonance concept is politically incorrect, and from quantum chemists, who found that MO offered a much more convenient approach for calculating the properties of large molecules.

The Soviet attack on resonance was part of the same movement that established Lysenko's theories in genetics and temporarily threatened the Copenhagen interpretation of quantum mechanics. According to dogmatic Marxist-Leninists, resonance theory was tainted with idealism because it described molecular structure in terms of forms that were physically impossible but had only a theoretical existence.

Although there was much rhetoric in the Soviet Union about the dangers of Pauling-Ingoldism, it appears that (unlike the Lysenko controversy) this particular attempt to impose the Communist Party line on science resulted in not much more than verbal modifications. Russian theoretical chemists could do their calculations however they thought best, as long as they did not use the word "resonance." In the meantime Pauling himself, far from being a stooge of capitalist imperialism, became an outspoken and effective leader of the nuclear disarmament movement in the West, and was awarded a second Nobel Prize—for promoting peace—in 1962.

In responding to the Soviet criticisms, Pauling tried to clarify his own philosophical views on resonance theory. Rather than treat it as a mathematical deduction from VB which was itself an approximation to quantum mechanics (Postulate III), he chose to regard it as an independent empirical theory in chemistry. It was an outgrowth of classical structure theory, and resonant structures were no more artificial than single and double bonds. Here Pauling disagreed with his colleague Wheland, who did consider resonance a man-made concept, with more human elements than chemical structure theory. But even Wheland's metaphor of resonance as a mule, a mixture of two real animals (horse and donkey) was beginning to be replaced, at Caltech and elsewhere, by the metaphor of the rhinoceros, a real animal, as a cross between two imaginary animals, the dragon and the unicorn.

For chemists in the West, the apparent absurdity of the Marxist ideological critique obscured a significant difference between VB and MO. VB could be interpreted in the spirit of Copenhagen: the wave function is a mixture of two states, $\Phi_A(1)\,\Phi_B(2)$ and $\Phi_B(1)\,\Phi_A(2)$, and it is not possible to say that the first electron is *really* in the state N_A (that is, in an orbital centered on nucleus A) or Φ_B. Only by a rather violent observation—which would involve pulling the molecule apart into two atoms—could one force the electron to decide which atom it belongs to. This is something like a Schrödinger cat paradox. In the case of the resonance of the benzene molecule between five different valence-bond structures the situation is even worse: there is no observation that could determine which structure is really present (or, which of the two supposed ortho-isomers of xylene is going to yield glyoxal and methyl glyoxal on ozonization and which will yield glyoxal and dimethyl glyoxal). Thus one has to abandon the idea that the electron is *in* a particular state.

The MO interpretation permits a more realist interpretation in this respect, since it assigns each electron to a particular state, though that state may be spread out over the entire molecule. As chemist Michael Dewar expressed it, "N^2 may be taken as the measure of density of an 'electron gas' representing the average electron distribution over a comparatively long period," and since this distribution represents all we can know about the electron, "the failure to localize the electrons more exactly is of little practical importance." So the shift from VB to MO was part of a more general movement from instrumentalist toward realist views in science after World War II. On the other hand, it could also be seen as a shift toward a more holistic description of the molecule, replacing the classical description in terms of bonds between pairs of atoms.

10.5 THE SHIFT TO MOLECULAR ORBITALS AFTER 1950

Several reasons have been suggested for the decision of quantum chemists to switch from VB to MO in the 1950s and 1960s. First, it was already recognized in the 1930s that properties such as spectra, which depend on determinations of excited electronic states, are more easily and accurately calculated by the MO approach; as more chemists became interested in molecular spectroscopy, and as more MO calculations were done, this advantage gradually gained more weight.

Chemists also became increasingly interested in physical properties of molecules such as magnetism. The failure of VB to give an adequate account of the paramagnetism of diatomic molecules was still regarded as a major defect.

Another reason is the increasing interest of chemists in large molecules, for which MO methods offer a significant computational advantage. The VB approach, which requires only five canonical structures for benzene, involves a rapidly increasing number of structures for larger aromatic molecules, and there is no simple reliable way to tell which ones will make the largest contributions to the hybrid, although in general the excited states become more and more important relative to the Kekulé structures. Hence this approach is not only impractical but loses one of its original advantages, the intuitive connection with classical bond theory. The MO approach allowed useful (and often visualizable) approximations so that large molecules could be treated successfully.

Prewar quantum chemistry calculations generally had to assume that the framework of atomic nuclei was given, in order to calculate the electronic wave functions. Postwar calculations, especially those done by the MO approach, were able to take bond lengths and angles as variable parameters, so that the size and shape of the molecule could be predicted rather than assumed.

The MO approach was also applied to the study of chemical reactions. Following the pioneering work of Wheland and Pauling in the 1930s, and of Coulson and

Longuet-Higgins in the 1940s, chemist Kenichi Fukui proposed his theory of frontier orbitals, and chemists R. B. Woodward and Roald Hoffmann developed their famous rules for predicting reactions. Fukui and Hoffmann shared the 1981 Nobel Prize in chemistry for this work (Woodward died in 1979 and was unable to share the prize). However, their theory did not contribute much to the understanding of benzene itself, but only to the reactions of some of its derivatives.

Finally, on the positive side, MO acquired advocates who could sell it to chemists as well as Pauling had presented VB. At this point I refer the reader to the comprehensive history of quantum chemistry by historians Kostas Gavroglu and Ana Simões, *Neither Physics nor Chemistry*. They describe the work of J. E. Lennard Jones, who became the first professor of theoretical chemistry in the world when he was appointed to an endowed chair at Cambridge University in 1932. In his lectures he asserted that "the object of theoretical chemistry [is] the explanation of chemical phenomena." But that did not preclude prediction: "He considered the prediction of directed valences as an additional advantage" of MO, which was "the more fundamental of the two, basically because the electrons were treated as belonging to the whole molecule rather than to atoms or localized parts of the molecule." Mulliken had perhaps lost some potential support for MO by remarking that VB followed the "ideology of chemistry" while MO followed the physicist's approach; Lennard-Jones "appeared to consider the electron-pair method [VB] as a physicist's method and the molecular orbital method as a chemical method."

Gavroglu and Simões describe in detail the success of chemists Charles Coulson, Alberte and Bernard Pullman, Raymond Daudel, and Per-Olov Löwdin in establishing MO as the preferred approach in quantum chemistry. "For Löwdin," they write, "the ultimate goal of quantum chemistry . . . was the accurate prediction of the properties of a hypothetical polyatomic molecule before its laboratory synthesis."

In 1950 Coulson claimed that it was now possible to predict bond lengths in many hydrocarbon molecules "by purely theoretical methods" using MO. By contrast, he noted, the success of resonance methods in predicting lengths is largely empirical and without theoretical foundation.

In 1952 he published Valence, a book that tried to do for MO what Pauling's *Nature of the Chemical Bond* had done for VB—not with the same spectacular triumph at first, but with substantial long-term success. It was easy for a theoretical chemist to criticize resonance theory because, as Coulson noted, "experimental organic chemists, having come across the idea of resonance, used it quite indiscriminately to justify any idea that came into their heads . . . by the end of 1945 resonance had become almost a 'dirty' word." Coulson complained that "the literature abounds in 'bastard mathematics'—by which I mean poor-quality mathematics which serves practically no purpose because it solves no problems." These are strong words, coming from the pen of a Methodist local preacher and author of books on science and religion.—although, characteristically, they are not explicitly directed against anyone in particular.

10.6 AROMATICITY

> Erich Hückel was a physicist who worked between two worlds. Because he was a physicist, organic chemists paid no attention to him, and because he worked in chemistry, physicists paid him no heed.
>
> —Andrew Streitweiser *(1997)*

"Aromaticity" means, in a general sense, benzene-like behavior. More specifically, we may start with the definition given in the textbook published by Erich Hückel's brother, Walter Hückel: Aromatic compounds are "substances derived from hydrocarbons containing considerably less hydrogen than the paraffins, and which, unlike other low hydrogen or 'unsaturated' compounds, are usually relatively stable and inert." As we will see, this definition is not stable and inert, but changes as chemists attempt to determine which compounds are or are not aromatic.

Some theorists expected that the aromaticity (benzene-like behavior) of cyclic compounds with four or eight carbon atoms, for which Kekulé-type structures with alternating double and single bonds could be written, would be a crucial test of resonance theory. According to Hückel's 4n+2 rule derived from his MO calculations, such compounds would *not* be aromatic if they have just one mobile electron per carbon atom, or if they are not planar.

The nonaromaticity of COT (cyclooctatetraene, C_8H_8) had been considered evidence against some classical theories. However, research in the 1940s indicated that the COT molecule is not planar like benzene but instead has a tub shape. In this case Pauling's VB theory does *not* predict that COT is aromatic, so it cannot be used as evidence against VB. It is "impertinent to the theory," according to chemists W. von E. Doering and L. H. Knox.

In several other cases the VB theory's predictions turned out to be wrong while MO's were correct. Chemist Roald Hoffmann recalls, "there was just a fantastic accumulation of organic synthesis and stability evidence on $C_3H_3^+$, $C_5H_5^-$, $C_7H_7^+$ being stable, $C_3H_3^-$, $C_5H_5^+$, and so on being unstable . . . the period 1950–1960 just convinced 99.9% of organic chemists that MO theory was right through continued, consistent successes of Hückel's rule."

Another obvious candidate for aromaticity was cyclobutadiene, C_4H_4, and the nature of this molecule was increasingly regarded as a crucial test case for MO and VB during the 1950s and 1960s. It proved to be very difficult to synthesize, and very unstable when it was finally constructed by two groups in 1965. By 1978 it was possible to make detailed predictions of its properties from improved versions of both MO and VB theory and test those predictions by experiment.

> **Prediction 13** (for cyclobutadiene). Most of the MO calculations indicated not only that there should be little or no aromaticity, but that it would be antiaromatic—*less* stable than the corresponding noncyclic compound. Moreover, the four carbon atoms would be in a rectangular rather than a square configuration, partly because of the Jahn-Teller effect. The ground state should be a singlet, but there should be a low-lying triplet state with square geometry.
> **Prediction 14**. VB calculations predicted that the ground state should be a square triplet, with the rectangular singlet lying slightly above.

Early evidence from the infrared spectrum was interpreted to support the square rather than rectangular shape. Thus MO theorists were in the classic Popperian situation of making a risky *contraprediction*: predicting a fact contrary to the empirical evidence known at the time, with sufficient confidence in the validity of their theory to hope that further experiments would vindicate them.

By 1978 it was clear that the experimental evidence favored the rectangular singlet ground state for cyclobutadiene, contrary to the VB predictions and to the reports of early observations, while confirming MO and Hückel's rule. During the following two decades this case could be cited as a successful prediction of MO and a refuted prediction of VB.

Important evidence came from charged molecules, as noted in the quotation from Hoffmann at the beginning of this section. For example, if you add an electron to C_5H_5, making $C_5H_5^-$, or take away an electron from C_7H_7, making $C_7H_7^+$, you should have just six electrons in the molecular orbital spread over the entire molecule (delocalized), as in benzene, so the resulting ion should be aromatic.

In 1954, W. Von E. Doering and L. H. Knox announced the synthesis of C7 H7+, known as tropylium, and presented evidence of its aromatic character. This has been called "the first verification of [a prediction from] Hückel's rule" by Shigeaki Kikuchi in his history of the subject. As a result, "Hückel's rule began to be regarded as effective, and the MO theory got considerable additional support in chemistry because Hückel's rule was derived from the MO theory."

Franz Sondheimer became a leader in the effort to synthesize the larger annulene molecules that should be aromatic according the Hückel's rule. He also proposed the annulene terminology for naming these molecules (from the Latin word *anulus* for a finger or signet ring.) In 1959 Sondheimer announced, in a paper with Reuven Wolovsky, the synthesis of [18] annulene. This is "the first cyclic polyene after benzene which might be expected to exhibit aromatic properties," they wrote. It satisfies Hückel's rule with n = 4.

But, you may ask, what about [10] annulus and [14] annulus, which satisfy the rule with n = 2 and 3, respectively? The answer is that if you try to draw the structure of these molecules using the accepted rules of organic chemistry, you find that the most stable arrangement of the carbon atoms is not a regular polygon with 10 or 14 sides, leaving a big empty space in the middle, but a more compact structure. However, that means that the hydrogen atoms can no longer be fitted into that space without excessive strain, and the molecule is not planar—hence not aromatic. This argument is attributed to a 1952 paper by Kurt Mislow.

At this point things get a little messy. In Hückel's derivation of his 4n+2 rule he assumed that the molecule is planar. But Sondheimer and other chemists assumed that a [4n+2] annulene could be aromatic even if it is only nearly planar. What does that mean? In 1960 Coulson and Golebiewski calculated that [18] annulene "cannot possibly be planar as a result of steric forces." But they seemed willing to accept the possibility that it could still "show aromatic-type behavior in its reactivity, in general agreement with Sondheimer and Wolovsky."

So is planarity a reliable prerequisite for aromaticity? Organic chemists dealt with this problem by finding other criteria for aromaticity. One is bond lengths: in benzene all the CC bonds should have equal length—they do not alternate between double and single bonds as one might have expected from some early theories of benzene.

Another criterion is the ring current: a delocalized system of B electrons corresponds to an electric current going around the framework of carbon atoms, as suggested by Linus Pauling in 1936. This current can be detected by nuclear magnetic resonance (NMR) measurements. Thus in 1961, Elvidge and Jackman asserted: "We can define an aromatic compound . . . as a compound which will sustain an induced ring current." Why? Because, as Jackman and several members of Sondheimer's group wrote, NMR is the "most convenient method for determining whether a compound is or is not aromatic." Moreover, as Sondheimer and his colleagues wrote in a 1962 article, there is

> no theoretical justification for equating "aromaticity" with "stability" or "benzene-like chemical behavior" as has often been done, despite the fact that many aromatic substances to exhibit such behavior. A cyclic substance is aromatic if all the B-electrons are delocalized around the ring . . . This is a property of the substance in its ground state, whereas chemical reactivity is dependent on the difference in free energy between that of the ground state and that of the transition state for the chemical change involved. Valid criteria for aromaticity are such properties as a lower energy content than would be predicted by comparison with an appropriate acyclic analog: the ability to sustain an induced ring current of B-electrons, and the presence of carbon-carbon bonds which are intermediate in length between those usual for single and double bonds.

Their conclusion is that [18] annulene is aromatic, despite the view of Mislow and of Coulson and Golebiewski that it cannot be planar. Jackman and the Sondheim group also concluded that [14] annulene is not aromatic (even though it satisfies the 4n+2 rule) because no ring current was detected by nuclear magnetic resonance.

The third criterion proposed by Sondheimer, Wolovsky, and Amiel in 1962 was "carbon-carbon bonds which are intermediate in length between those usual for single and double bonds." The usual situation for nonaromatic rings of carbon atoms is alternation of the CC bonds from single to double, which means that the *lengths* of the bonds will alternate. In 1960, H. C. Longuet-Higgins and L. Salem predicted that for values of n greater than eight (i.e., for $4n + 2 > 34$) the bond lengths will alternate. In the same year Sondheimer's group synthesized [30] annulene and concluded that it is not aromatic.

In 1969 Michael Dewar and Gerald Jay Gleicher developed a new treatment of the aromaticity of annulenes, and predicted that [26] and higher annulenes will not be aromatic even if they satisfy the 4n+2 rule. At that time it was believed that [22] is aromatic while [30] is not, so [26] would be a crucial test of their prediction. Thus at the end of our stated time period (1970) the critical value of n for onset of bond alteration had not yet been precisely established.

In a 1966 lecture to the Royal Society of London (published in 1967), Sondheimer stated that [10] and [14] annulenes "should be aromatic" by Hückel's rule but because of the "steric interaction of the internal protons" they "presumably cannot be planar" and therefore are not aromatic. So "[18] annulene was . . . the first member of this series to meet all the conditions for aromaticity." But when [18] was synthesized it was found to be "almost planar and contains (4n+2) B-electrons [so] it should be an aromatic substance." Is this actually borne out in practice? According to the classical definition, [18] annulene did not appear to be aromatic. Thus, the compound was not remarkably stable: "the fact that it can be nitrated under suitable conditions . . . points to the inherent danger in relating aromaticity to chemical reactivity. . . . The 'modern' definition, relating aromaticity to B-electron delocalization, is clearly preferable. According to the 'modern' definition, [18] annulene is without doubt aromatic. Thus . . . X-ray crystallographic analysis showed that there is no bond alternation . . . the n.m.r. spectrum represents the most convenient method for determining whether a given substance is aromatic . . . This spectrum clearly shows [18] to be aromatic."

In the case of [14] annulene, Sondheimer argues that is it aromatic "even in the temperature range in which the n.m.r. spectra show it to be 'nonaromatic.' Support for the aromatic nature of [14] annulene . . . is provided by the tentative finding that molecules of the crystalline substance appear to possess a centre of symmetry, which rules out bond alternation . . . this compound is aromatic, despite the fact that the carbon skeleton is presumably not coplanar."

By 1970, it was clear that Hückel's 4n+2 prediction had stimulated much valuable research, and the largely successful results of this research had provided important support for the MO theory. Here are some examples:

> Much of the interest in the field of nonbenzenoid aromatic compounds centers around attempts to verify theoretical predictions that certain conjugated systems will be especially stable, or "aromatic," while others will not.
>
> —Ronald Breslow *(1957)*

> "The major impetus leading to the current outburst of research in [aromatic chemistry] undoubtedly must be ascribed to the emergence of the molecular orbital theory of benzene, particularly as developed by E. Hückel."
>
> —E. Vogel *(1967)*

But the situation did not look so favorable in 1970, when the experts convened in Jerusalem for an international symposium on aromaticity, pseudoaromaticity, and antiaromaticity. Several participants complained that the term "aromaticity" had been burdened with too many different meanings and should now be abandoned. In particular, as one of the editors of the *Proceedings*, Ernst D. Bergmann, pointed out that "the chemical definition, which emphasizes the energy content of the molecule in the excited state," had been replaced by "the physical viewpoint which underlines the properties of the molecules in the ground state." Later he remarked: "we have not solved the problem of what is aromaticity . . . but we all agree that aromaticity can be defined only artificially, by convention, if we do not want to go to the extreme of abandoning the notion altogether."

Jean-Francois Labarre did want to go that far: because of the many different definitions of aromaticity,

> it seems, therefore, that the term "aromaticity" has lost all simple meeting, and, in fact, one finds oneself incapable of defining. Another difficulty arises from the fact that it is practically impossible to measure the aromaticity of a molecule . . . We believe that the term "aromaticity" should be eliminated from the scientific vocabulary.

D. Lloyd and D. R. Marshall argued that the meaning of aromaticity had changed so many times that "it would be better if the use of this term were discontinued."

E. Heilbronner proclaimed: "we should all realize that we are united here in a symposium on a nonexistent subject . . . aromaticity is not an observable property, i. e., it is not a quantity that can be measured and it is not even a concept which, in my experience, has proved very useful. . . . From the point of view of the chemist who wants to know the behaviour of his molecules, the question of whether a molecule is aromatic, pseudoaromatic, antiaromatic, or whatever aromatic is completely irrelevant. We want to know the properties of molecules, and we don't want to know whether a molecule falls into a certain abstract category that we have invented."

But these objections do not seem to have prevented chemists from continuing their attempts to prove (in both senses) old candidates and synthesize new candidates for the label of "aromatic." Much of this work was devoted to [18] annulene, as cataloged by Harold Baumann and Jean F. M. Oth in their 1982 survey.

In 1994 we find Vladimir Minkin and his colleagues writing that "there exists no unambiguous yardstick according to which one might assign a compound to the aromatic or antiaromatic class . . . one should strive to avoid the legendary situation when blind sages attempted to describe the elephant." Yet they also declare that "when Doering and Knox obtained the cycloheptatrienium (tropylium) cation in 1954, it was a vivid demonstration of the prognostic power of the Hückel rule."

Here is indeed a paradox: aromaticity is a meaningless concept, yet Hückel's theory, leading to his 4n + 2 rule predicting what compounds will be aromatic, is a great discovery which has enriched organic chemistry, according to Breslow and Vogel (quoted above). Hückel's rule "has been acclaimed one of the most fundamental rules of chemistry and has found wide applicability among (organic) chemists," according to D. J. Klein and N. Trinajstic in 1984.

Similarly, P. J. Garrett wrote in 1986:

> The idea that some compounds have unexpected stability arising from a common property has been historically valuable for the advance of chemistry and was the first area in organic chemistry in which a fruitful interaction between theoretical and experimental chemists occurred. Although many of the original ideas have been modified, many have been generally accepted, for example Hückel's Rule, and the interplay of experimental observation and theoretical deduction still continues to be fruitful in this area, leading to the synthesis of new nonnatural compounds and the generation of new theoretical concepts.

Even more extravagant praise came in 1999 from Jerome A. Berson :

> One must ask . . . whether any other theoretical advance ultimately has done more to enlighten the thinking of organic chemists than Hückel's brief, bright flare of cognition regrettably quenched too soon.

Roald Hoffmann (quoted at the beginning of this section) offers a larger context for the paradox in a 1974 article:

> By and large, theory has not predicted much chemistry. There have been some exceptions, some instances where theory made an impact on the experimental

field. I would call such an instance a credibility nexus—a place and time of interaction in which a group of experimentalists, otherwise sceptical of theory, suddenly found itself faced with the success of a simple theory. That set of specialists quickly became converts, often zealots. These episodes of doubt and conviction are important to identify because they play a crucial role in the peculiar symbiosis of theory and experiment. . . . It is important to note the counterproductive part of the story—the romance with Hückel's rule became an infatuation with stabilisation and resonance energies as derived from simple Hückel calculations. Many man-years were spent in attempts to make molecules that were supposed to be stable but in fact weren't.

10.7 THE REVIVAL OF PREDICTIVE CHEMISTRY

In 1933, N. V. Sidgwick wrote that the power of theories and models lies in their predictive character. Yet the quantum theory that he welcomed into chemistry seemed for many years, as chemist C. N. Hinshelwood warned, unable successfully "to predict new phenomena in organic chemistry"—it could only provide plausible explanations for known chemical facts. Pauling could predict the resonance energies and bond lengths of some aromatic molecules only with the help of parameters adjusted to agree with the observed properties of other molecules. Hückel's predictions about the aromaticity of other cyclic hydrocarbons such as COT and CBD could not be tested until those molecules were available for replicable experiments. Before the 1950s quantum chemists, whether they used VB or MO or some other approximation scheme, still almost always had to take the overall atomic structure of a molecule as given, and then try to calculate electronic wave functions for a fixed set of nuclear positions.

At the same time Pauling's resonance theory, treated as a set of qualitative or semiquantitative rules independent of its putative quantum-theoretical foundations, gained the reputation of being a successful technique for predicting and explaining chemical structure and reactions. Thus when M.J.S. Dewar published his MO-based book *The Electronic Theory of Organic Chemistry* (1949), some reviewers attacked it for failing to meet the standard of predictiveness set by resonance theory. Chemist Paul D. Bartlett complained that Dewar did not show the superiority of "the rather nebulous notation of molecular orbitals as an instrument of correlation and prediction, in comparison to the well-developed qualitative resonance scheme." Worse, one of its predictions (about the nonexistence of a bimolecular allylic rearrangement) had already been refuted by the time the book was published. Chemist M. S. Kharasch insisted that "part of the 'success' which the author claims as justifying theories based on such insecure ground is achieved by juggling 'secondary effects' to make 'predictions' fit observed facts . . . The 'acid test' of a new idea . . . is its predictive aspect. Thus far, the idea of B complex intermediates has been singularly barren in uncovering new facts."

Per-Olov Löwdin, one of the leaders in the postwar generation of quantum chemists, stated in 1957, as noted above, that the goal of his discipline is novel prediction, but admitted that enormous amounts of computation still had to be done to make quantitative predictions: "A theoretician is a person who knows how to solve a problem, but who cannot do it. This remark is certainly true concerning the quantum chemists."

At the Kekulé Symposium in 1958, Longuet-Higgins proclaimed that "a scientific theory must . . . be judged by its predictive as well as its explanatory power." By this criterion, VB failed when applied to the stability and aromaticity of molecules similar to benzene. Thus it predicted a greater resonance energy for cyclobutadiene, which is very unstable if it exists at all, than for benzene; and it cannot explain why the C_5H_5- ion should be more stable than the $C_5H_5^+$ ion. Hückel's MO theory and improved versions thereof are clearly better in those cases. Yet one cannot even test VB for most larger aromatic molecules because of "the enormous labour which is required to apply it," so MO wins by default.

But, as Longuet-Higgins recalled many years later, that victory did not yet constitute a return to the "golden age of valency theory" (the days of Lewis, Langmuir, and Sidgwick), when "it was taken for granted that the strength of any theory lay in its power to make strange predictions and cast doubt on existing evidence. Both Pauling and Mulliken . . . were happy to accept the reported structures of molecules at their face value and 'explain' them to the chemist."

This criticism was expressed more bluntly in a paper given at a 1963 meeting of the American Chemical Society by John R. Platt. He argued that some scientific fields progress faster than others because they use the method of "strong inference." This method—which is essentially the good old hypothetico-deductive method with a little Baconian induction, a dash of geologist T. C. Chamberlin's multiple working hypotheses, and a lot of Popperian falsificationism thrown in—is supposedly responsible for the success of molecular biology and nuclear physics. Other fields are "sick by comparison, because they have forgotten the necessity for alternative hypotheses and disproof." In particular he faults his own field, theoretical chemistry, in which he claims the theorist only tries to explain a given fact, not make testable predictions. The resonance theorists and MO theorists are equally to blame for failing to come up with crucial experiments that would let one or the other win out: "A failure to agree for 30 years is public advertisement of a failure to disprove."

To a chemist, this inability to resolve theoretical disputes by decisive experiments was especially frustrating because a rigorous scientific method was part of the discipline's proud history, and the nineteenth-century study of benzene and its isomers was still celebrated as a textbook example of scientific method. For Platt, organic chemistry has been the spiritual home of strong inference from the beginning: "Do the bonds alternate in benzene or are they equivalent? If the first, there should be five disubstituted derivatives; if the second, three. And three it is."

When Mulliken won the 1966 Nobel Prize in chemistry, Platt celebrated this event as a confirmation of the victory of MO over VB: their battle "has divided the chemical world for a generation. The theories are supposed to be formally identical when all high-order corrections are included, but in practice they are as different as night and day." MO, he said, "is now used far more widely, and simplified versions of it are being taught to college freshmen and even to high school students." But Platt did not name a single experiment that clearly proved the superiority of MO over VB.

In a 1974 article I criticized the accuracy of claims by Platt and others about the role of the hypothetico-deductive method in the progress of science, and questioned whether contemporary historians' accounts of important discoveries would support the lessons he wanted to teach about how successful science is done. On the other hand, as a former Coulson student who abandoned quantum chemistry in the late

1950s because of dissatisfaction with the scientific value of the results obtained by MO calculations, I am sympathetic to his view that the field at that time urgently needed a more Popperian attitude. But even in 1974, I did not realize that modern chemistry is more predictive than other physical sciences.

In order to be Popperian you must be able to deduce predictions from your hypothesis with reasonable confidence; otherwise, there is too great a temptation to blame any discrepancy between predicted and observed results on the inaccuracy of your approximations. In fact, the increasing popularity of MO in the 1950s and 1960s went along with increasingly accurate computational methods, and with the development of reliable ways to draw conclusions without excessive computation—those trends reinforced each other. The introduction of fast large-memory electronic computers allowed theorists to exploit those methods even more effectively.

Beginning in the late 1960s, MO theorists claimed that they could now calculate the chemical and spectral properties of molecules accurately enough to be useful to chemists. Coulson, in a review article published shortly before his death, emphasized that MO theory could predict as well as explain, citing as examples the successful proposals for stabilizing cyclobutadiene and cyclooctatetraene by using rare-earth atoms. Others also credited MO with many successful predictions.

10.8 RECEPTION OF MOLECULAR ORBITAL THEORY BY ORGANIC CHEMISTS

There is no doubt that MO methods became increasingly popular with theoretical chemists after 1945, and that by 1970 they had displaced VB methods as the preferred approach for calculations of molecular properties in general and for aromatic molecules in particular. Although from the viewpoint of fundamental physics these are just two different approximations to the same theory—quantum mechanics—from the viewpoint of chemistry there is a much more profound difference. As I argued above, the adoption of the resonance hypothesis for benzene was relatively painless since it could easily be described as just a more sophisticated version of a concept already familiar in chemistry; whereas the MO picture wiped out all the traditional models that postulated bonds between pairs of carbon atoms and replaced them by a delocalized cloud of electrons spread over the entire molecule. Thus, after 100 years, Kekulé's model of alternating or oscillating double and single bonds had to be abandoned as a description of the actual structure of benzene.

Why would chemists be willing to do this?

I have indicated above some of the reasons for this shift given by quantum chemists. Now I want to determine more precisely why (and to what extent) the entire community of organic chemists switched from resonance theory to molecular orbital theory for benzene—that is, what reasons did they give in their publications? Did they continue to use qualitative VB explanations for pedagogical purposes even though they agreed that MO came closer to an accurate description of the properties of the molecule?

As is generally the case in the physical sciences, authors of research papers do not often tell you why they use a theory; they just use it. So we must look at review articles, monographs, and textbooks where these reasons are more explicitly stated.

Most authors of review articles and monographs on quantum chemistry published from 1961 to 1980 gave preference to MO over VB but provided no specific reason for this preference aside from computational convenience (8 out of 11). Only one stated that MO results were in better agreement with empirical data than VB results for benzene. Successful predictions from Hückel's rule (generally understood to be based on a simple MO approach) that only those molecules with 4n+2 B electrons are aromatic were sometimes mentioned; these are directly relevant to the traditional problem of explaining the special properties of benzene (two authors). One monograph pointed to the Woodward-Hoffmann rules for pericyclic reactions as a success of molecular orbital theory.

A significant indicator of the consensus of quantum chemists is a 1968 volume dedicated to Linus Pauling by his students, colleagues, and friends in which only one author (M. Simonetta) defended VB, while five others discussed mostly MO calculations. (As the saying goes, with friends like that, who needs enemies?)

Pauling himself refused to use MO in later editions of his *Nature of the Chemical Bond*, explaining in correspondence with David P. Craig that he didn't know how to present it accurately at an elementary level suitable for students. But, according to Gavroglu and Simões, "in the abridged version . . . published in 1967 and specifically addressed to students, Pauling made a small concession . . . he introduced students to the molecular orbital approach."

10.9 ADOPTION OF MO IN TEXTBOOKS

In the period 1951–1960, MO concepts found their way into organic chemistry textbooks by 11 authors, but usually as a supplement rather than an alternative to VB and RH, and often with no recognition that there is any conflict between the two approaches. But two authors explicitly preferred molecular orbital theory to valence-bond theory.

In 1961–1970 textbooks by nine authors mentioned MO concepts, but only four preferred them. By the 1970s it was clear that MO was definitely the more correct picture; organic chemistry textbooks published in 1971–1980 by nine authors presented the MO approach even for qualitative discussions of structure and reactions. Valence-bond theory and the resonance hybrid hypothesis were retained only for heuristic purposes—or, as one book expressed it, because an MO picture is "cumbersome to draw and requires an artistic talent not possessed by most organic chemists."

Among the stated reasons for using MO the most popular was the success of the Woodward-Hoffmann rules for predicting the course of reactions from orbital symmetry and the associated "frontier orbital" method of Fukui. (But, in a remark recognizing the fallacy of affirming the consequent, Dewar argued that the success of the Woodward-Hoffmann rules did not imply the validity of the theory from which they were derived, and proposed an alternative derivation.) Also, the Hückel rule for aromaticity, derived from MO, was considered to have successfully predicted the instability and nonaromatic character of CBD, even before that strange molecule was found to be rectangular rather than square. When the MO prediction of its rectangular shape was confirmed by new experiments, this reinforced the support for MO in aromatic chemistry. A few authors cited the success of MO in explaining molecules in which a metal atom is sandwiched between two benzene rings.

There is explicit evidence that textbook authors were prepared to award extra credit for novel predictions. In 1976 chemist Norman Allinger and his colleagues wrote: "A good theory explains all the known facts in its province and can be used to predict new facts." In 1973, chemists Robert T. Morrison and Robert N. Boyd called the Fukui-Woodward-Hoffmann theory of reactions, based on MO, "one of the really giant steps forward in chemical theory" and remarked: "Woodward and Hoffmann made *predictions*, which have since been borne out by experiment. All this is the more convincing because those predictions were of the kind called 'risky': that is, the events predicted seemed unlikely on any grounds other than the theory being tested."

Similar views are expressed in other organic chemistry textbooks: MO theory is first presented as a general explanation of the chemical bond, with little or no mention of the VB alternative; the MO picture of benzene is given, with the RH picture being pushed increasingly into the background or eliminated altogether; while no claim is made that MO gives quantitatively better results for benzene itself than does RH (except possibly for excited states and spectra); the success of Hückel's rule (based on MO) in predicting whether similar compounds will be aromatic is noted; and, in advanced texts and even in some introductory texts, a substantial section on pericyclic reactions, featuring the predictively successful Woodward-Hoffmann theory, is added to more recent editions.

Just as in the case of the adoption of Mendeleev's periodic law (Chapter 5), we do not find that chemists adopted the MO theory *primarily* because of its successful novel predictions; the utility of the theory in organizing and explaining previously known facts is a more important factor. Nevertheless, it is clear that several chemists do agree that predicting a new fact counts more than explaining a known fact of similar importance.

Of course in science, no theory should be enshrined as the ultimate truth, never subject to correction or replacement—no matter how many successful predictions it can claim. During the past three decades, while MO was widely accepted by chemists, a small but vocal minority attempted to revive VB, arguing that, properly formulated, this approach could yield results just as accurate and reliable as those of MO, while at the same time preserving the qualitative advantages of the classical bond description. In particular, it was claimed that VB gives a more accurate model for benzene than MO and vindicates Kekulé. It was even claimed that contrary to MO, delocalization is not generally the driving force in chemistry responsible for the aromatic stability of benzene, but merely a byproduct of other factors.

10.10 A 1996 SURVEY

In 1996, in an attempt to go beyond the published record and ascertain the current views of the chemical community, I wrote to 133 chemists, selected because they were authors of books or review articles on quantum chemistry or textbooks on organic chemistry, or had published research papers on benzene or cyclobutadiene, since 1970. (A few historians of modern chemistry were also included.) The letter stated: "As part of my research in the history of science, I am comparing the reasons why theories are accepted or rejected in different fields. My current interest is in the structure and properties of benzene. . . . Most of the textbooks and articles I've examined indicate that the molecular orbital theory is now generally accepted by organic chemists, but do not clearly explain why it is considered superior to alternative theories such as the valence bond/resonance theory; and in particular what advantage it has for benzene." The letter then asked

1. Why do most chemists accept the molecular orbital theory?
2. In your opinion, which theory gives the best description of benzene and similar molecules? Why?

I received 43 replies, some of them several pages long with reprints enclosed, others only a few sentences. In some cases I followed up with a letter requesting more specific information.

Here is my analysis of the answers to the first question:

- (22 respondents) MO is easy to use, especially for larger molecules; good computer programs are available.
- (17) "It works"—it gives accurate results.
- (11) It is simpler, easier to visualize, and understandable without extensive mathematical training.
- (10) It explains Hückel's rule, which is successful in predicting aromaticity.
- (7) It provides the basis for the successful Woodward-Hoffmann rules and the Fukui frontier orbital theory.
- (6) It explained the properties of O_2 (two unpaired electrons, paramagnetism), which simple VB did not.
- (4) It has a better theoretical foundation.
- (3) Chemists have been brainwashed to prefer MO without any direct experience themselves; publicity for the Nobel Prize awarded to Woodward-Hoffmann rules and Fukui's frontier orbitals.
- (2) It avoids the (VB) pitfall of ascribing reality to resonance structures.

Five other reasons were mentioned only once. Six respondents rejected the premise of the question, insisting that most chemists have *not* switched from VB to MO.

The second question did not elicit as many different answers:

- (10 respondents) For benzene, both methods are equally good, it depends on which property is most important to you.
- (8) MO gives better results for excited states and spectra of benzene.
- (6) The case of cyclobutadiene shows that MO provides a better explanation of aromaticity and hence is preferable for benzene.
- (5) MO better explains benzene's stability and reactions.
- (5) VB (modern version) gives a better description of benzene.
- (2) MO gives the most accurate values for heat of formation, bond lengths, and so on.
- (2) The ring current effect in NMR is most easily explained by MO.

The predictivist hypothesis that cyclobutadiene would be considered a crucial experiment was not very well supported by these results. Of the nine respondents who had published papers on cyclobutadiene, only one mentioned it as a reason to accept the MO description of benzene, and then only in response to my follow-up letter. Two others, as mentioned above, stated that VB actually gives a better account of cyclobutadiene.

10.11 CONCLUSIONS

It is a commonplace that people often proclaim norms they do not follow in practice, and scientists do not seem to be unusually virtuous in this respect. Chemists, like other scientists, insist that theories must be tested by experiment and argue that the best theories are those that lead to successful predictions. The results of the present study are consistent with the hypothesis that organic chemists actually try to follow that prescription and are unhappy when it doesn't work. Of course these results by themselves do not verify the hypothesis, although it is fair to say that the structure of benzene has been considered one of the most important and difficult problems for the discipline ever since 1865.

To recapitulate: research on benzene in the late nineteenth century did follow the method of conjecture and refutation, and this period is still recalled fondly by modern chemists. Several initially plausible hypotheses such as Ladenburg's prism were definitely excluded by experimental research. But in the early twentieth century the method failed to work; it was not possible to use empirical tests to distinguish between alternatives such as Kekulé's oscillation (K-Osc), Thiele's partial valence (PV), and the Armstrong-Bayer centric hypothesis (ABC). Thus in the 1930s organic chemists were receptive to a new idea—the Pauling-Ingold resonance/mesomerism hypothesis based on the valence-bond (VB) approximation to the Schrödinger equation—which not only enjoyed some of the prestige of quantum physics, but could also be related to ideas that chemists themselves had found useful in explaining the properties of benzene (Postulate III).

Nevertheless, the acceptance of Pauling's resonance hybrid theory of benzene was not justified by the confirmation of any novel predictions or any direct quantitative calculations of the properties of benzene itself. Of course we should keep in mind that quantum mechanics, which provided the theoretical basis for RH, had been accepted by physicists because of its success in explaining previously known facts in a simpler, more accurate, and more coherent manner than its rivals, *not* because of successful novel predictions.

There was in fact one successful novel prediction that made a strong impression on organic chemists. In 1932, Levine and Cole announced that they had confirmed Kekulé's oscillation hypothesis by showing that the ozonization of a di-substituted benzene derivative, ortho-xylene, yielded products that must have come from the two different isomers of that derivative required by K-Osc. Their results were quantitatively confirmed and extended by Haaijman and Wibaut. Since RH represented benzene in a way that could be imagined as an oscillation between the two Kekulé structures, this experiment was taken to be also a confirmation of the resonance hybrid theory.

Some philosophers of science have argued that a new theory should explain all the facts explained by the old theory plus some others. Since RH was initially derived from VB, and VB was subsequently replaced by the alternative MO approach, we could ask those philosophers how MO explains the products of the ozonization of benzene. I do not doubt that a modern quantum chemist could provide such an explanation if necessary, but so far I have not found one in print. This crucial experiment has simply dropped out of the textbooks. We might compare this situation to the successful novel prediction of the positron from Dirac's relativistic wave equation (which also happened in the early

1930s). Although it was initially followed by widespread acceptance of Dirac's theory, that theory was later replaced by another one (the quantum electrodynamics of Feynman, Schwinger and Tomonaga) that did not claim to predict the positron but simply accepted the existence of antiparticles as a basic postulate (Section 9.3).

In the 1950s quantum chemists replaced the RH description of benzene by the MO description. The primary reason was not a successful novel prediction about benzene itself, but the greater computational convenience and accuracy of MO in dealing with *larger* molecules. It was generally believed that any adequate theory of molecular structure and reactions must be generally consistent with quantum mechanics, VB and MO were the two major approximation methods for deriving molecular properties from quantum mechanics, and MO was undoubtedly better *as an approximation to quantum mechanics*. Thus the success of Pauling's VB theory led chemists to accept quantum mechanics as a basis for chemical theory, and this in turn led them to *reject* VB because MO was a better way to use quantum mechanics.

In the 1970s and 1980s, organic chemists followed quantum chemists in taking MO as the basic approach to the chemical bond, and hence in particular as the correct picture of benzene. By this time, however, there were some successful novel predictions that encouraged them to replace VB by MO not just as a general foundation, but as a working tool. One was the MO theory of pericyclic reactions, developed under the name frontier orbitals by Fukui and his colleagues, and in an equivalent but more qualitative manner as orbital symmetry by Woodward and Hoffmann. Another, of more direct relevance to benzene theory, was the success of Hückel's rule (derived from MO) in predicting which of the various molecules similar in structure to benzene would share its aromatic properties. In particular, the MO prediction that cyclobutadiene (C_4H_4) would not be aromatic and would have a rectangular (though very unstable) structure was experimentally confirmed in the late 1970s. So the theory of benzene was changed not because of experiments on benzene itself, but because of experiments on other molecules that could be better explained by a general theory that also explained benzene. Again we have an analogy from the history of physics: the substance or caloric theory of heat, generally accepted around 1800, was abandoned by some physicists in the 1820s and 1830s in favor of a vibration theory of heat, not because of experiments on heat but because of experiments on *light* that favored a vibration (wave) theory over a substance (particle) theory, combined with a general belief that both heat and light must be different aspects of the same general phenomenon.

If a theory is good enough it may able to make successful *counter*predictions: to challenge the validity or accuracy of an accepted experimental result and force a replication of the experiment that vindicates the theory. In 1973, Charles Coulson delivered an inaugural lecture marking his appointment to the newly created chair of theoretical chemistry in the University of Oxford. He reported on the case of the bond energy in the hydrogen molecule. The theoretical value was 36117.7 cm^{-1};

> until recently the best experimental value was 36113 cm^{-1}. This could not be fitted into the theoretical values and so sure were people that the calculations were right that Herzberg in Ottawa was led to reconsider the experimental measurements. He found that there were corrections to be made.

The new experimental results, 36116.3 to 36118.3 cm^{-1}, agreed with the theoretical value.

Do chemists give more weight to novel predictions than other scientists? Eric Scerri argues that they do not, but he also points out that until recently little research has been done on the philosophy of chemistry, so many questions of this kind are still open.

My own view, based on the very limited evidence from only two case histories (periodic law and benzene), is that the predictiveness norm *is* stronger in modern chemistry than in modern physics. It was not strong enough to prevent a physics-based, plausible, but only weakly predictive hypothesis from dominating benzene theory for three decades. But when that theory was beaten at its own game by another physics-based theory that could also make successful novel predictions, chemists eagerly embraced the newer theory, even though doing so meant giving up (or treating as merely heuristic) one of their most treasured possessions, Kekulé's single/double bond hexagon.

Successful predictions may have a greater impact in the context of a crucial experiment: a situation in which two competing theories make very different predictions. This was the case with cyclobutadiene, and the result was a significant victory of molecular orbital theory over valence-bond theory.

One final remark: in the cases of relativity (special and general) and quantum (old quantum theory and new quantum mechanics), I found that when a theory is judged mostly on its *explanations* it may be accepted more quickly than when it is judged mostly on its *predictions*. This generalization seems to apply here: Kekulé's theory was accepted only slowly and reluctantly in the nineteenth century despite its successful predictions; VB was accepted quickly because of Pauling's persuasive explanations; MO, proposed about the same time as VB, was accepted much more slowly despite its successful predictions.

NOTES FOR CHAPTER 10

291 *The structure and reactions of atoms and molecules are determined by quantum mechanics* For details and references see my paper "Dynamics of Theory Change in Chemistry: Part 1. The Benzene Problem 1865–1945" and "Part 2. Benzene and Molecular Orbitals, 1945–1980," *Studies in History and Philosophy of Science* 30 (1999): 21–79, 263–302.

291 *A key concept in benzene theories after 1930 is resonance* General surveys: Costas Gavroglu and Ana Simões, *Neither Physics nor Chemistry: A History of Quantum Chemistry* (Cambridge, MA: MIT Press, 2012); Ana Simoes, "Chemical Physics and Quantum Chemistry in the Twentieth Century," in *The Modern Physical and Mathematical Sciences*, edited by Mary Jo Nye (Cambridge: Cambridge University Press, 2003), pp. 394–412; Buhm Soon Park, "The 'Hyperbola of Quantum Chemistry': The Changing Practice and Identity of a Scientific Discipline in the Early Years of Electronic Digital Computers," *Annals of Science* 60 (2003): 219–247; Ana Simões and Kostas Gavroglu, "Different Legacies and common Aims: Robert Mulliken, Linus Pauling and the Origins of Quantum Chemistry," in *Conceptual Perspectives in Quantum Chemistry*, edited by J.-L. Calais and E. S. Kryachko (Dordrecht: Kluwer, 1997), pp. 383–413; Simoes and Gavroglu, "Quantum Chemistry in Great Britain: Developing a Mathematical Framework

for Quantum Chemistry," *Studies in History and Philosophy of modern Physics* 31 (2000): 511–548.

291 *"he was the hare and the MO was the tortoise"* Philip Anderson, "Who or What is RVB?" *Physics Today*, 61, no. 4 (April 2008): 8–9.

Section 10.1

291 *A key concept in benzene theories after 1930 is* resonance In classical mechanics, resonance occurs in a mechanical system consisting of two harmonic oscillators that, if they did not interact, would each have the same frequency f_0. If the interaction is characterized by a small parameter, the potential energy for the system can be written in the form

$$V(x_1, x_2) = 2\pi^2 m f_0^2 x_1^2 + 2\pi^2 m f_0^2 x_2^2 + 4\pi^2 m \varepsilon x_1 x_2,$$

where x_1 and x_2 are the coordinates of the two oscillators, each of mass m.

If ε / f_0^2 is small, one finds that the interacting oscillators (coordinates x_1 and x_2) carry out approximately harmonic oscillations with frequency f_0 but with amplitudes that change with time, in such a way that the total energy of the system gradually shifts back and forth between them. The period of the resonance (time between successive maxima of x_1) is approximately f_0/ ε.

Whereas in classical mechanics resonance is relatively uncommon—real oscillators generally do not have exactly the same frequency, and their interaction is not usually determined by the linear form above—in quantum mechanics it turns out to be very important. This is because atoms and molecules consist of particles most of which are identical (at least the electrons are) and the basic equations of the theory are linear. In the simplest case of two-electron atoms, first discussed by Heisenberg from this viewpoint, the wave function of the system corresponding to one electron in state *n* and the other in state *m* can be written

$$\varphi = (1/\sqrt{2})\left[\varphi_n(1)\varphi_m(2) \pm \varphi_n(2)\varphi_m(1)\right]$$

Instead of saying that the first electron is in state *n* while the second electron is in state *m*, we have to say that this configuration is mixed with another in which the first is in *m* while the second is in *n*. The two possible mixtures corresponding to the + and – signs are called symmetric and antisymmetric; these two states have slightly different energies. Heisenberg attributed the splitting of spectral lines in helium to this effect, which he described as a continual interchange of the electrons. For further discussion see Eamon F. Healy, "Heisenberg's Chemical Legacy: Resonance and the Chemical Bond," *Foundations of Chemistry* 13 (2011): 39–49.

291 *one can construct a wave function for the system* Thus for the first electron, in atom a, one has

$$\varphi_a(1) = (1/\sqrt{\pi a_0^3}) \exp\{-r_{a1}/a_0\}$$
$$\left[a_0 = h^2/4\pi^2 me^2 = 0.529\text{Å}, \text{ 'Bohr radius'}\right]$$

where r_{a1} is the distance between the nucleus of atom a and the first electron. Since the system resonates between a state with the first electron in atom a and the second in b, and a state with the first in b and the second in a, the wave function for the system is written

$$\varphi(1,2) = c_1\varphi_a(1)\varphi_b(2) + c_2\varphi_b(1)\varphi_a(2)$$

The nuclei are assumed to be fixed at a distance R; the energy of the system is calculated for various values of R, and it is assumed that the equilibrium distance between nuclei in the hydrogen molecule is the one for which the energy is a minimum.

292 *When Heitler presented his theory* L. Rosenfeld, "Quantentheorie 1929: Erinnerungen an die erste Kopenhagenkonferenz," in *Niels Bohr 1885–1962, Der Kopenhagener Geist in der Physik* (Braunschweig/Wiesbaden: Friedr. Vieweg & Sohn, 1985), as quoted/translated by Andreas Karachalios, *Erich Hückel (1896–1980). From Physics to Quantum Chemistry* (Boston Studies in Philosophy of Science, 283), Berlin: Springer, 2010, p. 43.

292 *the Van der Waals attractive force . . . could be explained* See S. G. Brush, *Statistical Physics and the Atomic Theory of Matter from Boyle and Newton to Landau and Onsager* (Princeton, NJ: Princeton University Press, 1983, section 5.4.

292 *resonance between the two Kekulé structures* J. C. Slater, "Directed Valence in Polyatomic Molecules," *Physical Review* 37 (1931): 481–489, on p. 489.

292 *the molecular orbital theory (MO) . . . was better* Slater, Discussion Remark, *Papers and Discussions, International Conference on Physics*, London, 1934, Vol. II, p. 53. London: Physical Society, 1935. John C. Slater, *Solid-State and Molecular Theory: A Scientific Biography* (New York: Wiley, 1975), pp. 47, 105.

293 *"a tetrahedral arrangement of the four bonds of the quadrivalent atoms"* "The Shared-Electron Chemical Bond," *Proceedings of the National Academy of Sciences* 14 (1928): 359–362, on p. 361.

293 *acknowledged that Slater had independently discovered* Pauling, "The Nature of the Chemical Bond: Application of Results Obtained from the Quantum Mechanics and from a Theory of Paramagnetic Susceptibility to the Structure of Molecules," *Journal of the American Chemical Society* 53 (1931): 1367–1400, on page 1367.

293 *This term was first used by Mulliken in 1932* Mulliken, "Electronic Structures of polyatomic Molecules and Valence; II: General Considerations," *Physical Review* 41 (1932): 49–71; Mulliken, *Selected Papers*, edited by D. A. Ramsey and J. Hinze (Chicago: University of Chicago Press, 1975), p. 458.

293 *He suggested a function of the form* $\varphi(1) = \varphi_A(1) + \varphi_B(1)$ "The Electronic Structure of Some Diatomic Molecules," *Transactions of the Faraday Society* 25 (1929): 668–686.

294 *"the molecule must be viewed as a whole"* Hückel, "Interview with Erich Hückel," *Journal of Chemical Education* 49 (1972): 2–4, on p. 3. For a detailed account of Hückel's career and his theory of benzene see Karachalios, *Hückel*. On his earlier work see Helge Kragh, "Before Quantum Chemistry: Erich Hückel and the Physics-Chemistry Interface," *Centaurus* 43 (2001): 1–16.

294 *This became known as Hückel's rule* E. Hückel, "Grundzüge der Theorie ungesägten und aromatischen Verbindungen," *Zeitschrift für Elektrochemie* 43 (1937): 752–788, 827–849, on p. 778. Karachalius, *Hückel*, pp. 87–90.

295 *Pauling and George W. Wheland modestly described their theory* Pauling and Wheland, "The Nature of the chemical Bond, V. The Quantum-Mechanical Calculation of the Resonance Energy of Benzene and Naphthalene and the Hydrocarbon free Radicals," *Journal of Chemical Physics* 1 (1933): 362–374, on p. 363.

295 *Pauling and Wheland wrote the wave function for benzene* Wheland and Pauling, "A Quantum Mechanical Discussion of Orientation of Substituents in Aromatic Molecules," *Journal of the American Chemical Society* 57 (1935): 2086–2095.

295 *Some authors took care to distinguish resonance from* tautomerism N. V. Sidgwick, "Presidential Address delivered at the Annual General `Meeting, March 18th, 1937," *Journal of the Chemical Society, London* 151 (1937): 694–699; L. P. Hammett, *Physical Organic Chemistry* (New York: McGraw-Hill, 1940), on p. 27.

296 *Wheland explained the distinction with a biological analogy.* Wheland, *The Theory of Resonance and its Application to Organic Chemistry* (New York: Wiley, 1944), p. 3.

296 *VB "seems to give results in somewhat better agreement"* Wheland, "The Quantum Mechanics of unsaturated and aromatic Molecules: A Comparison of two Methods of Treatment," *Journal of Chemical Physics* 2 (1934): 474–481, on p. 474.

296 *MO is "often the more convenient"* Van Vleck and Sherman, "The Quantum Theory of Valence," *Reviews of Modern Physics* 7 (1935): 167–228, on p. 171.

297 *Mulliken himself admitted that MO departs from "chemical ideology"* "Electronic Structure of polyatomic Molecules and Valence, VI: On the Method of Molecular Orbitals," *Journal of Chemical Physics* 3 (1935): 375–378, on p. 376.

Section 10.2

297 *the power of theories lies in their predictive character Some Physical Properties of the Covalent Link in Chemistry* (Ithaca, NY: Cornell University Press, 1933).

297 *the theory of resonance "must now be taken seriously into account"* Sidgwick, "The Theory of Resonance and the Coordination of Hydrogen," *Annual Reports of the Progress of Chemistry* 31 (1934): 37–43, on pp. 37 and 40. A. I. Simoes, *Converging Trajectories, Diverging Traditions: Chemical Bond, Valence, Quantum Mechanics and Chemistry, 1927–1937* (PhD dissertation, University of Maryland, College Park, 1993), pp. 211–216.

297 *it is not even mentioned in his brother's book* W. Hückel, *Theoretische Grundlagen der organische Chemie* (Leipzig: Akademische Verlagsgesellschaft M.B.H., 2nd ed., 1934).

297 *because of the excellence of its presentation, VB would eclipse MO* Joseph Mayer, review of Pauling's *Nature of the Chemical Bond* in *Reviews of Scientific Instruments* 10 (1939): 290–291.

297 *Pauling's book retarded the progress of quantum chemistry* E. Hückel, *Ein Gelehrtenleben: Ernst und Satire* (Weinheim: Verlag Chemie, 1975), p. 176.

298 *"Pauling made a special point of making everything sound as simple as possible"* R. S. Mulliken, *Selected Papers*, edited by D. A. Ramsey and J. Hinze (Chicago: University of Chicago Press, 1975), pp. 9–10, based on Mulliken's interview with T. S. Kuhn, Transcript at Center for History of Physics, College Park, MD, in the Archive for History of Quantum Physics Collection.

298 *Pauling "was an eloquent teacher"* Hager, *Force of Nature: The Life of Linus Pauling* (New York: Simon and Schuster, 1995), p. 164.

298 *he cited their work in support of his claim* C. K. Ingold, "Significance of Tautomerism and of the Reactions of aromatic Compounds in the Electronic Theory of Organic Reactions," *Journal of the Chemical Society, London* 143 (1933): 1120–1127 (note added July 1927).

298 *calling it "a modernized version"* Ingold, "Principles of an Electronic Theory of Organic Reactions," *Chemical Reviews* 15 (1934): 225–274, on pp. 247 and 251.

298 *"the physical interpretation of mesomerism was developed independently"* Ingold, "The Structure of Benzene," *Proceedings of the Royal Society of London* A169 (1938): 149–173, on pp. 154 and 158.

299 *Sidgwick praised the success of Pauling's resonance theory* Sidgwick, "Presidential Address Delivered at the Annual General Meeting, March 18th, 1937," *Journal of the Chemical Society, London* 151 (1937): 694–699.

299 *"the modern development of the structural theory"* "Presidential Address. Delivered at the Annual General Meeting, April 16th, 1936," *Journal of the Chemical Society, London* 149 (1936): 533–538 on page 538.

300 *the need to escape from the "thought-forms of the physicist"* "Recent developments in Valence Theory," *Pure and Applied Chemistry* 24 (1970): 257–287, on p. 259.

300 *his now-familiar double-donut MO representation of benzene* Coulson, "Representation of simple Molecules by Molecular Orbitals," *Quarterly Review of the Chemical Society of London* 1 (1947): 144–178, on p. 168.

Section 10.3

300 *the two Kekulé structure of o-xylene "merely represented phases"* Levine and Cole, "The Ozonides of Ortho-xylene and the Structure of the Benzene Ring," *Journal of the American Chemical Society* 54 (1932): 338–341, on p. 339, 346. For a detailed discussion see Brush, "Dynamics, Part I," pp. 58–62.

Section 10.4

301 *Pauling disagreed with his colleague Wheland* See A. I. Simoes, *Converging Trajectories*, pp. 170–173.

302 *"the failure to localize the electrons more exactly"* M. J. S. Dewar, *Electronic Theory of Organic Chemistry* (Oxford: Clarendon Press, 1949, p. 1.

302 *part of a more general movement from instrumentalist toward realist views* Brush, "The Chimerical Cat: Philosophy of Quantum Mechanics in Historical Perspective," *Social Studies of Science* 10 (1980): 393–447; Kostas Gavroglu, "Can Theories of Chemistry Provide an Argument Against Realism?" in *Realism and Anti-Realism in the Philosophy of Science*, edited by R. S. Cohen, R. Hilpinen and Qiu Renzong (Dordrecht: Kluwer, 1996), pp. 149–170.

Section 10.5

303 *establishing MO as the preferred approach* A remarkable example of Coulson's strategy is a popular article he published in 1947. Superficially it looks like an enthusiastic account of the achievements of resonance theory. He notes that resonance is not a "real phenomenon" but an idea that "has proved extraordinarily successful in correlating, explaining, and even predicting an astonishingly large and varied body of chemical experience." But, having converted the reader to that theory, he points out that in the resonance picture of benzene "an electron originally round atom No. 1 is sometimes paired in a molecular orbit round No. 2, so that it can migrate to this atom" and thence to No. 3 and all the way around the ring. So "there are six electrons

in benzene which are able to flow as tiny currents round the central hexagon"—a model that implies magnetic effects, confirmed experimentally in the diamagnetic anisotropy of benzene. Without exactly mentioning MO theory, he has prepared the reader to accept it as a more detailed and accurate version of the wonderful theory of resonance. Coulson, "The Meaning of Resonance in Quantum Chemistry," *Endeavour* 6 (1947): 42–47, on pp. 47, 45.

Coulson adopted a different strategy when addressing other quantum chemists. At a Royal Society meeting on bond energies and bond lengths held on March 9, 1950, he first presented a critical survey of VB, arguing that despite its "astonishing success . . . in correlating a vast field of chemical knowledge and experience" it has no good theoretical basis; it is really just a semiempirical method, of little use for more accurate calculations. He then reported recent results from his group and from chemist A. Pullman in France, on the calculation of bond lengths in conjugated molecules. Coulson and chemist B. H. Chirgwin had just achieved a major computational breakthrough by showing that the MO method could be formulated in such a way that the overlap factor is completely irrelevant in calculating bond lengths and charges for most conjugated hydrocarbons; thus one does not have to assume that the atomic orbitals are orthogonal. Coulson, "Bond Lengths in conjugated Molecules: The Present Position," *Proceedings of the Royal Society of London* A207 (1951): 91–100.

303 *In 1952 he published Valence* Ana Simões, "A Quantum Chemical Dialogue Mediated by Textbooks: Pauling's *The Nature of the Chemical Bond* and Coulson's *Valence*," *Notes and Records of the Royal Society of London* 62 (2008): 259–269; Ana Simões and Kostas Gavroglu, "Quantum Chemistry *Qua* Applied Mathematics: The Contributions of Charles Alfred Coulson (1910–1974)," *Historical Studies in the Physical and Biological Sciences* 29 (1999): 363–406; Gavroglu and Simões, *Neither*, Chapter 4.

303 *"by the end of 1945 resonance had become almost a 'dirty' word"* Coulson, "Recent Developments in Valence Theory," *Pure and Applied Chemistry* 24 (1970): 257–287, on p. 259; "A-Electron Chemistry," *Nature* 195 (1962): 932–933, on p. 932.

Section 10.6

303 For a general survey of the CBD problem see M. P. Cava and M. J. Mitchell, *Cyclobutadiene and Related Compounds, with a Chapter on Theory by H. E. Simmons and A. G. Anastassiou* (New York: Academic Press, 1967).

303 *"a physicist who worked between two worlds"* Andrew Streitweiser, *A Lifetime of Synergy with Theory and Experiment* (Washington, DC: American Chemical Society, 1997), p. 181.

304 *the textbook published by Erich Hückel's brother* Walter Hückel, *Theoretical Principles of Organic Chemistry* (New York: Elsevier, 1955), Volume I, Chapter IX, p. 642.

304 *It is "impertinent to the theory"* W. von E. Doering and L. H. Knox, "The Cycloheptatrienylium (Tropylium) Ion," *Journal of the American Chemical Society* 76 (1954): 3203–3206, on p. 3203.

304 *"there was just a fantastic accumulation"* Hoffmann, Letter to SGB, February 8, 1997.

304 *carbon atoms would be in a rectangular rather than a square configuration* H. A. Jahn and E. Teller, "Stability of Polyatomic Molecules in Degenerate Electronic States. I. Orbital Degeneracy," *Proceedings of the Royal Society of London* A161 (1937): 220–235. They showed that a configuration of a polyatomic molecules whose electronic state function is orbitally degenerate cannot be stable with respect

to *all* displacements of the nuclei, unless in the original configuration the nuclei all lie on a straight line, as summarized by H. E. Simons and A. G. Anastassiou, "Theoretical Aspects of the Cyclobutadiene Problem," in *Cyclobutadiene and related Compounds*, edited by M. P. Cava and M. J. Mitchell (New York: Academic Press, 1967), pp. 368–422.

305 *W. Von E. Doering and L. H. Knox announced the synthesis of* $C_7H_7^+$ Doering and Knox, "The Cycloheptatrienylium (Tropylium) Ion," *Journal of the American Chemical Society* 76 (1954): 3203–3206.

305 *"the first verification of [a prediction from] Hückel's rule" by Shigeaki Kikuchi . . . "Hückel's rule began to be regarded as effective"* Kikuchi, "A History of the Structural Theory of Benzene: The Aromatic Sextet Rule and Hückel's Rule," *Journal of Chemical Education* 74 (1997): 194–201, on p. 199.

305 *Sondheimer announced, in a paper with Reuven Wolovsky* Sondheimer and Wolovsky, "The Synthesis of Cycloöcta Decanonaene: A New Aromatic System," *Tetrahedron Letters* 1, no. 3 (1959): 3–6, on p. 3.

305 *the COT molecule is not planar like benzene* Kurt Mislow, "Aromaticity of Conjugated Monocyclic Hydrocarbons," *Journal of Chemical Physics* 20 (1952): 1489–1490.

305 *[18] annulene "cannot possibly be planar"* C. A. Coulson and A. Golebiewski, "Molecular Deformation in Cycloöctadecanoene," *Tetrahedron* 11 (1960): 125–130.

305 *a delocalized system of B electrons corresponds to an electric current* Linus Pauling, "The Diamagnetic Anisotropy of Aromatic Molecules," *Journal of Chemical Physics* 4 (1936): 673–677.

305 *"We can define an aromatic compound"* J. A. Elvidge and L. M. Jackman, "Studies of Aromaticity by Nuclear Magnetic Spectroscopy, Part 1. 2-Pyridones and Related Systems," *Journal of the Chemical Society* 8 (1961): 859–866, on p. 859. See also L. M. Jackman, F. Sondheimer, Y. Amiel, D. A. Ben-Efraim, Y. Gaoni, R. Wolovsky, and A. A. Bothner-By, "The Nuclear Magnetic Resonance Spectroscopy of a Series of Annulenes and Dehydro-annulenes," *Journal of the American Chemical Society* 84 (1962): 4307–4312.

306 *NMR is the "most convenient method"* L. M. Jackman et al., "NMR of Annulenes," p. 4307.

306 *also concluded that [14] annulene is not aromatic* Jackman, Sondheimer et al., "NMR Spectroscopy."

306 *"no theoretical justification for equating 'aromaticity' with 'stability'"* Franz Sondheimer, Reuven Wolovsky, and Yakov Amiel, Unsaturated Macrocyclic Compounds. XXIII. The Synthesis of the Fully Conjugated Macrocyclic Polyenes Cycloöctadecanonaene ([18] Annulene), Cyclotetracosadodecaene ([24] Annulene), and Cyclotriacontapentadecaene ([30] Annulene)." *Journal of the American Chemical Society* 84 (1962): 274–284, on p. 281.

306 *The third criterion proposed by Sondheimer, Wolovsky, and Amiel in 1962* Sondheimer et al., "Unsaturated Compounds XXIII," on p. 281.

306 *values of n greater than eight . . . the bond lengths will alternate* H. C. Longuet-Higgins and L. Salem, "The Alternation of Bond Lengths in large Conjugated Molecules. III. The Cyclic Polyenes $C_{18}H_{18}$, $C_{24}H_{24}$, and $C_{30}H_{30}$," *Proceedings of the Royal Society of London* A257 (1960): 445–456.

306 *Sondheimer's group synthesized [30] annulene* Franz Sondheimer, Reuven Wolovsky and Yehiel Gaoni, "Unsaturated Macrocyclic Compounds. XII. Synthesis of Two Completely Conjugated Thirty-Membered Ring Cyclic Systems," *Journal of the American Chemical Society* 82 (1960): 754–755.

306 *predicted that [26] and higher annulenes will not be aromatic* Michael Dewar and Gerald Jay Gleicher, "Ground states of Conjugated Molecules. III. Classical Polyenes," *Journal of the American Chemical Society* 87 (1965): 692–696.

306 *[10] and [14] annulenes "should be aromatic"* Franz Sondheimer, "Review Lecture. The Annulenes." (Lecture delivered March 17, 1966—MS received July 29, 1966) *Proceedings of the Royal Society of London* A297 (1967): 173–204, on pp. 175–176, 179–182, 185–199, 201–202.

307 *"Much of the interest in the field of nonbenzenoid aromatic compounds"* Ronald Breslow, "Synthesis of the s-Triphenylcyclopropenyl Cation,' *Journal of the American Chemical Society* 79 (1957): 5318.

307 *"The major impetus leading to the current outburst of research"* Emanuel Vogel, "Aromatic 10 B-Electron Systems," in *Aromaticity: An International Symposium*, edited by W. D. Ollis et al. (London: Chemical Society, Special Publications no. 21, 1967), pp. 113–147, on p. 113.

307 *had been replaced by "the physical viewpoint"* E. D. Bergmann and I. Agranat, "Review of the Theoretical and Experimental Means for the Determination of Aromaticity," in *"Aromaticity, Pseudo-Aromaticity, Anti-Aromaticity," Proceedings of an International Symposium held in Jerusalem, 31 March–3 April 1970)*, edited by Ernst D. Bergmann and Bernard Pullman (Jerusalem: The Israel Academy of Sciences and Humanities, 1971), pp. 9–24, on pp. 9–10.

307 *Jean-Francois Labarre did want to go that far* Labarre, "Appendix: Some Considerations on the Term 'Aromaticity,' " in Bergmann and Pullman, *Aromaticity*, pp. 55–56, on p. 55.

307 *"it would be better if the use of this term were discontinued"* D. Lloyd and D. R. Marshall, "Quasi-Aromatic Compounds. Examples of Regenerative, or Meneidic, Systems," in Bergmann and Pullman, *Aromaticity*, pp. 85–88, on p. 87.

308 *"we are united here in a symposium on a nonexistent subject"* Heilbronner, discussion remark in Bergmann and Pullman, *Aromaticity*, p. 21–22, on p. 21.

308 *[18] annulene, as cataloged by Harold Baumann and Jean F. M. Oth* Baumann and Oth, "The Low-Temperature UV.VIS. Absorption Spectrum of [18] Annulene," *Helvetica Chimica Acta* 65 (1982): 1885–1893.

308 *"there exists no unambiguous yardstick"* Vladimir I. Minkin, Mikhail N. Glukhovstev, and Boris Ya. Simkin, *Aromaticity and Antiaromaticity: Electronic and Structural Aspects* (New York: Wiley, 1994), on p. 6.

308 *"a vivid demonstration of the prognostic power of the Hückel rule"* Ibid., p. 186.

308 *Hückel's rule "has been acclaimed one of the most fundamental rules"* D. J. Klein and N. Trinajstic, "Hückel Rules and Electron Correlation," *Journal of the American Chemical Society* 106 (1984): 8050–8056, on p. 8050.

308 *"The idea that some compounds have unexpected stability"* Peter J. Garratt, *Aromaticity* (New York: Wiley, 1986), p. 293.

308 *Even more extravagant praise came in 1999 from Jerome A. Berson* Berson, *Chemical Creativity: Ideas from the Work of Woodward, Hückel, Meerwein and Others* (Weinheim: Wiley-VCH, 1999), p. 70.

308 *offers a larger context for the paradox* Roald Hoffmann, "Theory in Chemistry," *Chemical & Engineering News* 52 (July 29, 1974): 32–34, on p. 32. I thank Ko Hojo for this reference; see his article "Theory, Synthesis, and Aromaticity—Planar or not Planar: [10] Annulene Problem," In *Chemistry in the Philosophical Melting Pot*, edited by D. Sobczy½ska (Frankfurt am Main: Peter Lang, 2004): pp. 241–260, on pp. 245, 255.

SECTION 10.7

309 *the power of theories and models lies in their predictive character* Sidgwick, *Some Physical Properties of the Covalent Link in Chemistry* (Ithaca, NY: Cornell University Press, 1933), pp. 12–14.

309 *unable successfully "to predict new phenomena"* C. N. Hinshelwood, "General and Physical Chemistry, I: General," *Annual Reports on Progress in Chemistry* 29 (1933): 13–21, on p. 20.

309 *Dewar did not show the superiority of "the rather nebulous notation"* "Review of Dewar's *Electronic Theory of Organic Chemistry*," *Journal of the American Chemical Society* 71 (1949): 3859–3860.

309 *"part of the 'success' which the author claims"* "Review of Dewar's *Electronic Theory of Organic Chemistry*," *Journal of Chemical Education* 26 (1949): 505–506.

309 *"a person who knows how to solve a problem, but who cannot do it"* "Present Situation of Quantum Chemistry," *Journal of Physical Chemistry* 61 (1957): 55–68, on p. 68.

309 *"be judged by its predictive as well as its explanatory power"* "The ground State of some A-electron Systems," in *Theoretical Organic Chemistry (Kekulé Symposium, London, September 1958)* (London: Butterworths, 1959), pp. 9–19, on pp. 11–12.

310 *the "golden age of valency theory"* "Robert Sanderson Mulliken, 7 June 1896–31 October 1986," *Biographical Memoirs of Fellows of the Royal Society* 35 (1990): 329–354., on p. 341.

310 *he faults his own field, theoretical chemistry* "Strong Inference," *Science* 146 (1964): 347–353, on pp. 350–351.

310 *organic chemistry has been the spiritual home of strong inference* Platt, "Strong Inference," p. 351.

310 *MO, he said, "is now used far more widely"* Platt, "1960 Nobel Laureate in Chemistry: Robert S. Mulliken," *Science* 154 (1966): 745–747, on p. 745.

310 *In a 1974 article I criticized the accuracy of claims by Platt and others* S. G. Brush, "Should the History of Science Be Rated X?" *Science* 183 (1974): 1164–1172.

311 *emphasized that MO theory could predict as well as explain* "The Influence of Wave Mechanics on Organic Chemistry," in W. C. Price et al., editors, *Wave Mechanics* (New York: Wiley, 1973), pp. 255–271.

SECTION 10.8

311 *Others also credited MO with many successful predictions* W. J. Noble, *Highlights of Organic Chemistry* (New York: Dekker, 1974), p. 260; M.J.S. Dewar, *A Semi-Empirical Life* (Washington, DC: American Chemical Society, 1992), pp. 56, 75, 134; William A. Goddard III, "Theoretical Chemistry Comes Alive: Full Partner with Experiment," *Science* 227 (1985): 917–923; J. A. Pople, "*A Priori* Geometry Predictions," in *Applications of Electronic Structure*, edited by H. F. Schaefer (New York: Plenum, 1977), pp. 1–27; G, Richards, "Third Age of Quantum Chemistry," *Nature* 278 (1979): 507.

312 *1968 volume dedicated to Linus Pauling by his students, colleagues, and friends* A. Rich and N. Davidson, Editors, *Structural Chemistry and Molecular Biology* (San Francisco: Freeman, 1968).

312 *Pauling himself refused to use MO in later editions* See Brush, *Dynamics,* Part 2, pp. 282–284.

312 *Pauling made a small concession* Gavroglu and Simões, *Neither*, p. 125.

Section 10.9

312 *because an MO picture is "cumbersome to draw"* C. D. Gutsche and D. Pasto, *Fundamentals of Organic Chemistry* (Englewood Cliffs, NJ: Prentice-Hall, 1975). For names of other authors see Brush, "Dynamics," Part 2, notes 97–103.

312 *the Woodward-Hoffmann rules did not imply the validity of the theory* M.J.S. Dewar, "Aromaticity and Pericyclic Reactions," *Angewandte Chemie, International Edition* 10 (1971): 761–776. On attempts to derive the Woodward-Hoffmann rules from VB see W. J. LeNoble, *Highlights of Organic Chemistry: An Advanced Textbook* (New York: Dekker, 1974), Chapter 14.

312 *"A good theory explains all the known facts"* Allinger et al., *Organic Chemistry* (New York: Worth, 2nd ed. 1976), on p. 13.

312 *"one of the really giant steps forward in chemical theory"* Morrison and Boyd, *Organic Chemistry* (Boston: Allyn & Bacon, 4th ed., 1983), pp. 1203, 1201.

313 *delocalization is* not *generally the driving force* S. S. Shaik et al., "Is Delocalization a Driving Force in Chemistry? Benzene, Allyl Radical, Cyclobutadiene, and their isoelectronic Species," *Journal of the American Chemical Society* 109 (1987): 363–374; see also Shaik et al., "A Different Story of Benzene," *Journal of Molecular Structure (Theochem)* 398–399 (1997): 155–167.

Section 10.11

316 *Coulson delivered an inaugural lecture* C. A. Coulson, *Theoretical Chemistry Past and Future. An Inaugural Lecture Delivered before the University of Oxford on 13 February 1973* (Oxford: Clarendon Press, 1974), p. 21.

PART THREE

Space and Time

11

Relativity

> Nature and Nature's Laws lay hid in night
>
> God said, *Let Newton Be!* And all was light.
>
> —Alexander Pope

> It did not last: the Devil howling "Ho!
>
> Let Einstein be!" restored the status quo.
>
> —Sir John Collings Squire (1926)

As noted in Chapter 2, Einstein's relativity is the only theory in the physical sciences for which a large number of reception studies have been done (thanks in part to the efforts of historian Thomas F. Glick), and it is therefore a good test case for various opinions about why scientists accept theories.

For simplicity I will divide the reasons for accepting a theory into three general categories: first, empirical tests; second, social or psychological factors; and third, aesthetic or mathematical evaluations of the theory. After some brief comments on each of these categories, I review the conclusions of historians, including a summary of my own research comparing the role of gravitational light bending and the advance of Mercury's perihelion on the acceptance of general relativity in the 1920s.

Many historians treat relativity as a single theory to be accepted or rejected, but it is possible to distinguish between the reception of the special and the general theory.

11.1 THE SPECIAL THEORY OF RELATIVITY

Until recently, textbooks and histories of modern physics introduced Einstein's theory of relativity as if it were an attempt to account for the negative result of the Michelson-Morley experiment done by physicist A. A. Michelson and chemist E. W. Morley in the 1880s. They failed to observe any motion of the earth relative to the ether, the hypothetical substance believed to fill all space and invoked to explain the propagation of light. The well-established motion of the Earth in its orbit around the sun and rotation around its own axis (Section 2.2) should have been detectable by their apparatus.

Earlier attempts to explain this puzzling result by physicists G. F. FitzGerald and H. A. Lorentz (separately) had led to the so-called FitzGerald-Lorentz contraction formula. FitzGerald and Lorentz suggested, around 1890, that the apparatus used in the experiment might actually decrease in length in the direction of motion in such a way as to cancel out the expected effect. The formula for this decrease is such that the change in length of the apparatus is extremely small unless the speed of the earth through space is nearly as great as the speed of light, but the length shrinks to zero as the speed approaches the speed of light.

The formula is $L_v = L_O\sqrt{[1 - (v^2/c^2)]}$ where L_v = length at speed *v*, L_O = length at rest (v = O), *c* = speed of light. Thus if an object travels at half the speed of light, its length would shrink to $\sqrt{[1 - (1/4)]}$ = ½(.866) = .43 of its original length.

Einstein's theory led to precisely the same formula, but with a different physical interpretation: the apparatus does not really shrink in an absolute sense, but rather it is inherent in the way we make measurements of the size of fast-moving objects that it must *appear* to shrink.

The notion that relativity theory was invented to account for an anomalous experimental result was convenient for physics teachers and philosophers of science, but did not stand up under detailed analysis of the available evidence about the origins of Einstein's ideas. Physicist and historian Gerald Holton showed in 1969 that the problem of the Michelson-Morley experiment played little if any part in Einstein's early development of his theory.

Einstein's first paper on relativity, published in 1905, appears to have emerged from a prolonged period of reflection on some aspects of Maxwell's electromagnetic theory. At the age of 16, as he reported later in his article "Autobiographical Notes," he had discovered that there is something peculiar about traveling at the speed of light:

> If I pursue a beam of light with the velocity c (velocity of light in a vacuum), I should observe such a beam of light as a spatially oscillatory electromagnetic field at rest. However, there seems to be no such thing, whether on the basis of experience or according to Maxwell's equations. From the very beginning it appeared to me intuitively clear that, judged from the standpoint of such an observer, everything would have to happen according to the same laws as for an observer who, relative to the earth, was at rest. For how, otherwise, should the first observer know, i.e., be able to determine, that he is in a state of uniform motion?

Thus the possibility of reducing the relative speed of light to zero for some observer seemed to be an absurdity, yet there was nothing in mechanics as then understood to exclude such a possibility.

In the opening paragraph of his 1905 paper, Einstein pointed out an inconsistency in the conventional interpretation of electromagnetic theory. It was customary in textbooks at that time to explain electromagnetic induction in two different ways depending on whether the magnet or the conductor was moving. Yet the effect—the current produced—really depends only on the *relative* motion of the magnet and the conductor.

According to Einstein, the textbooks introduced an artificial asymmetry in describing the phenomena by assuming the existence of an absolute space. Influenced by his reading of the critiques of Newtonian science by the eighteenth-century philosopher David Hume and the nineteenth-century physicist Ernst Mach, he decided to reject this assumption, and to build physics on a new assumption:

> In all coordinate systems in which the mechanical equations are valid, also the same electrodynamic and optical laws are valid.

Those coordinate systems are the inertial frames of reference: systems that are moving at constant velocity with respect to the fixed stars and within which Newton's laws of motion apply.

A terrestrial laboratory is not quite an inertial frame, since the Earth is rotating around its axis. But for most purposes we can ignore or easily correct for such discrepancies. Once we have identified a single inertial frame, we can immediately identify an infinite number of others, namely all those frames moving at constant velocity with respect to the first one.

Einstein continues:

> We shall raise this conjecture (whose content will be called "the principle of relativity" hereafter) to the status of a postulate, and shall introduce, in addition, the postulate, only seemingly incompatible with the former one, that in empty space light is always propagated with a definite velocity V which is independent of the state of motion of the emitting body. These two postulates suffice for arriving at a simple and consistent electrodynamics of moving bodies on the basis of Maxwell's theory for bodies at rest. The introduction of a "light ether" will prove superfluous, in accordance with the concept to be developed here, no "space at absolute rest" endowed with special properties will be introduced, nor will a velocity-vector be assigned to a point of empty space at which electromagnetic processes are taking place.

The first postulate means that the same law of nature (for example, Maxwell's equations of electromagnetism) holds in an inertial frame of reference that is at rest with respect to a fixed star as in one that moves at, say, nearly the velocity of light.

The second postulate requires that if a spaceship traveling away from earth at half the speed of light sends back a light ray signal to earth, we would find that the speed of this signal is precisely c, the speed of light we measure for stationary light sources; the astronauts in the spaceship would also find that the speed of that signal was c. This situation is obviously impossible in Newtonian physics, where the speed of the signal could be c relative to earth, or relative to the spaceship, but not relative to both.

From these two postulates follow a few of the startling consequences of Einstein's *special theory of* relativity:

1. *Relativistic contraction of moving objects*: If an inertial frame of reference, denoted by A, is moving at constant velocity v with respect to another inertial frame of reference, B, then an observer in either frame would find from his own measurements that the lengths of objects moving in the other frame appear to be contracted from what they would be if there were no relative motion. The amount of the contraction is given by the

same formula as the FitzGerald-Lorentz contraction, but the effect is now reciprocal: for each observer, lengths of objects in the other frame are contracted.

2. *Relativistic time dilation*: For an observer in either frame, time intervals between events in the other frame will be dilated (increased) by a factor similar to that occurring in the Lorentz-FitzGerald contraction; that is, the other observer's clocks will appear to be running more slowly. (The two events might be successive swings of a pendulum or successive ticks of an atomic clock.) The formula for the time dilation is $T_v = T_O/\sqrt{[1 - v^2/c^2]}$ where T_v = time interval between two events in frame moving at speed v, T_O = time interval between events in other frame.

A well-known predicted consequence of the time-dilation effect is the *twin paradox* (sometimes called the *clock paradox*). If one twin goes off in a spaceship traveling at a speed close to the speed of light, then turns around and returns to earth, he should find that he is younger than the other twin who stayed behind. Seen from the frame of reference of the Earth, biological processes of aging, as well as the operation of clocks (including the interval between heart beats) will run more slowly in the spaceship than on Earth.

The predicted outcome is difficult to believe intuitively—that's why it is called a paradox. After all, why is the situation not symmetrical in this case? Why couldn't the spaceship be taken as the frame of reference, so that the earthbound twin would be younger from the viewpoint of the traveler? One has the feeling that when the traveler returns, all time dilation effects on both sides should have cancelled out, since each cannot objectively be younger than the other when they meet again. But that argument assumes incorrectly that the spaceship is an inertial frame of reference all the time; in fact, unlike the Earth, it has to undergo large accelerations in order to start, turn around and stop; hence the time-dilation effects are *not* the same for both twins. The traveling twin will in fact have aged less than the stay-at-home twin—though the actual differences for today's travel speeds are quite small.

3. There is no *absolute space*, thus no justification for saying that one frame is really at rest or really moving. Only *relative* motions have physical reality. (This statement will have to be somewhat qualified when we come to the general theory of relativity.)
4. There is no *absolute time*; that is, two events that may be simultaneous for an observer in one frame of reference will not in general be simultaneous for an observer in another frame. It is even possible that event #1 may be found to occur *before* event #2 in frame A but *after* event #2 in frame B. So one cannot say that two events are simultaneous (unless they happen at the same *place* as well as the same time).
5. *Relativistic mass increase*: The observable mass of any object increases as it goes faster (relative to the observer). Like the dilation of time intervals, this effect is negligible at low speeds but becomes infinite as the speed approaches the speed of light. The mass at speed v is equal to

$$m_v = m_O / \sqrt{[1 - v^2 / c^2]}, \text{ where } m_O \text{ is the } \textit{rest mass}.$$

Since the amount of force needed to accelerate an object is proportional to its mass (Newton's second law), this means that it would take an infinite force (or infinite amount of energy) to make such an object travel at the speed of light. Thus it appears that the speed of light is a universal speed limit. (If $v > c$, the mass would be proportional to the square root of a negative number, which mathematicians call an imaginary number. As I mentioned in Section 8.3, although imaginary numbers are useful in quantum mechanics, Einstein does not tell us how to detect an imaginary mass.) If indeed it is impossible for anything to go faster than light, we don't have much hope of visiting any more than a handful of nearby stars in our galaxy, those that are only a few light years away. If it were possible to travel at a speed very close to that of light we could visit more distant stars, but by the time we got back home all our friends and family would long since be dead (because of the twin paradox).

6. Closely connected with the mass-increase effect is Einstein's famous formula $E = mc^2$: mass and energy are no longer separately conserved but can be interconverted. In particular, nuclear reactions in which the total mass of the products is less than the total mass of the reagents will release energy, in the form of electromagnetic radiation, neutrinos, and excess kinetic energy of the product nuclei. Examples are the *fission* (splitting) of heavy nuclei such as uranium or plutonium, as in the atomic bombs dropped on Hiroshima and Nagasaki by the United States to end World War II; nuclear reactors, also based on fission, that are now providing an increasing share of our energy supply; and *fusion* (combining together) of light nuclei such as hydrogen, as in the hydrogen bomb. Fusion provides the energy to keep the sun shining and thus is essential to the development and continued existence of life on earth. Einstein's theory of relativity provides part of the explanation for all these things, though it does not by itself tell us why certain reactions will actually provide energy under particular circumstances.
7. Because of #1, 2, 3 and 4, we can no longer treat space and time as independent entities. In a sense they are fused together into a four-dimensional *space-time manifold*. One speaks of a four-dimensional space-time interval between two events; for one observer the two events may be simultaneous, separated only by a spatial distance, while for another observer the events may be separated by a time interval as well; yet it is possible to define a four-dimensional space-time interval that has the same value for all observers (all frames of reference).

In two dimensions the distance between points with coordinates (x_1,y_1) and (x_2,y_2) is

$$d = (x_2 - x_1)^2 + (y_2 - y_1)^2$$

by the Pythagorean theorem. If you rotate the coordinate system the x's and y's will all be different but d remains the same. The four-dimensional interval is defined as

$$d = (x_2 - x_1)^2 + (y_2 - y_1)^2 + (z_2 - z_1)^2 - c^2(t_2 - t_1)^2.$$

Every one of these seven results of Einstein's special theory of relativity is contrary to common sense. Every one has survived all possible experimental tests. The theory has therefore been built into the core of all fundamental physical theories in the twentieth century. This does not mean that Newtonian physics is wrong, but rather that its region of validity is severely limited (e.g., it does not apply to objects moving at high speeds). On the other hand, Maxwell's electromagnetic theory turns out to be completely correct, though the ether model that originally inspired it, like all ether models, has had to be abandoned.

11.2 GENERAL THEORY OF RELATIVITY

The special theory of relativity has one significant limitation: it applies only to physical phenomena in which gravitational forces are not involved, and in which only inertial frames moving at constant speeds are considered. By 1915 Einstein had worked out the basic form of a more comprehensive theory, known as the *general theory of* relativity, which did include gravity and accelerated frames of reference. Although some aspects of this theory are still subject to experimental test and may have to be modified, the basic principles and crucial predictions can be summarized here.

In explaining his general theory, Einstein pointed out first a fact that was in a certain sense well known, yet never properly understood before: the *inertial mass* of every object is the same as its *gravitational mass*. It is just because of this fact that, as Galileo discovered, all objects are accelerated the same amount by the Earth's gravitational field at a given location.

Within the framework of Newtonian physics, it is at least conceivable that some objects might possess a kind of mass that was not acted on by gravitational forces yet would contribute to the inertia (resistance to acceleration) of the object; such an object would fall more slowly (be accelerated less) than others. Yet this does not happen. Somehow every bit of mass that contributes to a body's *weight* (interacts gravitationally with the Earth) also contributes the same amount to its *inertia*.

Einstein argues that this equality must mean that "the same quality of a body manifests itself according to circumstances as 'inertia' or as 'weight.'" The significance of this assertion may be illustrated by considering a large box (of negligible mass) in a part of empty space far away from a planet or other gravitating objects. Inside the box is an observer, who will be floating freely in space since there is no gravitational field present. Now suppose someone attaches a hook and rope to the outside of one end of the box and begins to pull it with a constant force. The box will undergo a uniform acceleration, which is transmitted to the observer inside the box by the floor when it scoops him up or hits him:

> He must therefore take up this pressure by means of his legs if he does not wish to be laid out full length on the floor. He is then standing in the box in exactly the same way as anyone stands in a room of a house on our earth. If he releases a body which he previously had in his hand, the acceleration of the box will no longer be transmitted to this body, and for this reason the body will approach the floor of the box with an accelerated relative motion. The observer will further convince himself that *the acceleration of the body toward the floor of the box is always of the same magnitude, whatever kind of body he may happen to use for the experiment.*

> . . . the man in the box will thus come to the conclusion that he and the box are in a gravitational field which is constant with regard to time. Of course he will be puzzled for a moment as to why the box does not fall in this gravitational field. Just then, however, he discovers the hook in the middle of the top of the box and the rope which is attached to it, and he consequently comes to the conclusion that the box is suspended at rest in the gravitational field.
>
> Ought we to smile at the man and say that he errs in his conclusion? I do not believe we ought to if we wish to remain consistent; we must rather admit that his mode of grasping the situation violates neither reason nor known mechanical laws.

Thus, even though to an *outside* observer the box may appear to be in accelerated motion in gravitation-free space, the observer *inside* the box may legitimately regard it as being at rest, but subject to a gravitational field. This discussion illustrates that physics can be formulated in such a way that frames of reference in uniformly accelerated motion relative to each other may be considered equivalent. This is done, in Einstein's *Principle of Equivalence*, by deciding that gravity is not a real force but rather a local property of space and time as described in a particular coordinate system.

Einstein thus accomplished, for gravitational forces, what some nineteenth-century physicists (Ernst Mach, Heinrich Hertz) had wanted to do for all forces: to eliminate force as a primitive concept and replace it by a description of its effects on moving objects. Yet Einstein did this by attributing much more reality to space and time than his predecessors would have liked.

In this same paper (1907) in which the equivalence principle was first proposed, Einstein showed that, by analogy with the Doppler effect, a spectral line produced by an atom in a strong gravitational field (when observed at a place with lower gravitational potential energy) has a greater wavelength. This is the *gravitational redshift*. Einstein predicted that light coming from the surface of the sun would have a wavelength longer by two parts in a million than light from the same atom on earth. But it was not until much later that the existence of this redshift (as distinct from various other kinds of redshifts) was established by observation. (The Pound-Rebka experiment, published in 1960, is now regarded as the best direct test.)

In 1911 Einstein used his principle of equivalence to predict that light passing near the sun would change its direction. The amount of deflection was the same as one might predict from the Newtonian particle theory of light; in fact, geodesist and astronomer Joseph Georg von Soldner had attempted a similar calculation in 1801. Einstein urged astronomers to try to detect this bending of light near the sun, but—perhaps fortunately for the popular reputation of his theory—this was not done until after Einstein had developed his general theory of relativity, from which a deflection twice as large was estimated. The first organized attempt to test the prediction was made in 1914, but World War I broke out just as the expedition left Germany for Russia, and its members were taken captive and prevented from making their observations (see Section 15.2).

During the years 1912–1914 Einstein worked with mathematician Marcel Grossman, who helped him translate his physical ideas into mathematical form. This effort involved a major shift in Einstein's attitude toward physical theory, for previously he had preferred to use the bare minimum of mathematical apparatus in working out his ideas. For example, he had initially shown little enthusiasm when

Hermann Minkowski, in 1908, proclaimed that relativity theory had fused space and time into a single four-dimensional continuum. But now Einstein adopted this approach, which had been originated by Riemann in 1854.

In 1915 Einstein completed the development of his basic general theory by giving a set of field equations. As it happened, the same equations were derived independently at the same time by the German mathematician David Hilbert. Although Hilbert started to work on the problem only after Einstein had already formulated it and never claimed any kind of priority as a discoverer of general relativity theory, his derivation of the final equations is in some respects better than Einstein's. So, since Einstein already has a wealth of discoveries to his credit, I think it is fair to call the field equations of general relativity the Einstein-Hilbert equations.

Testing the Predictions of General Relativity

Einstein showed that his theory now predicted that the gravitational field of the sum would bend a ray of starlight by about 1.7 seconds of arc, twice the value computed earlier from Newtonian gravitational theory. The bending would cause the apparent positions of stars to shift away from the sun. Of course this effect could be observed only during a total eclipse when the sun's own light was not preventing us from seeing those stars; moreover, the effect is so small that it would not be noticed unless one were looking very carefully for it. It was the successful test of this prediction by the British eclipse expedition of 1919 that provided the first convincing evidence for Einstein's theory. (The two observational results of 1919 were 1.98 ± 0.12 and 1.61 ± 0.30 seconds, compared to the theoretical value 1.751 seconds.)

The advance of the perihelion of Mercury is another well-known test of relativity theory; it is one that might be called an anomaly of the Newtonian theory, since it was known in the nineteenth century that there was a discrepancy between the observations and the perihelion (closest point of orbit to sun) motion calculated from perturbations by known planets. The discrepancy was first announced in 1859 by U.J.J. LeVerrier, known as the discoverer (by theoretical calculation) of the planet Neptune. By the time Einstein turned to the problem, the most refined astronomical calculations indicated that the discrepancy was 42.56 seconds per century.

LeVerrier had suggested that the difference might be due to the actions of several small planets between Mercury and the sun, but extensive searches failed to discover them (though there were a few false alarms). There were various attempts to explain it by modifying Newton's law of gravity—for example, to make the force between two masses depend on their relative speed (as one might expect if gravity is propagated with finite velocity)—but none of these were satisfactory.

In 1915, Einstein calculated from his general theory of relativity that the perihelion of the orbit of Mercury should advance by 43 seconds per century, extremely close to the observational value. His explanation depends on the idea that energy is associated with mass, and that even energy in empty space—for example, the energy represented by the sun's own gravitational field—may be affected by and in turn act on other objects that exert gravitational forces. As a planet gets close to the sun, it experiences a gravitational force greater than would be expected from Newton's law of universal gravitation in its original form; one way of expressing this is to say that the effective mass of the combination, sun plus its gravitational field, increases. This effect is significant only for planets that come in very close to the sun, such as

Mercury; they must also have eccentric orbits, so that there is a considerable variation in their distance from the sun and thus a variation in the effective mass in different parts of the orbit.

The advance of the perihelion of Mercury was not a novel prediction, contrary to what some philosophers have claimed (Section 3.2). Not only was Einstein aware of the effect before he published his theory, he used it (along with other arguments) in deciding which of two possible equations to select.

11.3 EMPIRICAL PREDICTIONS AND EXPLANATIONS

Table 11.1 lists a few of the many empirical consequences of relativity, including the ones most often mentioned before 1940, arranged in order of the date first discovered

Table 11.1. EMPIRICAL CONSEQUENCES OF RELATIVITY THEORY

E1. Advance of perihelion of planet Mercury (Le Verrier, 1859)
E2. Failure to detect the Earth's motion through space (Michelson and Morley, 1887)
E3. Mass increase with velocity, by factor $1/\sqrt{(1 - v^2/c^2)}$ (Bucherer, 1908; Neumann, 1914; Guye and Lavanchy, 1916)
E4. Effect on electron orbits and spectrum in Bohr model of atom as calculated by Sommerfeld (Paschen, 1916)*
E5. Gravitational bending of light (Eddington et al., 1919)
E6. Existence of positron predicted by Dirac from his relativistic wave equation (Anderson, 1932; Blackett and Occhialini, 1933) (see Chapter 9)
E7. Deviation of atomic weights from Prout's hypothesis, owing to binding energy of nuclear forces (Aston, 1919–1927; Bainbridge, 1933); partial mass-energy transformation in nuclear reactions (Cockcroft and Walton, 1932)
E8. Complete mass-energy transformation in electron-positron creation and annihilation (Blackett and Occhialini, 1933; Klemperer 1934)
E9. Time dilation (Kennedy and Thorndike, 1932; Ives and Stilwell, 1937)
E10. Gravitational redshift of spectral lines (Pound and Rebka, 1960)
E11. Geodetic drift and frame-dragging of satellite in orbit around Earth (Everitt et al., 2011)
E12. Gravitational radiation ("gravity waves") (indirect evidence found by Taylor and Hulse, 1975)

*Despite its confirmation by Paschen, this effect was later shown to be a consequence of electron spin in quantum mechanics (see Section 8.4). Since spin itself can be derived from the relativistic Dirac equation (Section 9.1), one may argue that E4 is still *indirectly* a consequence of relativity.

As we will see in Section 11.6, the special theory of relativity was accepted by leading physicists around 1911. There is no such definite time for the general theory; as a rough estimate, I suggest that it was accepted by most astronomers, mathematicians, and physicists who were qualified to judge the theory around 1925 ± 1. But then according to historian Jean Eisenstadt, most scientists tended to ignore it until the 1960s, when Pound and Rebka performed a successful laboratory test of the gravitational redshift (E10), and John Wheeler popularized cosmological applications and the black hole concept.

or established. Note that the names and dates refer to experiments that were considered confirmations at the time; in some cases (e.g., mass increase with velocity) later work has thrown doubt on their accuracy.

As explained in Chapter 3, I am especially interested in uncovering the role of confirmed *novel* (in advance) predictions as compared with deductions or explanations of known facts, in persuading scientists to accept a new theory. Relativity provides a good opportunity for making this comparison. Thus for the special theory of relativity (STR), proposed in 1905, was the Michelson-Morley experiment (E2) more or less important, other things being equal, than the evidence found after 1905 (E3, E4, E6, E7, E8, E9)? Similarly for the general theory (GTR): was the advance of Mercury's perihelion (E1), discovered in the nineteenth century, more or less important than E5, E10, and E11?

Those questions are a little misleading, because some of the evidence was not found until after the theory was accepted, and therefore could not have influenced the acceptance of the theory. To take a couple of well-known examples, the discovery of stellar parallax might have been a decisive proof of the heliocentric theory if it had been discovered in 1600, and the determination that light moves faster in air than in water might have been a decisive proof of the wave theory if it had been accomplished in 1800, but as it happened physicists had to assess those theories without these empirical observations. So we first need to find out *when* the theory was accepted.

11.4 SOCIAL-PSYCHOLOGICAL FACTORS

During the 1970s and 1980s, several sociologists argued that science should be studied by an approach that came to be known as social construction (Chapter 4). Although they differed among themselves on many points, they generally accepted a principle of *symmetry*: the same causes should be invoked to explain why scientists accepted a theory now considered correct, as those invoked to explain why they accepted a theory now considered wrong. In other words, one should not simply say that a correct theory was adopted *because* it is correct, and use social explanations only to account for the behavior of those scientists who failed to adopt it.

Unfortunately some of these sociologists described their approach in ways that seemed to imply scientific research does not discover, and is not even significantly constrained by, the natural world. To say that scientific knowledge is socially constructed was perceived by some scientists as debunking science.

Sociologist Stephen Cole argued that while social factors could affect the direction and rate of progress of scientific research, there was no evidence that such factors influenced the substantive content of established scientific knowledge. In response to Cole's challenge to social constructionists to produce even one example in which social causation had been demonstrated by the accepted methods of sociological research, sociologist David Bloor (one of the founders of social constructionism) pointed to historian Andrew Warwick's study of the reception of relativity at Cambridge University. Warwick found that mathematicians and physicists understood Einstein's theory in radically different ways, depending on their own disciplinary commitments.

As an example of social construction this is not very convincing, since Warwick's study is limited to the early period 1905–1911, but, according to Staley and other historians, special relativity had not yet been accepted by most physicists before the end of that period. This is certainly true of British physicists, who did not abandon their strong belief in the ether until much later, as shown by historian Stanley Goldberg. Warwick himself points out that before 1911, "there existed no well-defined special theory of relativity" that could be accepted or rejected.

Historian Loren Graham pointed out that social constructionists have up to now failed to make one of the most obvious tests of their thesis: to compare the development of science in radically different cultures such as the Soviet Union and Great Britain. He states that the relativity research of physicist V. A. Fock is a legitimate example of social construction, and shows the good effects of what Western scientists would generally consider a bad social influence, dialectical materialism.

Historian Lewis Pyenson argued that "physicists in Central Europe were particularly disposed to seeking revolutionary solutions for the scientific problems that they faced," and noted that words meaning "revolution[ary]" were often used by the early defenders of relativity. This would imply that physicists judged the new theory not just on its own merits, but also on the basis of their own like or dislike of radical change in general.

Why would a particular physicist tend to accept or reject an idea because it is revolutionary? This is a psychological question, and might find an answer in historian-psychologist Frank Sulloway's study of openness to scientific innovation. Based on analysis of 308 scientists whose positions on relativity before 1930 are known, Sulloway concluded that age is a strong predictor of tendency to accept Einstein's theories, while social attitudes and birth order are moderately good predictors: young, more liberal scientists who were the second or later child in their family were statistically more likely to support relativity.

11.5 AESTHETIC-MATHEMATICAL FACTORS

Although some historians and sociologists seem to believe that empirical (or positivist) and social factors are the only alternatives for explaining how scientists choose theories, other students of (and participants in) the scientific enterprise have stressed the importance of a third factor: the belief that a theory must be correct because it is mathematically elegant, aesthetically pleasing, and expresses a necessary truth about nature. Albert Einstein and Paul Dirac endorsed this view, in the passages quoted in Section 1.8.

According to historian Jeroen van Dongen, Einstein's preference for mathematically elegant theories was the result of his experience in developing general relativity. Previously he had given greater weight to empirical support from physics; thus in 1912, commenting on Max Abraham's theory of gravity, he asserted that Abraham had been led astray by mathematical esthetics and ignored physics. So when Einstein himself took up the new tensor methods (at the suggestion of mathematician Marcel Grossman) he worked "with a cautious attitude toward considerations of mathematical beauty and simplicity," according to Dongen. But his focus on a physical basis for his equations led him to the so-called Entwurf (sketch or outline) theory, which not only failed to have the desired invariance properties but also led to an erroneous value for the advance of the perihelion of Mercury. It was only by putting mathematical criteria first that he could derive the correct field equations and the correct value for Mercury's orbit-change.

Pyenson also argues that physicists and mathematicians in Germany accepted relativity because it satisfied their desire to believe in a "preestablished harmony between mathematics and physics." Philosopher James W. McAllister has reviewed the role of aesthetic factors in scientists' evaluation of theories; he argues that the frequent invocation of such factors as a reason for accepting relativity shows that this theory is *not* revolutionary.

Thomas Glick, in his study of the reception of relativity in Spain, found that many mathematicians were among the early advocates of Einstein's theory. Mathematics was the strongest discipline among the exact or physical sciences; "identification with relativity enhanced its prestige." Mathematicians were needed to explain the absolute differential calculus to physicists and to resolve conceptual problems in the general theory of relativity. Moreover, they had close connections with Italian mathematicians, who (as historian Judith Goodstein shows) were enthusiastic about relativity.

Of course the three factors—empirical, social, and aesthetic—are not independent. For example, a theory may initially be accepted by prestigious leaders of a scientific community for empirical or aesthetic reasons; other scientists may then be influenced to accept it not so much because of their own expert evaluation, but because they accept the judgment of the leaders and decide that it is in their own interests to jump on the bandwagon. In this case the social factor would amplify the empirical or aesthetic reason—in other words, it would be a factor in the mathematical sense, multiplying rather than just adding to the other factors. Conversely, the example of Spain suggests that mathematicians may be initially attracted to a theory for social reasons but find it worthy of intense study because of its mathematical-aesthetic virtues. I take up this subject again in Section 11.9.

11.6 EARLY RECEPTION OF RELATIVITY

The most important advocate of the theory was Max Planck. A later-born, Planck assisted at the birth of the relativity revolution. He presented the theory at the physics colloquium in Berlin during the winter semester 1905–1906 and published a paper on it in 1906 (the first publication on relativity other than Einstein's). While Planck did not yet believe that the truth of the theory had been demonstrated experimentally, he considered it such a promising approach that it should be further developed and carefully tested. Planck challenged the conclusions of physicist Walter Kaufmann, who claimed that the results of his experiments on the mass increase of accelerated electrons supported Abraham's theory rather than the Lorentz-Einstein formula, and showed that Kaufmann's data did not rule out relativity. (Eventually, experiments by physicist A. H. Bucherer and others agreed with the Lorentz-Einstein formula.) As a professor at Berlin (at that time one of the major centers of physics) Planck encouraged his students and colleagues to work on relativity theory. As editor of the prestigious journal *Annalen der Physik*, Planck saw to it that any paper on relativity meeting the normal standards would get published.

According to historian Stanley Goldberg, Planck was attracted to relativity theory because of "his philosophical and ethical convictions about the ultimate laws of reality." He liked the "absolute character with which physical law was endowed" by relativity theory, such as that the laws of nature are the same for all observers: "For Planck this represented the supreme objectivity toward which science was striving."

Other German physicists became interested in relativity because it promised a new approach to theoretical puzzles such as the rigidity of the electron. In this way Arnold Sommerfeld, Wilhelm Wien, Max Born, Paul Ehrenfest, and others, some of whom were initially skeptical for various reasons, made relativity theory useful and even fashionable among theoretical physicists. Historians Arthur I. Miller and Martin J. Klein concluded that Ehrenfest's analysis of the distortion of moving bodies was an important factor in special relativity's acceptance as a theory different from Lorentz's. Physicist Max Laue, who learned about the theory from Planck, was quickly converted to it and eventually published the first definitive monograph on relativity in 1911; he wrote that relativity "has found an ever growing amount of attention" despite its inadequate empirical foundation and puzzling assertions about space and time.

An interesting exception is Bucherer , who was apparently converted to relativity because of his own experimental results despite an earlier dislike of the theory.

Pyenson, as mentioned earlier, attributed the acceptance of relativity by German mathematicians and some theoretical physicists to its conformity with their convictions about the harmony of the world; they liked the theory itself even if they did not fully appreciate its physical meaning or experimental predictions. Pyenson quotes the physicists Wilhelm Wien and J. J. Laub, and the mathematicians Hermann Minkowski, Felix Klein, and Adolf Kneser, to show their liking for the aesthetic qualities of relativity.

Historian Richard Staley concluded that German physicists accepted special relativity around 1911, and that the most important empirical factor was its explanation of the failure to detect the motion of the Earth through the ether. Experiments to test the predicted increase of mass at high speeds were not accurate enough to confirm Einstein's theory until after 1911.

Aesthetic and mathematical evaluations of the theory, especially as formulated by Minkowski , in a famous 1908 lecture, played an important role in its acceptance. Historian Suman Seth writes that Arnold Sommerfeld was converted by Minkowski's lectures. Historian Scott Walter reports that Minkowski inspired mathematicians to study relativity; "between 1909 and 1915, sixty-five mathematicians wrote 151 articles on nongravitational relativity theory, one out of every four articles published in this domain."

Historian José Sánchez-Ron suggests that British physicists were interested in relativity because it offered a way to deal with problems arising from the new atomic physics, in which the magnetic interactions of electrons and their behavior at very high speeds need to be better understood. Sánchez-Ron also points to the interest of these physicists in symmetry considerations. But the most explicit statement he quotes about the reason for accepting relativity is that of an anonymous reviewer of Ludwik Silberstein's 1914 book: "without the result of Michelson and Morley's experiment there would be no justification for the theory at all. . . . [It] will only be when further experimental data of a crucial kind are obtained that the theory will run much chance of becoming definitely accepted as scientific knowledge."

In France, relativity was ignored or rejected by most physicists until the 1950s, despite the efforts of its major advocate, physicist Paul Langevin, according to historian Michel Paty. Langevin had been working on electrodynamic theory along the lines initiated by Lorentz and Poincaré; this work "made him receptive to the influence of relativistic ideas." He was also "sensitive to the formal perfection of the theory."

Historians V. P. Vizgin and G. E. Gorelik suggest that in Russia, the conditions for the reception of relativity theory were rather favorable because of the strong interest of physicists like P. N. Lebedev, N. A. Umov, and A. A. Eichenwald in light pressure and electrodynamics, and of mathematicians like V. F. Kagan in non-Euclidean geometry (continuing the tradition started by Lobachevskii). In 1907, K. K. Baumgart supported the Lorentz-Einstein theory because it was compatible with the negative result of the Michelson-Morley experiment. In 1909, O. D. Khvol'son praised relativity theory, in part because it replaced the mechanical with the electrical worldview. It appears that several Russian physicists had already become enthusiastic about the electromagnetic worldview before 1905, and saw Einstein's work as a major contribution to that worldview. The high prestige of German physics also inclined them to take an interest in a theory that was being widely discussed in Germany. Paul Ehrenfest, who resided in St. Petersburg from 1907 to 1912, aroused further interest in relativity with his discussions of the problem of electron rigidity. Mathematicians were attracted to relativity because of the opportunity to investigate new geometrical structures.

In Italy, according to historian Giulio Maltese, Planck's principle applies: the older generation (Righi, Volterra, etc.) never accepted relativity, with the exception of Levi-Civita; the younger generation (Fermi, Persico, both students of Levi-Civita) took it for granted.

11.7 DO SCIENTISTS GIVE EXTRA CREDIT FOR NOVELTY? THE CASE OF GRAVITATIONAL LIGHT BENDING

> If time were linked with space after the manner of the fourth dimension, relativity in electrodynamic fields would be secured . . . but the sources of the field could not be permanent particles of electrons. If physical science is to evolve on the basis of relations of permanent matter and its motions, time must be maintained distinct from space.
>
> —W. J. Johnston and Joseph Larmor, *September 12, 1919*

> This was written before it became known that the Greenwich and Cambridge astronomers, in their recent eclipse expeditions, had confirmed Einstein's prediction for the amount of the deflection of a ray of light by the influence of the sun. It must be recognized that the theory has come to stay in some form or other.
>
> —Joseph Larmor, *October 20, 1919*

In 1916, a new phase in the history of relativity began with the publication of Einstein's general theory and its confirmation by the calculation of the anomalous motion of Mercury's perihelion. Then came A. S. Eddington's sensational confirmation of Einstein's prediction that the path of light is deflected by the sun's gravitational force. How did these two empirical tests affect the acceptance of relativity?

The reception of general relativity theory in the 1920s offers a good opportunity to find out whether scientists are more likely to accept a theory that has made a successful novel prediction than one that has provided a successful explanation of a previously known fact. To review the situation: Einstein's theory made three testable predictions: the advance of the perihelion of Mercury (hereafter called Mercury's orbit), the bending of light by a gravitational field, and the gravitational redshift of spectral lines from a massive source. The first phenomenon had been known for several decades but not satisfactorily explained within the framework of Newtonian theory. Einstein was influenced by it in his choice of a gravitational field equation, so it was not a novel prediction, contrary to the claims of some philosophers (see Section 3.2). The second was first observed in 1919 for the purpose of testing Einstein's prediction. The status of the third prediction was still in doubt until after World War II.

There is a large amount of scientific and popular literature that mentions the eclipse test of Einstein's theory, and some of it allows us to judge the weight attributed to the prediction of light bending by comparison with the Mercury perihelion test. Did light bending provide better evidence *because* it was a novel prediction?

The comparison between Mercury's orbit and light bending is not quite fair, because the quality of the data and the fundamental significance of the two effects are not the same. That difficulty probably affects all attempts to test simple philosophical theses in complex historical situations.

The eclipse results created enormous publicity for relativity theory and helped make Einstein the most famous scientist of the twentieth century. In fact, he was declared by Time magazine to be person of the century. The public was impressed in part by his ability to forecast a striking new phenomenon; Einstein's fame or notoriety was enhanced by the suggestion that he possessed "secret and mysterious methods to harness enormous power and thus control, and maybe destroy, the ordinary person's life." But scientists, writing in technical journals and books addressed to other scientists, rarely ascribe such efficacy to any theory, although they often concede that the eclipse results brought Einstein's theory to their attention.

In the initial excitement caused by the announcement of the eclipse results, several scientists made extravagant statements that one might take to imply that general relativity should be considered favorably because Einstein had predicted the result in advance. But elsewhere these same physicists suggested that the prediction of light bending was not so important after all.

Most of the published comments by physicists during the first two or three years after the 1919 eclipse observation indicated that light bending and Mercury's orbit counted equally strongly in favor of general relativity. If light bending was more important that was not because it had been forecast in advance, but because the data themselves were believed to be more definitive.

For astronomers, especially in the United States, light bending played a more important role, but not (except in one case) because it was a novel prediction. Historian Jeffrey Crelinsten has studied the reception of general relativity among American astronomers in great detail. He finds that while the calculation of Mercury's perihelion "sparked great interest," the primary reason why California astronomers became advocates of relativity was that their own observations, especially those on light bending, turned out to confirm Einstein's predictions. But Crelinsten stresses that this was not a simple case of doing a crucial experiment in order to test a theory, and then accepting the theory because the result was favorable. Astronomers

W. W. Campbell, Charles St. John, and Heber D. Curtis did not understand general relativity theory and were initially skeptical of it, but had the necessary equipment and skills to test some of its predictions. Only astronomer R. J. Trumpler, trained in Europe, was knowledgeable about the theory. Yet by 1920 American astronomy was widely acknowledged to be the best in the world, especially on the observational side, and Europeans looked to the United States for definitive tests of relativity. After the analysis of the 1922 eclipse results confirmed Einstein, and after St. John concluded that his observations confirmed the gravitational redshift, Campbell, St. John, and Trumpler were forced to defend their results against the attacks of antirelativists to protect their own reputations and the prestige of their institutions (the Lick and Mt. Wilson observatories). They and other California astronomers such as R. G. Aitken, also an early skeptic, were struck by the reactionary, pathological, and anti-Semitic views of the most outspoken critics of Einstein, in America and Europe, and saw a need to make public statements about the strong empirical support for relativity. By the end of the 1920s they were asserting that "the theory had to be right because it had passed all the empirical tests" despite their "'slight antipathy' to its wider implications, and a continuing inability to master its mathematics."

Crelinsten identifies the turning point for American astronomers as Campbell's 1923 announcement that Einstein's light-bending prediction had been confirmed by analysis of the 1922 eclipse observations in Australia. It did not cause astronomers to change their minds about the validity of the theory, but "positions tentatively taken after the 1919 eclipse results came out, were entrenched after Campbell's corroboration"—this was "the seal of approval of American technology." Yet H. D. Curtis and others continued to reject relativity despite its empirical confirmation; "they just did not like relativity in the first place."

During the next couple of decades it became clear to the experts (including Einstein himself) that Mercury's orbit change was stronger evidence for general relativity than light bending. In part this was because the observational data were more accurate—it was very difficult to make good eclipse measurements, even with modern technology—and in part because the Mercury orbit calculation depended on a deeper part of the theory itself. That light bending was a forecast whereas the Mercury orbit was not seems to count for little or nothing in these judgments. In fact, one cosmologist, Willem de Sitter, asserted that all of the evidence previously found to support Newton's theory of gravitation also supports general relativity in those cases (the vast majority) where they have the same empirical consequences. This would imply that any such evidence that had been forecast by Newton's theory but not by Einstein's counts no more for the former than for the latter.

But the most significant argument (though it was not often explicitly stated) is that rather than light bending providing *better* evidence because it was predicted before the observation, it actually provides *less secure* evidence for that very reason. This is the case at least in the years immediately following the announcement of the eclipse result, because scientists recognized that any given empirical result might be explained by more than one theory (recall the fallacy of affirming the consequent). Because the Mercury orbit discrepancy had been known for several decades, theorists had already had ample opportunity to explain it from Newtonian celestial mechanics and had failed to do so except by making implausible ad hoc assumptions. This made Einstein's success all the more impressive and made it seem quite unlikely that anyone else would subsequently come up with a better alternative explanation. Light bending, by contrast, had not been previously discussed theoretically (with rare exceptions), but now that the phenomenon was known to exist one might expect that another equally or more satisfactory explanation would be found. It was only 10 years after the initial report of light bending observations that Einstein's supporters could plausibly assert, as did R. J. Trumpler, that

> no other theory is at present able to account for the numerical values of the observed displacements. The assumption that there is an actual curvature of space in the immediate surroundings of the sun, which is implied in Einstein's theory, seems indeed to furnish the only satisfactory explanation why the observed light deflections arc twice as large as those predicted on the basis of Newton's theory.

The view that Eddington's confirmation of the light-bending prediction "won wide acceptance for general relativity" continues to be expressed in popular and philosophical writings on science. It is difficult to find good evidence for the view in those writings.

In the post-1923 technical literature I have examined, only one physicist explicitly states that light bending was better evidence because it was a novel prediction. R. C. Tolman stated that the verification of Einstein's theory by "the three so-called crucial tests" was

> all the more significant, since the advance in the perihelion of Mercury was the only one of the three phenomena in question which was actually known at the time when Einstein's theory was developed, and the effects of gravitation both in determining the path and wavelength of light had not even been observed as qualitative phenomena prior to their prediction by the theory of relativity.

I have not searched the astronomy literature on this point; Crelinsten quotes one example that does give credit for novelty. Astronomer D. H. Menzel remarked, in a 1929 review article, that relativity's success in explaining results such as light bending and the gravitational red shift "is a strong argument for its reality; that it predicted many of them before any study had been made is even more convincing proof of its correctness."

I think that Tolman and Menzel represent only a small minority; in the case of gravitational light bending most scientists ascribed essentially no weight to the mere circumstance that the phenomenon was predicted before it was observed. The majority view is stated in a book by physicist Mendel Sachs: the Mercury orbit test was

> not as spectacular . . . because the theoretical result came after the experimental facts were known. But this test was certainly as important as the other two. The timing of experimental confirmation of a theory should have nothing to do with its significance for the scientific truth of that theory.

So the main value of a successful novel prediction (as compared to a successful deduction of a known fact) is favorable publicity. In this case the publicity helped to make Einstein the most famous (living) scientist in the world.

But what if plans to test his earlier light-bending prediction had been successful and had shown that the actual amount is twice that predicted by his theory? His revised theory, with an extra factor of two, would have looked suspiciously ad hoc to those not familiar with the theory. As it happened, however, the German astronomers who went to Russia to observe an eclipse in 1914, to test Einstein's earlier prediction, were prevented from doing so by the outbreak of the first World War (see Part Five).

As a cynic once said about the use of sex in advertising, "sex doesn't sell, it just gathers a crowd." The prediction itself, whether confirmed or not, may of course advance science by causing scientists to perform an experiment that might not otherwise have been done until much later. Even those physicists who rejected Einstein's general relativity theory had to admit that his prediction had led to the discovery of an important fact about nature. The confirmation of the light-bending prediction certainly did force scientists to give serious consideration to a theory that they might otherwise have ignored or rejected. This is by no means a negligible factor in a situation where many theories compete for attention, and those that seem to violate established ideas about the world can easily be dismissed. The eclipse results put relativity much higher on the scientific agenda and provoked other scientists to try to give plausible alternative explanations. But light bending could not become reliable evidence for Einstein's theory until those alternatives failed, and then its weight was independent of the history of its discovery.

11.8 ARE THEORISTS LESS TRUSTWORTHY THAN OBSERVERS?

The reason why some philosophers and scientists want to give more credit to forecasts is presumably their suspicion that theorists may be influenced in reaching their conclusions by knowledge of the phenomena to be explained. But is it not just as likely that observers will be influenced in reporting their results by knowledge of theoretical predictions of those results? Some opponents of relativity theory, and some contemporary physicists who are experts on relativity, have suggested that the astronomers who first observed light bending exaggerated the agreement between their results and Einstein's prediction because they were already supporters of the theory. Historian Norris Hetherington argues that the history of the third classical test of general relativity, the gravitational redshift, illustrates this influence. Conversely, Crelinsten notes that H. D. Curtis, who disliked relativity, found that the results from one plate taken during the 1918 eclipse "were in excellent agreement with the prediction based on general relativity" but chose to average those results with the null results from another plate and concluded that Einstein was wrong.

In effect, the preference for novel predictions implies a double standard for theorists and observers, based on a discredited empiricist conception of science. In view of the increasing evidence that (as suggested by the Einstein and Eddington statements quoted below) observations are not intrinsically more reliable sources of knowledge than theories, perhaps it would be just as reasonable (or unreasonable) to give more weight to observations performed before rather than after a theoretical prediction.

11.9 MATHEMATICAL-AESTHETIC REASONS FOR ACCEPTING RELATIVITY

Some eminent scientists and mathematicians have asserted that none of the empirical tests is as convincing as the coherence and beauty of the theory itself; some go so far as to say that even if all its predictions were falsified, the theory should still be retained. Conversely, some opponents of relativity assert that even if there were perfect agreement between its predictions and the results of observation, it still would not be an acceptable theory.

Einstein himself, though pleased by the eclipse results, gave them little weight as evidence for his theory. According to his student, Ilse Rosenthal-Schneider, after showing her a cable he received from Arthur Eddington about the measurements, Einstein remarked, "but I knew that the theory is correct." When she asked what he would have done if the prediction had not been confirmed, he said, "then I would have been sorry for the dear Lord—the theory *is* correct." Later, in 1930, he wrote: "I do not by any means find the main significance of the general theory of relativity to be the prediction of some tiny observable facts, but rather the simplicity of its foundations and its logical consistency."

Arthur Eddington, the person primarily responsible for carrying out the eclipse observation project, was already convinced of the truth of Einstein's theory before making the observations. According to astrophysicist S. Chandrasekhar, the project was undertaken not primarily because Eddington wanted to test relativity , but rather because he was a pacifist ; his friends and mentors wanted him to escape the disgrace of refusing to serve his country in wartime. Astronomer Royal Frank Dyson was the instigator of the project and arranged for Eddington to be deferred from military service on the condition that he organize the eclipse expedition if the war ended in time.

In his preliminary discussion of the results, Eddington stated that they confirmed only the law of propagation of light in a gravitational field—the mathematical formula for the interval *ds*—but not Einstein's general theory. He referred to both Mercury's orbit and light bending as predictions of the theory. Reflecting on the status of relativity in 1923, Eddington wrote:

> The present widespread interest in the theory arose from the verification of certain minute deviations in the theory from Newtonian laws. To those who are still hesitating and reluctant to leave the old faith, these deviations will remain the chief centre of interest; but for those who have caught the spirit of the new ideas the observational predictions form only a minor part of the subject. It is claimed for the theory that it leads to an understanding of the world of physics clearer and more penetrating than that previously attained.

As Eddington asserted in a famous dictum, one should not "put overmuch confidence in the observational results that are put forward *until they have been confirmed by theory.*"

P.A.M. Dirac, who constructed the first successful synthesis of special relativity and quantum mechanics (Section 9.1), expressed similar views in his recollections. Dirac stated that he first learned about relativity theory when it was widely publicized in England after World War I by Eddington. His initial interest in the theory was captured by the experimental evidence—the Michelson-Morley experiment, Mercury's orbit, and light bending. But he made it clear that this evidence was not the primary source for his subsequent belief in the theory:

> Suppose a discrepancy had appeared, well confirmed and substantiated, between the [general relativity] theory and observations. . . . Should one then consider the theory to be basically wrong? I would say the answer . . . is emphatically no. The Einstein theory of gravitation has a character of excellence of its own . . . a theory with the beauty and elegance of Einstein's theory has to be substantially correct. If a discrepancy should appear in some application of the theory, it must be caused by some secondary feature relating to this application which has not been adequately taken into account, and not by a failure of the general principles of the theory.

Thus, for some scientists, mathematical-aesthetic factors were more important than empirical factors in persuading them to accept relativity.

11.10 SOCIAL-PSYCHOLOGICAL REASONS FOR ACCEPTING RELATIVITY

> Our travellers tell us that there is complete ignorance in the public mind as to what relativity means. A good many people seem to think that the book deals with the relations between the sexes.
>
> —*Methuen Publishers in England to Robert Lawson, Einstein's translator (1920)*

> In November–December 1922, Einstein went on a six-week trip to Japan, where he was received with great enthusiasm. In addition to Einstein's other attributes, its citizens may also have been curious about him because the Japanese characters for "relativity principle" are very similar to those for "love" and "sex."
>
> —ALBRECHT FÖLSING *(1993)*

Frank Sulloway, in his analysis of the response of firstborn and later-born scientists to radical innovations, found that the correlation between birth order and acceptance of relativity was much weaker after 1915 than it had been before the publication of the general theory. He suggests that "the eclipse results of 1919 caused empirical arguments to uncouple from ideological ones." Although relativity still appealed to younger scientists with liberal social attitudes after 1919, and it is well known that opposition to Einstein was part of a strong anti-Semitic movement in German society, it seems to be difficult to find much evidence that physicists accepted relativity because of social or psychological factors.

The case of V. A. Fock, mentioned by Graham as an example of social construction, turns out on closer examination of Graham's own analysis to be not so relevant to our question. Fock first learned and accepted relativity as a physicist, then later connected it with Marxist ideology. In the introduction to his well-known monograph on gravitational theory (he rejected the label "general theory of relativity") Fock wrote:

> The philosophical side of our views on the theory of space, time and gravitation was formed under the influence of the philosophy of dialectical materialism,

> in particular under the influence of Lenin's "Materialism and Empiriocriticism." The teachings of dialectical materialism helped us to approach critically Einstein's point of view concerning the theory created by him and to think it out anew. It helped us also to understand correctly, and to interpret, the new results we ourselves obtained. We wish to state this here, although this book does not explicitly touch philosophical questions.

It remains to be determined whether any of his students who started out as committed Marxists were persuaded by Fock's lectures and writings to accept relativity *because* it was compatible with Marxism.

A dissertation by historian Maxim William Mikulak gives a detailed account of the interactions between relativity theory and Marxist ideology in the USSR. Mikulak concludes that "there is no universally accepted dialectical materialist interpretation of relativity physics in the Soviet Union." There was a lively debate in the 1920s and 1930s about whether relativity is compatible with Marxism; by the 1950s, an affirmative consensus had been reached. The major ideological obstacle to accepting the special theory of relativity was the idea that matter is equivalent to or can be transformed into energy, since this would seem to undermine materialism. But, Mikulak writes, "what was indeed striking about the special theory of relativity in the hands of Soviet physicists after 1954 was its singular interpretation that emphasized not the dialectical materialist analysis but the substantial empirical data that assured the theory's sound foundation." Thus social factors retarded but did not ultimately prevent the acceptance of the special theory by Soviet physicists.

The situation was different with the general theory, especially in regard to its cosmological applications; a universe limited in space and time was seen as clearly contradictory to Marxist philosophy. Mikulak reports that "most Soviet physicists suspended judgment on [general] relativity theory and simultaneously shunned any Marxist discussions involving Einstein's physics" in the 1920s. Little work was done in relativistic cosmology in the USSR until the 1950s. In 1953, V. A. Fock and D. Aleksandrov rejected Einstein's view that all inertial frames are equivalent and suggested that nature prefers a particular coordinate system; according to Mikulak, they "were groping for a dialectical materialist rendition of relativity theory without disturbing the theory's mathematical edifice." According to Mikulak their view was accepted, so that at least as of 1960 one could argue that the particular version of relativity theory adopted in the USSR was, in part, socially constructed.

One could also argue that in the West, the enthusiasm of mathematicians for relativity reflects not just an aesthetic judgment about the theory but a self-interested decision to exploit their own skills in order to become valued experts on a popular subject. According to Sánchez-Ron, the general theory of relativity gave them a "feeling of life" not easy to find in other areas of mathematical research. In this sense they were motivated to accept relativity by social-psychological factors. Similarly, the ability of theoretical and mathematical physics—disciplines hospitable to relativity—to acquire greater institutional acceptance and resources during the early decades of the twentieth century is a significant social factor. But the initial reception of a theory may be influenced by diverse factors—personal, disciplinary, geographical—while its eventual survival as established textbook knowledge probably depends on its meeting more universal criteria.

An adequate social explanation for the acceptance of relativity should tell us why the popularity of the general theory declined after the 1920s and was revived only in the decades after World War II. The empirical reasons are fairly clear: there were no new convincing tests of the theory's predictions until the development of better technology and advances in atomic physics such as the Mössbauer effect made possible the Pound-Rebka verification of the gravitational redshift. In the meantime quantum mechanics offered many more research opportunities for physicists. At the same time, the establishment of quantum mechanics and the introduction of Dirac's relativistic wave equation made *special* relativity part of the core theories of atomic physics.

The revival of general relativity began in astronomy, with renewed interest in cosmological models inspired partly by theoretical speculation, partly by developments in nuclear physics, and partly by the microwave technology that allowed the discovery of the cosmic black-body background radiation. The mathematical-aesthetic factor comes in again to revive Einstein's program for a unified field theory: a single theory should encompass gravity, electromagnetism, strong and weak interactions—"a theory of everything."

Social factors seem to have been relevant only in Communist countries. In the USSR, pressures to make scientific theories conform to Marxist ideology were strong in the 1930s and 1940s but weaker after the mid-1950s. In China, where Einstein had been popular since the 1920s, he was attacked as an idealist and reactionary in the 1960s. But Zhou Enlai, who had been one of Einstein's earliest supporters in the 1920s, gained political power in the 1970s and was able to restore support for relativity theory.

11.11 A STATISTICAL SUMMARY OF COMPARATIVE RECEPTION

My analysis of the reception studies in Glick's book shows that out of 191 scientists, mathematicians, and textbook authors in nine countries, 137 favored relativity (7 of them having converted from initial opposition) while 54 opposed it. Leaving out the countries in which the views of fewer than 10 scientists, mathematicians, and textbook authors were reported, I find that the highest rate of acceptance of relativity was in Poland (27 out of 29 = 93%). The others were Spain (18/21 = 86%), the United States (17-21 = 81%), Russia/USSR (30/43 = 70%), Italy (15/22 = 68%), Germany (15/26 = 42%), and France (7/18 = 39%).

It is difficult to draw reliable conclusions about national differences from these statistics because the studies were done by different authors using different methods, and in all cases the number of scientists for whom the authors report *reasons* for accepting relativity is small compared to the size of the scientific community. With these qualifications, we can say that according to the articles in Glick's book, scientists accepted relativity (1) because of the negative results of the Michelson-Morley and other ether-drift experiments; (2) their attraction to its mathematical aspects and its promise to unify physics; and (3) the confirmation of the light-bending prediction. These authors give practically no evidence for social-psychological reasons except in the case of Spain, where the most popular reason in this category was religious: the rejection of absolute space and time was deemed consistent with neoscholastic theology. Aesthetic-mathematical reasons were most often given in Italy and Russia/USSR.

Historian Danian Hu has comprehensively described the reception of relativity in China, and found similar results by a detailed analysis of the publications of six Chinese physicists. Reasons (1) and (3) above were the most popular, along with the advance of Mercury's perihelion and the increase of electron mass with speed. One mentioned the mathematical/aesthetic attraction of the theory. The most interesting difference was the assertion that Chinese scientists (unlike those in Europe and America) have no difficulty accepting the relativistic conception of time because they never acquired the Newtonian prejudice in favor of absolute time.

11.12 CONCLUSIONS

> Anyone who bets against Einstein better get out their wallet. That's because those supposedly faster-than-light particles that shook up the world of physics last September are now looking a lot slower.
>
> A second experiment deep in an Italian mountain times these subatomic particles, called neutrinos, traveling at precisely the speed of light and no faster, a team from the experiment, called ICARUS, announced Friday.
>
> —Journalist Brian Vastag *(2012)*

> I suspect . . . that both Einstein's theory and my boxer shorts are safe.
>
> *—physicist Jim Al-Khalili, who had promised to eat his boxers on live TV if neutrinos were shown to travel faster than light (Science, 2012)*

Why was relativity accepted? The historical studies reviewed in this chapter can be put together to suggest a three-stage answer. In the first stage, a few leading scientists such as Planck and Eddington adopted the theory because it promised to satisfy their desire for a coherent, mathematically sophisticated, fundamental picture of the universe. In the second stage, their enthusiastic advocacy persuaded other scientists to work on the theory and apply it to problems that were currently of great interest: the behavior of electrons, and Bohr's atomic model. The special theory was accepted by many German physicists by 1910 (strong opposition developed later) and had begun to attract some interest in other countries.

In the third stage, the confirmation of Einstein's light-bending prediction attracted so much attention among the general public as well as among scientists that no one could ignore it after 1919. Physicists who had not previously accepted relativity now had to take it seriously, and when they did they were persuaded of its validity by a combination of factors. Those who insisted on empirical evidence gave priority to the Michelson-Morley experiment, gravitational light bending, and the advance of Mercury's perihelion. For the experts who could appreciate Einstein's deductions, the Mercury orbit test was more convincing, even though it was not a novel prediction but an explanation of a long-known discrepancy. Those who demanded a coherent fundamental theory were converted to relativity but still found difficulty in giving up old concepts such as the ether, absolute space and time. Mathematicians, whose commitment to the concepts of Newtonian physics was weaker, were more eager to embrace a theory that demanded (and rewarded) expertise in abstract reasoning. For astronomers, especially those in the United States who were involved in successful tests of general relativity, defending the research of their own institutions (and their own discipline) made them defenders of relativity.

Younger scientists who had more liberal social attitudes and were the second or later child in their family were statistically more likely to accept relativity, especially before 1919 (according to Sulloway). There is little evidence that the technical content of relativity was socially constructed, except for general relativity and cosmology in the USSR; however, the theory's acceptance was facilitated by the prestige and institutional resources of mathematics and theoretical physics.

Although the authors of reception studies rarely address this point directly, I suspect that one result of the confirmations of the *general* theory of relativity was to persuade many more physicists to use the *special* theory in their research and teaching; while they accepted the general theory in a vague sense, it was of little use to them. Yet the status of general relativity as the approved way to extend the relativistic approach to problems involving gravity made it the first candidate for consideration when such problems again attracted the interest of scientists.

NOTES FOR CHAPTER 11

329 *Relativity* This chapter is based on my papers "Prediction and Theory Evaluation: The Case of light Bending," *Science* 246 (1989): 1124–1129, and "Why Was Relativity Accepted?" *Physics in Perspective* 1 (1999): 184–214. These papers should be consulted for further details and references. Einstein's publications and correspondence are easily available in *The Collected Papers of Albert Einstein* (Princeton, NJ: Princeton University Press, 1987–), edited by John Stachel et al. English translations are provided in separate volumes, indicated by ET.

329 *"Nature and Nature's Laws lay hid in night"* Epitaph XII for Sir Isaac Newton. Alexander Pope, *The Works of Alexander Pope, Esq. In Nine Volumes, Complete. With Notes and Illustrations by Joseph Warton, D.D. and Others.* London: Printed for B. Law, J. Johnson, C. Dilly [et al.] [1797] vol. 2, p. 403.

329 *"It did not last: the Devil howling 'Ho'"* Cited by Alice Calaprice, editor, *The Ultimate Quotable Einstein* (Princeton, NJ: Princeton University Press, 2011), p. 526. My own proposed addendum is: "God rolled his dice, to Einstein's great dismay: 'Let Feynman be!' and all was clear as day" (Calaprice, *Ultimate*, p. 526).

329 *a large number of reception studies have been done* Glick, editor, *The Comparative Reception of Relativity* (Dordrecht and Boston: Reidel, 1987). T. F. Glick, *Einstein in Spain: Relativity and the Recovery of Science* (Princeton, NJ: Princeton University Press).

Section 11.1

329 *the negative result of the Michelson-Morley experiment* Albert A. Michelson, "The Relative Motion of the Earth and the Luminiferous Ether," *American Journal of Science* 22 (1881): 120–129; Michelson and Edward W. Morley, "Influence of Motion of the Medium on the Velocity of Light," *American Journal of Science* 31 (1886): 377–386; "On the Relative Motion of the Earth and the Luminiferous Ether," *American Journal of Science* 34 (1887): 333–345. Loyd S. Swenson, *The Ethereal Aether: A History of the Michelson-Morley-Miller Aether-Drift Experiments, 1880–1930* (Austin, TX: University of Texas Press, 1972).

330 *Earlier attempts to explain this puzzling result* George Francis FitzGerald, "The Ether and the Earth's Atmosphere," *Science*, 13 (1889): 390; Hendrik Antoon Lorentz,

Versuch einer Theorie der electrische und optischen Erscheinungen in bewegten Körpern (Leiden: Brill, 1895). S. G. Brush, "Note on the History of the FitzGerald-Lorentz Contraction," *Isis* 58 (1967): 230–232. (FitzGerald didn't know that his paper, submitted to *Science,* had actually been published.)

330 *the Michelson-Morley experiment played little if any part* Holton, "Einstein, Michelson, and the 'Crucial' Experiment," *Isis* 60 (1969): 133–197; reprinted in his *Thematic Origins of Scientific Thought, Kepler to Einstein* (Cambridge, MA: Harvard University Press, 2nd ed. 1988), Chapter 8. There are a few counterexamples—statements in which Einstein does credit the Michelson-Morley experiment with some role in his early thinking—but these statements may have been simply diplomatic moves to please Michelson himself or his colleagues. For a recent example see the discussion of a 1921 lecture by Einstein in Chicago, in Jeroen van Dongen, *Einstein's Unification* (Cambridge: Cambridge University Press, 2010), p. 90. The text of the lecture (in German) is in Einstein, *Collected Papers,* vol. 12 (2009), Appendix D, p. 519: "Vortrag des Herrn Prof. Dr. Albert Einstein, Chicago, 5 Mai, 1921, Francis Parker School." Dongen, *Unification,* p. 90.

330 *"If I pursue a beam of light with the velocity* c*"* Einstein, "Autobiographical Notes," in *Albert Einstein Philosopher-Scientist,* edited by Paul Arthur Schilpp (Evanston, IL: Library of Living Philosophers, 1949), p. 52, ET on p. 53.

330 *Einstein pointed out an inconsistency in the conventional interpretation* "Zur Elektrodynamik bewegter Körper," *Annalen der Physik* [series 4] 17 (1905): 891–921, on p. 891; reprinted in Einstein, *Collected Papers,* vol. 2, pp. 276–306, on p. 276, with extensive editorial notes, pp. 253–274, 306–310; ET, "On the Electrodynamics of Moving Bodies," pp. 140–171 (quotation from p. 140).

331 *"We shall raise this conjecture"* Einstein, *Collected Papers,* vol. 2, p. 276; ET, pp. 140–141.

332 *A well-known predicted consequence of the time-dilation effect* Einstein, *Collected Papers,* vol. 2, p. 289; ET, p. 253. Paul Langevin, "L'Evolution de l'Espace et du Temps," *Scientia* 10 (1911): 31–54. Also known as the "clock paradox."

333 *mass and energy are no longer separately conserved* Einstein, "Ist die Trägheit eines Körpers von seinem Energieinhalt abhängig?," *Annalen der Physik* [Vierte Folge] 18 (1905): 639–641; Einstein, *Collected Papers,* vol. 2, pp. 312–314; ET, "Does the Inertia of a Body depend on its Energy Content?," 172–174.

Section 11.2

334 *By 1915 Einstein had worked out the basic form of* Einstein, "Zur allgemeinen Relativitätstheorie," *Königlich Preussische Akademie der Wissenschaften* (1915): 778–786; " . . . (Nachtrag)," 779–801; "Die Feldgleichungen der Gravitation," ibid., 844–847; Einstein, *Collected Papers,* vol. 6, pp. 215–223, 226–228; 245–248; ET, "On the General Theory of Relativity," pp. 98–107; "(Addendum)," pp. 108–110; "The Field Equations of Gravitation," pp. 117–120.

334 *"the same* quality of a body manifests itself*"* Einstein, *Über die spezielle und die allgemeine Relativitätstheorie (Gemeinverständlich)* (Braunschweig: Druck und Verlag von Riedr. Vieweg & Sohn, 1917), p. 45; Einstein, *Collected Papers,* vol. 6, pp. 420–534, on p. 469; ET, reprint of *Relativity: The Special and the General Theory,* translation by Robert W. Lawson (New York: Crown, 1961), pp. 247–420, on p. 317.

335 *"his mode of grasping the situation violates neither reason"* Einstein, *Spezielle und allgemeine.* pp. 45–46; Einstein, *Collected Papers,* vol. 6, p. 470; ET, reprint of *Relativity,* translated by Lawson, on p. 319. I have changed the translation of the word *Kasten* from "chest" to "box."

335 *In this same paper (1907) in which the equivalence principle was first proposed* Einstein, "Über das Relativitätsprinzip und die aus demselben gezogenen Folgerungen," *Jahrbuch der Radioaktivität und Elektronik* 4 (1907): 411–462, on p. 459; *Collected Papers*, vol. 2, pp. 433–484, on p. 481; ET, "On the Relativity Principle and the Conclusions, Drawn from it," pp. 252–311, on p. 307.

335 *to predict that light passing near the sun would change its direction* "Über den Einfluss der Schwerkraft auf die Ausbreitung des Lichtes," *Annalen der Physik* [series 4] 35 (1911): 898–908, on p. 908; Einstein, *Collected Papers*, Vol. 3, pp. 486–496, on p. 496; ET, "On the Influence of Gravitation on the Propagation of Light," pp. 379–387, on p. 387.

335 *Soldner had attempted a similar calculation in 1801* "Ueber die Ablenkung eines Lichtstrahl von seiner geradlinigen Bewegung, durch die Attraction eines Weltkörpers, an welchen er nahe vorbei geht," *Astronomisches Jahrbuch für das Jahr 1804* (1801): 161–172.

335 *from which a deflection twice as large was estimated* Einstein, "Erklärung der Perihelbewegung des Merkur aus der allgemeinen Relativitätstheorie," *Königlich Preussische Akademie der Wissenschaften* (Berlin), *Sitzungsberichte* (1915): 831–839, on p. 834; Einstein, *Collected Papers*, 6: 234–242, on p. 237. ET (by Brian Doyle, 1979), "Explanation of the Perihelion Motion of Mercury from the General Theory of Relativity," pp. 113–116, on p. 114. Einstein, "Die Grundlage der allgemeinen Relativitätstheorie, *Annalen der Physik* [Vierte Folge] 49 (1916): 769–822, on pp. 821–822; *Collected Papers*, vol. 6, pp. 284–337, on pp. 336–337; ET, "The Foundation of the General Theory of Relativity," pp. 147–200, p. 199.

336 *relativity theory had fused space and time Raum und Zeit. Vortrag gehalten auf der 80, Naturforscher-Versammlung zu Köln am 21. September 1908* (Leipzig: Teubner, 1909).

336 *the same equations were derived independently at the same time* Hilbert, "Die Grundlagen der Physik (Erste Mitteilung)," *Königliche Gesellschaft der Wissenschaften zu Göttingen, Mathematisch—physikalische Klasse, Nachrichten* (1915): 395–407. John Earman and Clark Glymour, "Einstein and Hilbert: Two Months in the History of General Relativity," *Archive for History of Exact Sciences* 19 (1978): 291–308. It was discovered in 1997 that Hilbert did *not* discover the field equations before Einstein; he altered the proofs of his article after Einstein's paper came out so while Hilbert's paper was published in March 1916, the dateline just said it was submitted November 20, 1915. See L. Corry, J. Renn, and J. Stachel, "Belated Decision in the Hilbert-Einstein Dispute," *Science* 278 (1997): 1270–1273.

336 *The advance of the perihelion of Mercury is another well-known test* N. T. Roseveare, *Mercury's Perihelion from Le Verrier to Einstein* (Oxford: Clarendon Press, 1982).

336 *the orbit of Mercury should advance by 43 seconds per century* "Erklärung der Perihelbewegung des Merkur aus der allgemeinen Relativitätstheorie," *Königlich Preussische Akademie der Wissenschaften* (Berlin), *Sitzungsberichte* (1915): 831–839; Einstein, *Collected Papers*, vol. 6, pp. 236–242; ET, "Explanation of the Perihelion Motion of Mercury from the General Theory of Relativity," pp. 113–116. For the earlier unsuccessful calculation based on incorrect field equations for general relativity (the "Entwurf" theory), see Einstein, *Collected Papers*, vol. 4, pp. 344–473. See the notes on Einstein's letter to Sommerfeld in Section 3.1.

SECTION 11.3

338 *later work has thrown doubt on their accuracy* For summary and references to experimental work related to special relativity, see Arthur I. Miller, *Albert Einstein's Special Theory of Relativity: Emergence (1905) and Early Interpretation (1905–1911)* (Reading, MA: Addison-Wesley, 1981); see also James A. Coleman, *Relativity for*

the Layman: A Simplified Account (New York: Mentor, 1958). Tests of general relativity are cited in the notes to Section 11.5. See also Magda Whitrow (editor), *ISIS Cumulative Bibliography 1913–1965* (London: Mansell & The History of Science Society, 1976), vol. 3, pp. 152–153. On gravity waves see Daniel Kennefick, "Controversies in the History of the Radiation Reaction Problem in General Relativity," in *Expanding Worlds of General Relativity*, edited by H. Goenner et al. (Boston: Birkhäuser, 1999), pp. 207–234.

337 *most scientists tended to ignore it until the 1960s Jean Eisenstadt, The curious History of Relativity: How Einstein's Theory of Gravity was Lost and Found Again* (Princeton, NJ: Princeton University Press, 2006).

337 *Pound and Rebka performed a successful laboratory test* R. V. Pound, "Weighing Photons," *Physics in Perspective* 2 (2000): 224–268; 3 (2001): 4–51.

337 *Wheeler popularized cosmological applications and the black hole concept* R. Ruffini and J. A. Wheeler, "Introducing the Black Hole," *Physics Today* (January 1971): 30–41 and front cover illustration.

Section 11.4

338 *no evidence that such factors influenced the substantive content* Stephen Cole, *Making Science* (Cambridge, MA: Harvard University Press, 1992); "Voodoo Sociology: Recent Developments in the Sociology of Science," in *The Flight from Science and Reason*, edited by P. R. Gross, N. Levitt, and M. W. Lewis (New York: New York Academy of Sciences, 1996), pp. 274–287.

338 *Warwick's study of the reception of relativity at Cambridge University.* D. Bloor, "Remember the Strong Program?" *Science, Technology and Human Values* 22 (1997): 373–355; A. Warwick, "Cambridge Mathematics and Cavendish Physics: Cunningham, Campbell and Einstein's Relativity 1905–1911," *Studies in History and Philosophy of Science* 23 (1992): 625–656; 24 (1993): 1–25.

339 *who did not abandon their strong belief in the ether until much later.* S. Goldberg, *Understanding Relativity: Origin and Impact of a Scientific Revolution* (Boston: Birkhäuser, 1984).

339 *before 1911, "there existed no well-defined special theory of relativity"* Warwick, "International Relativity: The Establishment of a Theoretical Discipline," *Studies in History and Philosophy of Science* 20 (1989): 139–149, on p. 140.

339 *relativity research of physicist V. A. Fock is a legitimate example.* L. Graham, "The Reception of Einstein's Ideas: Two Examples from Contrasting Cultures," in G. Holton and Y. Elkana, Editors, *Albert Einstein: Historical and Cultural Perspectives* (Princeton, NJ: Princeton University Press, 1982), pp. 107–136.

339 *"physicists in Central Europe were particularly disposed".* L. Pyenson, "The Relativity Revolution in Germany," in Glick, *Comparative Reception*, pp. 59–111, on p. 60.

Section 11.5

339 *Einstein's preference for mathematically elegant theories* Dongen, *Unification*, pp. 10, 26, 32, 43, 57.

340 *many mathematicians were among the early advocates* Glick, *Einstein in Spain*, p. 66.

340 *Italian mathematicians, who . . . were enthusiastic* J. R. Goodstein, "The Italian Mathematicians of Relativity," *Centaurus* 26 (1983): 241–261.

Section 11.6

340 *published a paper on it in 1906* Planck, "Das Prinzip des Relativität und die Grundgleichungen der Mechanik," *Deutsche Physikalische Gesellschaft, Verhandlungen* 8 (1906): 136–141.

340 *Kaufmann's data did not rule out relativity* For further discussion see Giora Hon, "Is the Identification of Experimental Error Contextually Dependent? The Case of Kaufmann's Experiment and its Varied Reception," in *Scientific Practice: Theories and Stories of Doing Physics*, edited by Jed Z. Buchwald (Chicago: University of Chicago Press, 1995), pp. 170–223.

340 *any paper on relativity meeting the normal standards would get published* Pyenson, "Physical Sense in Relativity: Max Planck Edits the Annalen der Physik, 1906–1918," in *Proceedings of the 9th International Conference on General Relativity*, edited by E. Schmutzer (Berlin: VEB Deutscher Verlag der Wissenschaften/Cambridge: Cambri dge University Press, 1983), pp. 285–302.

341 *Planck was attracted to relativity theory* Goldberg, *Understanding Relativity*, pp. 189, 191.

341 *Ehrenfest's analysis of the distortion of moving bodies* Miller, *Albert Einstein's Special Theory of Relativity: Emergence (1905) and early Interpretation (1905–1911)* (Reading, MA: Addison-Wesley, 1981), pp. 235–257. Klein, *Paul Ehrenfest*, Vol. 1, *The Making of a Theoretical Physicist* (Amsterdam: North-Holland Publishing Company, 1970), pp. 150–154.

341 *relativity "has found an ever growing amount of attention"* Laue, *Das Relativitätsprinzip* (Braunschweig: Vieweg, 1911), as quoted by Goldberg, *Understanding Relativity*, p. 203.

341 *Bucherer, who was apparently converted to relativity* Miller, *Einstein's Special Theory*, p. 349.

341 *German physicists accepted special relativity around 1911* Staley, *Einstein's Generation: The Origins of the Relativity Revolution* (Chicago: University of Chicago Press, 2008), pp. 23, 259, 315, 331, 339–340, 371.

341 *Minkowski, in a famous 1908 lecture* "Raum und Zeit," *Jahresberichte der deutschen Mathematiker-Vereinigung*, 18 (1909): 75–88. See Scott Walter, "Minkowski, Mathematicians, and the Mathematical Theory of Relativity," in *The Expanding Worlds of General Relativity*, edited by Hubert Goenner, Jürgen Renn, Jim Ritter and Tilman Sauer (Boston: Birkhäuser, 1999), pp. 45–86. See also Peter Galison, "Minkowski's Spacetime: From Visual Thinking to the Absolute World," *Historical Studies in the Physical Sciences* 10 (1979): 85–121.

341 *Sommerfeld was converted by Minkowski's lectures* Seth, *Crafting the Quantum: Arnold Sommerfeld and the Practice of Theory 1890–1926* (Cambridge, MA: MIT Press, 2010), p. 42. In 1910 he wrote to H. A. Lorentz: "Now I, too, have adapted to the relative theory; in particular, Minkowski's systematic form and view facilitated my comprehension"; letter dated January 9, 19[10], quoted by Walter, "Minkowski," p. 70.

341 *"sixty-five mathematicians wrote 151 articles"* Walter, "Minkowski," p. 73.

341 *"without the result of Michelson and Morley's experiment"* Sánchez-Ron, "The Role played by Symmetries in the Introduction of Relativity in Great Britain," *Symmetries in Physics (1600–1980)*, edited by M. G. Doncel et al. (Barcelona: Universitat Autònoma de Barcelona, 1987), pp. 166–193, on p. 170.

342 *Langevin had been working on electrodynamic theory* M. Paty, "The Scientific Reception of Relativity in France," in Glick, *Comparative Reception*, pp. 113–167, on p. 130.

342 *in Russia, the conditions for the reception of relativity were rather favorable* Vizgin and Gorelik, "The Reception of the Theory of Relativity in Russia and the USSR," in Glick, *Comparative Reception*, pp. 265–336.

342 *In Italy, according to historian Giuliio Maltese* Maltese, "The Late Entrance of Relativity into Italian Scientific Community (1906–1930)," *Historical Studies in the Physical and Biological Sciences* 31 (2000): 125–173.

Section 11.7

342 *"If time were linked with space after the manner of the fourth dimension"* W. J. Johnston and J. Larmor, "The Limitations of Relativity," presented September 12, 1919 at the British Association meeting, published in *Report of the Eighty-Seventh Meeting of the British Association for the Advancement of Science, Bournemouth 1919* (London: John Murray, 1920), pp. 158–159, as quoted by José M. Sanchez-Ron, "Larmor versus General Relativity," in Goenner, *Expanding Worlds*, pp. 405–430, on p. 420.

342 *"This was written before it became known"* Note added to his paper "On Generalized Relativity in Connexion with Mr. W. J. Johnston's Calculus," *Proceedings of the Royal Society of London* 96 (1919): 334–363, as quoted by Sanchez-Ron, ibid.

342 *Einstein was influenced by it . . . so it was not a novel prediction* Warren Z. Watson, "An Historical Analysis of the Theoretical Solutions to the Problem of the Advance of the Perihelion of Mercury," (PhD dissertation, University of Wisconsin, 1969); N. T. Roseveare, *Mercury's Perihelion from Le Verrier to Einstein*" (Oxford: Clarendon Press, 1982). See notes for Section 3.2 for references to the evidence that contradicts statements by Elie Zahar, Jarrett Leplin, and Ronald Giere that the advance of Mercury's perihelion did not play a role in the construction of Einstein's general theory of relativity.

343 *he was declared by* Time *magazine to be person of the century Time*, December 31, 1999, cover.

343 *he possessed "secret and mysterious methods".*

343 *suggested that the prediction of light bending was not so important after all* The scientists were J. J. Thomson, P. Langevin, M. Born, M. Von Laue, and H. A. Lorentz; for details and references see Brush, "Light Bending."

343 *the primary reason why California astronomers became advocates of relativity* "The Reception of Einstein's General Theory of Relativity among American Astronomers, 1910–1930" (PhD dissertation, Université de Montréal, 1981), p. xi. See also Crelinsten, *Einstein's Jury: The Race to Test Relativity* (Princeton, NJ: Princeton University Press, 2006).

344 *were forced to defend their results* Crelinsten, "Reception," p. xv.

344 *"the theory had to be right because it had passed all the empirical tests"* Crelinsten, "Reception," pp. xvi, 384–385, See also Crelinsten, "William Wallace Campbell and the 'Einstein Problem': An Observational Astronomer Confronts the Theory of Relativity," *Historical Studies in the Physical Sciences* 14 (1983): 1091.

344 *"positions tentatively taken after the 1919 eclipse results came out"* Crelinsten, "Reception," pp. 264, 269.

344 *Mercury's orbit change was stronger evidence for general relativity* Einstein, Letter to Robert W. Larson, April 22, 1920, included in a list of his "most important scientific ideas": "1915 [fundamental] field equations of gravitation. Explanation of the perihelion motion of Mercury." He does not mention the prediction of light bending in this list. Einstein, *Collected Papers*, volume 9, pp. 523–524; English translation by Ann Hentschel, ibid., p. 325. For additional references see Brush, "Why was Relativity," note 54, and Brush, "Light Bending," p. 1126 and notes 36, 37, 71, 72.

344 *asserted that all of the evidence previously found* W. De Sitter, *Kosmos: A Course of Lectures on the Development of our Insight into the Structure of the Universe* (Cambridge, MA: Harvard University Press, 1932), p. 111.

345 *theorists . . . had failed to do so* There is one small exception: In 1918, Ernst Gehrcke, one of the strongest opponents of relativity, argued that Paul Gerber in 1898 had explained it by postulating a specific relation between gravitation and the velocity of light. Einstein had to publish a note pointing out that the Gerber-Gehrcke idea was based on contradictory assumptions. See Helge Kragh, "The Fine Structure of Hydrogen and the Gross Structure of the Physics Community, 1916–1926," *Historical Studies in the Physical Sciences* 15, part 2 (1985): 67–125, on p. 112; Einstein, *Collected Papers*, volume 7, pp. 103–104, 346–347, 349.

345 *Einstein's supporters could plausibly assert, as did R. J. Trumpler* Trumpler, "The Relativity Deflection of Light," *Journal of the Royal Astronomical Society of Canada* 23 (1929): 208–218, on p. 218.

345 *Eddington's confirmation of the light-bending prediction "won wide acceptance"* E. L. Turner, "Gravitational Lenses," *Scientific American* 259, no. 1 (July 1988): 54–60, on p. 54; other references in Brush, "Light Bending"; Morris R. Cohen, "Einstein's Theory of Relativity," *New Republic* 21 (1920): 228, 341; Fritz Rohrlich, *From Paradox to Reality* (New York: Cambridge University Press, 1987), p. 18.

345 *only one physicist explicitly states that light bending was better evidence* Tolman, *Relativity, Thermodynamics and Cosmology* (Oxford: Clarendon Press, 1934), p. 213.

345 *"that it predicted many of them before any study had been made"* Menzel, "Progress of Astronomy," *Publications of the Astronomical Society of the Pacific* 41 (1929): 224–231, on p. 229.

345 *The majority view is stated in a book by physicist Mendel Sachs Sachs, Einstein versus Bohr: The continuing Controversies in Physics* (LaSalle, IL: Open Court, 1988), p. 193.

346 *"Sex doesn't sell, it just gathers a crowd"* Jim Twitchell, quoted by Ellen Goodman, "Forgettable Messages," *Boston Globe*, (July 7, 2002): p. D7.

Section 11.8

346 *exaggerated the agreement between their results and Einstein's prediction* According to physicist and historian C. W. F. Everitt, a detailed reading of the reports on the 1919 eclipse observations "leads only to the conclusion that this was a model of how not to do an experiment . . . It is impossible to avoid the impression—indeed Eddington virtually says so . . . that the experimenters approached their work with a determination to prove Einstein right. Only Eddington's disarming way of spinning a yarn could convince anyone that here was a good check of general relativity. The results of later eclipse expeditions have been equally disappointing." Everitt, "Experimental Tests of General Relativity: Past, Present, and Future," in *Physics and Contemporary Needs*, vol. 4, edited by Riazuddin (New York: Plenum, 1980), pp. 529–555. See also John Earman and Clark Glymour, "Relativity and Eclipses: The British Eclipse Expeditions of 1919 and their Predecessors," *Historical Studies in the Physical Sciences* 11, part 1 (1980): 49–85.

Defenders of Eddington include Matthew Stanley, "'An Expedition to Heal the Wounds of War': The 1919 Eclipse and Eddington as Quaker Adventurer," *Isis* 94 (2003): 57–89; Daniel Kennefick, "Testing Relativity from the 1919 Eclipse—A Question of Bias," *Physics Today* 62, no. 3 (March 2009): 37–42. Kennefick's argument has been rebutted by Samuel Schindler, "Theory-Laden Experimentation," *Studies in History and Philosophy of Science* 44 (2013): 89–101, on pp. 97.

346 *H. D. Curtis, who disliked relativity* Crelington, *Reception*, pp. 227–229.

Section 11.9

347 *Some eminent scientists and mathematicians* H. Weyl, E. T. Whittaker, P. G. Bergmann, L. Infeld, R. B. Lindsay, and H. Margenau, cited in Brush, "Light Bending"; A. S. Eddington and P. A. M. Dirac, quoted below.

347 *Some opponents of relativity assert that even if there were perfect agreement* William D. MacMillan, "The Fourth Doctrine of Science and its Limitations," in R. D. Carmichael et al., *A Debate on the Theory of Relativity* (Chicago: Open Court, 1927), pp. 117–127.

347 *"but I knew that the theory is correct"* Quoted by G. Holton, *Thematic Origins of Scientific Thought, Kepler to Einstein* (Cambridge, MA: Harvard University Press, revised edition 1988), on p. 255. According to Klaus Hentschel, this story is "highly questionable": Hentschel, "Einstein's Attitude towards Experiments: Testing Relativity Theory 1907–1927," *Studies in History and Philosophy of Science* 23 (1992): 593–624.

347 *"I do not by any means find the main significance of the general theory"* Quoted by A. Pais, *'Subtle is the Lord . . . ' The Science and the Life of Albert Einstein* (New York: Oxford University Press, 1982), p. 273.

347 *the project was undertaken . . . because he was a pacifist* Chandrasekhar, "The Richtmyer Memorial Lecture—Some Historical Notes," *American Journal of Physics* 37 (1969): 577–584.

347 *"The present widespread interest in the theory"* *The Mathematical Theory of Relativity* (Cambridge: Cambridge University Press, 1923), Preface.

347 *one should not "put overmuch confidence in the observational results"* *New Pathways in Science* (Cambridge: Cambridge University Press, 1934), p. 211.

348 *this evidence was not the primary source for his subsequent belief* Dirac, "The early Years of Relativity," in Holton and Elkana, *Albert Einstein*, pp. 79–80; "The Excellence of Einstein's Theory of Gravitation," *Impact of Science on Society* 29 (1979): 11–14, on p. 13.

Section 11.10

348 *"people seem to think that the book deals with the relations between the sexes"* Message from Methuen publishers in England to Robert Lawson, Einstein's translator, February 1920. Calaprice, *Ultimate*, p. 518.

348 *Japanese characters for 'relativity principle' are very similar* Albrecht Fölsing, *Albert Einstein, eine Biographie* (Frankfurt a. M.: Suhrkamp, 1993); *Albert Einstein*. translated by Edwald Osers (New York: Viking, 1997), p. 528; Calaprice, *Ultimate*, p. 438.

348 *"the eclipse results of 1919 caused empirical arguments to uncouple"* Frank Sulloway, *Born to Rebel*, pp. 40, 345.

348 *Fock first learned and accepted relativity as a physicist* Loren Graham, *Science and Philosophy in the Soviet Union* (New York: Knopf, 1972), Chapter IV; "Do Mathematical Equations Display Social Attributes?" *Mathematical Intelligencer* 22, no. 3 (2000): 31–36. V. J. Frenkel, "Fok, Vladimir Aleksandrovich," *Dictional of Scientific Biography* 17 (1990): 303–305; Alexei Kozhevnikov, private communication.

348 *"The philosophical side of our views"* *The Theory of Space, Time and Gravitation*, translated by N. Kemmer (New York: Pergamon Press, 1959), p. xviii.

349 *"there is no universally accepted dialectical materialist interpretation"* Mikulak, "Relativity Theory and Soviet Communist Philosophy (1922–1960)," PhD dissertation, Columbia University, 1965, on pp. ii, 230.

349 *they "were groping for a dialectical materialist rendition"* Mikuklak, "Relativity," pp. 238, 240–241. See also Mikulak,"Soviet Philosophic-Cosmological Thought," *Philosophy of Science* 25 (1958): 35–50. For a different view see Vizgin and Gorelik, "Reception."

349 *gave them a "feeling of life"* Sánchez-Ron, "The Reception of Relativity among British physicists and mathematicians (1915–1920)," in *Studies in the History of General Relativity*, edited by J. Eisenstaedt and A. J. Kox (Boston: Birkäuser, 1992), pp. 57–88, on p. 73.

350 *he was attacked as an idealist and reactionary in the 1960s* Danian Hu, *China and Albert Einstein: The Reception of the Physicist and his Theory in China, 1917–1979* (Cambridge, MA: Harvard University Press, 2005), pp. 143–165.

Section 11.11

351 *has comprehensively described the reception of relativity in China* Hu, *Einstein*, Chapter 3.

Section 11.12

351 *"Anyone who bets against Einstein better get out their wallet"* Brian Vastag, "Not so fast: Neutrinos clocked at light speed. Second team gets result Einstein predicted." *Washington Post*, March 17, 2012, p. A5.

Of course the situation is not quite that simple. According to Einstein, neutrinos could not travel at the speed of light unless their rest mass is zero. In fact that there is evidence that neutrinos (or at least kinds of neutrinos) do have nonzero mass. The important point is that neutrinos don't go *faster* than light.

351 *"my boxer shorts are safe"* *Science* 335 (2012): 1025. See following article by science journalist Edwin Cartlidge, "Loose Cable May Unravel Faster-Than-Light Result," *Science* 335 (2012): 1027.

12

Big Bang Cosmology

Oh, my husband does that on the back of an old envelope.

—*Elsa Einstein, after a host at Mount Wilson Observatory in California explained to her that the giant telescope is used to find out the shape of the universe*

12.1 THE EXPANDING UNIVERSE IS PROPOSED

By a remarkable coincidence, research in theoretical physics and cosmology and in stellar astronomy in the 1920s both led to the conclusion that the stellar universe—including perhaps space itself—is expanding. Galaxies are moving away from us; the more distant galaxies are moving faster. This implies that at some time in the past—the beginning of the expansion—they were all crowded together in a small space. Theory and observation left open questions such as:

1. Is the expansion proceeding at a constant rate, speeding up, or slowing down?
2. Was the universe actually created at the beginning of the expansion, or did it exist before that, and if so was it contracting or simply static for an indefinite time period?
3. Will it continue to expand forever, or will gravitational attraction eventually slow it down and turn the expansion into a contraction?
4. Is the universe oscillating between periods of expansion and contraction?
5. If there was a universe before the origin of the present expansion period, could stars, planets, and life have survived the passage through that origin?

Einstein's general theory of relativity, as worked out in detail by mathematician A. A. Friedmann, astronomer Willem de Sitter, physicist and astronomer Georges Lemaître, and others, allowed several answers to these questions. Einstein himself had assumed that the universe is stationary, and would eventually collapse as a result of the mutual gravitation of all its components unless a repulsive force acted to counteract gravitational attraction. This hypothetical repulsive force is represented in his equations by the famous *cosmological constant*. If the universe is actually expanding, at a rate continually reduced by gravity, then the constant is not needed. Indeed one might say that the theory (general relativity) predicts that the universe is expanding, a consequence that the theorist himself rejected until 1931. If the expansion is actually accelerating—a possibility not seriously considered until recently—then λ is needed, and might not even be constant.

Early in the twentieth century, astronomer Vesto Melvin Slipher observed the spectral lines of light coming from a number of galaxies outside our own Milky Way galaxy, and found that most of them were shifted toward the red end of the spectrum. At the same time, other astronomers were developing methods for estimating the distances of these galaxies using astronomer Henrietta Leavitt's discovery of an empirical correlation between the period and luminosity of a special kind of variable star, called the Cepheid variables (see the notes to this chapter).

Although some astronomers had proposed that the redshifts of galaxies increase with distance, their evidence was not very convincing, and it is generally agreed that astronomer Edwin Hubble was the first to establish the relationship in 1929. If the redshifts are due to the Doppler effect, they can be interpreted as motion away from us. Hubble enlisted astronomer Milton L. Humason in a project that established a "roughly linear relation between velocities and distances among nebulae."

The linear relation was later expressed by the equation

$$v = Hd$$

where v is the average velocity with which distant galaxies move away from us, d is their distance, and H is called the Hubble constant (first estimated by Lemaître in 1927). Since velocity has the dimension length/time, the inverse of the Hubble constant, $1/H$, will be a characteristic *time* for the expansion, called the Hubble time. (Astronomers usually measure H in units of velocity divided by distance—for example, kilometers per second divided by megaparsecs. One megaparsec [Mpc] is equal to about 3×10^{19} km, so H = [1 km/sec]/ Mpc corresponds to 1/H ~ 1,000 billion years.)

If the rate of expansion is constant (individual galaxies neither speed up nor slow down), then $1/H$ can be called the age of the universe: the time elapsed since all the galaxies were crowded together in a very small volume. (Since a galaxy moving away from us at constant speed v travels a distance $d = vt$ in time t, its distance was $d = 0$ when $t = 0$.)

The Hubble-Humason results were generally taken to indicate an age of about 1.8 billion years. But Hubble did not mention any inferences from his velocity-distance law concerning the timescale for expansion; in fact the phrase "expanding universe" (or any equivalent) does not appear in this paper.

Hubble pursued his research with Humason, and soon provided more extensive confirmations of the redshift-distance relations. But they quickly started to abandon the original assumption that the redshifts were caused by *motion* of the galaxies. Humason wrote two years later that "it is not at all certain that the large redshifts observed in the spectra are to be interpreted as a Doppler effect, but for convenience they are expressed in terms of velocity and referred to as apparent velocities."

In later years Hubble consistently avoided any definite statement that the redshifts of distant nebulae are due to motion (though he admitted that this interpretation did apply to nearby ones), or that the universe is actually expanding, and seemed to doubt the validity of this conclusion. Thus one may say that Hubble discovered the expanding universe in the same sense that Max Planck discovered the quantum: he established an empirical formula that seemed to imply the theory and indeed led others to adopt it (and later to assume that he must have adopted it himself)—yet he drew back from explicitly advocating it as a true statement about the world, and on some occasions even suggested that it is false (Section 7.2).

If Hubble did not really discover the expanding universe, then who did? Historians of science are often reluctant to answer such questions, which assume that discovery is a well-defined event. Moreover, it is almost impossible to persuade authors and editors to correct historical errors. Historians Helge Kragh and Robert W. Smith have nevertheless given it careful consideration. They conclude that the greatest share of the credit should be given to Georges Lemaître, with one important qualification:

> Although he explicitly predicted the expansion of the universe, he could not justify the prediction with observational data that convincingly supported the linear law he suspected. In so far as Lemaître did not establish observationally that the universe is in fact expanding, he did not make a discovery; but in so far as he gave theoretical as well as observational reasons for it, he did discover the expansion of the universe.

Mathematician Ari Belenkly disagrees, arguing that most of the credit should be given to Friedmann, who worked out more completely the solutions of Einstein's equations, including the one that has an initial singularity corresponding to the big bang. Fortunately I do not have to settle the priority dispute, since my project is to find out how and why astronomers eventually accepted the big bang, regardless of who invented the concept.

12.2 THE AGE OF THE EARTH

What's in a number? What difference does it make if the age of the Earth is 6,000 years, or 24 million years, or three billion years?

Estimates of the Earth's age can affect the relations between disciplines, undermine orthodox ideas, and encourage the development of alternative theories. For young-Earth creationists, any number significantly greater than Bishop Ussher's 6,000 years is a threat to the credibility of the Bible and invites the dangerous speculation that a day of the creation week is a metaphor for a geological age. For Darwinian evolutionists, who wanted several hundred million years of geological time, physicist Lord Kelvin's 24 million years (as estimated in the late nineteenth century) was a threat to the hypothesis that the slow process of natural selection could have produced present-day plants, animals, and humans; it encouraged revised evolutionary theories that could do the job more efficiently and complete it in the time allowed. And for cosmologists in the 1930s and 1940s, whose models assumed an expanding universe less than two billion years old, the radiometric dating of terrestrial rocks by geologist Arthur Holmes and others, showing that the age of the Earth is at least three billion years, was an anomaly so serious that it helped to inspire the competing steady-state theory.

Here I focus on the last of these controversies, asking how the age of the Earth influenced the history of cosmology between 1929 and 1952. The first of these dates marks Hubble's announcement of the linear relation between spectral redshifts and distances of nebulae, and the last is the announcement of the doubling of the distance scale by astronomer Walter Baade and others, the first step toward making the age of the universe definitely greater than the age of the Earth.

The problem faced by cosmologists came from their axiom (or at least implied assumption) that *the earth cannot be older than the universe.* This might seem obvious to scientists, but it is not accepted by those who believe in the literal truth of the Bible, such as the young-Earth creationists. If "the universe" means not just "the firmament" but includes the *lights* in the firmament of the heaven, then *Genesis* tells us that the Earth was created one day *before* the entire universe. In the United States, where creationists control public education in many communities and influence the curriculum in entire states such as California, Texas, Kansas, and Ohio, scientists must keep this fact in mind when speaking to the public. The creationists reject not only biological evolution, but also the big bang cosmology, plate tectonics, and any other scientific theory that assumes a timescale longer than a few thousand years.

A more subtle point must also be made about the axiom. It actually does *not* imply that the universe is older than the Earth. There is a third possibility: that they are the *same* age. That possibility seems to be of interest only to a mathematician, but remember in science that "the same" often means "approximately the same"; in geology or astronomy, "approximately" could mean "within a few million years."

12.3 THE CONTEXT FOR THE DEBATE: FOUR NEW SCIENCES AND ONE SHARED MEMORY

Our story involves the interactions of four scientific fields that arose, or were revived and reorganized, at the beginning of the twentieth century:

1. *Stellar astronomy, based on observations with large telescopes.* These telescopes were mostly in the United States, built as a result of the efforts of George Ellery Hale, E. C. Pickering, and Percival Lowell. Theories of the evolution of stars before 1925 assumed a gradual transformation of mass into energy in accordance with Einstein's equation $E = mc^2$; after 1930, as a result of the discovery by astronomer Cecilia Payne and confirmed by astronomer H. N. Russell that the sun and stars are mostly hydrogen, a theory based on the synthesis of elements by fusion of hydrogen was developed by physicist Hans Bethe and others. Important results were obtained by Americans or by European astronomers who came to the United States (see persons mentioned table at the end of this chapter.)
2. *Theoretical physics/cosmology.* Theoretical physics is an old field, but in the late nineteenth century it became possible to specialize in theoretical physics without having to be a competent experimenter as well. With the development of quantum theory and relativity, theoretical physicists acquired considerable prestige and resources in Europe, and later in the

United States; it also attracted a number of mathematicians. Cosmology (including cosmogony) had been a sideline for some astronomers who had to earn their living doing observations and calculational celestial mechanics, or a dilettante pursuit for amateurs; in the twentieth century it became a professionalized specialty.

3. *Atomic physics.* Atomic theories go back to antiquity, but only at the end of the nineteenth century did it become possible to study atoms and their components experimentally, with the discovery of X-rays, the electron, and radioactivity. The laboratory of Marie and Pierre Curie in Paris and the Cavendish Laboratory at Cambridge under J. J. Thomson became the major centers for atomic research. Ernest Rutherford quickly became the leader in research on radioactivity, transmutation, and the nuclear atom, starting at the Cavendish and returning as its Director after making major discoveries at McGill and Manchester.
4. *Planetary geology.* Here I mean the kind of geology that tries to understand the structure and evolution of the earth as a whole, and as part of the solar system. It was unfashionable in Anglo-American geology during most of the nineteenth century but was revived in the twentieth century and made possible the revolution in the earth sciences of the 1960s.

The cooperation of theoretical physics and cosmology and stellar astronomy led to the result that the universe is about 1.8 billion years or less, while the combination of atomic physics and planetary geology led to the result that the Earth is about two billion years old or more. When there is a conflict between the conclusions of different sciences, which prevails? A sociologist might expect that astronomy and theoretical physics, having higher prestige than atomic physics and planetary geology in the twentieth century, would win out or at least be able to ignore the contrary results. But here we have to recognize a shared memory: all these scientists were probably aware of the great controversy between Kelvin and the geologists in the nineteenth century. They knew that Kelvin's estimate of the ages of the Earth and sun based on the application of a simple physical model—a cooling fluid sphere—and ignoring geological evidence, had led to spectacularly wrong results. In 1923 astronomer Arthur Eddington, in a lecture to the Geological Society of London titled "The Borderland of Astronomy and Geology," stated:

> I am sure it will not be supposed that, in presenting the astronomical side of these questions which belong both to geology and astronomy, I have any intention of laying down the law. The time has gone by when the physicist prescribed dictatorially what theories the geologist might be permitted to consider. You have your own clues to follow out to elucidate these problems, and your clues may be better than ours for leading towards the truth. . . .Where, as in the new views of the age of the Earth, physics, biology, geology, astronomy, all seem to be leading in the same direction, and producing evidence for a greatly extended timescale, we may feel more confidence that a permanent advance is being made. Where our clues seem to be opposed, it is not for one of us to dictate to the other, but to accept with thankfulness the warning from a neighbouring science that all may not be so certain and straightforward as our own one-sided view seemed to indicate.

At that time there seemed to be no danger that the Earth could be older than the universe. Around 1915, ten years after the publication of the first radioactivity estimates of geological time by physicists Ernest Rutherford and Robert J. Strutt, the figure of 1.6 billion years was frequently proposed as the best estimate for the age of the oldest minerals and the Earth's crust. Geologist and astronomer T. C. Chamberlin, combining his planetesimal hypothesis (accretion of small solid bodies) for the formation of the Earth with radioactivity and theoretical biology, found about 4.26 billion years for its age. H. N. Russell arrived at a similar estimate for the age of the Earth by a different method, relying solely on the radiative decay of thorium and uranium: an upper limit of eight billion years and a lower limit of 1.1 billion, of which the average would be 4.550 billion. Geophysicist Harold Jeffreys estimated that the shape (eccentricity) of Mercury's orbit indicated an evolution over a period from one to 10 billion, the most probable figure being about 2.5 billion .

During the 1920s radioactivity estimates of the age of the crust based on lead and uranium ratios started to creep up to values around three billion years, but this increase was counteracted by the recognition that a significant part of the lead actually is either nonradiogenic or comes from the decay of thorium. Thus values for the age of the Earth in the range of 1.6 to 2 billion continued to be accepted by geochronologists as late as 1935. Rutherford concluded that the best value is about 3.4 billion years, although Arthur Holmes thought 3.2 billion is too high.

12.4 COSMOLOGY CONSTRAINED BY TERRESTRIAL TIME

During the 1930s, cosmologists tried to reconcile the conflict between the age of the Earth (2 to 3 billion years) and the time between the beginning of the universe's expansion and the present (less than 2 billion years) by assuming, first, that stars might have been formed before the beginning of the expansion; second, that the solar system including the Earth was formed by the encounter of two stars, as postulated by Chamberlin and astronomer F. R. Moulton in America, and by physicist and astronomer James Jeans and geophysicist Harold Jeffreys in England, an event that would been more likely when the stars were close together; and, third, that neither of these two ages was accurately known.

But by 1940, the shaky consensus supporting this theory had begun to fall apart, and choosing a suitable timescale had become, as historian John North wrote, "the nightmare of the cosmologist." As the accuracy of determinations of the ages of Earth and universe improved, the discrepancy between them could no longer be attributed to errors in either one.

Another escape route from the nightmare was also closed off. In 1935, the encounter theory of the origin of the solar system was subjected to devastating criticism by Russell and could no longer be used to justify a scenario in which the Earth was formed at about the same time as the beginning of the expansion of the universe. Astronomer Harlow Shapley rejected the idea that cosmic congestion two billion years ago, implied by the expanding universe theory, could be causally connected with the origin of the Earth just because the latter *also* is a few billion years old. We are all "uncomfortable" with the short time scale of the Universe, he wrote in 1944: "It does not seem sufficiently dignified that the uncompromisingly majestic universe measure its duration as scarcely greater than the age of the oldest rocks on this small planet's surface."

12.5 HUBBLE DOUBTS THE EXPANDING UNIVERSE

Since Hubble is generally (though unfairly) regarded as the founder of the expanding universe theory, one might ask what role he played in the timescale controversy. The answer is: he used it as one of several arguments *against* that theory. In fact Hubble withdrew his public support for the expanding universe theory in 1935, arguing that the evidence did not favor his original interpretation of red shifts as Doppler shifts and challenging theorists to provide a better interpretation. Even before that time he had rarely discussed the 1.8 billion year age of the universe that seemed to be implied by his redshift-distance law; perhaps the best example is a 1934 article for a popular science journal. He wrote that we should assume red shifts are due to motion "until evidence to the contrary is forthcoming," but the implication of this assumption is that the nebulae were "jammed together in our particular region of space, and at a particular instant, about two billion years ago, they started rushing away in all directions at various velocities . . . The time scale seems suspiciously short—a small fraction of the estimated age of some stars—and the apparent discrepancy suggests the advisability of further discussion of the interpretation of redshifts as evidence of motion." Pending further research, "the cautious observer refrains from committing himself to the present interpretation and employs the colorless term 'apparent velocity.'"

The expanding universe, Hubble wrote, "is the latest widely accepted development in cosmology," but it depends on assuming red shifts are velocity shifts: "The 200-inch telescope will definitely answer the question of the interpretation of redshifts, whether or not they do represent motions." Hubble's work with cosmologist Richard C. Tolman led to another way to answer the question. They looked at two simple models based on the assumption that the redshifts (a) are or (b) are not velocity shifts, and applied them to the observational data.

The conclusion of the Hubble-Tolman collaboration was that model (a) could be made to fit the data only with unrealistic assumptions about the density and size of the universe. They seemed to favor model (b), with the assumption that the universe is static and "photons emitted by a nebula lose energy on their journey to the observer by some unknown effect."

According to science writer George Gray, the Hubble-Tolman analysis "which casts doubt on the reality of the expansion . . . has come like a bombshell into the camp of the theorists and is providing a major topic of conversation among astronomers, cosmologists, mathematicians, and other universe explorers." Francis Johnson, in a book on the history of astronomy published in the same year (1937), compared the contest between static and expanding universes to the geocentric/heliocentric debate at the end of the sixteenth century; he had learned from Hubble that we still lack a physical reason for the redshift of nebulae, just as in 1600 no one had a plausible physical reason for the motion of the Earth. But the analogy did not prove to be valid: the Hubble-Tolman dissent did not disturb for more than a couple of years the widespread acceptance of the expanding universe theory.

In his later publications (and in earlier correspondence) Hubble took the position that it was not his responsibility as an observer to determine the correct theoretical interpretation of his data. But he went further than that, with rather explicit statements that the "expanding models are definitely inconsistent with the observations unless a large positive curvature (small, closed universe) is postulated," which gives an unreasonably high density as well as a "small scale . . . both in space and time." On the other hand, the nonexpanding model (b) gives a "rather simple and thoroughly consistent picture." It is "more economical and less vulnerable, except for the fact that, at the moment, no other satisfactory explanation [of the red shift] is known." Moreover, the expanding model (a) assigns a unique location to the observer, which "is unwelcome and a priori improbable."

In his popular book *The Realm of the Nebulae*, published in 1936, Hubble took a completely neutral position in regard to the two models, but in a technical paper the next year he asserted that even when observational data are "weighted in favour of the [expanding universe] theory as heavily as can reasonably be allowed, they still fall short of expectations." The expanding universe would have to be closed, "curiously small and dense, and, it may be added, suspiciously young." If red shifts are *not* primarily velocity shifts, we have a much more reasonable universe in which "the observable region may extend indefinitely both in space and time."

In Autumn 1936 Hubble came to Oxford to give the Rhodes memorial lectures (he was himself a Rhodes scholar) and gave perhaps his most explicit statement that the Earth is too old for the expanding universe theory to be valid. If it were expanding, deviations from the linear redshift-distance relations would be expected because the expansion is being slowed down by gravity, so that the true age of the universe would be less than the Hubble time 1.86 billion years; it would be at most only 1.5 billion years. So the

> initial instant . . . clearly falls within the life-history of the Earth, probably within the history of life on the Earth. . . . Some there are who stoutly maintain that the Earth may well be older than the expansion of the universe. Others suggest that in those crowded, jostling yesterdays, the rhythm of events was faster than the rhythm of the spacious universe today; evolution then proceeded apace, and, into the faint surviving traces, we now misread the evidence of a great antiquity.

But such speculations, Hubble declared, "sound like special pleading, like forced solutions of the difficulty," so we must again look for a different explanation of the red shifts: "some unknown reaction between the light and the medium through which it travels."

Did Hubble actually reject the expanding universe? One historian who has examined all the available Hubble material, Norriss Hetherington, argues that he continued to favor the relativistic expanding universe for philosophical reasons even though his own research and the age of the Earth refuted it. For example, unpublished notes for the Hubble-Tolman paper show that the evidence was even more unfavorable to the expanding-universe theory than their published paper admitted, a fact that Hetherington interprets to mean that Hubble let his bias in favour of the expanding model outweigh the empirical facts. But the statements in his publications, while sometimes ambivalent, give me the impression that Hubble did not support the expanding universe after 1935, or at the very least that he wanted his audience to know about the strong empirical evidence against it. Several astronomers who have written about Hubble agree that he rejected the velocity interpretation of redshifts and was sceptical about the reality of expansion. Perhaps, using Hubble's own term, we should say he *apparently* rejected the theory without explicitly saying so.

12.6 A RADICAL SOLUTION: STEADY-STATE COSMOLOGY

> The centres of the nebulae are of the nature of "singular points," at which matter is poured into our universe from some other, and entirely extraneous, spatial dimension, so that, to a denizen of our universe, they appear as points at which matter is being continually created.
>
> —JAMES JEANS *(1928)*

> Astronomer Fred Hoyle "compared the notion of the universe emerging out of nothing to a 'party girl' jumping out of a cake: 'it just wasn't dignified or elegant.'"
>
> —*journalist* DENNIS OVERBYE *(1991)*

By 1948 the timescale problem had become much more serious. Additional astronomical observations reconfirmed the 1.8 billion year Hubble time, implying an age of only 1.2 billion years for the simplest plausible model with the expansion slowed by gravity. More sophisticated techniques in geochronology, combined with new determinations of lead isotopic abundances by Alfred Nier, made 3.35 billion years a firm *lower* limit for the age of the Earth.

Even Albert Einstein feared that relativistic cosmology was in danger:

> The age of the universe, in the sense used here, must certainly exceed that of the firm crust of the Earth as found from radioactive minerals. Since determination of age by these minerals is reliable in every respect, the cosmologic theory here presented would be disproved if it were found to contradict any such results. In this case I see no reasonable solution.

What happened next looks very much like the beginning of a Kuhnian scientific revolution. Since it appeared that the timescale anomaly could not be resolved within the existing paradigm, three cosmologists boldly proposed a new theory that explicitly violated one of the most sacred laws of that paradigm: conservation of mass. Astronomers Fred Hoyle, Hermann Bondi, and Thomas Gold postulated a perfect cosmological principle: *the universe, in the large, always looks the same to an observer at any time or place.* Yes, distant galaxies are rushing away from us and will eventually become invisible even to our most powerful telescopes, but not to worry. We will not be left alone with only our own galaxy to look at. Other galaxies will appear in their place—formed from matter that is being continually created at a rate just sufficient to keep constant the average density of matter in the visible universe. In this steady-state model the universe has always existed in the past and will always exist in the future. The Hubble time describes the rate of expansion but has nothing to do with the *age* of the universe, which is infinite. By waving their magic wand, Hoyle, Bondi, and Gold made the age paradox disappear.

Not so fast, gentlemen. How can you violate the law of conservation of mass-energy without throwing away modern physics?

The reply, already prepared, comes back: isn't the orthodox cosmology—which Hoyle sarcastically called the big bang—also based on a violation of this law—but one that occurred 1.8 billion years ago, when all the matter and energy of the universe was supposedly created at one instant? Since, by hypothesis, no one except God was around to observe this illegal event, we have no way to prove scientifically that it really happened—whereas in the steady-state theory, the creation is happening *right now* at a definite rate that can easily be calculated, and even directly observed by a sufficiently delicate measurement. Thus steady-state, unlike big bang, makes a *testable prediction*. Isn't that the essence of science?

At this point Bondi and Gold deviated from the Kuhnian script and started to read from the one written by Karl Popper. Bondi, especially, asserted that a theory, if it is to be considered scientific, must make falsifiable predictions and the theory must be abandoned if those predictions are refuted. The steady-state theory predicted that the universe in the distant past, which we observe by looking far out in space, must be essentially similar to the present (nearby) universe.

When that prediction was refuted in the 1960s, Bondi gave up steady-state (as did Gold a few years later), while Hoyle, who had never endorsed falsifiability, tried to keep the theory alive with various modifications. So we learn from this episode that Popper's philosophy is valid only if you believe in it.

There was another contribution from the philosophy of science: in 1954, philosopher Michael Scriven won a prize offered by the *British Journal for the Philosophy of Science* for an essay titled "What Is the Logical and Scientific Status of the Concept of the Temporal Origin and Age of the Universe?" Two astronomers, E. J. Öpik and G. J. Whitrow, also entered the contest but failed to persuade the philosopher-judges to take seriously the *scientific* evidence, though their papers did get published in the journal. In the great tradition of philosopher Auguste Comte, who had asserted more than a century earlier that we would never know the chemical composition of the stars, Scriven argued that "no verifiable claim can be made either that the Universe has a finite age or that it does not. We may still believe that there is a difference between these claims: but the difference is one that is not within the power of science to determine, nor will it ever be."

12.7 ASTRONOMY BLINKS: SLOWING THE EXPANSION

The conflict between astronomical and geological time scales was eliminated in the 1950s when the astronomers (Baade, Thackeray, Sandage, and others) revised their distance scale. This was in part due to new observations made with the 200-inch telescope, which, as Hubble had predicted, resolved the redshift problem and put the expanding-universe theory on a firmer foundation. By doubling and then quadrupling (or more) the distances of most galaxies, these observations increased the age of the universe to more than 10 billion years, at least twice the age of the Earth even when the latter was increased to its current value of 4.5 billion years. Since the new astronomical time scale removed a major objection to the big bang cosmology without introducing continuous creation, it appeared to many astronomers that the steady-state paradigm switch was no longer necessary.

What was Hubble's reaction to the rescue of the expanding-universe theory? One biographer has suggested that he "must have been secretly pleased" when Baade's preliminary results were announced in 1944, indicating that Population II stars must be "much more distant from one another than Hubble's calculations suggested" since that removed the major objection to Hubble's theory by making the universe as old as the Earth. But astronomer Donald Osterbrock notes that Hubble didn't keep up with astronomical research during the war because of his own preoccupation with military projects, and concludes that he didn't read (or at least didn't understand the significance of) Baade's 1944 work. As late as 1951, Hubble doubted that the redshifts of distant nebulae are velocity-shifts. Even after the announcement of Baade's results in 1952, making the Hubble time about 3.5 billion years, a journalist reported that Hubble was still not satisfied with the expanding-universe theory. Since the apparent velocities indicated that the expansion (if real) is slowing down, the actual time since the beginning of the expansion would be less than that: "The timescale is still an uncertain and disturbing feature . . . even the extended 'age of the universe' is no greater than current estimates of the age of rocks in the crust of the Earth." I don't know if Hubble was aware at that time that the age of the Earth had recently been extended to 4.5 billion years (some of the participants in that research, including geochemist C. C. Patterson, were at nearby Caltech), or that the encounter theory of the origin of the solar system, which made it somewhat plausible that the Earth is about as old as the universe, had been generally rejected in favor of a more gradual formation.

By 1958 the age of the universe was thought to be about 13 billion years, with an uncertainty of a factor of two, so even the lowest reasonable value was comfortably greater than the age of the Earth. Unfortunately Hubble died in 1953, so we don't know whether he would have retracted his opposition to the expanding universe in the light of this new evidence. His last published statement on the subject, edited by astronomer Alan Sandage, from the manuscript of his George Darwin lecture on May 8, 1953, acknowledges the doubling of the distance scale and concluded that the age of the Universe is "likely to be between 3000 and 4,000 million years, and thus comparable with the age of rock in the crust of the Earth." According to Sandage, he still rejected the velocity interpretation of redshifts and thus did not accept expansion at the time of that lecture.

12.8 LEMAÎTRE'S PRIMEVAL ATOM AND GAMOW'S BIG BANG

As noted in Section 12.1, Georges Lemaître may be considered the founder of the expanding universe theory. He also proposed a specific model for the expanding universe: it originated as a single primeval atom. He arrived at this conception by considering how the second law of thermodynamics should be expressed using the concepts of quantum theory. Natural processes that correspond to *increasing* entropy are like the splitting of a high-energy quantum into two or more quanta of lower energy. Because of Einstein's relation between mass and energy ($E = mc^2$), such a process is equivalent to the division of a single large mass into two or more smaller masses. Conversely, if we trace the history of the universe backward in time we should arrive at an initial (zero-entropy) state in which the total mass-energy is concentrated in a single superquantum or superatom.

Lemaître pointed out, in 1931, that acceptance of his primeval-atom theory would require us to abandon once and for all the *determinism* inherent in Newtonian physics:

> Clearly the initial quantum could not conceal in itself the whole course of evolution; but according to the principle of indeterminacy, this is not necessary. Our

> world is now understood to be a world where something really happens; the whole story of the world need not have been written down in the first quantum like a song on the disc of a phonograph. The whole matter of the world must have been present at the beginning, but the story it has to tell may be written step by step.

Thus quantum randomness plays a positive role by supplying novelty to the world: something new and unpredictable is always coming into existence as the primeval atom subdivides into smaller and smaller portions of mass-energy. But this randomness does not prevent us from looking back to the beginning of time, with the help of Lemaître's theory:

> The evolution of the world can be compared to a display of fireworks that has just ended: some few red wisps, ashes and smoke. Standing on a well-chilled cinder, we see the slow fading of the suns, and we try to recall the vanished brilliance of the origin of worlds.

George Gamow learned about relativistic cosmology firsthand from Aleksandr Friedmann at the University of Leningrad (now St. Petersburg). But his plan to pursue research in that subject was frustrated by Friedmann's early death. Since there were no other experts on cosmology at the university, Gamow did his graduate work in physics.

In 1928 Gamow made a major breakthrough in understanding how radioactive decay works. An alpha particle may be held inside a nucleus by forces that would keep it there forever according to Newtonian mechanics, but quantum mechanics allows it to escape by tunneling through a barrier that it lacks the energy to surmount. A similar explanation was published at the same time by physicists Ronald Gurney and Edward Condon. After moving to the United States, Gamow's interests turned in the 1930s to the astrophysical and cosmological aspects of nuclear reactions, and in particular to the problem of explaining how the elements had originally been formed.

What are the building blocks for elements? In the early 1930s, Werner Heisenberg and other physicists concluded that the atomic nucleus is composed of protons and neutrons (Section 9.4). The nucleus is surrounded by enough negatively charged electrons to balance the positive charge of the protons, and these electrons determine the chemical characteristics of an element. As a result of research by Cecilia Payne-Gaposchkin and Henry Norris Russell, astronomers had concluded that the sun and most stars are composed predominantly of hydrogen and helium. The hydrogen nucleus is a proton, and the helium nucleus consists of two protons and two neutrons. Thus it was reasonable to assume that matter originally consisted of protons, neutrons, and electrons, and that helium is the first step in the formation of heavier elements from hydrogen.

In order to overcome the strong electrostatic repulsion of the positively charged protons, extremely high densities and temperatures are needed. Such conditions could be found in two places: inside stars, and in the initial state of the Friedmann-Lemaître universe. Gamow was one of the first to suggest, in 1935, that the process could take place in stars, although he later concentrated his efforts on the other alternative.

The modern theory of nuclear reactions in stars was initially developed by physicist Hans Bethe in 1938. He was primarily interested in explaining the production of energy by mass conversion, to answer the question of why the sun shines. His answer involved the combination of protons to make helium nuclei, and the addition of protons to the carbon nucleus to form heavier elements (the carbon cycle).

The nuclear reactions needed to create the other elements from hydrogen and helium were studied experimentally by William A. Fowler and his group at Caltech, with important theoretical suggestions from E. E. Salpeter and Fred Hoyle about how helium nuclei could come together to form carbon. By 1957, a comprehensive scheme for synthesizing most of the elements in stars from hydrogen and helium had been worked out by Fowler, Hoyle, Margaret, and Geoffrey Burbidge, now called B^2FH ; a similar scheme was proposed (independently) by A.G.W. Cameron. This scheme did not provide a satisfactory explanation for the cosmic abundance of helium itself. But it did, given the existence of helium, offer an explanation for the evolution of carbon and other elements needed for the evolution of living beings. Hoyle, in his autobiography, suggests that his suggestion of a possible way to make carbon from helium, subsequently confirmed in Fowler's lab, was an application of the anthropic principle. This principle, proposed by Brandon Carter, states that the reason physical constants have the values they do is that if they did not, we would not here to discuss and measure them. In particular, there must be some way to make carbon, or there would be no physicists to study the question. (Historian Helge Kragh has shown that this was not the basis for the original prediction, only a retrospective comment that has become a popular myth.)

Meanwhile, Gamow had suggested in 1948 that the elements could be synthesized at the beginning of the universe, rather than waiting for stars to form. He postulated a high-density, high-temperature gas of neutrons, which quickly started to break down into protons and electrons. He assigned the task of developing the theory to his physics graduate student Ralph Alpher; Alpher was later joined by physicist Robert Herman.

According to Gamow's theory as worked out by Alpher, all the elements were formed by the successive capture of neutrons. When a nucleus had acquired too many neutrons to be stable, it would undergo beta decay—in effect, a neutron would change to a proton, ejecting an electron and a neutrino. This increases the atomic number of the nucleus by one, making it a different element. (Addition of neutrons without beta decay produces a heavier isotope of the same element.) By repeating these processes of nucleosynthesis, the periodic table of elements is gradually filled up. Nucleosynthesis stops when the supply of neutrons is exhausted, the temperature drops to reduce the reaction rates, and particles disperse as the universe expands.

Alpher and Herman soon realized that the radiation pervading the universe in their model would maintain the spectrum characteristic of black-body radiation (Section 7.1) as it cooled. Moreover, they could calculate the changing density and temperature of this radiation during the expansion and cooling of the universe; they could estimate its present temperature if the present density of matter were known. They found that the radiation would now be about 5° K above absolute zero.

Victor S. Alpher, Ralph Alpher's son, has published a detailed account of the interactions between Ralph Alpher, Robert C. Herman, George Gamow, and other scientists involved in the prediction and discovery of the cosmic microwave background radiation (CMBR). His research in documents and interviews shows that even now there are several misunderstandings about the role of the three primary actors. In particular:

> For three years, from 1948 to 1951, Gamow did not consistently support Alpher and Herman's efforts [to estimate the temperature of the CMBR]. His opposition to their prediction of the CMBR on theoretical grounds is ironic since, even today it is often cited as being the work of "Gamow and colleagues." Moreover . . . Gamow began to publish his own estimates of the CMBR, which were usually mathematically inconsistent, and in which he did not cite Alpher and Herman's work and hence tended to detract from it.

On the other hand, the fact that Gamow (a respected if eccentric physicist) was publishing estimates of the temperature of the CMBR in scientific journals and in his popular book *The Creation of the Universe*, which was widely distributed in paperback editions in the early 1960s, makes it even more puzzling that astrophysicists continued to ignore such estimates before 1965.

The original big bang theory had two major drawbacks. First, it failed to explain the formation of the elements beyond helium. Helium's mass number (total number of protons and neutrons) is 4; there are no stable isotopes with mass numbers 5 and 8, so it is difficult to build larger nuclei by addition of neutrons one at a time. This problem could be solved only by invoking the rival theory of nucleosynthesis in stars, and indeed the modern big bang theory relies on the B^2FH scheme to construct the elements beyond helium.

The second objection to the big bang theory was that the age of the universe, as estimated from Hubble's law using astronomical data available in the 1940s, was only about two billion years. This was significantly less than the age of the Earth and its oldest rocks, as determined by radiometric dating. As noted in Section 12.7, this objection was eliminated in the 1950s when the estimated distances of the galaxies and the Hubble time were greatly increased.

12.9 ARGUMENTS FOR STEADY-STATE WEAKEN

According to the steady-state theory, the universe as a whole must look about the same at all times. In particular, galaxies formed recently must be similar to (and just as abundant as) those formed long ago. Consequently, as Bondi pointed out in 1960, if you look out into space and see that distant galaxies are different from nearby ones, "then the steady-state theory is stone dead."

During the 1950s and early 1960s several astronomical observations were reported that seemed to refute the steady-state theory in the way suggested by Bondi. Looking farther out into space means looking farther back in time, since light travels at a finite speed. Counts of distant radio sources by astronomer Martin Ryle seemed to show that the universe *was* different several billion years ago. There were fewer sources in the distant past (as one might expect if the universe began only 10 or 20 billion years ago) and some of them—the quasistellar objects or quasars—were remarkably different from closer (more recent) objects.

By 1959, when the company Science Service conducted a poll of astronomers, a majority voted against the hypothesis that matter is continuously created in space, though only a third of the respondents affirmed that the universe started with a big bang.

Steady-state was further undermined by new estimates of the cosmic abundance of helium. According to astronomer Donald E. Osterbrock, the consensus of astronomers around 1960–1966 was that "most of the helium must have been formed in the big bang." Its abundance Y (ratio of helium mass to total mass) was too high to be explained by steady-state unless one simply postulated that helium as well as hydrogen was continuously created, but that was apparently too ad hoc for the steady-staters: "The assumption $Y = 0$ agreed with Hoyle's interpretation of the steady state theory . . . Perhaps by that time Hoyle was already convinced that continuous creation was dead because he knew . . . that Y does *not* equal zero anywhere in our galaxy."

Hoyle himself, with astronomer Roger Tayler, was partly responsible for showing that big bang nucleosynthesis would yield 25% helium (by mass) in agreement with observations, as estimated by Osterbrock and Rogerson in 1961. Hoyle had expected a much different result, as recalled by astronomer John Faulkner.

But the steady-state theory was by no means dead in the early 1960s. In fact, its advocates took heart from the fact that several earlier refutations had turned out to be spurious, and this made them suspicious of any new attacks. Moreover, as noted by sociologist B. R. Martin, the conflict between the two theories stimulated and helped gain resources for new observations that might not have been possible if everyone had supported the big bang.

12.10 THE TEMPERATURE OF SPACE

Since the contest between big bang and steady-state cosmologies was eventually settled by the discovery of the temperature of space—that is, by the detection of black-body radiation characterized by the Planck spectral distribution corresponding to a particular temperature—we need to say a little about that concept.

We are accustomed to thinking of our bodies, the surrounding air, fluids and solids in our environment as having definite temperatures, not necessarily all the same. But what is the temperature of space? Space contains electromagnetic radiation that pervades the regions between planets and stars, and can be observed by instruments on Earth. Much of this radiation is confined to specific frequencies determined by the physical and chemical characteristic of astronomical sources; it is not accurately characterized by giving a single temperature. But is there another kind of radiation that is actually in thermal equilibrium at a particular temperature? What kind of thermometer could register that temperature?

To say that radiation is in thermal equilibrium at a particular temperature means that its energy is continuously distributed over different frequencies according to Max Planck's distribution law (Section 7.2). The law has a characteristic shape for every temperature; for our universe, today, the background radiation corresponds to a temperature slightly less than 3 degrees above absolute zero on the Kelvin scale. The distribution peaks at a wavelength of about 0.07 inches, which is in the microwave region, hence this radiation is frequently referred to as the cosmic microwave background.

Obviously our own bodies and our ordinary thermometers are not in equilibrium with this radiation; its energy is much too small to be noticed in comparison with local sources of radiation such as the sun or gas molecules in the atmosphere. The microwave background can be observed only by instruments that are especially sensitive in the region of wavelengths from.01 to.1 inches, and which are insulated from interfering radiation in that region.

Eddington pointed out in 1926 that one could estimate the temperature of space by computing the total amount of light coming from all the stars. He did not assume that the energy of starlight follows Planck's law for the distribution of energy over different frequencies, but an equivalent temperature could still be computed from its total energy density and came out to be a little more than 3 degrees above absolute zero (3.2 °K). Eddington asserted that a black body in space would acquire this temperature, but he did not propose a specific procedure for measuring it.

About 15 years later, Andrew McKellar did suggest a practical way to measure an effective temperature of space. McKellar was one of the first astronomers to propose that molecules as well as atoms could exist in interstellar space. He identified spectral lines corresponding to electronic transitions in the CN (cyanogen) molecule, and pointed out that the relative intensities of two of those lines should depend on the temperature. According to a standard formula of quantum statistical mechanics, the number of electrons in higher-energy states should increase with temperature. Since the intensities of spectral lines depend on the populations of those states, the temperature can be estimated from the observed intensities and comes out to be about 2.3 °K .

The microwave radar equipment developed at MIT during World War II made it just barely possible to detect the background radiation directly—if one had a motivation to look for it. In 1946 a group led by physicist Robert Dicke reported atmospheric radiation measurements with a microwave radiometer; they noted that "there is very little (< 20 °K) radiation from cosmic matter at the radiometer wavelengths," but did not follow up this observation. Dicke later recalled that "at the time of this measurement we were not thinking of the 'big bang' radiation but only of a possible glow emitted by the most distant galaxies in the universe."

In 1955 astronomer É. Le Roux and colleagues used radar equipment developed by the Germans in World War II to measure the temperature of background radiation in the sky. They estimated an average of about 3° K from three measurements at a wavelength of about 13 inches. But they did not recognize the cosmological significance of these measurements or try to test the Planckian character of the radiation by extending their measurements to other wavelengths, and their results attracted no attention from other scientists.

In 1962 astronomer William K. Rose used a heterodyne receiver with a maser preamplifer to measure the cosmic background radiation. He estimated its temperature to be between 2.5° K and 3° K, but was not able to use the equipment to make further confirmatory measurements and did not publish his result.

12.11 DISCOVERY OF THE COSMIC MICROWAVE BACKGROUND

Alpher and Herman asked several other scientists about the possibility of actually observing the cosmic microwave radiation they had predicted, and were told that it couldn't be distinguished from other kinds of energy arriving at the earth. Gamow advised them that the microwave background would be masked by starlight and cosmic rays, so that observations could only put an upper limit of about 5 °K on the residue of the original heat of the universe. The crucial point that the primordial radiation could be distinguished by its characteristic Planck spectrum was not recognized. After about 1955 they abandoned their pursuit of an experimental test of their prediction.

Physicist Steven Weinberg, in his book *The First Three Minutes,* speculated about why no one made a systematic search for the background radiation before 1965. He suggested, first, that the credibility of the big bang theory was diminished by its failure to explain the origin of heavy elements, so it didn't seem important to test its other predictions. By contrast, the B^2FH theory of nucleosynthesis in stars—a theory that was associated with the steady-state cosmology—did seem to provide a satisfactory explanation for the construction of the heavy elements from hydrogen and helium, even though it didn't provide enough helium to start with.

Second, Weinberg pointed to a breakdown of communication between theorists and experimentalists: the theorists didn't realize that the radiation could be observed with equipment already available, and the experimentalists didn't realize the theoretical significance of what they had observed. From this perspective it is significant that Robert Dicke, a physicist who is both a theorist and an experimentalist, played a major role in the discovery.

The most remarkable missed opportunity was a breakdown in communication between the two theorists, Gamow and Hoyle. While each criticized the other's theory, they could still have friendly discussions. In the summer of 1956 Gamow was hired as a consultant by General Dynamics; they paid him enough money to buy a Cadillac convertible, the main requirement being that he remain on call in La Jolla, California. So Gamow invited Hoyle, who was at that time visiting Fowler and the Burbidges at Caltech, to visit him in La Jolla. Hoyle recalls being taken for a ride by Gamow, who told him that the universe must be filled with microwave radiation.

As it happened, Hoyle was familiar with Andrew McKellar's proposal that the temperature of space is about 3 °K, because McKellar's report on interstellar molecules had enabled Hoyle to publish a theoretical paper on that topic which would otherwise have been rejected. So Hoyle could point out that the temperature could not be as high as Gamow was then claiming (tens of degrees above zero) because of McKellar's work: "Whether it was the too-great comfort of the Cadillac, or because George wanted a temperature higher than 3 °K, whereas I wanted a temperature of zero °K, we missed the chance of spotting the discovery made nine years later by Arno Penzias and Bob Wilson."

A different kind of communication problem did lead to the actual discovery of the cosmic microwave background. To achieve optimum accuracy in transmitting information by microwaves using artificial satellites, it was necessary to design an antenna that could minimize the noise from all sources. Bell Laboratories had been involved in microwave technology since World War II, because of the need to develop radar equipment. A Bell Labs engineer, Arthur B. Crawford, built a 20-foot horn reflector at Bell's facility near Holmdel, New Jersey in 1960, originally to receive signals bounced from a plastic balloon high in the atmosphere. After it had served its purpose in Bell's communication satellite project, the antenna was available for research in radio astronomy—just in time for the arrival of astrophysicist Arno Penzias and astronomer Robert Woodrow Wilson.

Wilson recalled, decades later, that

> my only cosmology course [as a graduate student] at Caltech was taught by Fred Hoyle. . . . Philosophically, I liked his Steady State theory of the universe except for the fact that it relied on untestable new physics.

Penzias and Wilson wanted to use the Bell Labs antenna for radioastronomy, but first they had to get rid of the excess noise that had been found in the antenna. They failed to identify the source of this noise despite extensive experimentation and analysis. Finally, in January 1965, Penzias happened to be using the means of communication that paid his salary: the telephone. In a phone conversation with Bernard Burke, he learned about a theory proposed by physicist P.J.E. Peebles, which might explain the origin of the microwave noise.

Peebles was working with Dicke at Princeton, a few miles away. Dicke had become interested in cosmology, and was working on a cyclic version of the Friedmann model. He did not accept the hypothesis that the universe began with a big bang; he didn't believe that all the matter and energy could have suddenly appeared at an instant of time. He thought it more likely that the universe went through phases of expansion and contraction. But what happened to the elements produced in previous phases? Astronomical observations of old stars indicated that they were relatively free of the heavier elements, whereas stars formed more recently contained a higher proportion of those elements. That was what one might expect if the heavy elements were created in a first generation of stars that collapsed and scattered their atoms through space in a supernova explosion, providing raw material for the second generation of stars (such as our sun). But it implied that matter was born again as hydrogen in each phase of expansion. It must have gone through a fiery furnace, hot and dense enough to break down the heavier nuclei into protons and neutrons.

Thus, although Dicke's universe did not *start* with a big bang, each cycle of it must begin and end in conditions that are very similar to those of the Gamow's big bang theory. Moreover, Dicke's cosmology implied a primeval fireball of high-temperature radiation that retains its Planckian black-body character as it cools down, and he estimated that the present temperature of the radiation would be 45 °K. He had forgotten his own 1946 measurement that suggested the existence of background radiation at a temperature less than 20 °K. Peebles made further calculations from Dicke's theory and obtained an estimate of about 10 °K.

Dicke and Peebles, together with two graduate students, P. G. Roll and D. T. Wilkinson, then started to construct an antenna at Princeton to measure the cosmic background radiation. Before they had a chance to get any results, Dicke received a call from Penzias suggesting that they get together to discuss the noise in the Bell Labs antenna, corresponding to a temperature of 3.5 ± 1.0 °K. It was soon apparent that Penzias and Wilson had already detected the radiation predicted by Dicke and Peebles (and earlier by Alpher and Herman, whom they did not mention). The reports of the Bell Labs and Princeton groups were sent to the *Astrophysical Journal* in May 1965 and published together in the July 1 issue.

12.12 IMPACT OF THE DISCOVERY ON COSMOLOGISTS

> It is a bit surprising how quickly our results were accepted among the astronomers I talked to. It probably helped that the Steady State theory was failing to fit observations and Bell Labs had a reputation for doing good science. There were only a couple of occasions where I was challenged about the correctness of our measurement. More often, paradigm changes of that magnitude are resisted much more by established scientists.
>
> —ROBERT W. WILSON *(2009)*

Even before the official publication of the discovery of the background radiation and its theoretical interpretation, the *New York Times* published a front-page article under the headline "Signals Imply a 'Big Bang' Universe." Reports in other newspapers and magazines soon followed, celebrating the triumph of the big bang over the steady-state cosmology. Scientific journals started to receive a flood of papers about the microwave background and its cosmological significance.

Hoyle reviewed the situation in a lecture on September 6, 1965 to the annual meeting of the British Association for the Advancement of Science, published a month later in *Nature*. He admitted that the Penzias-Wilson discovery, along with other evidence, showed that the universe had been much denser in the past. Hence the steady-state theory, at least in its original form, "will now have to be discarded." He went on to argue for a modified theory in which finite regions of an infinite universe expand and contract, while the universe as a whole remains in a steady state. An observer within one region would not see the overall structure but would simply assume that the universe is expanding or oscillating.

Hoyle's statement that the steady-state theory must be abandoned was widely publicized. His attempt to retain a modified steady-state theory was ignored or treated with contempt; as Newsweek pointed out, it violated his own rule to keep theories "simple and free of unobservable, untestable ingredients." Bondi's emphasis on the falsifiability of the steady-state theory had come back to haunt its proponents, and any attempt to twist the theory to explain the new discoveries risked the stigma of Popper's pseudoscience label.

But Popper's criterion, despite its endorsement by Bondi and some other scientists, is not always an accurate description of how science works. As Robert Wilson told a reporter in the summer of 1965, steady-state proponents would not simply abandon their theory, but would try to account for the background radiation in a way consistent with their own theory. And indeed they were soon hard at work doing just that. Hoyle pointed out that as Eddington had shown in 1926, the total energy flowing through all regions of space from starlight and other sources is of the same order of magnitude as the cosmic microwave background. All one needs is a mechanism to thermalize it so that it will have the Planck frequency distribution. The steady- staters promptly set about searching for types of interstellar grains that might do this job. Although they have continued their efforts up to the present time, they have failed to persuade other astronomers that this is a plausible alternative to the big bang explanation of the microwave radiation.

Although the press was quick to conclude that the prediction from the big bang theory had been confirmed by the Penzias-Wilson measurement, scientists realized that these results were limited to only a few wavelengths, all of them on one side of the Planck curve. Other explanations of the background radiation, such as a combination of radio sources, could explain those data points but would have difficulty accounting for a spectrum that agreed with Planck's law over a wide range of frequencies. It was not until the mid-1970s that enough measurements at different frequencies had been made to convince the skeptics that the background radiation really follows Planck's law. The spectrum of the CN molecule played an important role here, as astronomers resurrected and built on the earlier work of McKellar. Until then it was quite reasonable, and in fact scientific, for scientists to seek other explanations of the 3 °K radiation.

More importantly, even *after* the background radiation had been shown to follow Planck's law, it would have been foolish for *all* scientists to adopt the big bang theory and abandon the search for other explanations. The fact that the radiation had been predicted in advance by the big bang theory did not prove that this theory was correct (remember the fallacy of affirming the consequent), although (in view of the other objections to steady state cosmology) it was clearly the best buy for scientists looking for a working hypothesis to guide further research. Moreover, if another theory can give an equally plausible explanation, the big bang does not automatically deserve preference just because it predicted the radiation *in advance*. Indeed, although it is widely believed that the present universe evolved from an earlier hot dense state, it is possible that the big bang theory in its original form will be modified to provide a better explanation of specific features of the cosmic background radiation.

Although the discovery of the cosmic microwave background did not prove the validity of the big bang theory, it did effectively demolish the steady-state theory. Scientists did not accept any of the various hypotheses devised to explain the radiation in the steady-state framework. By 1975, nearly all of those who had supported the steady-state had explicitly abandoned it or simply stopped publishing on the subject. I begin with two of the cofounders:

- Bondi, in his 1990 autobiography, said that he had not worked in cosmology since the mid-1950s, and acknowledged that evidence from microwave radiation and the abundance of helium "made most astronomers accept the Big Bang model though this too has some problems."
- Gold, in a 1968 paper with Pacini, argued that the microwave background could be explained by postulating a set of previously unresolved sources with specific properties; but he did not publish anything more on the subject. Nevertheless, he still considers the big bang theories "very unsatisfactory because they fail to account in a serious way for much observational evidence."

Other steady-state supporters are listed in order of their date of conversion:

- Hidekazu Nariai, Kenji Tomita, and Shoji Kato, 1966
- Dennis Sciama, 1966
- M. J. Rees, 1966
- W. H. McCrea, 1968
- Geoffrey Burbidge, 1971 (though his conversion was only temporary)

Table 12.1. Shift in Opinions of Astronomers Following Discovery of Cosmic Microwave Background

	1959		1980	
	% Favorable	Unfavorable	Favorable	Unfavorable
Big bang	33	36	69	7
Steady-State	24	55	2	91

- J. V. Narlikar, 1974 (he joined Hoyle's later attempt to revive the steady-state theory)
- V. C. Reddish. 1974

Others who had supported the steady-state theory simply stopped publishing on cosmology (with the exception of Hoyle and Wickramasinghe): R. L. Agacy, A. Giao, R. L. Liboff, R. A. Lyttleton, I. W. Roxburgh, P. G. Saffman, and A. O. Zupancic.

We also have statistical evidence for the theory-change, in the form of opinion surveys (predominantly of American astronomers) conducted before (1959) and after (1980) the crucial experiment (see Table 12.1).

The only major holdout was Hoyle, who recovered from the initial blow (the Penzias-Wilson discovery "knocked the stuffing out of me" psychologically, he recalled) and tried to revive a modified steady-state cosmology in the 1980s. He was even able to rally some of the former steady- staters to publish a joint paper denouncing the big bang in 1990.

12.13 CREDIT FOR THE PREDICTION

> Thank you for sending me your paper on 3° K radiation. It is very nicely written except that "early history" is "not quite complete." The theory of, what is now knows [sic] as "primeval fireball" was first developed by me in 1946 (Phys. Rev. *70*, 572, 1946; *74*, 505, 1948; Nature *162*, 680, 1948). The prediction of the numerical value of the present (residual) temperature could be found in Alpher & Hermann's paper (Phys. Rev. *75*, 1083, 1949) who estimate it as *5°K*, and in my paper (Kong. Dansk. Ved. Sels. 27, no. 10, 1953) with the estimate of 7°K. Even in my popular book "Creation of the Universe" (Viking 1952) you can find (on p. 42) the formula $T = 1.5\ 10^{10}/t^{½}$ °K, and the upper limit of 50°K. Thus, you see the world did not start with almighty Dicke. Sincerely G. Gamow
>
> —*letter to Arno Penzias, September 29, 1965*

Did the discovery of the cosmic microwave background bring any recognition to those who had predicted it, or to their theory?

A curious feature of the astronomy literature in the years immediately following the Penzias-Wilson discovery is the reluctance to credit Alpher, Herman, and Gamow with the original prediction of cosmic background radiation. Instead, the 1965 paper by Dicke's group at Princeton was usually cited. The first writers to call attention to the earlier predictions were R. J. Tayler and Ya. B. Zeldovich, in review articles published in 1966. In the same year Isaac Asimov's popular book The Universe recounted the dramatic struggle between two protagonists, Gamow and Hoyle, culminating in Gamow's victory through the confirmation of his prediction.

In 1967 Alpher, Gamow, and Herman, annoyed that their early work was still being ignored, published an article in the *Proceedings of the National Academy of Sciences* to set the record straight. In the meantime *Scientific American* had published an account by Peebles and Wilkinson (June 1967) giving minimal credit to Alpher, Gamow, and Herman for a theory which implied the 3 °K radiation. Eventually scientists began to cite the original prediction, and in 1971 Peebles gave full credit to the pioneers in his widely read monograph *Physical Cosmology*.

After Penzias and Wilson received the 1978 Nobel Prize in physics for their discovery, some of Dicke's colleagues at Princeton complained that he should have shared the prize, because it was his prediction that was confirmed. But why Dicke rather than Alpher, Herman, and Gamow?

Clearly the latter deserved more credit for their successful prediction than they received. Gamow died in 1968 and thus was not eligible for the Nobel Prize in 1978, but Alpher and Herman were still alive. One reason for the relative neglect of their theory even after the Penzias-Wilson discovery may be that it was not presented as a cosmology, but as a hypothesis about the origin of the elements. As such it was not generally successful; nucleosynthesis in the big bang is needed to explain the cosmic abundance of helium, but nucleosynthesis in stars is needed to account for the formation of the heavier elements. The discovery of the microwave background suggested that there had been a high-temperature high-density state billions of years ago (contrary to the steady-state theory), but did not make it any more likely that elements heavier than helium were formed at that time. So the big bang theory was not the final answer to the problem of the origin of the elements, only an essential first step toward the answer.

12.14 CONCLUSIONS

Scientific theories are supposed to be tested by empirical observations. Rarely does a theory stand or fall on the basis of a single experiment—things are usually much more complicated than that. In this case there were several pieces of evidence against the steady state theory before 1965, but the theory's advocates could explain them away. The discovery of the cosmic microwave radiation changed the situation almost overnight. Nearly all astronomers adopted the big bang theory, and within a few years most steady-staters had either switched to the big bang or stopped publishing on cosmology.

In 1962, O. Heckmann concluded a review of competing cosmological models by citing Planck's view that scientific controversies are settled only when proponents of one theory die out. "If, on the other hand," he wrote, "we are less pessimistic and believe that, to a large extent, we still speak the same language, it should be possible to come to a better understanding than we have today." His optimistic view was justified in this case. The rapid conversion of many cosmologists from the steady-state to some kind of evolutionary universe (or to silence on the subject) provides a counterexample to Planck's dictum. The discovery of the cosmic microwave radiation, combined with arguments about helium abundance and observations of distant radio sources and quasars, was responsible for this conversion. No need to wait for the older generation of scientists to retire from the field.

Did the universe really begin at the big bang, or was there a previous contraction phase—a "big crunch"—that led to the high temperature and density? According to physicist Peter G. Roll, Dicke believed that the detection of cosmic microwave background radiation was evidence for a closed universe that would one day contract back on itself; Gamow and others were also willing to entertain the idea of an oscillating universe.

Does the creation of the universe involve quantum theory in a fundamental way? Can one infer the development of material structures (galaxies, etc.) after the big bang from inhomogeneities in the microwave background? Can different parts of the universe, widely separated in space and time, influence each other despite the apparent prohibitions of relativity theory? Will the universe continue to expand forever at an increasing rate, or will it eventually collapse into a black hole? The fact that scientists now consider such questions worthy of serious investigation is itself largely a consequence of the discovery of the cosmic microwave radiation.

NOTES FOR CHAPTER 12

361 *Big Bang Cosmology* Sections 12.1 through 12.7 are based on my article "Is the Earth too Old? The Impact of Geochronology on Cosmology, 1929–1952," in C.L.E. Lewis and S. J. Knell, editors, *The Age of the Earth: From 4004 BC to AD 2002* (London: Geological Society Special Publications, no. 190, 2001): 157–175, which may be consulted for details and references. Other useful works are P. J. E. Peebles, L. A. Page Jr., and R. B. Partridge, editors, *Finding the Big Bang* (New York: Cambridge University Press, 2009); Michael J. Crowe, *Modern Theories of the Universe from Herschel to Hubble* (New York: Dover, 1994); N. S. Hetherington, editor, *Cosmology* (New York: Garland, 1993) and *Encyclopedia of Cosmology* (New York: Garland, 1993); Robert W. Smith, *The Expanding Universe: Astronomy's 'Great Debate' 1900–1931* (Cambridge: Cambridge University Press, 1982; Helge Kragh, *Cosmology and Controversy: The Historical Development of Two Theories of the Universe* (Princeton, NJ: Princeton University Press, 1996). Harry Nussbaumer and Lynda Bieri, *Discovering the Expanding Universe* (New York: Cambridge University Press, 2009).

361 *"Oh, my husband does that on the back of an old envelope" The Ultimate Quotable Einstein*, Collected and Edited by Alice Calaprice (Princeton, NJ: Princeton University Press, 2011), p. 502.

SECTION 12.1

361 *predicts . . . a consequence that the theorist himself rejected until 1931* John D. Barrow, *Cosmic Imagery: Key Images in the History of Science* (New York: Norton, 2008), pp. 81–82. Einstein. "Kosmologische Betrachtungen zur allgemeinen Relativitätstheorie," *Sitzungsberichte der Königlich Preussischen Akademie der Wissenschaften,*: 6 (1917): 142–152, as cited by Nussbaumer and Bieri, *Discovering*, p. 212.

362 *Henrietta Leavitt's discovery of an empirical correlation* Distances of nearby stars could be estimated by the parallax method (used by Bessel and others in the nineteenth century): accurately measure the position of the star relative to those near it, then measure again six months later when the Earth has moved halfway along its orbit around the sun. The shift in apparent position of the star will be inversely

proportional to its distance, which can be calculated by trigonometry using the (known) diameter of the Earth's orbit. But some other method is needed when the star is too far away to have a measurable parallax, especially when it is outside our own Milky Way galaxy. The problem was partly solved after Leavitt analyzed observations of a class of stars whose brightness varies periodically, known as Cepheid variables. She discovered in 1908 that brighter variables have longer periods. Since in one particular group all of the stars could be assumed to be at the same distance, this meant that the intrinsic brightness (luminosity) was correlated with the period of variation. If the distance of one of these stars could be determined by the parallax method, and if more distant Cepheids have the same relation between period and luminosity, one can estimate the distance of the more distant Cepheids and of the galaxy in which they are found. (This was first done by astronomers Ejnar Hertzsprung and Harlow Shapley.) Leavitt, "1777 Variables in the Magellanic Clouds," *Annals of Harvard College Observatory* 60, no. 4 (1908), as cited by Owen Gingerich in his *Dictionary of Scientific Biography* article on Leavitt; Leavitt, "Periods of Twenty-Five Variable Stars in the Small Magellanic Cloud," *Harvard College Observatory Circular* no. 173: 1–3, as cited by Kragh, *Controversy.*

362 *Edwin Hubble was the first to establish the relationship* Hubble, "A Relation Between Distance and Radial Velocity among the Extra-Galactic Nebulae," *Proceedings of the National Academy of Sciences of the USA* 15 (1929): 168–173. See also N. S. Hetherington, "The Measurement of Radial Velocities of spiral Nebulae," *Isis* 62 (1971): 309–313; Smith, *Expanding Universe.* According to Marco Mamone Capria, "Several authors have pointed out that Hubble's observations suggested a quadratic rather than a linear relation." See his article "The Rebirth of Cosmology: From the static to the expanding Universe," in *Physics Before and After Einstein*, edited by Capria (Amsterdam: Ios Press, 2005), pp. 129–162, on p. 148.

362 *H is called the Hubble constant* Nussbaumer and Bieri, *Discovery*, p. 108. They note that this section of his paper was omitted from the English translation published in 1931: Lemaître, "A Homogeneous Universe of Constant Mass and Increasing Radius Accounting for the Radial Velocity of Extra-Galactic Nebulae," *Monthly Notices of the Royal Astronomical Society* 91 (19310: 483–490.

362 *H = [1 km/sec]/Mpc corresponds to 1/H ~ 1,000 billion years* In this chapter some ages are given in billions of years, which is anachronistic because in Great Britain, until the late twentieth century, the word "billion" was used to mean 10^{12} (a million million) while in the United States it meant 10^{9} (a thousand million). To avoid confusion, scientists in both countries and elsewhere used the metric prefix "giga" for 10^{12}. Thus in the scientific literature one finds the abbreviation "Ga" (giga-annum) for 10^{12} years. Since this book is intended for general readers, I have used billion years (US usage).

362 *wrote two years later that "it is not at all certain"* Humason, "Apparent Velocity-Shifts in the Spectra of faint Nebulae," *Astronomical Journal* 74 (1931): 35–42, on p. 35. Hubble and Humason's caution is echoed by some modern cosmologists who argue that the effect should really be called the cosmological redshift. According to Nussbaum and Bieri, *Discovering*, "The Cause of the cosmological redshift is not the relative velocity, but the change of the [spacetime] metric." Similarly, Kragh states that it "is not a Doppler redshift: the cosmological expansion is not a motion of galaxies *through* space, but an expansion of space, carrying the galaxies with it. It is only for relatively small velocities that the cosmological redshift coincides with the Doppler redshift" (Kragh, *Controversy*, p. 19).

According to my colleagues at the University of Maryland, physicists Charles Misner and Dieter Brill, these statements are not quite accurate. Brill writes: "I have never understood why people argue about this, because what you call it has always seemed to me to be a question of what coordinates you use. But apparently this is not a matter of indifference to some. I recall a talk I gave as a young postdoc in Berlin, with Max von Laue in the audience. I treated the cosmological redshift as one thing (I think, expansion) and v. Laue insisted that it is the other (Doppler, if I recall correctly). . . . The view that galaxies are actually moving is more intuitive if you consider on galaxy; the view that more space is opening up works better if you consider radiation apart from its sources, as in the cosmic black-body radiation" (e-mail to SGB, May 24, 2012).

Misner writes: "The cosmological redshift wouldn't be there if the galaxies weren't moving apart, but calling it a Doppler shift is only clearly O.K. for relatively nearby galaxies such as those Hubble studied. For those galaxies the curved spacetime that Einstein, Friedmann and Lemaitre gave us is adequately approximated by flat Minkowski spacetime with the galaxies moving away from the Earth. But to describe a larger Universe containing some very distant galaxies that are now routinely observed, a curved spacetime is needed." In a two-dimensional projection of the curved spacetime universe, with one space and one time dimension, showing the motion of galaxies and light rays, they "move on geodesics of the curved spacetime . . . Since the galaxies are moving apart one is free to call the redshift a 'Doppler shift,' but this extends the meaning of that phrase to something consistent with, but not available to, its usage in flat spacetime. But another verbalization of the same insight . . . is to say that the curvature of spacetime is having a significant effect. Following a wavelength of light forward in time . . . shows that the length of that wavelength increases in exactly the same proportion as does the distance between galaxies." For more details and illustrations see Misner's paper "Infinity in Physics and Cosmology," in *Infinity*, edited by D. O. Dahlstrom et al. (*Proceedings of the American Catholic Philosophical Association*, vol. 55, Washington, DC, 1981) pp, 59–72.

My own conclusion is (1) Whether the observed redshift is simply a Doppler effect depends on what coordinate system you are using to describe curved four-dimensional space-time, and how far away from us the source of light is. (2) if the source is fairly close (e.g., as close as the galaxies Hubble and Humason could observe) it is essentially a Doppler effect, and one would expect a randomly chosen source to be either blue-shifted or red-shifted. For example, the first Doppler shift Slipher found for a spiral galaxy was a blue shift: the Andromeda Nebula is approaching the sun at about 300 kilometers per second. Hubble's 1929 plot of the apparent recessional velocities of galaxies as a function of distance shows negative values (blueshifts) for just a few nearby galaxies, positive values for the rest (reproduced by Kragh, *Controversy,* p. 18). (3) If the source is farther away, the effect of the expansion of space-time will produce an additional red shift (I think this is what Nussbaumer and Bieri meant by their phrase "the change in the size of the Universe between the emission and absorption of the photon"). The combination (not just simple addition) of the emitter's motion relative to space and the motion of space relative to use produces the observed cosmological shift, which is much more likely to be red than blue.

363 *"he did discover the expansion of the universe"* Kragh and Smith, "Who Discovered the Expanding Universe?" *History of Science* 41 (2003): 141–162, on page 153; Nussbaumer and Bieri, *Discovering*, Chapter 9; Lemaître, "Un Universe homogène de Masse constante et de Rayon Croissant, rendant compte de la Vitesse radiale

des Nebuleuses extra-galactiques," *Annales de la Société Scientifique de Bruxelles.* Serie A, 47 (1927): 45–59; English translation [partial], "A Homogeneous Universe of Constant Mass and Increasing Radius Accounting for the Radial Velocity of Extra-Galactic Nebulae," *Monthly Notices of the Royal Astronomical Society* 91 (1931): 483–490. Cf. For the Planck analogy, section 7.2, and S. G. Brush, "Cautious Revolutionaries: Maxwell, Planck, Hubble," *American Journal of Physics* 70 (2002): 119–127.

363 *most of the credit should be given to Friedmann* Belenkly, "Alexander Friedmann and the origins of modern cosmology," *Physics Today* 65, no. 10 (October 2012): 38–43.

363 *almost impossible to persuade authors and editors to correct historical errors* Here is my own attempt to do so, in the following letter to the editor of the *New York Times*: "The article '3 Win Nobel for Work on Accelerating Universe' (October 5, 2011) is a fascinating account of the prize-winning research of Saul Perlmutter, Adam Riess and Brian Schmidt, who discovered that the expansion of the universe is speeding up rather than slowing down. But reporter Dennis Overbye repeated the belief that 'Cosmic expansion was discovered by Edwin Hubble.' Expansion at a constant rate would mean that galaxies are generally moving away from us at a speed proportional to their distance. Hubble did show that the change in the spectrum of a galaxy—known as the 'red shift'—is increasing with its distance from us. But he doubted that this shift is due to speed, thus he did not claim that the universe is expanding. Historians of astronomy have concluded that the discoverer of the expanding universe was not Hubble but the Belgian astronomer Georges Lemaître. Details may be found in the article by Helge Kragh and Robert W. Smith, 'Who Discovered the Expanding Universe?' in the journal *History of Science*, vol. 41 (2003), pages 141–162." The *Times* did not publish this letter.

I have discussed elsewhere the success (or lack thereof) of two eminent historians of science (Gerald Holton, Thomas Kuhn) in persuading physics textbook writers to correct longstanding errors: the claim that Einstein developed special relativity in order to account for the negative result of the Michelson-Morley experiment, and the claim that Planck introduced physical quantization of energy in 1900: Brush, "Thomas Kuhn as a Historian of Science," *Science & Education* 9 (2000): 39–58.

Section 12.2

364 *What difference does it make if the age of the Earth is 6,000 years . . .* For a general survey of this topic see S. G. Brush, *Transmuted Past: The Age of the Earth and the Evolution of the Elements from Lyell to Patterson* (New York: Cambridge University Press, 1996).

Section 12.3

365 *"The time has gone by when the physicist prescribed"* "The Borderland of Astronomy and Geology," *Nature* 118 (1923): 18–21, on p. 21.

366 *found about 4.26 billion years for its age* Chamberlin, "Diastrophism and the Formative Process. XIII. The Bearings of the Size and Rate of Infall of Planetesimals on the Molten or Solid State of the Earth," *Journal of Geology* 28 (1920): 665–701; Brush, *Transmuted Past*, p. 73.

366 *Russell arrived at a similar estimate* Russell, "A Superior Limit to the Age of the Earth's Crust," *Proceedings of the Royal Society of London* A99 (1921): 84–86.

366 *Jeffreys estimated . . . about 2.5 billion* Jeffreys, "Age of the Earth," *Nature* 108 (1921): 24; *The Earth* (Cambridge: Cambridge University Press, 2nd ed., 1929), pp. 58–59.

366 *Rutherford concluded that the best value is about 3.4 billion years* Rutherford, "Origin of Actinium and the Age of the Earth," *Nature* 123 (1929): 313–314; Holmes, "The Problem of geological Time," *Scientia* 42 (1927): 263–272. See also Brush, *Transmuted Past*, p. 74.

Section 12.4

366 *"the nightmare of the cosmologist"* North, *The Measure of the Universe* (Oxford: Clarendon Press, 1965), p. 125.

366 *was subjected to devastating criticism by Russell* Russell, *The Solar System and its Origin* (New York: Macmillan, 1935); see S. G. Brush, *Fruitful Encounters: The Origin of the Solar System and of the Moon from Chamberlin to Apollo* (New York: Cambridge University Press, 1996), pp. 75–86.

366 *"It does not seem sufficiently dignified"* Shapley, "Trends in the Metagalaxy," *American Scientist* 32 (1944): 65–77, on p. 74.

Section 12.5

367 *"the cautious observer refrains from committing himself"* Hubble, "The Realm of the Nebulae," *Scientific Monthly* 39 (1939): 193–202, on p. 199.

367 *"the latest widely accepted development in cosmology"* Hubble, *Realm*, p. 202.

367 *"photons emitted by a nebula lose energy on their journey"* Hubble and Tolman, "Two Methods of Investigating the Nature of the Nebular Red-Shift," *Astrophysical Journal* 82 (1935): 302–337, on p. 304.

367 *"the reality of the expansion . . . has come like a bombshell" Gray, The Advancing Front of Science* (New York: Whittlesey House/McGraw-Hill, 1937), pp. 66–67.

367 *compared the contest between static and expanding universes Johnson, Astronomical Thought in Renaissance England* (Baltimore: Johns Hopkins University Press, 1937), p. 215.

367 *"expanding models are definitely inconsistent with the observations"* Hubble, "Effect of Red Shifts on the Distribution of Nebulae," *Astrophysical Journal* 84 (1936): 517–554, on pp. 517, 554; *Proceedings of the National Academy of Sciences* 22 (1936): 621–627, on pp. 624–626.

368 *The expanding universe would have to be closed, "curiously small"* Hubble, "Red Shifts and the Distribution of Nebulae," *Monthly Notices of the Royal Astronomical Society* 97 (1937): 506–513.

368 *"we now misread the evidence of a great antiquity"* Hubble, *The Observational Approach to Cosmology* (Oxford: Oxford University Press, 1937), pp. 42, 44, 45l; see also Hubble, "Our Sample of the Universe," *Scientific Monthly* 45 (1937): 481–493.

368 *evidence was even more unfavorable to the expanding-universe theory* N. S. Hetherington, "Edwin Hubble and a Relativistic Expanding Model of the Universe," *Astronomical Society of the Pacific, Leaflet* no. 509 (1971); "Philosophical Values and Observation in Edwin Hubble's Choice of a Model of the Universe," *Historical Studies in the Physical Sciences* 13 (1982): 41–67; "Geological Time versus Astronomical Time: Are Scientific Theories Falsifiable?" *Earth Sciences History* 8 (1989): 167–169; "Hubble's Cosmology," *American Scientist* 78 (1990): 142–151.

368 *he rejected the velocity interpretation of redshifts* G. J. Whitrow, "Hubble, Edwin Powell," *Dictionary of Scientific Biography* 6 (1972): 528–533, on p. 532. Alan Sandage, "Edwin Hubble, 1889–1953," *Journal of the Royal Astronomical Society of Canada* 83 (1989): 351–362, on p. 357; Sandage, "Foreword," in Nussbaumer and Bieri, *Discovering*, pp. Xi–xvii; Osterbrock, J. A. Gwinn and R. S. Brashear, "Edwin

Hubble and the Expanding Universe," *Scientific American* 269, no. 1 (July 1992): 84–89, on p. 88; J. Gribbin, *The Birth of Time: How Astronomers Measured the Age of the Universe*, (New Haven: Yale University Press, 2000). See also Brush, "Earth Too Old?" p. 167.

Section 12.6

369 *"matter is poured into our universe from some other . . . dimension"* Jeans, *Astronomy and Cosmogony* (Cambridge: Cambridge University Press, 1928), p. 352, as cited by Nussbaumer and Bieri, *Discovering*, p. 161.

369 *"Hoyle 'compared the notion of the universe emerging out of nothing'"* D. Overbye, *Lonely Hearts of the Cosmos* (New York: Harper-Collins, 1991), quoted by Graham Farmelo, *The Strangest Man: The Hidden Life of Paul Dirac, Mystic of the Atom*: (New York: Basic Books, 2009), p. 379.

369 *"In this case I see no reasonable solution"* Einstein, *The Meaning of Relativity*, reprint of the second edition, with a new Appendix (Princeton, N J: Princeton University Press, 1945), p. 132.

369 *very much like the beginning of a Kuhnian scientific revolution* Thomas S. Kuhn, *The Structure of Scientific Revolutions* (Chicago: University of Chicago Press, 1962).

369 *Hoyle, Bondi, and Gold made the age paradox disappear* H. Bondi and T. Gold, "The Steady State Theory of the Expanding Universe," *Monthly Notices of the Royal Astronomical Society* 108 (1948): 252–270; F. Hoyle, "A New Model for the Expanding Universe," ibid.: 372–382. On Bondi's approach see George Gale and John Urani, "Milne, Bondi, and the 'Second Way' to Cosmology," in *The Expanding Worlds of General Relativity*, edited by Hubert Goenner, Jürgen Renn, Jim Ritter, and Tilman Sauer (Boston: Birkhäuser, 1999), pp. 343–375. According to Helge Kragh, "Almost all of the technical development of the steady-state theory took place within Hoyle's version, which was much closer to general relativity than the Bondi-Gold version." For details see Kragh, *Controversy.*

370 *the orthodox cosmology—which Hoyle sarcastically called the big bang* Hoyle, *The Nature of the Universe* (New York: Harper, 1950), p. 113. He called the big bang "an irrational process that cannot be described in scientific terms." But advocates of the theory eventually adopted the name.

370 *to be considered scientific, must make* falsifiable predictions Bondi, "Setting the Scene" in *Cosmology Now*, edited by L. John (London: British Broadcasting Corporation, 1973), pp. 11–22, on p. 11; "The Philosopher for Science," *Nature* 358 (1992): 363. For more on Popper see chapters 1 and 3.

370 *Scriven argued that "no verifiable claim can be made"* Scriven, "The Age of the Universe," *British Journal for the Philosophy of Science* 5 (1954): 181–190, on p. 190.

370 *also entered the contest but failed to persuade the philosopher-judges* Öpik, "The Age of the Universe," *British Journal for the Philosophy of Science* 5 (1954): 203–214; Whitrow, "The Age of the Universe," ibid. 215–225.

Section 12.7

370 *he "must have been secretly pleased"* G. Christiansen, *Edwin Hubble: Mariner of the Nebulae* (New York: Farrar, Straus and Giroux, 1995), p. 293.

370 *Hubble didn't keep up with astronomical research* D. E. Osterbrock, "Walter Baade, observational Astrophysicist, (3) Palomar and Göttingen, 1948–1960 (Part A)," *Journal for the History of Astronomy* 28 (1997): 283–316.

371 *Hubble doubted that the redshifts of distant nebulae are velocity-shifts* "Explorations in Space: The Cosmological Program for the Palomar Telescopes," *Proceedings of the American Philosophical Society* 95 (1951): 463–470, on p. 463.

371 *"The timescale is still an uncertain and disturbing feature"* G. W. Gray, "A Larger and Older Universe," *Scientific American* 188, no. 6 (June 1953): 56–66.

371 *the age of the Earth had recently been extended to 4.5 billion years* C. C. Patterson, "The Isotopic Composition of Meteoric, Basaltic and Oceanic Leads, and the Age of the Earth," in *Proceedings of the Conference on Nuclear Processes in Geologic Settings, Williams Bay, Wisconsin, September 21–23, 1953*, pp. 36–40; for further details, see Brush, *Transmuted Past*, pp. 82–85.

371 *By 1958 the age of the universe was thought to be about 13 billion years* A. Sandage, "Current Problems in the Extragalactic Distance Scale," *Astrophysical Journal* 127 (1958): 513–526. For a summary of Sandage's later research see Donald Lynden-Bell, "Allan Sandage (1926–2010)," *Science* 330 (2010): 1763.

371 *he still rejected the velocity interpretation of redshifts* E. P. Hubble, "The Law of Red-Shifts" (George Darwin Lecture, May 8, 1953, edited by A. R. Sandage), *Monthly Notices of the Royal Astronomical Society* 113 (1954): 658–666, on p. 666. Sandage, "Edwin Hubble," p. 357.

Section 12.8

371 *Lemaitre's primeval atom and Gamow's Big Bang* Sections 12.8 through 12.11 are based on my article "Prediction and Theory Evaluation: Cosmic Microwaves and the Revival of the Big Bang," *Perspectives on Science* 1 (1993): 565–602, by permission of the University of Chicago Press; which may be consulted for details and references.

371 *"Clearly the initial quantum could not conceal in itself the whole course"* "The Beginning of the World from the Point of View of the Quantum Theory," *Nature* 127 (1931): 706.

372 *"The evolution of the world can be compared to a display of fireworks"* Lemaître, "L'Expansion de l'Espace," *Revue des Questions Scientifiques* 100 (1931): 391–410.

372 *In 1928 Gamow made a major breakthrough* Gamow, "Zur Quantentheorie des Atomkerne," *Zeitschrift für Physik* 51 (1928): 204–212.

372 *A similar explanation was published at the same time* Gurney and Condon, "Wave Mechanics and Radioactive Disintegration," *Nature* 122 (1928): 439.

373 *a comprehensive scheme . . . now called B^2FH* E. M. Burbidge, G. P. Burbidge, W. A. Fowler, and F. Hoyle, "Synthesis of the Elements in Stars," *Reviews of Modern Physics* 29 (1957): 547–650.

373 *a similar scheme was proposed (independently) by A. G. W. Cameron* "Nuclear Reactions in Stars and Nucleogenesis," *Publications of the Astronomical Society of the Pacific* 69 (1957): 201–222.

373 *an application of the anthropic principle* Fred Hoyle, *Home is where the Wind Blows: Chapters from a Cosmologist's Life* (Mill Valley, CA: University Science Books, 1994). Brandon Carter, "Large Number Coincidences and the Anthropic Principle in Cosmology," in *IAU Symposium 63: Confrontation of Cosmological Theories with Observational Data* (Dordrecht: Reidel, 1974), pp. 291–298. Lee Smolin argued that this is not a legitimate use of the anthropic principle—all it does is show there is "some process whereby all that carbon got made. The fact that we and other living things are made of carbon is unnecessary to the argument." *The Trouble with Physics: The Rise of String Theory, the Fall of a Science, and What Comes Next* (Boston: Houghton Mifflin, 2006, paperback edition 2007), p. 165.

373 *Kragh has shown that this was not the basis* Helge Kragh, "An Anthropic Myth: Fred Hoyle's Carbon-12 Resonance Level," *Archive for History of Exact Sciences* 64 (2010): 721–751.

373 *the elements could be synthesized at the beginning of the universe* Ralph A. Alpher, Hans Bethe, and George Gamow, "The Origin of chemical Elements. *Physical Review* [series 2] 73 (1948): 803–804. The order of names and inclusion of Bethe (who did not actually participate in this research, which was nevertheless based in part on his earlier work) was chosen to form a memorable team "∀∃γ".

373 *the radiation would now be about 5 °K above absolute zero* Alpher and Herman, "Evolution of the Universe," *Nature* 162 (1948): 774–775; "Remarks on the Evolution of the Expanding Universe," *Physical Review* [series 2] 75 (1949): 1089–1095; "Theory of the Origin and Relative Distribution of the Elements," *Reviews of Modern Physics* 22 (1950): 153–212.

373 *a detailed account of the interactions* Victor S. Alpher, "Ralph A. Alpher, Robert C. Herman, and the Cosmic Microwave Background Radiation," *Physics in Perspectives* 14 (2012): 300–334, quotation from p. 311.

374 *was publishing estimates of the temperature of the CMBR* Gamow, *The Creation of the Universe* (New York: Viking, 1952, reprinted by Bantam, 1965), p. 40. His estimate of the present temperature was 50 °K. The error of this estimate (it is more than 10 times as high as the current value) is less important than the fact that astrophysicists in the 1960s seemed completely unaware that such an estimate had been published in both the professional and the popular literature by one of the founders of nuclear physics.

SECTION 12.9

374 *"the steady-state theory is stone dead"* Bondi, "The Steady-State Theory of the Universe," in *Rival Theories of Cosmology* by Bondi et al. (London: Oxford University Press, 1960), pp. 12–21, on p. 19.

375 *a majority voted against the hypothesis Science News Letter* 76, no. 2 (July 11, 1959): 22.

375 *"Perhaps by that time Hoyle was already convinced"* Osterbrock, "The Helium Content of the Universe," in Peebles, *Finding* pp. 86–92, on p. 90. Osterbrock also suggests that Bondi and Gold could not believe that the helium was created in a steady state universe.

375 *responsible for showing that big bang nucleosynthesis would yield 25% helium* Hoyle and Tayler, "The Mystery of the Cosmic Helium Abundance," *Nature* 203 (1964): 1108–1110.

375 *Hoyle had expected a much different result* Faulkner, "The Day Fred Hoyle Thought he had Disproved the Big Bang Theory," in Peebles, *Finding*, pp. 244–258.

375 *conflict between the two theories stimulated and helped gain resources* Martin, "Radio Astronomy Revisited: A Reassessment of the Role of Competition and Conflict in the Development of Radio Astronomy," *Sociological Review* 26 (1978): 27–56.

SECTION 12.10

376 *Eddington pointed out in 1926 that one could estimate the temperature of space The Internal Constitution of Stars* (Cambridge: Cambridge University Press, 1926), p. 371; "Bakerian Lecture: Diffuse Matter in Interstellar Space," *Proceedings of the Royal Society of London* A111 (1926): 424–456.

376 *a practical way to measure an effective temperature of space . . . about 2.3 °K* "Evidence for the Molecular Origin of some hitherto unidentified interstellar Lines," *Publications of the Astronomical Society of the Pacific* 52 (1940): 187–192; "Molecular Lines from the Lowest States of Diatomic Molecules Composed of Atoms Probably Present in Interstellar Space," *Publications of the Dominion Astrophysical Observatory* 7 (1941, published 1949), pp. 25–272.

376 *physicist Robert Dicke reported atmospheric radiation measurements* R. H. Dicke et al., "Atmospheric Absorption Measurement with a Microwave Radiometer," *Physical Review* [series 2] 70 (1946): 340–348.

376 *They estimated an average of about 3 °K* J.-F. Denisse, É. LeRoux, and J. C. Sternberg, "Observations du Rayonnement Galactique sur la Longeur d'Onde de 33 cm," *Comptes Rendus hebdomadaires des Séances de l'Academie des Sciences, Paris* 240 (1955): 278–280.

376 *He estimated its temperature to be between 2.5 °K and 3 °K* Rose, letter to S. G. Brush, March 17, 1992.

Section 12.11

377 *observations could only put an upper limit of about 5 °K . . .* Letter to Alpher (1948), quoted by Penzias in his Nobel Lecture; see "The Origin of the Elements," *Reviews of Modern Physics* 51 (1979): 425–431, on p. 430.

377 *they abandoned their pursuit of an experimental test* Alpher, "The Development of the Big Bang Model and its present Status," in *Proceedings, GIREP '91, International Conference on Physics Education. Teaching about Reference Frames: From Copernicus to Einstein*, edited by H. Kühnelt et al., (Torun: Nicholas Copernicus University Press, 1992), pp. 52–79.

377 *a breakdown of communication between theorists and experimentalists The First Three Minutes: A modern View of the Origin of the Universe* (New York: Basic Books, 1977), Chapter VI.

377 *"we missed the chance of spotting the discovery"* Hoyle, "The Big Bang in Astronomy," *New Scientist* 92 (1981): 521–524, 526–527.

378 *"my only cosmology course . . . at Caltech was taught by Fred Hoyle"* Robert W. Wilson, "Two Astronomical Discoveries," in Peebles, *Finding the Big Bang,* pp. 157–176, on p. 159.

378 *Dicke's cosmology implied a primeval fireball* Interview with Dicke in A. Lightman and R. Brawer, *Origins: The Lives and Worlds of Modern Cosmologists* (Cambridge, MA: Harvard University Press, 1990), p. 205f.

378 *The reports of the Bell Labs and Princeton groups were sent* R. H. Dicke et al., "Cosmic Black-Body Radiation," *Astrophysical Journal* 142 (1965): 414–419; A. A. Penzias and R. W. Wilson, "A Measurement of Excess Antenna Temperature at 4080 Mc/s," ibid. 419–421.

Section 12.12

379 *"Signals Imply a 'Big Bang' Universe"* Walter Sullivan, "Signals Imply a 'Big Bang' Universe," *New York Times,* May 21, 1965: 1. Penzias writes: "Walter Sullivan . . . apparently had a 'mole' in *The Astrophysical Journal* editorial office"; Penzias, "Encountering Cosmology," in Peebles, *Finding*, pp. 144–157, on p. 152.

379 *Hoyle reviewed the situation in a lecture* Hoyle, "Recent Developments in Cosmology," *Nature* 208 (1965): 111–114, on p. 113.

379 *as* Newsweek *pointed out, it violated his own rule Newsweek* (October 25, 1965): 77.

379 *would try to account for the background radiation* D. Cohen, "Did the Universe Ever Begin?" *Science Digest* (August 1965): 40–44, on p. 42.

379 *The steady-staters promptly set about searching for types of interstellar grains* Hoyle, "Recent Developments in Nucleosynthesis," *Observatory* 86 (1966): 217–223; J. V. Narlikar and N. C. Wickramasinghe, "Microwave Background in a Steady-State Universe," *Nature* 216 (1967): 43–44; F. Hoyle, G. Burbidge and J. V. Narlikar, *A Different Approach to Cosmology* (New York: Cambridge University Press, 2000).

380 *measurements at different frequencies had been made to convince the skeptics* Steven Weinberg, *Gravitation and Cosmology* (New York: Wiley, 1972); E. I. Robson et al., "Spectrum of the Cosmic Background Radiation between 3 mm and 800 μm," *Nature Physical Science* 251 (1974): 591–592; Alpher and Herman, "Big Bang Cosmology and the Cosmic Black-Body Radiation," *Proceedings of the American Philosophical Society* 119 (1975): 325–348; M. S. Rothenberg, "Cosmic Background Radiation Reveals its Blackbody Shape," *Physics Today* 28, no. 7 (July 1975): 17, 20; D. P. Woody et al., "Measurement of the Spectrum of the Submillimeter Cosmic Background," *Physical Review Letters* 34 (1975): 1036–1039. Peebles, *Finding*, pp. 280–360, presents recollections of several participants in this effort.

380 *"most astronomers accept the Big Bang model"* Bondi, *Science, Churchill and Me: The Autobiography of Herman Bondi, Master of Churchill College, Cambridge* (Oxford: Pergamon Press, 1990), p. 65. Jayant V. Narlikar recalls a radio interview during the 1990s when "I had asked him what he felt about the steady state theory in the light of the observations of the CMBR, especially by COBE. He replied . . . That the CMBR spectrum had turned out to be so close to the Planckian was, in his opinion, a very difficult observation for the steady state theory to explain. So he had felt that the theory was no longer viable." Narlikar, "Some Comments on the early History of the CMBR" in Peebles, *Finding*, pp. 272–275, on p. 274.

380 *argued that the microwave background could be explained* T. Gold and F. Pacini, "Can the Observed Microwave Background Be Due to a Superposition of Sources?" *Astrophysical Journal* 152 (1968): L115–L118.

381 *statistical evidence for the theory-change, in the form of opinion surveys* C. M. Copp, "Relativistic Cosmology. 1. Paradigm Commitment and Rationality," *Astronomy Quarterly* 4 (1982): 103–116.

381 *Others who had supported the steady-state theory simply stopped publishing* According to Helge Kragh, several of the astronomers listed in Brush, "Cosmic Microwaves," p. 587, were not supporters of the steady-state theory but "only expressed a technical interest in, had a brief flirtation with, or wrote on aspects related to the steady-state theory" see Kragh, *Controversy*, p. 444, note 197. I have deleted those names from the list, and thank Dr. Kragh for this information.

381 *the Penzias-Wilson discovery "knocked the stuffing out of me"* Hoyle, "The Big Bang in Astronomy," *New Scientist* 92 (1981): 521–524, 526–527.

381 *He was even able to rally some of the former steady-staters* H. C. Arp, F. Burbidge, F. Hoyle, J. V. Narlikar, and N. C. Wickramasinghe, "The Extragalactic Universe: An Alternative View," *Nature* 346 (1990): 807–812.

Section 12.13

381 *"Thank you for sending me your paper on 3°K radiation"* Arno Penzias, "Encountering Cosmology," in Peebles, *Finding*, pp. 144–157, on p. 154. Dicke explained ten years later why he failed to give proper credit for the prediction: "There is one unfortunate and embarrassing aspect of our work on the fire-ball radiation. We failed

to make an adequate literature search and missed the more important papers of Gamow, Alpher and Herman . . . I had heard Gamow talk at Princeton but I had remembered his model universe as cold and initially filled only with neutrons." "A Scientific Autobiography" (1975), unpublished manuscript held by National Academy of Sciences, quoted in J. D. Barrow, *Cosmic Imagery* (New York: Norton, 2008), note 87, p. 554. But the continued neglect of the prediction continued into the next century: see letters by Ralph de Blois and Charles Kaufman, "Alpher and Herman's Work Often Forgotten," *APS* [American Physical Society] *News*, October 2002, p. 4.

382 *The first writers to call attention to the earlier predictions* Tayler, "The Origin of the Elements," *Reports on Progress in Physics* 29 (1966): 489–538; Zeldovich, "The 'Hot' Model of the Universe," *Soviet Physics Uspekh*, 9 (1967): 602–617, translated from *Uspekhi Fizicheskikh Nauk* 89 (1966): 647–668.

382 *Asimov's popular book* The Universe *recounted the dramatic struggle* Isaac Asimov, *The Universe, from Flat Earth to Quasar* (New York: Walker, 1966), p. 301.

382 *In 1971 Peebles gave full credit to the pioneers Physical Cosmology* (Princeton, NJ: Princeton University Press, 1971), pp. 125–128. Peebles was annoyed by "the myth that our paper did not refer to earlier work." He admits this was true in the paper by Dicke and Peebles ("Gravitation and Space Science," *Space Science Reviews* 4 (1965): 419–460). But he asserts "we remedied that pretty quickly in the paper by Dicke et al. (R. H. Dicke, P. J. E. Peebles, P. G. Roll, and D. T. Wilkinson, "Cosmic Black-Body Radiation," *Astrophysical Journal* 142 (1965): 414–419) where 'the citations are normal and proper.'" Peebles, "How I Learned physical Cosmology," in Peebles, *Finding*, p. 192.

382 *Dicke's colleagues at Princeton complained that he should have shared the prize* Bernard F. Burke, "Radio Astronomy from First Contacts to the CMBR" in Peebles, *Finding*, pp. 176–183, on p. 181; Peter G. Roll, "Recollections of the second Measurement of the CMBR at Princeton University in 1965," in Peebles, *Finding*, pp. 213–221, on p. 216, disagrees.

Section 12.14

382 *Heckmann concluded a review of competing cosmological models* "General Review of Cosmological Theories," in *Problems of Extra-Galactic Research*, edited by G. C. McVittie (New York: Macmillan, 1962), pp. 429–438, on p. 437.

383 *that the detection of cosmic microwave background radiation was evidence* Roll, "Recollections of the second Measurement of the CMBR at Princeton University in 1965," in Peebles, *Finding*, pp. 213–221, on p. 218.

383 *were also willing to entertain the idea of an oscillating universe* Helge Kragh, *Higher Speculations: Grand Theories and Failed Revolutions in Physics and Cosmology* (Oxford: Oxford University Press, 2011), Chapter 8.

PART FOUR

Heredity and Evolution

13

Morgan's Chromosome Theory

> The charge is often made that the geneticist finds it impossible to present that which he has to say in terms so simple that anyone may read and understand. . . . It would seem to be far easier to discuss, in everyday language, the origin and nature of the universe than to explain the origin and genetic nature of a new coat colour in the rabbit.
>
> —F.A.E. Crew *(1931)*

> The earlier theory of pangens (which nobody could see) made virtually no specific predictions by which its merits could be judged. On the other hand, the chromosome theory offered abundant predictions, the testing of which led to the rapid growth of genetics as an exact science.
>
> —G. G. Simpson and William S. Beck (1965)

13.1 INTRODUCTION

In the second decade of the twentieth century, biologist Thomas Hunt Morgan and his colleagues proposed that the fundamental carriers of heredity are chromosomes (colored bodies visible with a microscope), which contain genes (then-invisible entities arranged like beads on a string). The genes are factors that determine in some way the characters of an organism. The theory can be used to map their (relative) locations on the individual chromosomes. Morgan's theory came to be widely accepted in the 1920s and is the direct ancestor of modern molecular genetics.

13.2 IS BIOLOGY LIKE HYPOTHETICO-DEDUCTIVE PHYSICS?

Is the method of biology different from that of the physical sciences? Some biologists argue that theories in the life sciences cannot be judged by the predictive hypothetico-deductive method supposedly employed in the physical sciences (see Chapters 1 and 3). It would appear that theories such as Darwinian evolution cannot be tested this way, since one cannot control the relevant environmental factors for thousands of years, and even if that were possible, the experiment could not tell us *now* whether to accept the theory. So Ernst Mayr and other biologists concluded that predictiveness is not a requirement for a biological theory, contrary to Karl Popper who once claimed that Darwinism is not a *scientific* theory because it cannot make testable predictions (see Chapter 14). Instead biology, according to some philosophers of science, should be treated as a descriptive and historical science that advances by discovering facts about specific cases rather than by discovering general laws.

The notion, apparently widespread until the late twentieth century, that biological theories such as evolution are not testable has created an impression that biology is a soft science whose conclusions are not as reliable as those in the hard sciences like physics and chemistry, and this has undermined support for evolution in the creation-evolution controversy. Genetics was one of the first subfields of biology to win respect by its predictions. Thus in the Mendelian-biometric controversy of the early twentieth century, biologist W. E. Castle claimed in response to biometrician and statistician Karl Pearson's attack on Mendelism "that he could predict the specific results of crosses using Mendelism, and Pearson could not do so using the law of ancestral heredity. . . . Castle, according to historian Will Provine, tended to publish bold hypotheses and regarded their refutation as an advance in knowledge"; others, like geneticist Sewall Wright, "preferred to get the hypothesis right the first time and dispense with the retractions." Castle's successful prediction of the color of guinea pigs was, Provine writes, one reason why the Mendelians defeated the biometricians.

According to historian Jane Maienschein, the increasing popularity of experimental research in the early twentieth century led biologists to call for testable predictions. Not surprisingly, we find examples of the fallacy of affirming the consequent: the success of a prediction confirms the truth of the hypothesis.

Morgan's chromosome theory of heredity did make novel predictions as well as deductions of known facts. One might argue that genetics is not typical of biology in this respect. Philosopher Mary Williams notes that when biologists are asked to give an example of prediction in biology, they are most likely to cite only the Hardy-Weinberg law in genetics. But the Morgan theory is not just an important theory in genetics, it is an interfield theory involving both genetics and cytology (the study of cells). (The nature of interfield theories in biology has been discussed by philosopher and historian Lindley Darden.) Moreover, Morgan's chromosome mapping is ancestral to today's human genome project, the biggest single project in recent biology. Evidence that cytogenetics *does* follow the hypothetico-deductive method would thus seriously undermine any general claims that biology does not.

13.3 PRECURSORS

The study of chromosomes began in the nineteenth century with microscopic observations of the structure of cells and their behavior during reproduction. A major step forward was the proposal by cytologist and zoologist Clarence E. McClung, in 1901–1902, that the accessory chromosome discovered by zoologist Hermann Henking might play an important part in sex determination. Then, in 1902, biologists Walter S. Sutton and Theodor Boveri proposed, independently, that chromosomes could provide the basis for Mendelian inheritance. In 1905 cytogeneticist Nettie M. Stevens discovered that the male Tenebrio (darkling) beetle has a small chromosome (now called Y) corresponding to the large accessory chromosome (now called X) in the female. Shortly afterward cytologist and embryologist Edmund B. Wilson made a similar discovery for Hemiptera (winged insects), with the significant difference that there is no chromosome in the male corresponding to the X in the female.

Stevens and Wilson established that a specific character—sex—was associated with a factor located on a particular chromosome, suggesting that other characters might also be localized on particular chromosomes. Although they did not claim to have discovered a *universal* mechanism of sex determination—there were too many exceptions—their results did provide a natural transition to the theory Morgan subsequently proposed. As geneticist and zoologist Leonard Doncaster wrote in 1914, if we accept sex-determination by chromosomes it is "almost impossible to reject the belief in a similar connection with Mendelian factors." According to Maienschein, "it was sex determination that led Wilson into the nucleus, pushed Morgan to genetics, and illustrated the separability of the related problems of sex inheritance and sex development or production. Complex problems sorted out for sex determination pointed the way to understanding more general problems of heredity and development."

Morgan's early work on the chromosome theory dealt with other characters linked to sex in inheritance and thus presumably also located on one of the sex chromosomes. In fact, it was this work that converted him from being an opponent of the chromosome theory to being its most powerful advocate.

13.4 MORGAN'S THEORY

I will state the Morgan theory as it was developed and presented by Morgan and his collaborators, and as it was described by authors of genetic textbooks and review articles in the period 1915–1930. Although these authors could have obtained their knowledge of Morgan's theory from many of the numerous publications of this prolific group, it appears that the single most important source was the monograph The Mechanism of Mendelian Heredity (1915), coauthored by Morgan, Alfred Henry Sturtevant, Herman Joseph Muller, and Calvin Blackman Bridges.

Morgan was a full generation older than the other three (see Table 13.2); in fact when the group was formed in 1910, Bridges and Sturtevant were undergraduate students at Columbia University, where Morgan was a full professor. Hence the group, together with those who came later, was often referred to as Morgan's school. I have listed in Table 13.1 those who were authors or coauthors of papers from Morgan's laboratory from about 1911 to about 1926. Their median age in 1915 was 26, not counting Morgan himself (49) and his wife Lillian Vaughan Morgan (45).

Mechanism presents nearly all of the major arguments for Morgan's chromosome theory that contributed to its early acceptance by geneticists. Its basic postulates are:

1. *The hereditary material is carried primarily, if not entirely, by the chromosomes*. These contain factors or genes that persist through the reproductive process. The alternative, cytoplasmic inheritance is of little importance except in a few special cases.
2. *The genes are arranged in a linear sequence on the chromosomes, in such a way that a gene for a particular character is located at a particular place on a particular chromosome*. During the process of reproduction the

Table 13.1 Morgan's School

Name	Date of Columbia Ph.D.	Institutions	Age in 1915
Bridges, Calvin B.	1916	Carnegie Inst. Washington; Calif. Inst. Tech.	26
Browne, Ethel Nicholson see Harvey, Ethel Browne			
Dexter, John S.	1914	U. of Saskatchewan, Northwestern U., U. Puerto Rico	30
Goodale, H. D.	1913	Mass. Agric. Expt. Station; Mt. Hope Farm	36
Gowen, John W.	1917	Maine Agric. Expt. Station; Rockefeller Inst.	22
Harvey, Ethel Browne	1913	Cornell, New York U.	30
Hyde, Ida Henrietta	[1896]	Kansas State Agric. College	58
Hyde, Roscoe R.	1915	Terre Haute Vet. College, Johns Hopkins U.	31
Lancefield, Donald E.	1921	Oregon, Columbia, Queens Coll. Flushing	22
Lynch, Clara Julia	1919	Rockefeller Inst. (New York City)	33
McEwen, Robert S.	1917	Oberlin College (Oberlin, Ohio)	27
Metz, Charles W.	1916	Cold Spring Harbor; Johns Hopkins U.	26
Morgan, Lilian Vaughan		Columbia U.	45
Morgan, Thomas H.	[1890]	Columbia U., Calif. Inst. Tech.	49
Muller, Hermann J.	1916	Columbia U., U. Texas, Leningrad U., Moscow U.	25
Payne, Fernandus	1909	Indiana U.	34
Plough, Harold H.	1917	Amherst College (Amherst, MA)	23
Plunkett, Charles R.	1926	New York U.	23
Safir, [Dean] Shelley Ray	1920	City College of New York; Yeshiva U. (New York City)	24
Schrader, Franz	1919	Bryn Mawr College (Bryn Mawr, PA), Columbia U	24

(*Continued*)

Table 13.1 (Continued)

Name	Date of Columbia Ph.D.	Institutions	Age in 1915
Shull, Aaron Franklin	1911	U. of Michigan	34
Stern, Curt	[1923]	U. Rochester (NY), U. Calif. Berkeley	13
Strong, Leonell Clarence	1922	St. Stephens College, Michigan; Jackson Lab. (Bar Harbor, ME)	21
Sturtevant, Alfred H.	1914	Carnegie Inst, Wash. (Columbia), Calif. Inst. Tech.	24
Weinstein, Alexander	1920	Carnegie Inst. Wash., U. Illinois	22
Zeleny, Charles	[1904]	U. Illinois	37

Date of Ph.D. is in brackets if not from Columbia.

chromosomes interact with each other with the possibility that segments of one chromosome will be exchanged with the corresponding segment of the other chromosome of the same kind. During this crossing over event, the probability that two genes originally on the same chromosome will be separated is an increasing function of their distance apart.

3. *In a given species both sexes have the same set of chromosomes except for two special chromosomes, called X and Y.* Thus in humans the male has one X and one Y, while the female has two X's. In a few anomalous cases, referred to as nondisjunction, the two X chromosomes of a female fail to separate during reduction division, leading for example to XXY females.

As evidence for these postulates, Morgan and his collaborators listed the following:

1a. *The behavior of the chromosomes during reduction division is exactly the same as required by the Mendelian laws.* The absence of cytoplasmic inheritance is best supported by Boveri's experiment with sea urchins: when the amount of cytoplasm is reduced there is no effect on the amount of parental influence.

They mentioned also the argument that since only the nucleus of the sperm penetrates the egg, yet both parents contribute equally to inheritance, there must not be any non-nuclear inheritance; they wrote that the argument is weak because there may be a thin enveloping protoplasm around the nucleus.

1b. Some genes do not segregate independently *in reproduction but are linked together; for example several different characters are linked to the sex chromosomes.* Detailed study of a single organism, *Drosophila melanogaster* (vinegar or fruit fly with dark belly) shows that all the characters can be divided into four linkage groups. *Drosophila melanogaster* happens to have four different chromosome pairs. This is exactly what one would expect if each chromosome carried a specific set of genes that determined characters that are inherited together. In the other organisms studied, there is not enough data to identify all the linkage groups, but in no case have more linkage groups than chromosome pairs been found.

I will call this the linkage group theorem: the number of linkage groups is less than or equal to the number of chromosomes (i.e., the number of pairs of identical chromosomes).

2a. *If the crossover frequencies for three genes A, B, C on the same chromosome (same linkage group) are measured, then the locations of these genes can be determined.* For genes close together the distances follow the relation AB + BC = AB. Conversely if one measures two of the cross-frequencies and uses those values to determine the locations of the genes, one can then use the above relation to predict the third frequency. The prediction is confirmed for small distances.

For larger distances where the results are less than predicted from summing the distances between intervening genes, one can account for the results by postulating multiple crossovers.

3a. *We would expect that the white-eyed XXY female produced by nondisjunction would "repeat the process and be nondisjunctional. This is what actually occurs, for all white females that are the product of such a cross do, in fact, give nondisjunction in the next generation."* Other predictions from the hypothesis are also confirmed.

Nondisjunction, predicted and discovered by Bridges, was stated to be "the most direct evidence yet obtained concerning the relations of particular characters and particular chromosomes." Bridges proposed the hypothesis in July 1913 to explain the fact that a few percent of the daughters of *Drosophila* matings did not possess the expected sex-linked character of their fathers but instead looked like their mothers. He argued that the observed deviations from the expected results could be explained if the two X chromosomes from the mother failed to separate and were then joined by fertilization to a Y chromosome from the father. He did not use the word "prediction" in this paper, but did say there was not yet any direct cytological evidence for nondisjunction in *Drosophila* (although something like it had been observed in other organisms).

A year later, Bridges did explicitly mention the "prediction . . . that half the daughters of a nondisjunctional female would be found to contain in addition to the two X chromosomes a supernumerary chromosome which is a Y," and reported that it had now been confirmed by cytological investigation.

13.5 THE PROBLEM OF UNIVERSALITY

Morgan's chromosome theory of heredity was generally regarded as a theory that should be valid for all organisms. One might think that its acceptance would depend on its confirmation for a number of different animals and plants. As geneticist E. M. East noted, biologists tend to be hypercritical; they "often require a large amount of affirmative data before assenting to a proposition which is in reality a simple corollary of one already accepted." Many philosophers of science argue that "there is not much generality in biology."

For example, the linkage group theorem was predicted to be true for all organisms. In 1915 there were not enough data on organisms other than *Drosophila* to test this prediction. Geneticist William Bateson and others gave the absence of such data as a reason not to accept Morgan's theory, and Orland E. White stated that the theory would be disproved if an organism were discovered for which the linkage group theorem is violated. Even those who accepted the theory saw a need to test it on other organisms. But according to cytogeneticist and evolutionist Cyril Dean Darlington, geneticists who studied *Oenothera* (evening primrose) thought that plants might be different from animals in this or other respects.

Morphologist and geneticist R. C. Punnett devoted considerable effort to testing the linkage group theorem for the sweet pea (*Latharus odoratus*), for which a large amount of data had been collected. Acknowledging the "brilliant researches of Morgan and his colleagues" who "showed beyond a doubt" that the theorem is valid for *Drosophila*, Punnett "decided to carry on with the programme of the sweet pea work, especially as it seemed at one time not unlikely that the number of independent groups of characters might prove to be greater than that of the chromosomes." He tentatively concluded that the theorem is indeed valid for the sweet pea, though some years of work were still needed to show that.

According to Punnett, no other confirmations of the linkage group theorem had yet been found four years later, when he reported that as a result of decades of research by himself and others, the sweet pea *Lathyrus* has seven linkage groups and seven different chromosomes: "We now have a plant as well as an animal in which this fundamental postulate of the chromosome theory is fulfilled."

There were, of course, very good reasons why the theorem could not easily be tested for other organisms. Most of them have more chromosomes that *Drosophila*; more importantly, their rate of reproduction is much slower (the span of a generation is measured in months or years rather than in minutes as for *Drosophila*.) Thus the huge amount of work and time required to establish linkage groups precluded quick testing of the theorem.

13.6 MORGAN'S THEORY IN RESEARCH JOURNALS

In studying the reception of a new theory or discovery by the scientific community, one usually examines the impact of one or a few important publications as reflected in a relatively large number of publications by other scientists. But in this case the published articles of the originators and developers of the new idea form a significant part of the total pool of articles in the relevant specialty, and their authors were beginning to dominate the profession, at least in the United States. While Morgan was clearly the leader of the group or school, his principal associates in the 1910s—Bridges, Muller, and Sturtevant—made some of the most important discoveries. His was one of the first scientific laboratories where students were treated as colleagues and encouraged to be coauthors or even sole authors of papers. In addition, 20 or 25 other students and collaborators published at least one paper using or testing

Morgan's theory and could thus be considered part of Morgan's school in the broader sense (see Table 13.1). Their influence on American biology is suggested by the fact that 11 were starred in the 1933 edition of American Men of Science, meaning they were considered especially eminent. Historian Hamilton Cravens estimated that by 1928, 24 of the Columbia department's 84 PhDs had written genetics dissertations primarily with Morgan, and entered the profession between 1914 and 1928; he also noted the expansion in the number of geneticists after 1910.

Morgan's policy of freely sharing research material with other scientists often converted them to supporters as well. He was known for his generosity in other ways: loaning eight dollars to the British statistician Ronald Fisher when he ran out of cash on a trip to the United States; and posting a two thousand dollar bond for one of his rivals, the German biologist Richard Goldschmidt, who was interned as an enemy alien at the end of World War I, even though Morgan declined to recommend that Goldschmidt be released merely on the grounds of his scientific eminence.

A rather different picture of Morgan's personality and behavior is presented by science writer James Schwartz, who tells the story of the chromosome theory from the perspective of H. J. Muller. According to Schwartz, Muller accused Morgan of stealing credit from his three student collaborators for their discoveries, and creating an atmosphere in which the students stole from each other. Schwartz implies that Muller, rather than Morgan, was the most important geneticist in the first half of the twentieth century. Since the validity of this claim does not affect my conclusions concerning the reception of the chromosome theory, I will leave that issue to be settled by other historians.

Morgan's institutional resources must be given some weight in explaining the rapid acceptance of his chromosome theory. Although he did not have ample funds to pay research expenses until after he started getting funds from the Carnegie Institution of Washington in 1915, he enjoyed more important—if less tangible—advantages. He had already established his reputation as a biologist before starting work on the theory; he was a professor at a major university who could attract several bright and energetic students. He was on the editorial boards of several journals (including *Biological Bulletin, Genetics, Journal of Experimental Zoology*) and was a respected adviser to the editor of the *American Naturalist*, a journal widely read by biologists and influential among geneticists. So he apparently had no difficulty in quickly publishing his work and that of his school, especially after establishing *Genetics* in part for this purpose.

With the help of these resources and the good fortune of a very fruitful experimental organism (*Drosophila melanogaster*), the Morgan school flooded the major American biology journals with so many papers that any attentive reader would have gotten the impression that his school's research dominated the mainstream of cytogenetic research in the 1910s. I have scanned some of these American journals for the crucial decade 1913–1922, a period that begins with Sturtevant's first chromosome map and Bridges's discovery of nondisjunction of the sex chromosomes and ends with the capitulation (even though temporary) of their principal rival, Bateson.

In the two journals that published the largest number of papers on chromosomes, *American Naturalist* and *Genetics*, during the period 1913–1917, Morgan's group published 40 papers, out of a total of 127 on chromosomes. In the period 1918–1922, Morgan's group published 28 out of a total of 146 on chromosomes. So faithful readers of those two journals were certainly somewhat familiar with Morgan's theory.

What about the *authors* of papers who were not members of Morgan's group? In the issues of the two journals published in 1913–1917, 30 papers by those authors favored Morgan's theory or at least cited it. For 1918–1922, that number increased to 46, while the total number of chromosome papers by non-Morgan authors increased from 58 to 67. So the average rate of acceptance (in a minimal sense) had increased from 52% to 69%, suggesting that the tipping point (50%) had been reached around 1916 ± 1, only about a year after the publication of *Mechanism.*

In Great Britain the major publication for papers in this field was the *Journal of Genetics* (which started in 1916). Using a similar procedure for estimating the rate of acceptance, I found that it increased from 36% in the period 1923–1927 to 53% in 1928–1932, suggesting a tipping point around 1929 ± 1.

13.7 IMPORTANT EARLY SUPPORTERS

In discussing the response of individual scientists to Morgan's theory, I start with the leaders in genetics—members of the National Academy of Sciences and scientists starred in *American Men of Science*, and fellows of the Royal Society of London—since their public statements were likely to influence the rest of the biology community.

See Table 13.2 for the following reasons were most frequently mentioned by the 30 US Leaders and 15 British leaders.

Note that nondisjunction, which impressed American geneticists more than British, was a very specific confirmed prediction about an infrequent phenomenon, but was not often referred to as such. The linkage group theorem, on the other hand, was a very general statement about heredity in all animals and plants; its universal validity could have been called a prediction but usually was not. Moreover, the theory was accepted when the linkage groups had been established for only a few organisms. The word "prediction" was more often used to characterize the connection between crossover frequencies and gene locations.

Table 13.2 Reasons Why US and British Biologists Accepted Morgan's Chromosome Theory of Heredity

	US	British
Number of *linkage groups* is less than or equal to the number of chromosome pairs	11	6
Mapping of chromosomes	11	5
Nondisjunction	8	2
Explains *Mendelian genetics*	7	6
Successful predictions (unspecified but probably refers to prediction of crossover frequency from positions on chromosome map, and conversely)	6	4
Sex-linked inheritance (e.g., recessive white-eyed Drosophilia are predominantly male)	6	2
Interference effect in crossing over	1	0
Chromosome length is proportional to number of genes	0	1

Doncaster's response is perhaps the most significant because it pinpoints a reason why many geneticists accepted Morgan's theory, while providing a counterexample to the view that British geneticists, under the influence of William Bateson, were slow to recognize its merits. The first prominent non-American supporter of the theory, he was influenced by Wilson's work on sex determination (though his own theory was somewhat different); as Bateson wrote (anonymously), "the element of apparent fundamentality which he found in cytology very strongly appealed to Doncaster's analytical mind, and he was therefore from the first greatly attracted by the theory of linkage propounded by Morgan."

In a review article on heredity and sex determination, Doncaster began with the view that Morgan's crossing-over hypothesis "must be regarded as entirely speculative," but could not resist coming back to it again and again. He pointed out that "there is one difficulty, however, which makes this hypothesis doubtful, in spite of its beautiful simplicity. It is that exceptions to sex-limited transmission occur in almost all the known cases." But in a note added before publication, he cited Bridges's suggestion that these exceptions are caused by nondisjunction of the sex chromosomes: "This explanation, if substantiated, would account for most of the recorded exceptions, but not that of the tortoise shell male cat." He considered that objections by a few researchers were either contradicted by other work or not relevant, so the net impression left by the article was generally favorable to Morgan's theory.

By 1920 Doncaster was fully converted to Morgan's theory, though he failed to carry Bateson with him. His book on cytology thoroughly reviewed the evidence and pointed out how it supported the theory on almost every point. Despite some objections, he concluded that Morgan's hypothesis is clearly the best available and is "the only one which seems at all nearly to approach the truth." Reviewing the latest publication of the Morgan group in *Nature*, he declared: "It is impossible to read the facts presented in this volume without being impressed by the great strength of the evidence for Morgan's theory that Mendelian factors are borne by chromosomes and arranged in definite sequence within them. Difficulties remain, but a theory which enables predictions to be made and verified cannot lightly be disregarded."

By this time Doncaster had more than enough reasons to accept the theory, but it is important to recall the one that first converted him—nondisjunction—because the same evidence also seemed decisive to several other geneticists. Why?

As often happens in biomedical research, one learns how a system works only when it malfunctions, or when an essential part is destroyed. (The strange story of Phineas Gage, discussed in detail by Malcolm Macmillan, is one of the first examples.) The fact that the observed normal behavior of chromosomes during reproduction corresponds closely to the observed Mendelian inheritance was not quite enough to convince geneticists that the latter is *caused* by the former. Bridges's 1913 discovery that abnormal chromosome behavior (failure to separate at the appropriate time) corresponds to abnormal inheritance in a completely deterministic, predictable fashion nailed down the argument in a dramatic way. The research was, in the words of Harry Federley , "idealisch schönen " (absolutely beautiful); it confirmed the theory's prediction, according to E. Guyénot, "d'une façon éclatante" in a brilliant manner. Bridges himself called it a proof of Morgan's theory when he published his thesis three years later, and many other geneticists (though not all) agreed with him or at least mentioned it as one of several strong pieces of evidence.

Geneticist Edward M. East, in his 1914 Harvard lectures, argued that while the correspondence between normal chromosome behavior and Mendelian genetics was strong evidence for a causal relation, nondisjunction—a confirmed novel prediction—and the linkage group theorem left no room for doubt: "the facts fit perfectly all that is known of chromosome behavior. It seems impossible, therefore, that there should be so many coincidences." Moreover, "it was proven cytologically *after the prediction had been made from the breeding facts* that these females resulted from the nondisjunction of the X chromosomes . . . This appears to be definite proof that sex-linked genes are borne by the X chromosomes." (Emphasis in original.) (Novelty does count.) The number of linkage groups provides a crucial test of the theory: "*If one single character should be found that did not fit into one of these four groups, the whole theory would break down. But no such character has appeared.*" (Emphasis in original.) In later papers he briefly reaffirmed the validity of the Morgan theory.

E. B. Wilson was invited to give the Croonian Lecture at the Royal Society of London in 1914 and took the opportunity to promote Morgan's theory to British biologists. He summarized the work of Morgan's school, including the "very convincing evidence" from nondisjunction found by Bridges, and also discussed chromosome maps constructed from crossing-over data. Alluding perhaps to the reason why Bateson and his colleagues were skeptical about Morgan's theory, he remarked: "To those not actually engaged in such [cytological] investigations this hypothesis will, perhaps, seem of highly speculative character." Cytology is an unfamiliar field in which "hypothesis and speculation have continually run far in advance of observation and experiment," so one cannot be blamed for being skeptical. But for those familiar with both cytology and genetics, the theory is "irresistible."

Seven years later, in a lecture in Boston, Wilson insisted that the experiments of Morgan and his colleagues " have almost revolutionized the study of heredity"; their result that genes have a definite serial order on the chromosome is "indeed staggering—to a certain type of mind even harder to assimilate than those [results] which physicists are now asking us to accept concerning the structure of atoms. *Nevertheless they are probably true* . . . They make possible precise quantitative prediction concerning the outcome of new experiments. In these respects they are employed in the same way as the exact concepts of the chemist or the physicist and they may, I think, lay claim to a validity of the same kind even if it be not yet quite of the same degree." Yet despite all its successes, some biologists were still hostile to the theory, Wilson complained. But other enthusiastic endorsements of Morgan's theory by prominent American biologists appeared in the next few years. Historians of biology seem to agree that Morgan's chromosome theory was widely accepted by American geneticists by 1920.

13.8 BATESON AND THE MORGAN THEORY IN GREAT BRITAIN

In Great Britain, Morgan's theory was initially rejected by the leading geneticist, William Bateson. He had proposed his own presence and absence theory—which had previously been "generally accepted by students of these matters," according to his colleague R. C. Punnett, though other geneticists found it seriously flawed. By 1914, according to Morgan, Bateson had "dissociated himself from the movement already in full swing that identified the genetic elements with the stable materials of the chromosomes."

In his review of *Mechanism of Mendelian Heredity*, Bateson conceded that Bridges's work on disjunction was an impressive confirmation of a prediction from the theory, but still regarded it as a "complex web of theory . . . so exceedingly elastic that it can be fitted to any facts." Four years later, delivering the Croonian Lecture to the Royal Society of London, he argued: " There is no body of evidence that the number of linkage-systems agrees with that of the chromosomes, a primary postulate of Morgan's theory. *Drosophila* is the only example which has been adequately investigated. The cytological appearances are not readily consistent with the other postulate of Morgan's case, that crossing-over is effected by . . . exchange of materials between [chromosomes] . . . Without personal familiarity with cytology no one can have a confident opinion. . . . The appearance of chromosomes is not to me suggestive of strings of bands of extreme heterogeneity, but rather of strands of some more or less homogeneous substance."

But in December 1921, Bateson visited Morgan and had the opportunity to see for himself in the laboratory the phenomena on which Morgan's theory was based. He was forced to admit that chromosomes were in some way involved in heredity, though he also said that "the details of the linkage theory still strike me as improbable."

In an address a few weeks later to the Toronto meeting of the American Association for the Advancement of Science, Bateson said that the validity of the main thesis could no longer be doubted. Yet he continued to criticize Morgan's theory, and as late as 1926 published a lecture (originally delivered in 1922) defending his presence and absence theory. He again claimed that the evidence supported his own theory, and resented the fact that prominent geneticists like Johannsen had instead adopted what he called the "American view." He did however concede that the Bateson-Punnett reduplication hypothesis "fades inevitably into the background" because of the superior explanation provided by Morgan's theory.

Although Bateson continued to be skeptical about the theory, his 1922 concession on its major postulates freed Punnett to support it (if he had previously been constrained by loyalty to his colleague) and probably broke down the resistance of other British biologists, thus effectively ending the period during which professional geneticists could reasonably oppose Morgan's theory.

Perhaps the best indication that Morgan's theory had been (at least tacitly) accepted by leaders of the British biological community came in December 1924, when the Royal Society of London awarded him its Darwin Medal. According to the citation, his results "throw light on the relation of the factors of heredity to the chromosomes . . . these researches mark an advance in the science of heredity with which the name of T. H. Morgan will always be associated."

13.9 THE PROBLEM OF UNIVERSALITY REVISITED

Bateson's research group at the Innes Institution did not seem to feel any obligation to support their boss's presence and absence theory, though they did not publicly attack it while he was still alive. But his body was still warm in his grave when younger geneticists who owed their research support to his patronage began trying mightily, in accordance with Planck's principle, to confirm the theory of his nemesis, Morgan. In January 1927, Aslaug Sverdrup published a remarkable article in Volume 17 of the *Journal of Genetics*, a volume with Bateson's photograph as its frontispiece but now edited solely by Punnett. Sverdrup expressed the geneticists' frustration at trying to confirm the linkage group theorem for *Pisum* (the pea), "a well-known classical object of genetical research," with "a comparatively great many factors . . . showing simple mendelian segregation." Despite intensive efforts to reduce the number of groups to seven (to agree with the theorem), this goal had not yet been attained.

Did this failure cast doubt on the validity of Morgan's theory, given that its fundamental theorem had been verified only for a few species of *Drosophila* and had "so far not been established for a certainty within any other genetical material"? Not at all. The theorem had become an essential part of a Kuhnian paradigm or (if you prefer) a Lakatosian research programme (see Section 3.2), and failure to confirm it could no longer be regarded as a crucial test of the validity of that paradigm or program—it was instead a reflection on the competence of the experimenter or his or her ingenuity in imagining suitable supplementary hypotheses to explain away the discrepancy. As Sverdrup explained, "one can only state that so far the number of factors and the number of chromosomes in *Pisum* fail to show agreement. To my mind this need not necessarily affect the theory of chromosomal inheritance but does perhaps indicate a looser structure of the chromosomes within certain species."

Another student at the Innes Institution, C. D. Darlington, simply ignored the lack of confirmation of the linkage group theorem, asserting that "studies have supported the original conclusion [of the Morgan theory] with reasonable consistency. We may therefore assume the general truth of the theory."

One might ask how supporters of Morgan's theory got away with such statements, while Bateson was criticized by other scientists and by later historians for asserting the general validity of his presence and absence theory on the basis of only two experiments and vague references to "all observations" and "recent experience." The answer is that Morgan's group really did have a large amount of evidence for their theory, even if much of it was based only on *Drosophila*, and moreover their theory was internally consistent, while Bateson's was not, and led to absurd consequences when applied to evolution. After Punnett's strong endorsement of Morgan's theory in the seventh edition (1927) of his authoritative treatise on *Mendelism*, it was difficult even for the British to continue defending Bateson's theory.

Punnett's report that the linkage group theorem is valid for the sweet pea *Lathyrus* (7 chromosome pairs), also published in 1927, seemed to provide the long-sought confirmation for the plant kingdom, though it was based on a rather small amount of evidence (as compared with that for *Drosophila*) and only a few geneticists recognized its significance. Another eight years passed before the group led by agriculturist and geneticist Rollins Adams Emerson could state confidently that the postulate also holds for maize, one of the most intensively studied plants: "It can now be said, with a considerable degree of assurance, that ten independent linkage groups corresponding to the ten chromosomes are known."

For another perspective on the difficulty of confirming the linkage theorem for several organisms, it should be noted that in some cases the problem was not just in establishing the number of linkage groups but in counting the number of chromosomes. For example, during the 1930s and 1940s it was generally believed that humans have 48 chromosomes (24 pairs), a number now believed to be too large by two.

13.10 BOOKS AND REVIEW ARTICLES ON GENETICS, EVOLUTION, AND CYTOLOGY

John A. Moore surveyed several books on inheritance published in the years just before 1913, as an indication of the views of their authors around 1910. He found considerable skepticism about the idea that chromosomes could be the primary carriers of inheritance. He found the most favorable statement about that idea in a book by the British biologist R. H. Lock. Morgan's theory began to reach a wider audience of biologists through books and review articles published soon after the four-man work of 1915. Many of these were written by the early supporters already mentioned, and their reasons for accepting the theory do not need to be quoted here insofar as they were similar to the views in their research articles. But from my viewpoint it is significant that while Doncaster's review article for practitioners (see Section 13.7) stressed the importance of nondisjunction, his textbook on cytology omitted this item from a list of four main arguments in favor of Morgan's theory, though he had mentioned it earlier as another piece of evidence that further strengthened the theory. Similarly, A. F. Shull, who had called attention to the importance of nondisjunction in a research paper, stressed instead the linkage group theorem in his textbook on heredity. Books on genetics by several prominent American or Canadian biologists strongly endorsed Morgan's theory, noting its success in locating genes on chromosomes and in explaining Mendel's rules. One of them, Herbert Eugene Walter, also praised the predictiveness of the theory and called Morgan's analysis "one of the fundamental principles of heredity." He asserted that " the work of gene-localization is quite comparable to that done by mathematicians and astronomers in determining the distances that separate the stars in the heavens from each other." Horatio Hackett Newman called Morgan's analysis of the heredity machine " so detailed as to be almost unbelievable. It seems too good to be true, yet the keenest critics of the work have failed to find any real flaws."

L. W. Sharp gave a favorable but balanced evaluation of Morgan's theory in his textbook on cytology. The predictions of crossing-over frequencies have been confirmed: "such an agreement of the results of new crosses with predictions made on the basis of known linkages has occurred over and over again in the experiments of Morgan and his students. The . . . hypothesis as thus elaborated fits the observed facts remarkably well." Sharp's final assessment of the theory, which seems to be all the more convincing because of his careful effort to evaluate all its deficiencies and alternative explanations, is that it reduces to order "a huge number of the observed facts of inheritance." It "seeks to make use of known cell mechanisms rather than entirely hypothetical processes"; even those parts of the theory "which are as yet unsupported by the results of direct cytological observation, though not contradicted thereby, at least have the virtue of affording a useful and graphic representation of the mutual behavior of hereditary characters," and "it is scarcely to be doubted that the chromosome theory of heredity in some form will turn out to be in accord with the truth."

Many other American textbook authors presented Morgan's theory with similar phrases in the 1920s and presumably influenced students and colleagues at their own institutions. To conclude this survey of American textbook authors I quote from zoologist A. Franklin Shull's 1926 book. As a result of the work of Morgan and his students, he wrote: "far greater progress has been made in our understanding of heredity in the last fifteen years than in all the centuries before that time. As a further result, knowledge of heredity has proceeded further toward the condition of an 'exact science' than any other division of biology."

In Great Britain, following Bateson's death in 1926, Punnett could proclaim his strong support for Morgan's theory and assert that it "is now generally accepted by geneticists." After that, few British authors publicly opposed the theory. At least 11 publicly supported it before 1927. They most frequently cited chromosome maps, agreement with Mendelian rules, and the linkage group theorem as reasons for accepting the theory.

Most of these biologists gave moderate or qualified support. But by the end of the 1920s there were only three dissenters among book authors who held academic positions in Great Britain. Ernest William MacBride strongly attacked the theory, despite an earlier favorable attitude. J. Henry Woodger rejected the theory as incompatible with his empiricist methodology. Alexander Meek favored the cytoplasm rather than the chromosomes as carrier of heredity. But several other authors did accept the theory and called attention to its practical applications.

13.11 BIOLOGY TEXTBOOKS

In the United States, college texts in botany, zoology, and general biology were almost as quick to adopt Morgan's theory as the more specialized books discussed in a previous section. Several authors favored the general view that chromosomes are carriers of heredity in the 1910s, without discussing the details of crossing-over and linkage or the *Drosophila* experiments. The first I have found that mentions the Drosophila research and sex-limited inheritance was published in 1914 by James Francis Abbott; he credits this research with "remarkable results," but doesn't say exactly what they are. Abbot thinks the chromosomes may be involved in heredity, but isn't sure how.

Another textbook published before 1920 was by A. S. Pearse. He endorsed the theory for no specific reason except the authority of T. H. Morgan (who "speaks with enough assurance . . . to be able to assign characters to particular positions on the chromosomes"), and concluded with a three-page quotation from Morgan (including the provocative phrase "the problem of heredity has been solved").

Two prominent biologists endorsed the Morgan theory in their general biology textbooks. Gary N. Calkins called Morgan's research "a brilliant confirmation of Weismann's hypothesis of the constitution of the germ plasm" and asserted flatly that "each set of chromosomes contains all the factors necessary for the complete individual." L. L. Woodruff stated that chromosomes, including those involved in sex determination, behave just like Mendelian factors; the phenomena of linkage and crossing over made it possible to assign relative positions of genes on chromosomes.

Morgan's theory was discussed in seven other zoology or general biology textbooks published in the 1920s. The evidence most frequently mentioned in these books was chromosome mapping and the linkage group theorem.

One might wonder whether botanists would be enthusiastic about a theory based primarily on experiments with animals. Gilbert M. Smith and his colleagues reviewed the evidence for the identity of the number of linkage groups and chromosome pairs in plants. In the pea, only six out of the expected seven linkage groups had been recognized; in corn, eight out of 10 had been found. (They ignored the results of Punnett.) The most they could claim was that "in no species of plant or animal have the genes been found certainly to constitute a number of linkage groups greater than the number of chromosome pairs." Yet they seemed to consider the Morgan theory well-established.

13.12 AGE DISTRIBUTION OF SUPPORTERS AND OPPONENTS

According to Max Planck, "a new scientific truth does not triumph by convincing its opponents and making them see the light, but rather because its opponents eventually die, and a new generation grows up that is familiar with it." Planck's principle is often interpreted to mean that a new theory is more likely to be supported by younger scientists. I have collected data to test this hypothesis using the publications of 132 American and British biologists (not including any members of Morgan's school), in the period 1913–1932. (For the Americans the search was less comprehensive after 1922, because almost all the Americans had accepted the theory before then.) I was able to determine for 132 of them their age in 1915 when the first comprehensive statement of the Morgan theory was published. For each one I looked at their publications to see whether they (1) supported or (2) opposed the theory; the others (3) wrote on chromosomes but did not express any view on the Morgan theory.

The median ages in 1915 of the people in these three categories are 34, 47, and 32, respectively. There were 87 biologists in the first category, 26 in the second, and 29 in the third. This does not pretend to be a statistically significant confirmation of Planck's statement, but is at least roughly consistent with it.

13.13 CONCLUSIONS

Why was Morgan's chromosome theory of heredity accepted in the United States and Great Britain? Since the evidence presented above shows that it was widely if not universally accepted in the United States before 1925 and in Great Britain by 1930, we can exclude some of the reasons later cited as proofs of the theory: the rediscovery of giant chromosomes in the salivary glands and their use in directly identifying certain genes (T. S. Painter); the cytological demonstration of crossing-over for maize (Harriet B. Creighton and Barbara McClintock) and for *Drosophila* (Curt Stern); Muller's X-ray research; and the final proof of the equality of number of linkage groups and chromosome pairs for *Lathyrus odoratus* (sweet peas) by Punnett and for maize by R. A. Emerson's group. As important as these achievements were in confirming some of the plausible hypotheses of Morgan's group and in the development of cytogenetics, they came too late to convert the community to the Morgan theory.

We are left with six major reasons for accepting the Morgan theory. None by itself was a crucial experiment compelling assent, but their cumulative effect was overwhelming:

1. *Sex-linked inheritance* of certain characters such as white-eyes in DM is easily explained by the assumption that the genes for those characters

are located on a sex chromosome. This was the first clear support for that assumption, and was mentioned by eight of the 45 supporters of the theory.

2. More generally, the Morgan theory provides a plausible *mechanism for Mendelian heredity*. This factor was mentioned by 13 supporters. It might have been even stronger evidence, according to Darlington, if it had been a novel prediction: "it has been the misfortune of the chromosome theory of heredity that the chromosomes were discovered before the theory . . . Had genetical theory, based on linkage experiments, advanced to its limits without any knowledge of the chromosomes, the later discovery of these actual organs of transmission would have verified prediction so completely that all objections to the principle would have at once dissolved."
3. The fact that *the number of linkage groups is equal to* (or no greater than) *the number of chromosome pairs* (linkage group theorem) confirms the connection between cytology and genetics, whether or not one accepts the detailed calculations leading to precise locations of genes on chromosomes. Since this number varies for other species than *Drosophila*, one might have thought that the linkage group theorem should have been verified for more than just one species (or the several species of *Drosophila*), but that did not seem to be crucial to accepting the universal validity of the Morgan. Seventeen supporters identified the connection between the number of linkage groups and the number of chromosome pairs as evidence in favor of the theory.
4. *Nondisjunction* was the strongest evidence for professionals familiar with both cytology and genetics, because it provided a clear connection between those two fields: a visible variation in the observable behavior of cells was directly associated with a significant variation in inheritance. Even Bateson was impressed by that. According to Sturtevant, with the publication of Bridges's paper on nondisjunction, "that phase of the history was closed, for since that time no informed geneticist has taken seriously any of the attempts that have been made to discredit the chromosome interpretation." The fact that only 10 supporters mentioned nondisjunction suggests that this was a fairly technical phenomenon, not considered suitable for discussion in elementary texts.

Nondisjunction was a confirmed prediction, but for a physicist reading this literature that fact would not seem to have accounted for much of its evidential value, since East was the only influential supporter who stressed the fact that the prediction came before the confirmation. However, I think that biologists, at least at that time, used the word "prediction" to mean *novel* prediction. (As will be seen in the next chapter, by the mid-twentieth-century biologists were beginning to moved toward the physicists' usage of the word "prediction.")

5. The *ability to predict crossing-over frequencies* and other observable phenomena from the model of genes arrayed on chromosomes was impressive. Morgan himself asserted that "this ability to predict would in itself justify the construction of such maps, even if there were no other facts concerning the location of genes." Other biologists stressed the importance

of predictions. But one might have expected that those biologists who considered predictiveness (the ability to make successful *novel* predictions) an important property of a theory would have waited until the linkage group-chromosome number prediction had been confirmed for a few species—at least for one plant.

Again, the fact that only a few (no more than 10) supporters mentioned it may indicate that predictiveness is more important to researchers than to other biologists and textbook writers.

6. The ability to construct a *map of chromosomes showing the location of individual genes*—even before this could be confirmed by microscopic observations—seemed to have great persuasive power. It was noted, sometimes with great surprise or wonder, by 16 supporters. It is known from studies of other sciences that the use of visual representations is correlated with the "hardness" of a discipline.

All six reasons were present as early as 1920. In his article "Biology Since Darwin" summarizing post-Darwinian developments for a general audience in a book on *Recent Developments in European Thought*, Leonard Doncaster wrote: it is "almost indubitable that inherited characters are in some way borne by the *chromosomes* in the nuclei of the germ cells. The work of Morgan and his school has shown that the actual order in which these inherited 'factors' are arranged in the chromosomes can almost certainly be demonstrated, and his results go far to suggest the conception of the organism . . . as a combination or mosaic of independently inherited features."

Aside from the scientific reasons for accepting the Morgan theory, I have noted some social and psychological factors. The theory was proposed not by an individual scientist, but by a group led by a senior scientist who had access to resources for developing and promoting it. The other members of the group were younger and energetic. The Morgan theory had the best of both worlds: it had the support of influential older members of the American scientific establishment, plus the advantage of being seen as a radical new idea popular with the rising generation of geneticists and cytologists. I have identified a total of 116 American and British biologists who supported the theory during the first two decades (1913–1932), but found only 15 opponents.

Finally it may be noted that the decade 1905–1915 was propitious for a mechanistic theory based on particulate structure and quantitative reasoning, features that seemed compatible with cultural trends in Europe and America; 20 years earlier, or 20 years later, such a theory would have been opposed by fashionable idealism or holism.

Although social and psychological factors may have accelerated the acceptance of the theory I see no evidence that they affected the final outcome. Morgan's theory became part of the core of scientific knowledge not by being socially constructed, but because it provided a very good explanation for empirical facts that could be replicated in many laboratories and a method for successfully predicting new facts. Unlike the cases in the physical sciences discussed earlier, there were hardly any serious attempts to show that an alternative theory could have yielded the same predictions.

NOTES FOR CHAPTER 13

397 *Morgan's Chromosome Theory* For details and additional references see Brush, "How Theories Became Knowledge: Morgan's Chromosome Theory of Heredity in America and Britain," *Journal of the History of Biology* 35 (2002): 471–535.

397 *"The charge is often made that the geneticist finds it impossible"* F. A. E. Crew, "Mammalian Genetics," *Nature* 127 (1931): 847–848, on p. 848.

397 *"the chromosome theory offered abundant predictions"* G. G. Simpson and W. S. Beck, *Life: An Introduction to Biology* (New York: Harcourt, Brace & World, second edition 1965), p. 175. Later in the book they discuss Bridges's nondisjunction prediction, which was "a landmark in the history of genetics" pp. 182, 184.

Section 13.2

398 *Biology . . . advances by discovering facts about specific cases* John Beatty, "The Synthesis and the Synthetic Theory," in *Integrating Scientific Disciplines*, edited by William Bechtel (Dordrecht: Nijhoff, 1986), pp. 125–135, on p. 133; John Beatty, Robert Brandon, Elliott Sober, and Sandra D. Mitchell, "Symposium: Are there Laws of Biology," in PSA 1996 *[Proceedings of the 1996 Biennial Meeting of the Philosophy of science Association]*, Part II, *Supplement to* [the journal] *Philosophy of Science* 64, edited by Lindley Darden (1997), pp. S432–S479.

398 *Castle's successful prediction* William B. Provine, *Sewall Wright and Evolutionary Biology* (Chicago: University of Chicago Press, 1986), pp. 37–38, 53.

398 *led biologists to call for testable predictions* Maienschein, "Arguments for Experimentation in Biology," *PSA 1986 [Proceedings of the 1986 Biennial Meeting of the Philosophy of Science Association]* vol. 2 (Philosophy of Science Association, 1987), pp. 180–195. See Brush, "Morgan's Chromosome Theory," p. 486, notes 18, 19, 20.

398 *an interfield theory involving both genetics and cytology* For definition and discussion of interfield theories see Lindley Darden and Nancy Maull, "Interfield Theories," *Philosophy of Science* 10 (1977): 87–106; for application of the concept to cytogenetics see Darden, *Theory Change in Science: Strategies from Mendelian Genetics* (New York: Oxford University Press, 1991).

Section 13.3

398 *the male* Tenebrio *(darkling) beetle has a small chromosome Studies in Spermatogenesis with especial Reference to the "Accessory Chromosome"* (Washington, DC: Carnegie Institution of Washington, 1905), Publication No. 36. See S. G. Brush, "Nettie Stevens and the Discovery of Sex Determination by Chromosomes," *Isis* 69 (1978): 163–172; Scott F. Gilbert, "Sex Determination and the Embryological Origins of the Gene Theory," M. A. essay, Johns Hopkins University (1975).

399 *Wilson made a similar discovery for* Hemiptera "Studies on Chromosomes. I. The Behavior of the Idiochromosomes in Hemiptera," *Journal of Experimental Zoology* 2 (1905): 371–405.

399 *it is "almost impossible to reject"* "Chromosomes, Heredity and Sex: A Review of the Present State of the Evidence with Regard to the Material Basis of Hereditary Transmission and Sex-Determination," *Quarterly Journal of Microscopical Science* 59 (1914): 487–521.

399 *According to Maienschein, "it was sex determination"* Maienschein, "Preformation or New Formation—Or Neither or Both?" in *A History of Embryology*, edited by

T. J. Horder et al. (New York: Cambridge University Press, 1983), pp. 73–112, on pp. 98–99.

399 *The nature of interfield theories in biology has been discussed* Darden, *Theory Change in Science: Strategies from Mendelian Genetics* (New York: Oxford University Press, 1991).

Section 13.4

399 *the monograph* The Mechanism of Mendelian Heredity Morgan et al., *The Mechanism of Mendelian Heredity* (New York: Holt, 1915). Second edition, 1922; French translation, 1923.

399 *Its basic postulates are* Morgan, *Mechanism*, pp. 135–139, 48–61, 80–81.

399 *As evidence for these postulates* Morgan, *Mechanism*, pp. 110–113; 5 and Chapter III; p. 149.

399 *Some genes do not segregate independently* J.B.S. Haldane wrote, in his "Autobiography in Brief," that "at a seminar for zoology students in 1911, I announced the discovery, from data published by others, of what is not called linkage between genes in vertebrates. My evidence was considered inadequate, and I began breeding mice . . . In 1915 my mouse work with [A. D.] Sprunt and my sister, Mrs. Mitchison, was published." *What I Require from Life: Writings on Science and Life from J. B. S. Haldane* edited by Krishna Dronamraju (Oxford: Oxford University Press, 2009), pp. xxix–xxxv, on p. xxx. He also published two papers on the theory of linkage in *Journal of Genetics* (1919).

402 *Bridges proposed the hypothesis in July 1913* Bridges, "Non-Disjunction of the Sec Chromosomes of Drosophila," *Journal of Experimental Zoology* 15 (1913): 587–606; quotation from Bridges, "Direct Proof through Non-Disjunction that the Sex-linked Genes of Drosophila are Borne by the X-Chromosomes," *Science* 40 (1914): 107–109; see the more extensive discussion in Bridges, "Non-Disjunction as Proof of the Chromosome Theory of Heredity," *Genetics* 1 (1916): 1–52, 107–163.

Section 13.5

403 *they "often require a large amount of affirmative data"* East, "The Genotype Hypothesis and Hybridization," *American Naturalist* 45 (1911): 160–174.

403 *Many philosophers of science agree that "there is not much generality"* R. M. Burian, "Against Generality: Meaning in Genetics and Philosophy," *Studies in History and Philosophy of Science* 27 (1996): 1–29 on p. 2, footnote 6. See also Beatty, "Laws."

403 *gave the absence of such data as a reason not to accept Morgan's theory* William Bateson, "Croonian Lecture—Genetic Segregation," *Proceedings of the Royal Society of London* B91 (1920): 358–368, on p. 361. See also E. B. Babcock, "Crepis—A Promising Genus for Genetic Investigation," *American Naturalist* 54 (1920): 270–276, on p. 271; Louis Blaringham, *Principes et Formules de l;'Hérédité Mendélienne* (Paris: Gauthier-Villars, 1928), p. 134.

403 *White stated that the theory would be disproved* White, "Studies of Inheritance in Pisum. II. The Present State of Knowledge of Heredity and Variation in Peas," *Proceedings of the American Philosophical Society* 56 (1917): 487–588.

403 *plants might be different from animals* Darlington, "The Evolution of Genetic Systems," in *The Evolutionary Synthesis*, edited by E. Mayr and W. B. Provine (Cambridge, MA: Harvard University Press, 1980), pp. 70–80. On Darlington see Robert C. Olby, "Darling, Cyril Dean," *Dictionary of Scientific Biography* 17 (1990): 203–209.

403 *He tentatively concluded that the theorem is indeed valid* "Linkage in the Sweet Pea (Lathyrus orodatus)," *Journal of Genetics* 13 (1923): 101–123.

403 *"this fundamental postulate of the chromosome theory is fulfilled"* "Linkage Groups and Chromosome Number in Lathyrus," *Proceedings of the Royal Society of London* B102 (1927): 236–238, on p. 238.

Section 13.6

404 *starred in the 1933 edition of* American Men of Science Stephen Sargent Visher, *Scientists starred 1903–1943 in "American Men of Science"* (Baltimore: Johns Hopkins Press, 1947).

404 *by 1928, 24 of the Columbia department's 84 PhDs* Cravens, *The Triumph of Evolution* (Baltimore: Johns Hopkins University Press, second edition, 1988), pp. 163, 160.

404 *Morgan's policy of freely sharing research material* G. E. Allen, *Thomas Hunt Morgan: The Man and his Science* (Princeton, NJ: Princeton University Press, 1978).

404 *the story of the chromosome theory from the perspective of H. J. Muller* Muller accused Morgan of stealing credit: Schwartz, *In Pursuit of the Gene: From Darwin to DNA* (Cambridge, MA: Harvard University Press, 2010), p. 186. He also quotes Edgar Altenburg's 1941 list of Muller's 11 "major accomplishments," starting with "the discovery of linear linkage (including interference)" and "the theory of genic interaction and the proper interpretation of the relationship of genes and traits." Muller received the 1946 Nobel Prize in physiology of medicine or medicine for showing that genetic mutations can be induced by exposing chromosomes to X-rays.

Section 13.7

405 *"the element of apparent fundamentality which he found in cytology"* Obituary signed "B" (probably by Bateson), "Leonard Doncaster, 1877–1920," *Proceedings of the Royal Society of London* B92 (1921): xli–xlvi, on p. xliv.

406 *Morgan's crossing-over hypothesis "must be regarded as entirely speculative"* "Chromosomes, Heredity and Sex: A Review of the present State of the Evidence with Regard to the Material Basis of hereditary Transmission and Sex-Determination," *Quarterly Journal of Microscopical Science* 59 (1914): 487–521, on pp. 492, 510.

406 *"a theory which enables predictions . . . cannot lightly be disregarded" An Introduction to the Study of Cytology* (Cambridge: Cambridge University Press, 1920), p. 229. Doncaster, "Genetic Studies of Drosophila," *Nature* 105 (1920): 405–406, on p. 406.

406 *The strange story of Phineas Gage, discussed in detail by Malcolm Macmillan* Lashley, *Brain Mechanisms and Intelligence: A Quantitative Study of Injuries to the Brain* (Chicago: University of Chicago Press, 1929); Macmillan, *An Odd Kind of Fame: Stories of Phineas Gage* (Cambridge, MA: MIT Press, 2000). "In September 1848, an explosion in Vermont drove a 100-centimeter tamping iron through Gage's skull . . . Amazingly, Gage recovered, albeit with a radical change in personality, from personable and conscientious to irascible and rude." Constance Holden, "A Face for Phineas Gage," *Science* 325 (2009): 521, reproduced the only known image of Gage. Anthony Gottlieb wrote: "Much of what we know about the brain comes from seeing what happens when it is damaged." "A Lion in the Undergrowth," *New York Times Book Review*, January 30, 2011, p. 12.

406 *leaders in genetics* See Brush, "Morgan's Chromosome Theory," pp. 490–496.

406 *nailed down the argument in a dramatic way* Elof Axel Carlson, *The Gene: A Critical History* (Philadelphia: Saunders, 1960), p. 67.
406 *The research was, in the words of Harry Federley, "idealisch schönen"* Federley, "Chromosomverhältnisse bei Mischingen," *Verhandlungen de V. Internationale Kongresses für Vererbungswissenschaft*, Berlin, 1927, pp. 194–222, on p. 195.
406 *according to E. Guyénot, "d'une façon eclatante"* Guyénot, "L'Oeuvre de T. H. Morgan et le Mécanisme de l'Hérédité," *Revue Generale des Sciences pures et appliquées* 79 (1918): 262–269, on p. 266.
406 *Bridges himself called it a* proof *of Morgan's theory* "Non-Disjunction as Proof of the Chromosome Theory of Heredity," *Genetics* 1 (1916): 1–52, 107–163.
406 *many other geneticists (though not all) agreed with him* See Brush, "Morgan's Chromosome Theory," note 83.
407 *the linkage group theorem left no room for doubt* East, "The Chromosome View of Heredity and its Meaning to Plant Breeders," *American Naturalist* 49 (1915): 457–494, on p. 474; further references in Brush, "Morgan's Chromosome Theory," note 85.
407 *for those familiar with both cytology and genetics, the theory is "irresistible"* Wilson, "Croonian Lecture: The Bearing of Cytological Research on Heredity," *Proceedings of the Royal Society of London* B88 (1914): 333–352, on pp. 345, 352.
407 *"have almost revolutionized the study of heredity"* Wilson, *The Physical Basis of Life* (New Haven: Yale University Press, 1923), on pp. 15, 29.
407 *other enthusiastic endorsements of Morgan's* See Brush, "Morgan's Chromosome Theory," pp. 498–503.

Section 13.8

407 *which had previously been "generally accepted by students of these matters"* Punnett, *Mendelism* (New York: Macmillan, 1911), p. 39; R. G. Swinburne, "The Presence-and-Absence Theory," *Annals of Science* 18 (1962): 131–145.
407 *Bateson had "dissociated himself from the movement"* Obituary by T. H. Morgan, "William Bateson," *Science* 63 (1926): 531–535.
408 *Bridges's work on disjunction was an impressive confirmation* Bateson, review of Morgan et al., *Mechanism of Mendelian Heredity*, in *Science* 44 (1916): 536–543.
408 *"There is no body of evidence that the number of linkage-systems agrees"* "Croonian Lecture—Genetic Segregation," *Proceedings of the Royal Society of London* B91 (1920): 358–368, on pp. 361, 363.
408 *was forced to admit that chromosomes were in some way involved in heredity* "Evolutionary Faith and Modern Doubts," *Science* 55 (1922): 35–61, on p. 57.
408 *Yet he continued to criticize Morgan's theory* R. T. Gregory, D. De Winton and W. Bateson, "Genetics of *Primula sinensis*," *Journal of Genetics* 13 (1923): 219–253; Bateson, "Segregation: Being the Joseph Leidy Memorial Lecture of the University of Pennsylvania, 1922," *Journal of Genetics* 16 (1926): 201–235, on p. 215.
408 *the Royal Society of London awarded [Morgan] its Darwin Medal* Charles S. Sherrington, "Address of the President at the Anniversary Meeting," *Proceedings of the Royal Society of London* B97 (1925): 254–267, on p. 267.

Section 13.9

409 *expressed the geneticists' frustration at trying to confirm the linkage* Sverdrup, "Linkage and independent Inheritance in Pisum sativum," *Journal of Genetics* 17 (1927): 221–251, on p. 239.

409 *"To my mind this [failure] need not necessarily affect the theory"* Sverdrup, "Linkage," pp. 239–240.

409 *"We may therefore assume the general truth of the theory"* Darlington, "Chromosome Behaviour and structural Heredity in the Tradescantiae," *Journal of Genetics* 21 (1929) 207–286, on p. 208.

409 *on the basis of only two experiments and vague references* Swinburne, "Presence-and-Absence Theory," p. 135.

409 *After Punnett's strong endorsement Mendelism*, 7th ed. (London: Macmillan, 1927), pp. v, 429.

409 *Punnett's report that the linkage group theorem is valid for the sweet pea* "Linkage Groups and Chromosome Number in Lathyrus," *Proceedings of the Royal Society of London* B102 (1927): 236–238.

409 *"ten independent linkage groups corresponding to the ten chromosomes"* R. A. Emerson, G. W. Beadle and A. C. Fraser, "A Summary of Linkage Studies in Maize," *Cornell University Agricultural Experiment Station Memoirs* 180 (1935): 1–83, on p. 29.

410 *it was generally believed that humans have 48 chromosomes (24 pairs)* "In 1955, Eva Hansen-Melander and her colleagues in Sweden . . . counted 46 chromosomes "In 1955, Eva Hansen-Melander and her colleagues in Sweden . . . counted 46 chromosomes [in human liver tissue]. Eventually . . . [they] challenged in print the long-established notion that humans have 48 chromosomes. By 1960, after many confirmations of the number 46, it was agreed that the older count of 48 [attributed to T. S. Painter] was wrong." Garland Allen and Jeffrey Baker, Biology: Scientific Process and Social Issues (Bethesda, MD: Fitzgerald Science Press, 2001), p. 43.

For details of this strange history see Jan-Dieter Murken and Hubertus von Wilmowsky, *Die Chromosomen des Menschen: Die Geschichte ihrer Erforschung* (Munich: Werner Fritsch, 1973); Aryn Martin, "Can't Any Body Count? Counting as an Epistemic Theme in the History of Human Chromosomes," *Social Studies of Science* 34 (2004): 923–948.

Section 13.10

410 *Moore surveyed several books on inheritance* Moore, "Thomas Hunt Morgan—The Geneticist," *American Zoologist* 23 (1953): 855–865.

410 *the most favorable statement about that idea in a book* Lock, *Recent Progress in the Study of Variation, Heredity, and Evolution* (New York: Dutton, 1906).

410 *his textbook on cytology omitted this item from a list of four main arguments* Leonard Doncaster, *An Introduction to the Study of Cytology* (Cambridge: Cambridge University Press, 1920), pp. 223–224, 232, 228–229.

410 *stressed instead the linkage group theorem in his textbook on heredity* Shull, *Heredity* (New York: McGraw-Hill, 1926), p. 83.

410 *"the work of gene-localization is quite comparable"* Walter, *Genetics: An Introduction . . . to the Study of Heredity*, revised edition (New York: Macmillan, 1922), pp. 234, 246.

410 *"so detailed as to be almost unbelievable"* Newman, *Evolution, Genetics, and Eugenics*, 2nd ed. (Chicago: University of Chicago Press, 1925), p. 447.

410 *L. W. Sharp gave a favorable but balanced evaluation of Morgan's theory* Sharp, *An Introduction to Cytology* (New York: McGraw-Hill, 1921), pp. 367, 389, 396.

411 *"far greater progress has been made in our understanding of heredity"* Shull, *Heredity*, p. 83.
411 *it "is now generally accepted by geneticists"* Punnett, *Mendelism*, pp. V, 149.
411 *At least 11 publicly supported it before 1927* See Brush, "Morgan's Chromosome Theory," p. 512 and note 159.
411 *MacBride strongly attacked the theory* McBride, *Evolution* (New York: Cape & Smith, 1929).
411 *Woodger rejected the theory as incompatible with his empiricist methodology* Woodger, *Biological Principles* (London: Routledge & Kegan Paul, 1929).
411 *Meek favored the cytoplasm rather than the chromosomes* Meek, *The Progress of Life: A Study in Psychogenetic Evolution* (London: Arnold, 1930).
411 *several other authors did accept the theory* See Brush, "Morgan's Chromosome Theory," notes 161, 162, 163.

Section 13.11

411 *favored the general view that chromosomes are carriers of heredity* See Brush, "Morgan's Chromosome Theory," note 164.
411 *The first I have found that mentions the Drosophila research* Abbott, *The elementary Principles of General Biology* (New York: Macmillan, 1914), on p. 234.
411 *He endorsed the theory for no specific reason* Pearse, *General Zoology*, 2nd ed. (New York: Holt, 1917), pp. 355, 357.
411 *Calkins called Morgan's research "a brilliant confirmation"* Calkins, *Biology*, 2nd ed. (New York: Holt, 1917), pp. 232, 218.
411 *chromosomes . . . behave just like Mendelian factors* Woodruff, *Foundations of Biology*, 2nd ed. (New York: Macmillan, 1922), pp. 288, 292, 294.
411 *was discussed in seven other zoology or general biology textbooks* See Brush, "Morgan's Chromosome Theory," note 169.
412 *"in no species of plant or animal have the genes been found"* Smith et al., *A Textbook of General Botany*, revised edition (New York: Macmillan, 1929), p. 464.

Section 13.12

412 *I have collected data to test this hypothesis* The data are presented in Brush, "Morgan's Chromosome Theory," pp. 490–494, along with the reasons (if any) given for supporting the theory.

Section 13.13

412 *We are left with six major reasons* See Brush, "Morgan's Chromosome Theory," Table 4, pp. 490–496.
413 *"it has been the misfortune of the chromosome theory"* C. D. Darlington, "Chromosome Behavior and Structural Heredity in the Tradescantiae," *Journal of Genetics* 21 (1929): 207–286, on p. 207.
413 *"that phase of the history was closed"* Sturtevant, "The Relation of Genes and Chromosomes, as Studied Through Linkage," in *Genetics in the 20th Century*, edited by L. C. Dunn (New York: Macmillan, 1950), pp. 101–110, on p. 104.
413 *Morgan himself asserted that "this ability to predict"* "The Relation of Genes to Physiology and Medicine," Nobel Lecture, June 4, 1934. In *Nobel Lectures including Presentation Speeches and Laureates' Biographies. Physiology or Medicine, 1922–1941* (Amsterdam: Elsevier, 1965), pp. 313–328, on p. 317.

414 *the use of visual representations is correlated* Laurence Smith et al., "Scientific Graphs and the Hierarchy of the Sciences: A Latourian Survey of Inscription Practices," *Social Studies of Science* 30 (2000): 73–94.

414 *Doncaster wrote: it is "almost indubitable that inherited characters"* "Progress in Biology during the last Sixty Years," in *Recent Developments in European Thought* (New York: Oxford University Press, 1920), pp. 229–246, on pp. 235–236.

414 *such a theory would have been opposed by fashionable idealism or holism* See Brush, "Morgan's Chromosome Theory," note 191.

414 *empirical facts that could be replicated in many laboratories* Ibid., note 192.

14

The Revival of Natural Selection 1930–1970

No scientific theory is worth anything unless it enables us to predict something which is actually going on. . . . There are still a number of people who do not believe in the theory of evolution. Scientists believe in it, not because it is an attractive theory, but because it enables them to make predictions which come true.

—J.B.S. Haldane (1940)

The fact that natural selection offers a general explanation of adaptation is one of the chief reasons for the rapid acceptance of Darwinian theory among biologists. For adaptation is very widespread, and some of it is very remarkable. So abundant is it, and so marvelous are parts of it, that many naturalists have come to feel that adaptation is the outstanding feature of life requiring an explanation.

—A. F. Shull et al. (1941)

What makes evolution a scientific explanation is that it makes testable predictions.

—*mathematician and biologist* Eric Lander *(2007)*

14.1 INTRODUCTION

When did biologists accept Charles Darwin's theory of evolution? One might argue that in its original form, his theory was never accepted by more than a handful of biologists. The theory that was accepted in the twentieth century was based on a different understanding of natural selection, of the variations on which it acts, and on the source of those variations. Nevertheless, Darwin is now given credit for proposing a theory that is *essentially* correct when the necessary corrections are made. In particular, biologists came to believe that natural selection acting on small mutations, combined with Mendelian principles of heredity, is the major cause of evolutionary adaptation. Additional factors, such as random genetic drift, geographical isolation, and so forth, may be involved in some cases (especially in the creation of new species) and at some levels, but they would accomplish little by themselves. As modified in the twentieth century, Darwin's theory is generally believed by historians of biology to have been established in the 1940s. It was the core of an evolutionary synthesis that brought previously separate disciplines such as genetics, zoology, botany, and paleontology together into a unified science of biology.

Why did biologists accept the modern version of Darwin's theory? I discuss in this chapter the reception by evolutionists and other biologists of what I will call the *natural selection hypothesis (NSH):* the hypothesis that natural selection, with an ample supply of variation in heritable characters, is not only the major process involved in evolution; Lamarckian effects, random genetic drift, and macromutations also have essentially no evolutionary significance. The NSH was accepted by a bare majority of evolutionary biologists for only a short time (in the 1950s and 1960s). In particular, I ask what the reasons were that they gave for adopting that hypothesis—and for rejecting, for example, the argument that drift plays a significant role in evolution.

What weight did biologists give to the confirmation of predictions, supposedly the essential feature of the hypothetico-deductive scientific method and, according to Karl Popper, the basis of the criterion for demarcating science from pseudoscience (Section 1.2)? That question is especially appropriate in the case of evolutionary biology, since Popper's own critique of Darwinism provoked a long-running debate among philosophers and biologists about whether biological theories should be expected to make testable predictions, and whether more generally the standards for biology should be different from those of the physical sciences.

This chapter does not attempt to give a comprehensive account of topics such as the nature and fate of the evolutionary synthesis, Sewall Wright's shifting balance theory, or—especially—the definition of Darwinism (see Section 2.4).

I start the story at the beginning of the 1930s, with the publication of the three canonical works of modern evolutionary theory by Ronald A. Fisher, J. B. S. Haldane, and Sewall Wright. At that time natural selection was generally recognized as a factor in evolution, but was not considered sufficient by itself to account for the observed facts. It was sometimes argued that natural selection is only a negative factor: it can eliminate bad genes but does nothing to create good genes. Some biologists favored botanist Hugo De Vries's theory of large mutations, also called saltations. Lamarckism (inheritance of acquired characteristics) was losing credibility, but was not quite dead. Several versions of orthogenesis (evolution in a single predetermined direction) were proposed.

There was no widely accepted alternative to natural selection, but instead a need for that process to be clarified and elaborated in order to serve as the major process in evolution. H. J. Muller's work on radiation-induced mutations and Dobzhansky's research on natural populations helped to broaden the definition of natural selection to include variation as a basic component rather a separate factor.

In their development of the natural selection hypothesis, Fisher, Haldane, and Wright used different mathematical approaches based on similar assumptions to arrive at results that were, with some minor exceptions, in good agreement with each other. Fisher and Wright disagreed on the relevance of random genetic drift to the process of evolution as it actually occurred in nature. But by 1960 Wright had retreated from his earlier prodrift position, and asserted that he had not earlier and did not now deny the primacy of natural selection.

In this chapter I focus on the competition between natural selection and drift. The NSH considered here was rather simplistic compared with later discussions of levels of selection. I avoid many of the more sophisticated arguments made about natural selection by biologists and philosophers in recent decades, because they did not seem to play an important role in the acceptance of natural selection itself before 1970.

By the mid-1960s the other major alternatives to selectionism had been rejected, so one may say that the strong selectionist view had become established knowledge among the leaders of evolutionary theory. (This is what biologist Stephen Jay Gould referred to as the "hardening" of the synthesis.) So we can then ask the question: when, and to what extent, did this knowledge (the primacy of natural selection) infiltrate the writing and teaching of other biologists?

Unlike some of the other theories I have studied, which remained orthodox doctrine for at least a couple of decades, selectionism was challenged from several directions only a few years after it was established. This study is limited in space as well as in time: it covers only the Anglo-American biological community. Of course this community includes prominent evolutionists such as Theodosius Dobzhansky and Ernst Mayr, who emigrated from other countries in the 1930s.

14.2 FISHER: A NEW LANGUAGE FOR EVOLUTIONARY RESEARCH

The early reception—or perhaps nonreception—of the modern theory of natural selection was strongly conditioned by the presentation of the theory in a language incomprehensible to most Anglo-American biologists. I am not referring to the fact that some important features of the evolutionary synthesis were originally published in Russian by Sergei S. Chetverikov and his colleagues. Rather, I mean that Fisher especially, and to a lesser degree Haldane and Wright, reasoned and wrote in the language of nineteenth-century theoretical physics. That language cannot easily be translated into the language of early twentieth-century biology because, as in understanding any language, one cannot simply look up individual words in a dictionary; one must learn something about the background knowledge and style of thought of those who use the language.

A successful theoretical physicist often chooses a simple model, deliberately ignoring the rich complexity of the world so dear to the naturalist, in favor of a set of postulates whose mathematical consequences can be reliably computed. It may be helpful to note that a theoretical *model* does not make a statement about the world, but a *scientist* may make the empirical claim that a particular phenomenon is more or less well described by the model.

In Fisher's case it is essential to note one biographical fact: he studied theoretical physics, including the kinetic theory of gases, with physicist and astronomer James Jeans. Hence the following sentence, which appears halfway through the first chapter of his 1930 monograph:

> The particulate theory of inheritance resembles the kinetic theory of gases with its perfectly elastic collisions, whereas the blending theory resembles a theory of

> gases with inelastic collisions, and in which some outside agency is required to be continually at work to keep the particles astir.

In the particulate theory, elastic interactions tend to maintain a statistical distribution (of speeds, or genes). Fisher had already pointed out that in Darwin's theory based on blending inheritance,

> bisexual reproduction will tend rapidly to produce uniformity [and therefore] if variability persists, causes of new variation must be continually at work; [and hence] the causes of the great variability of domesticated species, of all kinds and in all countries, must be sought for in . . . changed conditions and increase of food.

Fisher introduced another important feature of the modern theory of natural selection when he called attention to the misleading use of the phrase "struggle for existence" in the evolutionary literature. This struggle was usually blamed on the "excessive production of offspring, supposedly to be observed throughout organic nature." This manner of speaking, which places the emphasis on overpopulation as a cause of competition between individuals, diverts attention away from the importance of understanding the quantitative aspects of reproduction:

> There is something like a relic of creationist philosophy in arguing from the observation, let us say, that a cod spawns a million eggs, that *therefore* its offspring are subject to natural selection; and it has the disadvantage of excluding fecundity from the class of characteristics of which we may attempt to appreciate the aptitude. It would be instructive to know not only by what physiological mechanism a just apportionment is made between the nutriment devoted to the gonads and that devoted to the rest of the parental organism, but also what circumstances in the life-history and environment would render profitable the diversion of a greater or lesser share of the available resources toward reproduction.

Fisher's view of the operation of selection foreshadowed the general shift in the conception of evolutionary fitness away from survival of the individual organism in a brutal battle with other organisms, to a definition of fitness in terms of differential reproductive success. This change made natural selection more acceptable. Thus, according to historian Peter J. Bowler, T. H. Morgan opposed Darwinism because of "his moral objections to the picture of a world dominated by struggle."

The shift away from a struggle for survival to the struggle to reproduce was to prove useful in combating the harsh image of natural selection, blamed by some critics for the excesses of the Nazis as well as for social Darwinism. Tennyson's famous line "Nature, red in tooth and claw" was frequently quoted. But in the 1960s some American textbook writers invoked the counterculture slogan "make love, not war" by calling natural selection a peaceful force that promoted sex rather than a destructive one promoting death.

Moreover, sexual rather than asexual reproduction is quantitatively important to evolution according to Fisher's theory. He pointed out that while natural selection would allow an asexual organism to evolve, an otherwise similar sexual organism would evolve much faster, depending on "the number of different loci in the sexual species, the genes in which are freely interchangeable . . . even a sexual organism with only two genes would apparently possess a manifest advantage over its asexual competitor . . . from an approximate doubling of the rate with which it could respond to natural selection."

The current explanation for the prevalence of sexual rather than asexual reproduction was developed after 1970 and therefore will not be discussed here. For a brief summary, see the 2011 perspective by Michael Brockhurst introducing a report by Levi T. Morran and colleagues (see the notes to this section).

A result that illustrates more sharply Fisher's mathematical sophistication is that large mutations are generally nonadaptive or lethal, so only small mutations can contribute to evolution. He proved this (for his model) by using a well-known (to mathematicians) formula for the volume of a sphere in n-dimensional space.

As Fisher pointed out, the idea that most mutations are harmful is not original, but was "regarded as obvious by the naturalists" who believed that organisms are generally very well adapted to their environments. It was the popularity of the De Vries large-mutation theory that made it necessary to reinforce that view.

If natural selection acts primarily on very small mutations, it might seem that it must be a very slow process. But Fisher showed that a rare gene with only a small selective advantage—a small mutation that produces no observable effect in the first generation—can spread rather quickly through a population if the same mutation is repeated (randomly but at a finite rate) in every generation. For example, if the selective advantage is 1% (and the mutation rate is one in a thousand million per generation), then "in a species in which 1,000,000,000 come in each generation to maturity, a mutation rate of one in a thousand million will produce one mutant in every generation [on average], and thus establish the superiority of the new type in less than 250 generations, and quite probably in less than 10, from the first occurrence of the mutation, whereas, if the new mutation started with the more familiar mutation rate of one in 1,000,000 the whole business would be settled, with a considerable margin to spare, in the first generation." Natural selection will be much less effective if the population is very small.

Fisher's theoretical conclusion that small mutations—those whose effects are not easily observable in one generation—are more important in evolution than large ones was confirmed by later experiments . An example was the study by Kenneth Mather (Fisher's first genetical assistant when he became Galton Professor at University College London) and L. G. Wigan. They found that the several small mutations that combine to determine a polygenic character such as the number of chaetae (hair or bristles) on the abdomen of a *Drosophila* fly, produce an effect that is initially masked by fluctuations. Under the influence of selection they may produce a continuous change or a sudden jump that mimics a macromutation.

Fisher also discussed situations in which stable gene ratios are maintained by a balance of selective forces: "one gene has a selective advantage only until a certain gene-ratio is established, while for higher ratios it is at a selective disadvantage. In such cases the gene ratio will be stable at the limiting value, for the selection in action will tend to restore it to this value whenever it happens to be disturbed from it in either direction." Of course the conditions of stability must themselves be transient during the course of evolutionary change, but the temporary stability provides a good opportunity to study in detail how natural selection works. Theory indicates that " a single factor may be in stable equilibrium under selection if the heterozygote has a selective advantage over both homozygotes." This situation, later known as a balanced polymorphism, was investigated extensively by Fisher's colleague E. B. Ford and his students, and provided important evidence for the validity of the theory.

In developing a new field it is important to be able to publish one's results quickly, even when the editors and referees of established journals are indifferent or hostile. In 1947, together with cytologist Cyril Darlington, Fisher cofounded *Heredity: An International Journal of Genetics.* Although Darlington did most of the editorial work, Fisher (according to Darlington's biographer O. S. Harman) "was content that Darlington was providing in Heredity a ready platform for the selectionist crusade of his protégé E. B. Ford and Ford's students Bernard Kettlewell, Arthur Cain, and Philip Sheppard."

In summary, Fisher was the most important of the three founders, not just because he came first alphabetically or chronologically (his first major contribution was published in 1918) but because he established a simple proposition, subject to certain assumptions: natural selection acting on small mutations can produce a significant change in the genetic composition of a population much faster than was previously believed. Thus a theory based on natural selection *may* be able to explain evolution, and is therefore worth pursuing. Biologists criticized Fisher's model (as defined by his assumptions) for being unrealistic: "bean bag genetics." But from the viewpoint of theoretical physics that is not a criticism at all. It is often by starting with an idealized model whose properties can be accurately calculated that we can start to understand how nature works. Of course it takes a while for other scientists to be persuaded that this approach can work in biology, and so it was very fortunate that Fisher attracted followers like E. B. Ford who could apply and propagate his ideas.

The above paragraphs may give the impression that Fisher was an armchair theorist who left it to other scientists to do the empirical work needed to test his theories. But he did take an active role in empirical research, by himself and in collaboration with E. B. Ford and others, as may be seen from the list of his papers on genetics, evolution, and eugenics at the beginning of his *Collected Papers.* An early example is the confirmation of a prediction, suggested by Charles Darwin and quantitatively derived from Fisher's theory of natural selection, that the variability of a species should be proportional to the size of its population.

Fisher's fame as a leader in the development of statistical methods in research undoubtedly helped call attention to his development of the NSH; his method of double-blind random trials is often called the gold standard for testing new substances and procedures in medicine and agriculture. For some historians and philosophers of science, his critical scrutiny of Mendel's data showed that one can never assume that empirical results are independent of the theoretical bias of the observer; Fisher's statistical analysis suggested that Mendel's published results are too good to be true.

14.3 WRIGHT: RANDOM GENETIC DRIFT, A CONCEPT OUT OF CONTROL

Sewall Wright, in his comprehensive 1931 article, showed that many of his results were consistent with those of Fisher, but stressed the importance of population size in evolution. His earlier work on inbreeding, as an employee of the US Department of Agriculture (USDA), indicated that random fluctuations are important in small populations. They will in general tend to decrease the *heterozygosity* (presence of different genes, such as *A* and *a*, at the same locus on corresponding chromosomes of an organism) at a rate inversely proportional to the population size *N*. This means that traits determined by the homozygous genes (e.g., *A* and *A*) will tend to become fixed. In such a population there is "little variation, little effect of selection and thus a static condition modified occasionally by chance fixation of new mutations leading inevitably to degeneration and extinction." On the other hand, in a very large interbreeding population, "there is great variability but such a close approach to complete equilibrium of all gene frequencies that there is no evolution under static conditions."

The most favorable situation for evolution is

> a large population , divided and subdivided into partially isolated local races of small size . . . Complete isolation in this case . . . originates new species differing for the most part in nonadaptive respects, but is capable of initiating an adaptive radiation as well as parallel orthogenetic lines, in accordance with the conditions.

At the sixth International Congress of Genetics in 1932, Wright presented a shorter, more qualitative version of his theory. This, according to his biographer William Provine, " was probably the most influential paper he ever published." He introduced diagrams to show various hypothetical "fields of gene combinations." The species will generally occupy a region around a peak representing the optimum combination of gene frequencies in a particular environment. Isolated small local populations will evolve by random genetic drift;

> The chances are good that one at least will come under the influence of another peak. If a higher peak, this race will expand in numbers and by crossbreeding with the others will pull the whole species toward the new position. The average adaptiveness of the species thus advances under intergroup selection, an enormously more effective process than intragroup selection.

Wright's general theory was called the shifting balance process of evolution (later to be called adaptive landscape). It is another example of introducing the language of nineteenth-century theoretical physics into twentieth-century biology: it invokes the fields used qualitatively by Michael Faraday to describe electrical and magnetic phenomena, but without the mathematical sophistication of James Clerk Maxwell or R. A. Fisher. This mechanism for evolution following isolation, which Wright called "an essentially nonadaptive one" although it involved selection as well as genetic drift, became known as the Sewall Wright effect.

Despite Wright's frequent statements that he did not think genetic drift by itself would produce evolution in the absence of selective forces, his effect and his term "drift" were most often invoked to explain (supposedly) nonadaptive characters. Wright himself may have encouraged this incorrect usage by statements such as:

> In the human species, the blood group alleles are neutral as far as is known. The frequencies vary widely from region to region and in such a way as to indicate that the historical factor (*I. e.*, partial isolation) is the determining factor. The

> frequency distribution indicates a considerable amount of random differentiation even among the largest populations.

The refutation of these statements was later used (perhaps unfairly) as a refutation of Wright's entire theory of evolution.

Wright's shifting balance theory was fairly popular in the United States before 1950, partly because it was adopted by Theodosius Dobzhansky. According to historian Jonathan Hodge, Wright's "diagrams of adaptive landscapes . . . have appeared . . . routinely in textbooks especially since the Second World War." The shifting balance theory "originated as his extrapolationary projection onto nature in the wild and the long run of his prior conclusions about optimum breeding strategies on the farm." His earlier research at the USDA was indeed useful in improving agricultural production. For example, poultry breeders were able to increase the number of eggs laid by hens, an achievement that makes Wright indirectly responsible for the fact that eggs are relatively cheap in the United States today.

But the shifting balance theory itself, however helpful it may have been in visualizing how evolutionary processes *might* work, turned out to be useless—or even misleading—in finding out how they actually do work. The only disagreement between experts is whether, as Will Provine asserted, his diagrams are incoherent and meaningless , or as philosopher Rob Skipper argues, they are simply wrong. At the same time, Wright's genetic drift, considered apart from its role in the shifting balance theory, did prove to be a valuable concept as a foil to natural selection because it had empirical consequences.

The career of genetic drift reached a high point in 1954 when Wright published a series of three papers reporting experiments done by Warwick E. Kerr. Very small populations of *Drosophila melanogaster* (four males and four females in each) were isolated and followed through 10 or more generations. Selection was assumed to be present but estimated to be not strong enough to overwhelm the effects of drift in such small populations. In each of the three experiments, the frequencies of a specified mutation competing with its allele were determined for about 100 lines. The results (including the eventual fixation of one mutation) confirmed the theoretical predictions.

The Kerr-Wright papers were an important contribution to population genetics, but they did not correct the misunderstanding of Wright's theory by other biologists. Despite the fact that Wright had explicitly included the effects of selection in designing and analyzing the experiments, even his supporters continued to assume that he believed drift to be an *alternative* to natural selection. For those who read the Kerr-Wright papers more carefully, the extremely artificial conditions of this laboratory experiment probably reinforced the criticism that genetic drift, while *theoretically* present in any population, is unlikely to play a significant role in evolutionary processes in nature.

In surveying the response to this controversy by other biologists, I found it necessary to distinguish between two hypotheses, both known as genetic drift or the Sewall Wright effect:

1. Evolution goes fastest in medium-size or large populations, divided into several partially isolated smaller populations, in which certain

characters first appear in one subgroup as a result of genetic drift and then incorporated into others by natural selection. In this case genetic drift has evolutionary significance only *in combination* with natural selection.

2. A nonadaptive character may be explained as the result of genetic drift *rather than* natural selection.

Wright usually meant (1) and rejected the ascription to himself of (2), even by those who (like Motoo Kimura) thought they were supporting Wright's theory.

Perhaps his best explanation of (1) was his analogy of Mexican jumping beans on a surface with several hollows of varying depths. The hollows correspond to adaptive peaks, but with gravitational force replacing selective force. The only way the bean can move from a shallow place to a deeper place is by randomly jumping around; gravity (selection pressure) will just keep it where it is. Perhaps if he had used the terminology of the kinetic theory of gases (more precisely, statistical mechanics) to formulate this physical analogy, with thermal fluctuations playing the role of genetic drift in making possible the transition from a shallow potential well to a deeper one, he could have communicated his idea to Fisher more effectively. But Fisher would still have objected that in a multidimensional space of gene frequencies there is no reason to think there is even one position that represents an absolute minimum (meaning maximum fitness) for *all* gene ratios.

Hypothesis (2) was much more popular among biologists not just because it was simpler, but because it seemed to solve one of the outstanding difficulties of natural selection theory: the fact that the most prominent visual differences between closely related species seem to have no adaptive value. How, then, could natural selection explain the differences between them? This objection was made most effectively by O. W. Richards and G. C. Robson, and was probably a major reason for the reluctance of some biologists to accept natural selection as an explanation for speciation in the 1930s. Since the theory of Fisher, Haldane, and Wright implied that there must be genetic drift in sufficiently small populations, those who did not understand the various ramifications of the theory assumed that it was legitimate to invoke genetic drift to explain any nonadaptive character. But the result was often unfair to Wright, since his theory lost credibility when it was judged by the outcome of predictions that he himself did not make (and probably would not have made).

14.4 HALDANE: A MATHEMATICAL-PHILOSOPHICAL BIOLOGIST WEIGHS IN

While the fundamental works of Fisher and Wright seem to have had the greatest effect in persuading biologists that natural selection could account for evolutionary history, the earlier papers of J.B.S. Haldane, starting in 1924, have been credited with first establishing that natural selection could account for the evolutionary change going on now. According to biologist Sahotra Sarkar, Haldane was the first to propose a true *synthesis* of population genetics with classical genetics, cytology, paleontology, and evolutionary theory: "Fisher's and Wright's theories were retrospective in intent; Haldane's was purely prospective [intended to predict processes in the future] though it allowed retrodiction. To the extent that an *evolutionary theory* should be one that attempts future prediction, only Haldane's work qualifies."

Haldane's 1932 book *The Causes of Evolution* addressed the nonmathematical reader with a lively style, providing a large number of biological examples of natural selection. At the same time he was anxious to take advantage of his own prestige as a mathematician, asserting " I can write of natural selection with authority because I am one of the three people who know most about its mathematical theory." But his many other activities in science and politics kept him from devoting his full attention to evolutionary theory after 1932.

As for the creationists who believe in (what is now called) intelligent design, their doctrine, Haldane asserted, is refuted by two hard facts: first, the extinction of many species; second, echoing Darwin, the evolution of parasites that lost their faculties and inflict pain on other animals. Could a morally perfect God have made the tapeworm?

14.5 EARLY RECEPTION OF THE THEORY

The works of Fisher, Wright, and Haldane published at the beginning of the 1930s attracted little attention before 1935 except among a small group of their colleagues in Great Britain and the United States, notably E. B. Ford at Oxford. According to Fisher, this was because of the unfortunate influence of William Bateson and his wrongheaded ideas about evolution. Fisher was also annoyed that T. H. Morgan could not see that his own work had made a genetic theory of evolution based on selection of small mutations not only possible, but irresistible.

But the standard arguments—that natural selection without large (perhaps *directed*) mutations could not originate new species—were still prevalent among biologists through the 1930s. It was only dimly recognized that H. J. Muller's experiments on the production of small mutations by X-rays might provide important evidence on the nature of such mutations. Mathematical derivations were admired, but did not convince many readers that they entailed real biological consequences. Evidence from genetics was not convincing to, or even known by, naturalists who considered the field more important than the laboratory.

In 1935 publications, I found a few indications that the Fisher-Haldane-Wright theory has become part of the discourse of evolutionists about mutations but little real acceptance of the theory. Four major publications in 1936 show that views on the causes of evolution were in a state of flux.

A. F. Shull, in his textbook *Evolution*, declared that after years of neglect "the theory of natural selection is coming back," but acknowledged that traditional and more recent objections still have some validity. It almost seems that we are forced to believe in natural selection because all the other known alternatives are even less credible. Robson and Richards, in their monograph *Variation of Animals in Nature*, also asserted that most of the traditional evidence for natural selection was weak, but they were much less optimistic than Shull about the prospects for rescuing the theory.

A discussion at the Royal Society of London (also in 1936) on "The Present State of the Theory of Natural Selection" began with the following statement by D.M.S. Watson:

> The theory of natural selection is the only explanation of the production of adaptations which is consonant with modern work on heredity, but it suffers from the

> drawback that by the introduction of related subsidiary hypotheses it becomes capable of giving a theoretical explanation of any conceivable occurrence.

Despite Fisher's vigorous protest, Watson's skepticism about natural selection set the tone for the discussion. Gates and MacBride attacked natural selection, favoring instead mutations or Lamarckism, respectively. Other participants (Timofeeff-Ressovsky, Carpenter, Haldane) pointed out evidence and arguments favorable to the natural selection hypothesis.

In the previous year, 1935, most publications that discussed natural selection expressed an unfavorable view of it. In 1936 and 1937 the pros and cons were about equally balanced, without counting the publications of Fisher and Haldane. What tipped the balance was a remarkable book by Dobzhansky (to be discussed in the next section). Beginning in 1938 and continuing through the 1940s, a majority of the articles and books that discussed natural selection were favorable.

During the 1940s the most prominent opponent of the NSH was the respected German biologist Richard Goldschmidt. Goldschmidt accepted natural selection as an adequate explanation for microevolution but argued that it could not explain macroevolution, in particular the formation of new species. For this he invoked his own theory of macromutations, using the notorious phrase "hopeful monsters."

Beginning in 1942 with the publication of Julian Huxley's *Evolution the Modern Synthesis* and Ernst Mayr's *Systematics and the Origin of Species*, there was a stream of books and articles by leading biologists articulating and strongly supporting the evolutionary synthesis. It is now time to ask: what were the reasons for this groundswell of support, and to what extent was it focused on a single coherent theory such as the natural selection hypothesis?

14.6 DOBZHANSKY: THE FARADAY OF BIOLOGY?

In Russia, where Theodosius Dobzhansky was born and educated, there was much more support for Darwin's theory of natural selection (see Section 2.4) and much less separation between genetics and natural history than in Great Britain and the United States. When he came to the United States in 1927 to work with T. H. Morgan's group, first at Columbia and then at Caltech, Dobzhansky enhanced his already-substantial knowledge of genetics and used it to his continue his field research in entomology. His famous series of papers on the genetics of natural populations established a strong link between the established knowledge about *Drosophila* in the laboratory and new knowledge about *Drosophila* in the wild.

Dobzhansky's research in the mid-1930s provided strong indirect support for the NSH by showing that speciation may proceed in small steps. He became interested in Donald Lancefield's discovery that a newly discovered strain of *Drosophila*, later called *D. pseudoobscura*, when crossed with other strains produced sterile males, but some fertile females. This "may represent an earlier step in the evolutionary process" as the first step in the origin of a new species, Dobzhansky said. He was inspired by the possibility of creating a new species in the laboratory, which he later learned had actually been achieved by in the 1920s by G. D. Karpechenko, who crossed broccoli with a radish.

Dobzhansky's work suggested that speciation was not necessarily a sudden discontinuous process involving a macromutation, thereby challenging the views of those evolutionists who still followed de Vries, and preemptively undermining Richard Goldschmidt's theory. Instead, he postulated that there is no qualitative difference between microevolution and macroevolution, as a pragmatic methodological assumption and then gathered empirical evidence for this postulate; thus, research on small changes at the level of genes and chromosomes is directly relevant to the general problem of the origin of species. Speciation is not a discrete event but a process that occurs over several generations. Wright agreed with Dobzhansky that we have examples of every conceivable intermediate step, and therefore there is no justification for assuming the "'bridgeless gap' between species which is crucial" for Goldschmidt's theory. Likewise, there is no justification for the more recent creationist claim that while micromutation has been demonstrated in the laboratory, natural selection cannot produce macromutation.

Dobzhansky initially relied on Sewall Wright to tell him the conclusions of mathematical population genetics and was influenced to favor Wright's views on the importance of genetic drift, but his own results led him to change those views in the 1940s. Thus Dobzhansky's career in America illustrates in microcosm the shift toward selectionism in the 1940s and 1950s. It is comparable to the career of physicist Michael Faraday, who started as an assistant to chemist Humphry Davy and then made qualitative experimental discoveries in electricity and magnetism, leaving it to physicist James Clerk Maxwell to provide a mathematical theory based on those results.

The publication of Dobzhansky's *Genetics and the Origin of Species* in 1937 may be taken as the first announcement of the evolutionary synthesis. Dobzhansky presented a comprehensive review of the biological evidence on mutations, variation, selection, mechanisms for speciation (including polyploidy and isolation), patterns of evolution, and species. He called attention to Muller's X-ray work, which he called the first conclusive evidence for an external cause of mutations. He mentioned, and emphasized in the second edition (1941), the "rediscovery of the giant chromosomes in the larval salivary gland of flies by Heitz and Bauer (1933) and the application of these chromosomes as a tool of genetic research by T. S. Painter (1934)," which "enormously facilitated the comparison of the gene arrangements in different strains." This tool provided the basis for Dobzhansky's research in the 1940s on natural populations of *Drosophila*.

Genetics and the Origin of Species went through three editions under that title (1937, 1941, and 1951), and a further revision was published under the title *Genetics of the Evolutionary Process* (1970). It is generally considered to have been very influential in persuading biologists to adopt the evolutionary synthesis. One reason for this influence, according to scholar Leah Ceccarelli, is the rhetorical approach its author used. Rather than claim that one field of science should be reduced to another, more fundamental field, Dobzhansky made it clear that knowledge and research methods from several fields must be integrated in order to understand how evolution works. In particular, geneticists and naturalists would have to cooperate, as they had not done so in the past (at least in the Anglo-American scientific community). At the same time, as Lindley Darden has pointed out, the relations between fields in the evolutionary synthesis were not symmetrical, because the synthetic theory had to deal with three levels in a hierarchy: genes and chromosomes, populations, and species: "Since genetics can proceed without taking into account the populational level, it retains more relative independence than do the other fields in the synthesis"; population genetics depends on "new findings of mutational processes in genetics," and " population processes . . . can continue whether or not isolation" (producing new species) has occurred.

In the first chapter of the third edition of *Genetics and the Origin of Species*, Dobzhansky distinguished two approaches to evolutionary problems: (1) "unraveling and describing the actual course which the evolutionary process took in the history of the earth" (this is phylogeny); and (2) "studies on the mechanisms that bring about evolution, causal rather than historical problems, phenomena that can be studied experimentally rather than events which happened in the past." Roughly speaking, these two approaches lead to explanation and prediction, respectively.

14.7 EVIDENCE FOR NATURAL SELECTION, BEFORE 1941

For our purposes the most relevant part of Dobzhansky's *Genetics and the Origin of Species* is the survey of evidence for natural selection. I have used the extensive list in the second edition (1941).

Laboratory Experiments

In *The Origin of Species*, Charles Darwin discussed the remarkable fact, discovered by Thomas Vernon Wollaston a few years earlier, that many of the beetles inhabiting the Madeira Islands in the Atlantic Ocean are wingless, unlike related beetles elsewhere. He suggested that

> the wingless condition . . . is mainly due to the action of natural selection, but combined probably with disuse. For during thousands of successive generations each individual beetle which flew least, either from its wings having ever so little less perfectly developed or from indolent habit, will have had the best chance of surviving from not being blown out to sea; and, on the other hand, those beetles which most readily took to flight will oftenest have been blown to sea and thus have been destroyed.

This suggestion is equivalent to a *prediction* that if you start with a population of beetles with varying wing sizes and put them in an environment where those that have the largest wings are most likely to be blown away by strong winds, then (assuming wing size is an inherited character) after many generations the population will consist of a much larger proportion of beetles with small or absent wings. I suggest that we call it an *implied prediction.*

In the following decades, Darwin's discussion of the wingless flies of Madeira was often given as an example of how an observation could be *explained* by natural selection: the insects with wings were blown away, while the others remained. But it was not clear whether selection acted on large or small mutations, or even whether wind is the most important selective factor. According to the (Popperian) scientific method, we should do an experiment in which the intensity of a mechanically produced wind is varied while other factors are kept constant; our hypothesis predicts that over the course of many generations, the proportion of wingless flies should increase if the wind is strong, but not if it is weak or absent.

A critic unfamiliar with modern biology might scoff at this proposal, because everyone supposedly knows that evolution is such a slow process that it can't be directly observed within a human lifetime, much less in a laboratory experiment lasting only a few weeks or months. But the critic has ignored the fact that geneticists in the early twentieth century developed a model organism, the *Drosophila* fly, which reproduces so rapidly that such experiments can be done in the laboratory in real time. By 1930 any good geneticist could do *Drosophila* research, and even assign the classical experiments as practical exercises for students.

In 1937, in France—where Lamarckism still dominated evolutionary biology—the test was done. Philippe l'Heritier, Yvette Neefs, and Georges Teissier carried out an explicit test of Darwin's prediction. They used a population cage, which allowed one to study an isolated population under controlled conditions:

> We decided to study experimentally the problem posed by Darwin and have, for this purpose, used *Drosophila melanogaster*. We have realized the conditions that Darwin assumed to exist originally: a mixed population of some insects with normal wings and others having the recessive character *vestigial,* which is manifested by an atrophy of the wings, so that they have become inadequate for flight.

And indeed, in a trial of less than two months they were able to confirm the results predicted from Darwin's hypothesis.

Dobzhansky, who gave this as his first example of natural selection in the laboratory, did not mention Darwin's role, nor did he use the language of hypothetico-deductivism and prediction-testing. Instead he quickly moved on to the next category.

Historical Changes in the Composition of Populations

At the beginning of the twentieth century, attempts to eradicate red scale insects that attack citrus trees in California by fumigation with hydrocyanic (HCN) gas were initially successful, but after a few years it was observed that fumigation no longer worked. Experiments by H. J. Quayle showed that there were two races (subspecies) of red scale, one of which is resistant to the gas, the other not. By eliminating the nonresistant insects, the fumigation allowed the resistant race to take over the population. Dobzhansky called this " probably the best proof of the effectiveness of natural selection yet obtained."

But where did the HCN-resistant insects come from? Dobzhansky suggested that they resulted from mutations rather than being introduced from elsewhere, but considered this question largely academic—a rather inappropriate pejorative for a professor to use. In a later revision of the book he seemed unsure whether the mutations "are present in pest populations before the insecticide is applied, or arise after the application. This question is usually insoluble, and it is of no particular importance anyway." The context for these statements is the shift in the views of many evolutionists toward the position (supported by Dobzhansky's own experiments) that there is so much hidden genetic variability in any present-day population that new mutations are not really needed to drive evolution.

After mentioning other examples of the evolution of agricultural pests, Dobzhansky discussed industrial melanism: the appearance of darker forms of moths in areas near industrial cities in England and other European countries. He seemed to accept E. B. Ford's explanation that

> melanics [darker forms] are superior to the light-colored types in vigor, and . . . their spread in populations is normally prevented because they are not protectively colored. In industrial areas this disability is removed by the general darkening of the landscape.

Protective and Warning Coloration

Examples of protective and warning coloration, including mimicry had been given as evidence for natural selection in the nineteenth century. Dobzhansky acknowledged that much of it is " uncritical and valueless speculation . . . bringing only disrepute to the whole theory. . . . Some of the alleged protections and warnings have been shown to be armchair protection and museum mimicry." But the theory was revived in the 1930s, and became quite respectable. There was still not much careful experimental work; Dobzhansky mentioned only the research of biologist Francis Sumner showing that "fishes whose color contrasts with their surroundings are caught by predators more easily than those with harmonizing colors."

14.8 HUXLEY: A NEW SYNTHESIS IS PROCLAIMED

Julian Huxley, according to E. B. Ford, was an inspirational mentor and " the most powerful force in developing the selectionist attitude at Oxford in the 1920s." Huxley himself recalled:

> One of the greatest changes in biological outlook during my active life has been the re-erection of natural selection as the main and perhaps the sole agency of significant evolutionary change and adjustment . . . perhaps owing to a certain familiarity with natural history, I clung firmly to a belief in the principle of natural selection and its efficacy during the period up to the early or middle 1920s, when it was being neglected or attacked by most of the biologists who thought of themselves as advanced, and when even eminent geneticists could seriously assert that without natural selection all existing types could have come into being, and a vast number of others as well!

Huxley's edited book, *The New Systematics* (1940), helped to publicize the work of specialists such as Darlington, Ford, and Timofeeff-Ressovsky. In his introduction he wrote:

> There is still a widespread reluctance, especially among some of the younger experimental biologists, to recognize the prevalence of adaptation and the power

> of selection. This is doubtless in large part a natural reaction against the facile arm-chair reasoning of a certain school of evolutionists.

Huxley introduced the term "synthesis" and popularized his own version of it with the title and readable style of his influential 1942 book *Evolution the Modern Synthesis*. He asserted that thanks to the mathematical work of Fisher, Haldane and Wright, we can now make "quantitative prophecies with much greater fullness than was possible to Darwin." But it is difficult to find any such prophecies in his book, whose pages are remarkably free of numerical data. Huxley's book is nevertheless a comprehensive qualitative survey of the arguments and evidence for the selectionist view. As such, it is not a synthesis of different *theories* but of *observations* in support of one theory. Will Provine argues that it is a constriction which eliminated most of the factors previously believed to affect evolution.

What kind of evidence did Huxley himself consider most persuasive for a biologist who was already familiar with the older evidence for natural selection but reluctant to accept it as the *dominant* process in evolution? Huxley explicitly answers this question: First, the mathematical work of Fisher and Haldane, combined with modern genetics, definitively refutes the three major alternatives to the natural selection hypothesis. Small mutations are much more likely to produce adaptive improvements that large ones, overwhelmingly so when (as now seems probable) "the harmonious adjustment of many independently varying characters is required"—hence the de Vries large mutation theory is untenable. Similarly, orthogenesis and Lamarckian effects, even if present, cannot be important in evolution as long as selection operates. But

> perhaps the most important single concept of recent years is that of the adjustment of mutations through changes in the gene-complex. Before this had been developed by R. A. Fisher and his followers, notably E. B. Ford, the effect of a mutation was assumed to be constant. A given mutation, we may say, made an offer to the germplasm of the species, which had to be accepted or declined as it stood. . . . Today we are able to look at the matter in a wholly different way. . . . the offer made by a mutation may be merely a preliminary proposal, subject to negotiation. Biologically, this negotiation is effected in the first instance by recombination and secondarily by mutation in the residual gene-complex.

Huxley also recognized the Sewall Wright phenomenon of drift in small populations. This effect is involved in speciation, and Huxley called it

> one of the most important results of mathematical analysis applied to the facts of neo-mendelism. It gives accident as well as adaptation a place in evolution, and at one stroke explains many facts which puzzled earlier selectionists, notably the much greater degree of divergence shown by island than mainland forms, by forms in isolated lakes than in continuous river-systems.

In later publications, Huxley became increasingly skeptical about the evolutionary significance of genetic drift. In 1951, he asserted that the establishment of non-adaptive characters by drift in small populations is much less frequent than Wright thought; Ford, Muller, and others have shown that genes and gene combinations that are truly nonadaptive are exceedingly rare. He was not asserting that Wright's theory made incorrect predictions, but rather that the theory was not relevant because adaptation is much more effective than drift.

In 1952 Huxley published an article in the American pictorial magazine *Life*, featuring a sensational example of natural selection: the crab *Dorippe japonica*, which bears what looks like the face of a Samurai warrior. According to a Japanese legend, after the Heike warriors were defeated by the Genji in 1155 A.D., they "committed mass suicide by throwing themselves into the sea." Later, fishermen noticed crabs that bore a slight resemblance to the face of a warrior, and threw them back, since they did not want to eat the supposed reincarnation of a Heike warrior. As a result, the crabs that looked more like a warrior's face were more likely to survive and reproduce, so the resemblance became more nearly perfect in later generations. Huxley argued that such mimicry is common; this one just happens to be "more curious and surprising than others."

I find it also curious and surprising that Huxley's example of the Heike crab was not used in any of the books I examined from the period 1953–1970. Perhaps, after skeptical biologists had become fed up with the plethora of just-so stories invented by earlier Darwinians, they thought this one was just too good to be true. But astronomer and science popularizer Carl Sagan used it in his television series Cosmos and the associated book.

Scholars have recently questioned whether evolutionary synthesis is an accurate label for the theory developed in this period. Philosopher Richard G. Delisle notes that while Huxley provided a name for the evolutionary synthesis he denied one of its basic premises: that macroevolution is a cumulation of microevolutionary steps, not a sudden change. Huxley also asserted that evolution is progressive, contrary to the views of other leaders of the synthesis. Dobzhansky "had to severely distort . . . the concept of natural selection" to fit his understanding of evolution. So "the evolutionary synthesis was not a movement from which all the synthesists sprung, but was rather a meeting place from which each extracted evolutionary mechanisms to insert them into distinct and quasi-incommensurable . . . frameworks."

Historian Joe Cain argues that we should abandon the unit concept of "the evolutionary synthesis" and focus on studies of "the nature of species and the process of speciation (variation, divergence, isolation, and selection); experimental taxonomy," as well as larger trends such as the shift "towards process-based biologies and away from object-based naturalist disciplines." The object to process shift can also be observed in other sciences, as discussed by philosopher Xiang Chen.

14.9 MAYR: SYSTEMATICS AND THE FOUNDER PRINCIPLE

Modern evolutionary genetics is a highly mathematical subject, yet many biologists in the twentieth century disliked mathematics and resisted using it. How did the subject become established? One might conjecture that the simultaneous emergence of three mathematical biologists—Fisher, Haldane, and Wright—was a very unlikely event that happened to occur in the early twentieth century. These three founders, though geographically separated, were able to interact with each other thanks to modern communications and transportation technology, and thus to start a new specialty within the discipline of biology. That specialty might have died out, except that Fisher and Wright were able to team up with two biologists who were not expert mathematicians but very good at doing empirical work and publicizing their results: E. B. Ford and Theodosius Dobzhansky. Those collaborators attracted bright students who carried on their research programs (at Oxford and Columbia, respectively) and thereby sustained the new specialty. (Haldane's influence was somewhat more diffuse, in part because he worked on several other topics as well as evolutionary theory, but he did inspire another theorist, John Maynard Smith.)

This scenario conveniently illustrates what zoologist Ernst Mayr called the founder principle. It is a special case of random genetic drift, in which a new species is formed by a very small and very nonrepresentative subset of a larger population.

Mayr is known as one of the leaders of the evolutionary synthesis. His book *Systematics and the Origin of Species*, published in 1942 with an introductory endorsement by Dobzhansky, formulated the synthesis from the viewpoint of taxonomy and ornithology. In 1963 Huxley called him "undoubtedly the best all-round evolutionary geneticist we now have in the world." At the time of his death in 2005 at age 100, he was widely considered to be the father of modern evolutionary biology. As a distinguished biologist-turned-historian, he had considerable influence not only on evolutionary theory but also on how its development was to be retrospectively interpreted.

In the opening pages of *Systematics*, Mayr welcomed the new "mutual understanding between geneticists and systematists " exemplified by "[Bernhard] Rensch and [Alfred C.] Kinsey among the taxonomists, [Nikolai] Timoféeff-Ressovsky and Dobzhansky among the geneticists, and Huxley and C. Diver among the general biologists." As for the mathematical population geneticists, he mentioned Fisher only briefly and did not cite Haldane at all, while making it clear that he found Wright's theory the most congenial. Mayr's own empirical work and general knowledge of the subject had already convinced him that geographic isolation is the key factor in the speciation of birds. Now, thanks to Dobzhansky, he had a theory to support that view: Sewall Wright said evolution should be faster in small populations, "and this is exactly what we find"—in the West Indies, Solomon Islands, Galapagos Islands, and Hawaii. Alfred Kinsey found this for gall wasps, W. F. Reinig for bumblebees. But now Mayr put his own stamp on the idea with a further twist:

> The reduced variability of small populations . . . [is due] sometimes to the fact that the entire population was started by a single pair or by a single fertilized female. These "founders" of the population carried with them only a very small proportion of the variability of the parent population. This "founder" principle sometimes explains even the uniformity of rather large populations.

As examples of the founder principle Mayr cited his own work on the reef heron (*Demigretta sacra*), whose color is always gray on some islands but both gray and white on others. Mayr's support for isolation as the major cause of speciation in birds did not prevent him from endorsing the importance of natural selection in other cases, while cautioning readers to be skeptical about the adaptive significance of some differences invoked by extreme selectionists.

Mayr took a firm stand in favor of another feature of the natural selection hypothesis: he asserted that there is no qualitative difference between macroevolution and microevolution. He wrote:

> All the properties and phenomena of macroevolution and of the origin of the higher categories can be traced back to intraspecific variation, even though the first steps of such processes are usually very minute.

But Mayr's early support for Wright's general theory was much less firm, according to his recollections half a century later. In a letter to Michael Ruse dated November 20, 1991, Mayr wrote:

> In my 1942 book, in order to be "modern," I quote Sewall Wright copiously. However, in my actual thinking and working I was very much opposed to him. And I fought Dobzhansky all along when he wanted to believe in the neutrality of the human blood group genes and the *Drosophila* gene arrangements.

During the 1940s Mayr expressed increasingly strong support for the natural selection hypothesis. In 1945, discussing Epling's data on gene arrangements in *Drosophila* (see Section 14.12), he made what he later called a prediction that the evolution of these arrangements would be found to be influenced by natural selection.

Will Provine sees a turning point toward selectionism in a 1950 paper by Mayr and Erwin Stresemann, which abandoned the position in *Systematics and the Origin of Species* that "most conspicuous color polymorphisms in birds and other animals were nonadaptive" and that they are "accidents of variation and without selective significance." Instead, Mayr and Streseman wrote, "since recent genetic evidence indicates that alleles involved in a balanced polymorphism have different selective values, it seems probable that many subspecies and species characters that have heretofore been considered as 'neutral' are controlled by genes which differ in their selective values."

In his 1963 book *Animal Species and Evolution*, Mayr included a vigorous attack on the use of the genetic drift concept in evolutionary biology. The term is used in so many contradictory ways that according to Mayr, it had been discredited:

> During the period from about 1935 to 1955 it was fashionable to attribute puzzling evolutionary changes to "drift" or to the "Sewall Wright effect" in the same manner in which the preceding generation of evolutionists had explained similar changes as due to "mutation."

Drift is often postulated in cases of supposed selective neutrality, but Mayr asserted that such cases are actually quite rare:

> Selective neutrality can be excluded almost automatically wherever polymorphism or character clines [gradients] are found in natural populations. This clue was used to predict the adaptive significance (previously denied) of the distribution pattern of the gene arrangements in *Drosophila pseudoobscura* (Mayr 1945) and of the human blood groups (Ford 1945). Virtually every case quoted in the past as caused by genetic drift due to errors of sampling has more recently been interpreted in terms of selection pressures.

Why did Mayr change his position on drift? We have at least two possible answers. According to his published remarks, it was because of the evidence. According to his letter to Ruse, he had always rejected drift, so we may conjecture that he went along in 1942 because he did not want to challenge what appeared to be scientific orthodoxy; later, when he had established his reputation, he could afford to do so. Here, as in other cases, we cannot prove why a particular scientist expressed a particular view, at least not without much fuller evidence.

Paradoxically, Mayr continued to maintain the validity of his own founder principle (which could be regarded as a special case of genetic drift), perhaps because it did not depend on the assumption of selective neutrality.

14.10 SIMPSON: NO STRAIGHT AND NARROW PATH FOR PALEONTOLOGY

In his 1944 book *Tempo and Mode in Evolution*, Simpson presented the evolutionary synthesis as a theory that is compatible with , if not derivable from, the findings of paleontology. In particular, he showed that those findings did not require the invocation of orthogenetic rectilinear trends. This kind of trend "is a product of the tendency of the minds of scientists to move in straight lines [rather] than of the tendency for nature to do so." The only specific evidence for natural selection, admittedly indirect, was the example of the horse. Evidence that paleontologists were abandoning orthogenesis in favor of natural selection may be found in the contributions of D.M.S. Watson and Alfred Romer to the 1947 Conference on Genetics, Paleontology, and Evolution.

Simpson favored Wright's drift over a purely selectionist explanation of evolution. He proposed a quantum theory of evolution based on Wright's shifting balance picture, to explain abrupt changes where there was no fossil evidence of intermediate forces. In later publications he downgraded the importance of drift.

According to Stephen Jay Gould and others, Simpson singlehandedly brought paleontology into the evolutionary synthesis with his book *Tempo and Mode*, followed by a more popular exposition, *The Meaning of Evolution*, and a series of articles. At the same time he strengthened the synthesis itself by adding a quantitative *temporal* dimension for evolution, which laboratory and field studies of natural selection could not supply.

But Gould, a paleontologist who had deviated from the evolutionary synthesis by proposing a theory of punctuated evolution, complained that the synthesis had evolved into a rigid dogma that stressed natural selection to the exclusion of all other alternative or contributory factors. As a paleontologist, Gould was especially concerned with the role of his childhood hero Simpson, who had argued that fossil evidence, especially for horses, that had been used to support orthogenesis, could be better explained by natural selection. Gould wrote that Simpson's 1953 book *The Major Features of Evolution*

> differs from *Tempo and Mode* . . . it . . . displays some subtle but important shifts in theoretical emphasis and content. These shifts mirror some general trends in the modern synthesis, as its theory won adherents, gained prestige, and (unfortunately in some respects) hardened. In particular, increasingly exclusive reliance on selection-toward-adaptation (for Simpson, in the gradual, phyletic mode),

> coupled with a greater willingness to reject alternatives more firmly than the evidence warranted, marks both Simpson's new book and the growing confidence of the synthetic theory in general . . . Though he chides others (quite properly) for assuming that structures are inadaptive because they cannot imagine a use for them, he often constructs adaptive scenarios, in the speculative mode and on the opposite (and equally invalid) assumption that prominent features must have some immediate use.

Moreover, Simpson had rejected Wright's genetic drift (which he had favored in *Tempo and Mode)*, concluding that (in Gould's words) " it could not trigger any major evolutionary event." For Gould this was tantamount to devaluing the profession of paleontology; Simpson had

> unified paleontology with evolutionary theory, but at a high price indeed—at the price of admitting that no fundamental theory can arise from the study of major events and patterns in the history of life. Why be a paleontologist if all fundamental theory must arise elsewhere?

Gould, although extremely interested and well read in the history of evolutionary biology, was certainly not an unbiased interpreter of primary sources. The acceptance of his own theory of punctuated equilibrium would require the rejection of extreme selectionism. Reinstating random genetic drift as a significant factor in evolution might help his case, although it would have been risky to face quantitative scrutiny by population geneticists versed in the mathematical theories of drift developed by Fisher and Wright. More to Gould's advantage would be a resurrection of Goldschmidt's macromutations, but as empirical phenomena without Goldschmidt's theoretical speculations. With this in mind, let's look at two of Simpson's views *not* quoted by Gould.

In 1967 Simpson stated that the theory of sudden major changes (Schindewolf, Beurlen, Goldschmidt) was "conclusively refuted by Rensch. In fact there is now almost no support for that view except by a few philosophers not sufficiently acquainted with scientific data on evolution." In particular, Goldschmidt's systemic mutations have

> never been observed and . . . need not be taken seriously if, as is the case, the phenomena that they were postulated to explain can be explained in terms of known processes and forces. . . . In some of his latest work Goldschmidt implicitly retreated from his position.

But at the same time he was rejecting macromutations and orthogenesis, by again allowing a role for genetic drift. Having (to Gould's dismay) hardened his view against drift in the 1950s, he restored it as a secondary factor in the 1960s. Drift "plays a role, perhaps a minor one," although Mayr's founder principle had more influence on evolution. It provided a useful weapon against the orthogenesis and teleology that still contaminated the writings of some of his fellow paleontologists.

Gould's criticism of Simpson has been called "ritual patricide" by historian Joe Cain. Cain suggests that it was motivated by his desire to promote the punctuated equilibrium (PE) theory proposed by himself and Niles Eldredge. Cain writes: "if

PE was going to have any claim to novelty, Gould needed some way to negate Simpson"—that is, to show that Simpson himself abandoned his quantum evolution theory and therefore it is not relevant to the later debates about punctuated equilibrium. But Cain found, in the American Philosophical Society Library, a 1980 letter from Simpson to John Bucher of *Discover* magazine, who had asked about his views on the Gould-Eldredge theory. Simpson wrote:

> They tend to overstate both the novelty and the generality of their ideas. . . . their main point had long ago been stated in other words as a part of the synthetic theory. The idea that their views approach a general theory of evolution that contradicts and replaces the synthetic theory . . . is not justified in my opinion. What they call "punctuation" involves the origin of new species . . . by changes that are either instantaneous . . . or occur at rapid rates of evolution, followed by either slower rates or no further change . . . it was already stated by Darwin in 1859 that rates of evolution demonstrable from the fossil record vary greatly . . . In *Tempo and Mode* . . . I showed, without claiming particular originality, that although most rates fall into a more or less normal distribution, some are very slow or for long periods nil and others are exceptionally rapid, resulting in seemingly abrupt evolutionary changes . . . I called the latter "quantum evolution."
>
> In *Major Features of Evolution* 91953) p. 389, I further generalized this concept. . . . quantum evolution is essentially the same as the "punctuation" of Eldredge and Gould.

14.11 STEBBINS: PLANTS ARE ALSO SELECTED

The process of bringing the major biological disciplines into the evolutionary synthesis was completed, according to most historians of the subject, by the publication of *Variation and Evolution in Plants* by botanist G. Ledyard Stebbins in 1950. Research in botany had been responsible for two major obstacles to the acceptance of the modern natural selection hypothesis in the early twentieth century: de Vries's mutation theory, based on his experiments with the evening primrose (*Oenothera)*, and Johannsen's pure line theory, based on research on beans. Yet Stebbins asserted that "with the exception of the fact that genes lie on chromosomes . . . every other significant fact about genetics needed for the synthetic theory was worked out from research on higher plants." Thus by 1950, a major statement about evolutionary theory from a botanist was long overdue.

Addressing the assertion (Section 14.1) that natural selection is only negative and cannot create anything new, Stebbins invoked a persuasive metaphor: it is negative only in the sense that a sculptor's creation of a statue by removing chips from a block of marble is negative. The sculptor metaphor was also invoked by other authors. The assertion that natural selection is only a negative factor is no longer taken seriously by biologists, since there is good evidence of ample variation in natural populations and enough favorable new mutations to allow progress by selection. But the issue is still being debated by philosophers.

The direct botanical evidence for natural selection was still meager. The only substantial evidence for higher plants was that of W. B. Kemp for mixtures of grass and clover, separated into nearby fields subjected to different conditions (used for grazing or for hay).

While Stebbins had defended the Sewall Wright effect in 1944, by 1950 he had lost his enthusiasm for it. In most cases, he believed, the population is *not* small enough for it to be significant. It is likely, but not proved, that it applies to some plants in Hawaii (*Gouldea, Cyrtandia, Bidens*), and very likely that it affects land snails: "The only plant example known to the writer in which the action of random fixation or drift seems to have taken a prominent part in the differentiation of small isolated populations is that of the complex of *Papava alpinum* [Alpine poppy] in the Swiss Alps, as described by Fabergé (1943)."

14.12 CHROMOSOME INVERSIONS IN *DROSOPHILA*

During the 1940s and 1950s, several biologists obtained empirical results that seemed to show that certain characters, previously considered nonadaptive and therefore attributable to the effects of random genetic drift, were instead primarily controlled by selection. First and perhaps most important in its impact on the views of evolutionists was the series of observations conducted by Dobzhansky and his colleagues (including Carl Epling and Sewall Wright) on the genetics of wild populations of *Drosophila pseudoobscura*. According to historian Will Provine, Dobzhansky originally wanted to find a definitive empirical confirmation of Wright's theory of genetic drift, but ended up doing just the opposite.

Epling's detailed observations on the third chromosome of *Drosophila pseudoobscura* suggested that different gene arrangements had evolved from a single ancestor by inversion of chromosome segments. His results were discussed in a 1945 symposium by Ernst Mayr, G. G. Simpson, and G. L. Stebbins. Epling had assumed that the gene arrangements were equal in selective value, but Mayr argued that " each gene arrangement . . . may have different selective values at each locality. This is a consequence of the fact that the reduction of crossing over in the inverted segments prevents the free recombination of genes." Stebbins agreed that some gene combinations would have a

> selective advantage under certain ecological conditions, an assumption for which Dobzhansky . . . has obtained some indirect evidence. I agree with Mayr that . . . the maintenance of any distribution pattern over periods of thousands or millions of years would be improbable unless the dispersal of the various chromosomal types were restricted by the selective activity of the environment . . . Epling . . . now holds the same belief.

Dobzhansky himself had already come to the same conclusion, as he explained later. The chromosome inversions in *Drosophila pseudoobscura* are examples of *adaptive polymorphism*, which can be observed in field studies but also produced under laboratory conditions where the relevant variables can be controlled:

> Two species, A and B, can be sympatric [occupy the same territory] only provided that the environment in a territory which they inhabit is heterogeneous. The heterogeneity may be spatial or it may be temporal . . . A may be better adapted than B in summer, while B is superior to A during the winter season.

By shifting the balance back and forth between A and B in response to seasonal changes in the environment, natural selection allows the organism to exploit the available resources more efficiently than if only one of the two genotypes, A or B, were allowed to survive.

These observations made use of the giant chromosomes in the salivary glands, which have a visible pattern reflecting the gene arrangement in the chromosomes. There are various possible rearrangements, or inversions, of the chromosome segments, producing a polymorphism. For example, the sequence ABCDEFGHI may be broken between A and B, and between E and F, giving A BCDE FGHI. Then the middle sequence BCDE is inverted, giving AEDCB FGHI. Finally the sequence DCBFGH is inverted, leading to AEHGFBCDI:

> The first can arise from the second or give rise to the second through a single inversion. The same is true for the second and the third. But the third can arise from the first, or vice versa, only through the second arrangement as the probable intermediate step in the line of descent. If we find in natural populations of some species only the first and the third arrangements, it is probable that the second remains to be discovered, or at least that it existed in the past. . . . The existence of previously unknown gene arrangements in *Drosophila pseudoobscura* and *Drosophila azteca* was predicted with the aid of the theory of overlapping inversions, and most of these predictions were subsequently verified by discovery of the requisite inversions in nature.

Note that we have here an example of a confirmed novel prediction, based on the hypothesis that the chromosomes have evolved, but not necessarily by natural selection. It is comparable to the prediction of missing links from the general hypothesis that later organisms have evolved (by some process) from earlier ones, repeatedly confirmed in the late nineteenth and twentieth centuries by discoveries of fossils.

Dobzhansky recalls that "it seemed at first that the chromosomal polymorphism had no adaptive significance. It is now known that the contrary is the case." The importance of this realization is reflected in the reorganization of the third edition of *Genetics and the Origin of Species* (1951): the inversion experiments are described in a chapter now titled "Adaptive Polymorphisms" (instead of the noncommittal "Chromosomal Changes" in the second edition), which is now placed immediately after the chapter on selection rather than before it.

According to Provine, the first public indication of Dobzhansky's shift toward the natural selection hypothesis was in 1943, where he

> breaks rather abruptly with the past: because of cyclic (seasonal) and year-to-year changes in the frequencies of . . . gene arrangements . . . he concludes that these gene arrangements affect the adaptive values of their carriers. Cyclic changes ruled out chance, and mass migration was ruled out by the studies on dispersal. Thus the proportions of these gene arrangements in a population were said to be subject to a surprisingly intense natural selection.

But, Provine notes, Dobzhansky "had concluded that the inversions were under selection two years before the appearance" of this paper, as is shown by his letter to Sewall Wright dated May 4, 1941. As sometimes happens in a fast-moving research project the public record of a scientist's position does not always keep up with changing views expressed in private. For our purposes the public position as well as the private view is important since for a scientist as influential as Dobzhansky the public position is likely to persuade others.

In a 1946 paper, Dobzhansky asserted on the basis of his *Drosophila* experiments that natural selection could cross the supposed barrier between micro- and macroevolution:

> Some of the chromosomes obtained by crossing over between the three ancestral wild chromosomes have properties very different from the latter. It is, therefore, possible to "select" products of recombination of the gene complexes that deviate greatly from the ancestral types, being completely outside the limits of variability of these ancestors.

This allows "great advances in rebuilding the organism in directions favored by artificial or natural selection."

In 1947, Dobzhansky claimed that his own work, along with that of geneticist N. V. Timofeeff-Ressovsky, who first discovered cyclic seasonal changes in 1940, and the more recent publications of N. P. Dubinin and G. G. Tiniakov, disproved the previous view that "adaptive evolution in nature is too slow a process to be observed within a human lifetime." This is an important conclusion, not only because it illustrates the efficiency of natural selection, but also because it casts doubt on the assumption that human behavior can be explained by a changing social environment interacting with a fixed human nature.

According to Dobzhansky's own testimony, it was the surprising result of his own experiments that led him to change his opinion about the relative importance of natural selection and genetic drift: in these experiments drift clearly had a negligible effect, contrary to what was previously believed.

14.13 FORD: UNLUCKY BLOOD GROUPS

E. B. Ford, as the leader of the Oxford School of Ecological Genetics, conducted and sponsored many experiments testing the roles of natural selection and genetic drift and developing the concept of balanced polymorphism. By 1940, he could claim that this research confirmed the importance of natural selection. Here I discuss just one of his contributions: the prediction, based on Fisher's theory, that human blood groups would be found to be correlated with certain diseases.

In 1940, Ford pointed out that variations in the fitness of one member of a polymorphism, caused by changes in the environment, should be correlated with changes in its variability. This can be observed, for example, in the butterfly *Papilio dardanus*: fitness depends on mimicry of another species. Conversely, if we find changes in the distribution of members of a polymorphism we may infer that fitness is due to some as-yet-unknown environmental factor. Ford proposed to apply this idea to the human blood group polymorphism, by predicting a connection between the frequencies of different groups and susceptibility to specific diseases.

There had already been hundreds of reports claiming to find statistical correlations between the ABO groups and various medical conditions or psychological characteristics (including criminality), going back to 1921. But these reports were not considered reliable by modern (i.e., 1940s) standards, because the investigators did not realize that very large samples must be used to obtain significant correlations. Within the community of experts on population genetics no such correlations were believed to have been demonstrated or likely to be established.

Ford made his prediction more explicit in 1945:

> Individuals belonging to the different blood groups are not equally viable, and we may expect elimination to fall upon the AB class . . . A valuable line of enquiry which does not yet seem to have been pursued in any detail would be to study the blood group distributions in patients suffering from a wide variety of diseases. It is possible that in some conditions infectious or otherwise, they would depart from their normal frequencies, indicating that persons of a particular blood group are unduly susceptible to the disease in question.

Ford's prediction was confirmed in the 1950s, beginning with a 1951 report by D. Struthers, who studied bronchopneumonia in babies and found the incidence higher with group A than group O. Two years later, Aird, Bentall and Roberts stated that persons with blood group A are significantly more likely than those with O to have cancer of the stomach. In a more comprehensive article in 1954, which also reported an association with peptic ulceration, these authors, together with Mehigan , credited Fisher and Ford with a prediction relevant to their results. In 1956 Buckwalter's group published a survey of early results, noting that Ford had hinted at a relationship between blood groups and diseases "in suggesting the possible relationship of blood groups to natural selection."

Ford was of course entitled to claim victory for his own theory, which he did in 1957 and again at the Darwin centennial celebrations in 1959. His students and colleagues were also quick to point out the new support for selectionism in the ongoing controversy with Sewall Wright. But Wright had already changed his views by 1951, when he suggested that the distribution of blood group frequencies may be the result of "intergroup selection . . . a certain balance between local isolation and crossbreeding." Nevertheless, I could not find any evidence that other biologists outside Ford's group were persuaded to accept his selectionist views *because* of this confirmation, nor that it made any difference whether it was a *novel* prediction rather than a theoretical deduction of known facts.

Before 1950, Dobzhansky had supported Wright's view that drift plays an important role in evolution, as an adjunct though not a substitute for natural selection, and in particular in the evolution of human blood groups. In 1955 in his book *Evolution, Genetics, and Man,* Dobzhansky still wanted to explain the differences in blood groups among different populations as a result of genetic drift, ignoring the recent discoveries concerning their connection with diseases. In a 1958 survey article in *Science*, he stated that "the functional significance of the blood groups in man is still full of uncertainty." In another survey article he still found the evidence inconclusive. In his 1962 book *Mankind Evolving* he cited the evidence for blood group-disease connections (now including plague, syphilis and smallpox), but tried to explain them by genetic drift.

By 1970 Dobzhansky had revived his affection for drift, especially Mayr's founder principle, which he now said was supported by the work of his own group. He denounced the "hyperselectionism" of other evolutionists, although he admitted that the new interest in neutral or non-Darwinian evolution was perhaps just another "swing of the pendulum."

The association between blood groups and disease is undoubtedly real, according to a 2000 review by Garraty, but biologists in the 1960s were not able to give a satisfactory explanation for the association. It was pointed out by several experts on population genetics that theoretically the blood group polymorphism should not be stable because the heterozygotes do not have a selective advantage over the homozygotes—unlike in the well-known sickle cell-malaria polymorphism, where they do.

14.14 RESISTANCE TO ANTIBIOTICS

Another application of natural selection to medicine appeared around the same time as Ford's prediction. René J. Dubos, who was one of the first to discover an antibiotic drug (gramicidin), warned that the microbes attacked by such drugs would develop resistance to them. This phenomenon had already been observed with other drugs and was known as training or fastness, but its cause was not understood. It might have been considered analogous to the well-known phenomenon of resistance to insecticides, although there are significant biological differences between the two.

In a comprehensive treatise on *The Bacterial Cell*, Dubos reviewed experiments made "to establish whether the resistant bacteria always occur in small numbers during normal growth in the presence of the drug, or whether they are produced only as a response to the presence in the medium of the substance with reference to which resistance develops." In other words, is resistance a Darwinian or a Lamarckian phenomenon?

According to historian Angela Creager, before the 1940s bacteriologists favored the Lamarckian explanation. But the notorious Lysenko affair in the Soviet Union cast suspicion on Lamarckian theories and may have made some scientists more receptive to selectionist theories. Creager considers the 1943 experiment by Salvador Luria and Max Delbrück to be "one of the first clear demonstrations that inheritance in bacteria was not Lamarckian."

Dubos concluded that the evidence favors the Darwinian explanation of resistance: it is due to selection acting on the preexisting variability of the microbes. Those that happened to be resistant to the drug could survive and pass on this attribute to their descendants.

By 1944 it was known that strains of bacteria resistant to various drugs, including penicillin, could be obtained by growing bacteria in media containing increasingly higher concentrations of the drugs. What was not established, despite the Luria-Delbrück experiment, was whether the resistance was induced by the action of the antibiotic or was already present in some of the bacteria as a result of earlier mutations. This question was answered in a classic 1944 experiment by Milislav Demerec. Demerec, using strains of *Staphylococcus aureus*, found evidence that "makes it probable that the second [Darwinian] alternative is correct": resistance of bacteria to antibiotics is a result of selection.

Dobzhansky published, in *Scientific American*, a good popular explanation of Demerec's experiment as a prominent example of "evolution in the laboratory." For those readers who were following a little too recklessly the neo-Darwinian command "make love, not war," he noted that "in certain cities penicillin-resistant gonorrhea has become more frequent." Dobzhansky supported the Darwinian view over the Lamarckian in the third edition of his influential book Genetics and the Origin of Species (1951), which according to Creager "made antibiotic resistance a crucial—and observable—example of natural selection at work."

Further confirmation came from an experiment by Joshua L. Lederberg and Esther M. Lederberg: they showed that bacteria could display an inherited resistance to penicillin even if they had never been directly exposed to it. The mutations were not *caused* by the penicillin but were already present, thus excluding a Lamarckian interpretation. Although they don't say they were confirming a prediction by Dubos, one might surmise this from the fact that Joshua Lederberg studied with Dubos at Columbia and later recalled that Dubos's *Bacterial Cell* "is the work from which I can say I learned most of the microbiology I know . . . this work was the launching pad for my own investigations."

Although bacterial resistance to antibiotics cannot be called a novel prediction from natural selection, the experiments mentioned above did provide important support for evolutionary theory, especially for those scientists and physicians who were interested in practical applications of the theory. Dubos himself seemed to regard it not so much as an important scientific discovery but more as an illustration of his general philosophy: humans should not try to conquer nature (in this case by using powerful drugs), but should try to live in harmony with it.

For the twenty-first century we need an addendum: if humans want to live in harmony with nature, they must understand how nature works. In the last few years there have been several newspaper and magazine articles about methicillin-resistant *Staphylococcus aureus* (MRSA). These articles explain that use of antibiotics such as methicillin by physicians and their patients and by those involved in agriculture and meat processing have led to the development of resistant strains of the bacterium. As a result, the methicillin is no longer very effective. Sometimes the articles mention that this has happened before—for example with the former miracle drug penicillin, which can no longer perform miracles.

What the articles generally *don't* explain is that this is a simple example of Darwinian selection: the bacteria found in nature already have a rich variety of mutations, including ones that confer resistance to the antibiotic. When you use the antibiotic the nonresistant bacteria are killed, and the resistant strain quickly takes over, just like the wingless insects on the windy island.

Why don't the journalists explain this? It might help persuade some readers to avoid unnecessary use of antibiotics. At least one could point out, as does *New York Times* reporter Erik Eckholm, that " resistance can evolve whenever drugs are used against bacteria or other microbes because substrains that are less susceptible to the treatment will survive and multiply." Instead we get pseudo-Lamarckian explanations, such as "staph learned how to resist methicillin " or "staph aureus is a very smart bug, it figured out every antibiotic we humans have thrown at it and has developed resistance mechanisms to them one by one."

The only reason I can think of—given the fact that the last two quotations came from people with "Dr." in front of their names—is the journalist's or editor's fear that creationists will attack them and perhaps threaten to boycott advertisers. As we know, self-censorship can be quite insidious. In any case, we must conclude that the creationist movement is a threat not only to public education but also to public health.

14.15 TWO GREAT DEBATES: SNAILS AND TIGER MOTHS

One of the chapters in Julian Huxley's book *The New Systematics* (1940) was a report by Cyril Diver on his studies of two closely related species, *Cepaea nemoralis* (land or woods snails) and *Cepaea hortensis* (garden snails), living in the same areas in England and elsewhere in the United States and Europe. He attributed the split of the presumed earlier single species to "random differentiation in small partially isolated populations" as described by Wright's theory, presented in the same book, and *not* to geographical isolation or natural selection.

After the end of World War II, Arthur J. Cain and Philip M. Sheppard collaborated in a new investigation of *Cepaea nemoralis,* designed to test Diver's conclusion and perhaps to settle the Ford-Wright debate of the late 1940s about the relative importance of natural selection and random drift. This was the beginning of what Provine, in his biography of Wright, calls the great snail debate. Since he has described it in great detail, I need only summarize his account and his conclusions.

Cain and Sheppard, in their first paper on *Cepaea nemoralis*, found (contrary to Diver) that color and banding patterns "have definite selective value" in connection with predation by thrushes. They concluded that speciation should be ascribed to natural selection, not genetic drift, and that other cases ascribed to genetic drift should be reexamined. According to Provine, this result did not play the role in the selectionist/drift debate that might have been expected. Since Dobzhansky and Mayr had used Diver's results to support their prodrift views in the early 1940s, Fisher, Ford, and other selectionists would have anticipated that this evidence against the importance of drift would be considered evidence against Wright's theory of evolution, and that Wright would therefore try to defend his theory by arguing that gene frequencies were strongly influenced by drift. But in fact, Provine reports, "even before Wright read the Cain and Sheppard [1950] paper on polymorphism in *Cepaea*, he was prepared to believe that the gene frequencies . . . were governed primarily by natural selection rather than by random genetic drift." Even after Maxime Lamotte published his own results on *Cepaea* populations in France, which he interpreted as supporting Wright's theory of genetic drift, Wright insisted that while genetic drift is a real phenomenon, it is not important for evolution except as it may interact with natural selection. He was not really interested in invoking drift to explain neutral characters, which other biologists seemed to think provided its most useful application.

Cain also started a mini-debate in *Nature* magazine when he criticized biologists, such as G. S. Carter, who postulate drift just because they personally could not see any possible selective value in a variation. Then, Cain asserted: " so far, every supposed example of random variation [or genetic drift] that has been properly studied has been shown to be nonrandom."

Carter accused Cain of claiming "that selective value should be assumed in all characters until the contrary is proved." Cain denied taking that position, maintaining instead that one should not assume either randomness or selection without proof. He did not claim that there is *no* evolution of nonadaptive characters. Carter replied: "I am glad that Cain admits that genetic drift may be a real factor in evolution . . . We should . . . regard it and selection as equally possible explanations when neither is proved." In a 1957 book on evolution, Carter omitted genetic drift entirely.

According to Sir Cyril A. Clarke, the American opposition to the natural selection hypothesis, in particular the belief that polymorphisms are adaptively neutral, was still strong in 1959. Sheppard's paper, presented at the Cold Spring Harbor Symposium in that year, was, according to his biographer C. A. Clarke, "the only contribution which gave facts and reasons against neutrality, [and] was deleted by the editor [of the published proceedings]. This injudicious act roused Philip to fury."

Although the great snail debate continued through the next few years, it was considered in the 1960s to have produced a victory for natural selection over genetic drift. As Stephen Jay Gould wrote in 2002, it was a triumph for adaptationism .

Fisher and Ford conducted another test of Wright's genetic drift theory using two colonies of the scarlet tiger moth (*Panaxia dominula*). They measured fluctuations in the frequency of a heterozygous form, *medionigra*, easily recognizable by the colored wing pattern it produces. They reported in 1947 that these fluctuations were much larger than estimated from random drift theory. Hence the fluctuations must be caused by changes in the selective forces. Wright, in his reply, complained that his theory had been misrepresented; furthermore, that it was not legitimate to assume that any effect not due to drift must be due to selection. There were further replies back and forth, leading not to a resolution of the controversy but to great animosity among its participants and some stimulus to useful further research.

14.16 SELECTION AND/OR DRIFT? THE CHANGING VIEWS OF DOBZHANSKY AND WRIGHT

In 1983 Stephen Jay Gould extended his complaint about hardening beyond Simpson to Dobzhansky and Wright, the two founders of the synthesis who had tempered natural selection by genetic drift within small isolated populations in their explanations of speciation.

By 1957 Dobzhansky, contrary to what one might expect from Gould's account, was launching his own campaign against the hardening of the synthesis. In a paper with Olga Pavlovsky, he attacked the view that natural selection is all-important while drift can be ignored:

> As soon as a gene is shown to have any effect whatever on fitness, the conclusion is drawn that its distribution in populations must be determined solely by selection and cannot be influenced by random drift. But this is a logical nonsequitur. The important work of Aird et al. (1954) and of Clarke et al. (1956) disclosed that the incidence of certain types of gastro-intestinal ulceration is significantly different in persons with different blood groups. This is, however, far from a convincing demonstration that the observed diversity in the frequencies of the blood group genes in human populations is governed wholly, or even partially, by selection for resistance to ulcers. To make such a conclusion tenable it would have to be demonstrated that the environments in which human racial differences have evolved actually favored greater resistance in certain parts of the world and lesser resistance in certain other parts. Thus far no evidence has been adduced to substantiate any such claim.

In a carefully designed experiment, Dobzhansky and Pavlovsky tested the prediction that because of genetic drift, a small population will evolve with greater genetic diversity than a large one starting with the same frequencies even when both are subject to strong selective forces. It was also a test of Mayr's founder principle. Ten small populations (20 founders each) and ten large populations (4,000 founders each) of *Drosophila pseudoobscura* were constructed to have a 50–50 ratio of PP (Pikes peak) and AR (arrowhead) gene arrangements in their third chromosome, and kept in population cages under uniform conditions for 18 months (about 19 generations). The results showed that the small populations had significantly greater variation in the PP and AR frequencies. "Although the trait studied . . . is subject to powerful selection pressure," as found in Dobzhansky's earlier experiments (Section 14.12), "the outcome of the selection in the experimental populations is conditioned by random genetic drift. The either-selection-or-drift point of view is a fallacy." Mayr's founder principle was also verified by the experiment.

The Dobzhansky-Pavlovsky experiment is regarded as a classic demonstration of the phenomenon of genetic drift, and it certainly shows that predictions from the Fisher-Haldane-Wright theory can be tested under controlled conditions. But does it have any relevance to evolution in nature? Dobzhansky and Pavlovsky argued that it explains the fact that populations of the butterfly *Maniola jurtina*, "rather uniform throughout Southern England, despite some obvious environmental diversity in different parts of the territory . . . show quite appreciable divergence on the islands of the Scilly archipelago, although these islands are within only a few miles of each other and their environments appear rather uniform." This divergence was attributed entirely to selection by Dowdeswell and Ford. Dobzhansky withdrew his earlier statements that the situation arose "through random drift in populations of continuously small size, or frequently passing through narrow 'bottlenecks.'" Instead, "in the island population we are observing the emergence of novel genetic systems moulded by interaction of random drift with natural selection."

But Ford strongly rejected the view that his results from Scilly could be ascribed to the founder principle or any other kind of drift, insisting that they are completely due to natural selection.

Dobzhansky's change from prodrift to antidrift has already been discussed (Section 14.12). It remains to note only that by 1970 he was beginning to suspect that it was time to reverse his position again, in the light of recent "theories of non-Darwinian evolution by random walk." These theories, which were to become more popular in the 1970s, he characterized as " a new swing of the pendulum in the opposite direction." He told Jeffrey Powell in 1974, " I began as a drifter and then became a selectionist. Now, in my old age, I find myself becoming a drifter again."

The credibility of Mayr's founder principle did not seem to suffer from the refutations of genetic drift by the Ford group. While Wright and others considered the founder principle a special case of genetic drift, it came to be regarded as a distinct hypothesis, perhaps the only exception to the proselectionist consensus circa 1960.

At the same time, Gould was jumping on a different pendulum: the one that swings from micro to macro mutations. In a 1966 paper he expressed a strongly selectionist view: "Our ultimate goal in the study of a phyletic lineage is the explanation of each morphological change in terms of its selective advantage." Only six years later he rejected the Darwinian paradigm in favor of a new theory of punctuated equilibrium, somewhat akin to the macromutations of deVries and Goldschmidt.

14.17 THE VIEWS OF OTHER FOUNDERS AND LEADERS

To begin with the founding fathers: Fisher, as far as I know, never deviated from his strong selectionist position. Wright more or less abandoned his drift theory, while others adopted what they thought was his theory. Haldane, in the early 1950s, seemed to favor Wright's earlier view that drift assists selection. Later he asserted that "chance effects . . . will rarely matter to a whole species" except in the case of very small populations where the founder principle may come into play, or "the simultaneous establishment of several factors which are harmful singly but adaptive in combination." In a review of the status of natural selection for one of the many publications celebrating the centennial of Darwin's *Origin of Species*, Haldane briefly mentioned the idea that evolution may result from genetic drift in small tribes. He wrote that "the weakest point in Wright's argument is that he has not adequately considered what happens when this tribe starts hybridizing with others."

Ford, like his colleague Fisher, always insisted that genetic drift has no important role in evolution but went further in his public criticism of Wright, saying the theory of adaptive peaks "is wholly unrealistic and has done much harm in biology; assuming, as it apparently does, that ecological conditions are so stable that a given type of genetic adaptation . . . is persistent." He admitted that when the theory of genetic drift was proposed it seemed reasonable because selective advantages were thought to be small; but we now know, Ford insisted, "that large selective advantages are in fact common in natural populations," so selection overwhelms drift. The verdict: experiments by Ford and his colleagues proved that natural selection explains everything while drift explains nothing.

Julian Huxley, an influential advocate of the Synthesis, did not show much sympathy for genetic drift after 1942. Returning from a two-week visit to the USSR in 1945, Huxley reported with approval that despite Lysenko's influence, Russian evolutionists were even more selectionist than those in Great Britain and the United States. Reviewing Simpson's *Tempo and Mode in Evolution*, Huxley noted that gaps in the fossil record might be explained away by Wright's hypothesis that evolution is more rapid in small isolated populations, but geneticists may doubt if such populations could exist for the millions of years needed for nonadaptive evolution. Later he wrote: "it is now clear that selective advantages so small as to be undetectable in any one generation, are capable, when operating on the scale of geological time, of producing all the observed phenomena of biological evolution . . . Evolutionary change is almost always gradual and is almost wholly effected through selection." He asserted that the discovery that human blood groups are maintained in a morphism by a balance of selective advantages and disadvantages (Section 14.13) confirms Fisher and refutes Wright's suggestion that the ratios originated by drift. In later writings he simply omitted genetic drift as a significant factor in evolution, until 1963 when he again changed his mind and stated that drift "can have definite evolutionary consequences."

In an abridged edition of his 1963 book published seven years later, Mayr took notice of the proposed revival of genetic drift under the name "non-Darwinian evolution" (inappropriate and misleading, he asserted, because Lamarckism, orthogenesis, and other heresies could also be described by that term). He attempted to refute the new ideas in a couple of pages, concluding that they should not replace natural selection as the primary cause of evolution.

Stebbins asserted that the burden of proof lies on the shoulders of anyone who wants to invoke genetic drift to explain nonadaptive differences: "if we cannot see why a difference should be connected with differential adaptation, we are not justified in concluding from this fact that the difference is non-adaptive." Contrary to early assumptions that genetic drift should be relatively more important in small populations,

> no example of differentiation between populations is known which can be ascribed solely or principally to [chance]. One reason is that the effects of selection, both direct and indirect, are very strong, particularly in populations which are becoming smaller because of a worsening environment. Consequently, no gene remains unaffected by selection for a sufficient number of generations so that its frequency can be permanently altered by chance alone.

Many phenomena *can* be explained by a *combination* of chance and natural selection; however, "chance factors by themselves probably have little effect on evolutionary processes, but in combination with reduction in population size and the accompanying natural selection, they may play an important role in guiding the early stages of adaptive differentiation." This is essentially Wright's later view. But three years later, Stebbins restated Wright's concept of adaptive peaks and valleys by proposing that mutations rather than drift trigger the jump from one peak to another. Presumably he meant small mutations. This was a nice balancing act on the part of Stebbins, since he insisted at the same time that the Goldschmidt macromutation theory had been definitely excluded by modern molecular biology and developmental genetics.

Why did the synthesis harden? David J. Depew and Bruce H. Weber write:

> Gould professes not to know the causes of this "hardening." He mentions the prestige of solid adaptationist explanations in the field . . . These seemed to confirm selectionist experiments conducted in the laboratory, where gene frequencies of different phenotypes could slowly be shifted by controlled environmental variation. We would like to suggest, however, that standing behind this prestige is the deepening influence of empiricist models of science on the synthesists, their successors, and their philosophical defenders. A clean series of inferences linking a mathematically expressible set of laws to laboratory work, and extending laboratory results to the field, was precisely what would be needed to argue that evolutionary theory now stood on a solid, quantifiable, and potentially axiomatic-deductive basis.

14.18 THE PEPPERED MOTH

In addition to the evidence presented by Dobzhansky (Section 147), and the more recent research on chromosome inversions (14.12), blood groups (14.13), drug resistance (14.14), and snails and tiger moths (14.15), many other observations and experiments were cited in books published during the four decades covered by this survey. I will mention here only one, which became quite well known in the 1960s.

Industrial melanism was frequently mentioned as a probable example of natural selection even before the research by biologist H.B.D. Kettlewell on the peppered moth (*Biston betularia*). It was widely noted in the late nineteenth century that in industrial areas of England and other countries, moths that had previously had light-colored wings were being replaced by moths of the same species with dark-colored wings. It seemed plausible that this phenomenon was somehow connected with the darkening of the bark of trees produced by smoke from factories, but there was considerable disagreement about the exact nature of the selective factor. E. B. Ford argued that the dark (melanic) form was more viable, but its numbers were ordinarily limited because dark moths were more likely to be seen and eaten by birds when the trees were a lighter color. Others suggested a more direct connection between the chemical compounds in polluted air and the physiological mechanism that determined wing color.

Starting in 1953 Kettlewell, a member of Ford's group, tested the hypothesis that the moths were being eaten by birds, and that they were more likely to be eaten when the color of their wings contrasted with the color of the trees on which they rested. He marked and released more than 200 moths, mostly the dark form (*carbonaria*), in a heavily polluted area near Birmingham, then tried to recapture them. A significantly higher proportion of the dark moths disappeared, and several were observed being taken by robins and sparrows. Biologist Niko Tinbergen took motion pictures of Redstart birds (old-world thrushes) taking and eating the moths. As a control, the experiment was repeated in unpolluted countryside in Dorset. Kettlewell claimed to have established not only that predation by birds was the selective agent, but that human judgments of which moths were more conspicuous against a light or dark background correlated well with the effectiveness of the birds in catching the moths.

Kettlewell's first results were published in 1955, with a full report in 1956. It began to be mentioned as an example of natural selection in British publications in the late 1950s and in American publications in the early 1960s. It became a familiar piece of evidence for natural selection in many textbooks, most of which reproduced Kettlewell's pair of photographs showing moths against similar and contrasting backgrounds.

Did Kettlewell's research on the peppered moth help to persuade evolutionists to accept natural selection as the primary cause of evolution? No. Most had already adopted that position. But through its appearance in many textbooks, popular articles, and even a novel by Margaret Drabble, it may have reinforced the evolutionary beliefs of the next generation of biology students—although by that time the evidence for evolution, and for the natural selection hypothesis, was already overwhelming.

14.19 THE TRIUMPH OF NATURAL SELECTION?

According to Ernst Mayr, at a meeting in Princeton in January 1947,

> there was universal and unanimous agreement with the conclusions of the synthesis. All participants endorsed the gradualness of evolution, the preeminent importance of natural selection, and the populational aspect of the origin of diversity. . . . Not all other biologists were completely converted. This is evident from the great efforts made by Fisher, Haldane, and Muller as late as the late 1940s and 50s to present again and again evidence in favor of the universality of natural selection, and from some reasonably agnostic statements on evolution made by a few leading biologists such as Max Hartman.

By the 1960s, the founders and leaders of the evolutionary synthesis in the United States and the United Kingdom had decided that random genetic drift by itself, while theoretically expected to occur under certain conditions, has essentially no significant role in evolution except perhaps to facilitate natural selection, so that natural selection acting on small variations is all-important. This is what Stephen Jay Gould called the "hardening of the synthesis." The views of founders and leaders have been discussed in earlier sections. I now turn to four other biologists whose work was often cited as evidence for natural selection: N. P. Dubinin, Francis B. Sumner, W. C. Allee, and David L. Lack.

Dubinin, who had earlier proposed an idea similar to Wright's genetic drift, started to publish his research in Western journals in 1945. In a series of papers with G. G. Tiniakov, he reported that natural selection rapidly produces measurable changes in the genetic structure of populations of *Drosophila funebris* near Moscow. No mention of genetic drift appears in these papers. The selective factors are seasonal changes in climate and industrialization.

Sumner, whose work on protective coloration of the deer mouse (*Peromyscus*) and the mosquito fish (*Gambusia patruelis*) was often cited as evidence for natural selection, was originally a Lamarckian, and also contemplated the possibility of nonadaptive evolution by drift. He was persuaded by his own results to accept an adaptationist interpretation. Since Lamarckism had been refuted, and "inner perfecting tendencies" (orthogenesis) "do not belong in the realm of science," natural selection was the only hypothesis worth considering.

In his autobiography Sumner wrote that he decided that the color of the mice was a result of selection by predators, in accordance with Mendelian laws, despite his dislike for particulate theories in science and for the faddish character of the Mendelian movement. According to Provine, "because he enjoyed a well-deserved reputation as an exceedingly careful experimentalist, his reinterpretation of the adaptive value of the differences between geographical races carried much . . . weight with both systematists and geneticists."

Allee, who followed Wright's theory of evolution in his 1938 book on the social life of animals, published (with four other authors) a massive treatise on animal ecology in 1949. Although he still seemed to favor Wright's theory, it was now overshadowed by a huge amount of evidence for natural selection.

Lack, an ornithologist known for his research on Darwin's finches, is an example of a scientist who, like Dobzhansky, changed from drift to natural selection as a result of his own research. In his monograph on the Galapagos finches, Lack rejected Darwin's view that natural selection would favor different varieties in different islands, because he could find no evidence that the differences have adaptive significance. Instead, like "most geographical forms which have been described in birds," their nonadaptive evolution as "small and isolated populations" is best described by the Sewall Wright effect. But in a 1973 autobiographical memoir, Lack wrote:

> Since . . . Arthur Cain asked me, in the mid-1950s, why I postulated that various subspecific differences in the finches are nonadaptive, it is worth stressing that, before my work, almost all subspecific differences in animals were regarded as non-adaptive (hence the importance of Sewall Wright's theory of genetic drift). . . . "The variation of animals in nature" by G. C. Robson and O. W. Richards in 1936

> fairly reflected current opinion . . . I reached my conclusion, that subspecific and specific differences in Darwin's finches are adaptive, and that ecological isolation is essential for the persistence [sic] of new species, only when reconsidering my observations five years after I was in the Galapagos.

When *Darwin's Finches* was reprinted in 1961, he added a preface, in which he explained that the burden of proof was now on anyone who claimed that species differences are nonadaptive:

> This text was completed in 1944 . . . in the interval, views on species-formation have advanced. In particular, it was generally believed when I wrote the book that, in animals, nearly all of the differences between subspecies of the same species, and between closely related species in the same genus, were without adaptive significance . . . Sixteen years later, it is generally believed that all, or almost all, subspecific and specific differences are adaptive, a change of view which the present book may have helped to bring about. Hence it now seems probable that at least most of the seemingly non-adaptive differences in Darwin's finches . . . would, if more were known, prove to be adaptive.

In his 1968 book on *Ecological Adaptations*, Lack did not mention drift but wrote that

> this book, like every other work of natural history, shows, at least for those who have eyes to see, the immense power of natural selection and the complex and subtle adaptations to which it can give rise.

14.20 RESULTS OF A SURVEY OF BIOLOGICAL PUBLICATIONS

Given this consensus of the founders and leaders of the evolutionary synthesis in favor of natural selection, denying any significant role for genetic drift or macromutations, what about the followers? That is, what about other biologists who published books and technical review articles on evolution and related topics during the 1940s, 1950s, and 1960s? Here the picture is not as clear-cut.

In the decade 1941–1950, less than 40% of those publications accepted the natural selection hypothesis (natural selection acting on small mutations) as the primary factor in evolution, indicating that drift is insignificant or not mentioning it at all. Almost 30%, while accepting natural selection as an important factor, stipulated that drift may also have a significant influence on evolution. Some would also allow a role for other factors: large mutations (13%) or orthogenesis (13%). These numbers add up to more than 100% because some writers mentioned more than one factor. About 13% rejected natural selection entirely.

In the next decade (1951–1960), about 60% accepted natural selection acting on small mutations, without drift, as the primary or only factor in evolution. Support for drift (along with natural selection) remained just below 30%.

In the last decade surveyed, 1961–1970, support for natural selection without drift dropped slightly but remained about 50%, while support for drift increased to about 35%. By the second half of this decade (1966–1970) large mutations, Lamarckism, and orthogenesis had completely disappeared. So the views of these other evolutionary biologists in the late 1960s reflect the views held by the founders and leaders in the 1940s, after rejecting other factors but before the hardening of the synthesis (rejection of drift).

Another group of publications includes textbooks, popular books, and articles (again excluding those written by founders and leaders). The value of textbooks is succinctly expressed in a statement by professional scientist and amateur historian Stephen Jay Gould:

> To learn the unvarnished commitments of an age, one must turn to the textbooks that provide 'straight stuff' for introductory students . . . surveys of textbooks provide our best guide to the central convictions of any era. . . . This field of vernacular expression has been neglected by scholars, though the subject would yield great insight.

In my survey I have included popular books and articles in the category of textbooks, in order to get a reasonably large sample; a more comprehensive study would probably reveal significant differences.

In the decade 1941–1950, if we first disregard the question of drift, we find that whereas about two-thirds of the evolutionary biologists accepted natural selection acting on small mutations in their books and technical reviews, about half of the textbooks did so. Fewer than 15% of evolutionary biologists supported large mutations, compared to nearly 45% of authors of texts. I would identify this decade as the time when evolutionary biologists as a community rejected de Vries (and Goldschmidt) and accepted the hypothesis that evolution proceeds by natural selection acting on *small* mutations, while many other biologists continued to support evolution by large mutations. In the same decade, while drift was gaining support among evolutionary biologists it was gaining a foothold in textbooks but was not quite as popular as orthogenesis.

In the next decade, 1951–1960, the proportion of textbooks that endorsed natural selection (with or without drift) jumped from half to about three quarters, while the support for drift rose from 13% to about 20%. Large mutations received the support of about 20%, considerably less than in the previous decade.

Finally, in the decade 1961–1970, the textbooks were similar to the books and technical reviews in that they had almost 60% support for natural selection without drift, and about 35% for natural selection with drift. The major difference is that about 10% of the textbooks still endorsed large mutations, which had almost disappeared from the monographs and technical reviews.

So our conclusion—that the natural selection hypothesis became established knowledge in the 1960s—is just barely (and probably only temporarily) valid. The strong proselection consensus of the founders and leaders did not translate into equally strong support for natural selection among other biologists during the period of this study, a consequence of the familiar fact that knowledge presented in textbooks generally lags behind the research frontier by a decade or so.

One advantage of looking at textbooks popular enough to be frequently revised is that one can sometimes see an author changing his or her mind in successive editions, and thereby get an idea of what reasons were persuasive. Here we have four cases, all American:

1. James Watt Mavor, professor of biology at Union College, changed from the de Vries large-mutation theory to the NSH (small mutations) between the second (1941) and third (1947) editions of his *General Biology* text, but forgot to remove his endorsement of de Vries in making this revision.
2. Alfred M. Elliott, professor of zoology at the University of Michigan, was completely selectionist in the first and second editions of his zoology text (1952, 1957), but in later editions he discussed one example of genetic drift. At the same time he gradually abandoned his support for orthogenesis.
3. William Beaver, professor of biology at Wittenberg University, presented a negative view of natural selection in the fifth edition of his general biology textbook (1958), which he dropped in the sixth edition (1962), where he also mentions the modern synthesis. The seventh edition (1966), coauthored by George B. Noland of the University of Dayton, gives a fuller treatment of evolution, stating that the major factors are (presumably in order of importance) mutation, genetic drift, natural selection, and migration. No examples of natural selection are given.
4. One author presented one view in a popular book and another in a textbook published a few years later. Biologist Garrett Hardin discussed several cases in which genetic drift is important in *Nature and Man's Fate* (1959). But in the second edition of his textbook on biology (1966), he presented the theory of natural selection with no mention of drift.

14.21 IS EVOLUTIONARY THEORY SCIENTIFIC?

> Students often hold a variety of erroneous views of science, including for example a narrow definition of science as strictly experimental. Only claims that derive from hypothesis testing, manipulation of variables, etc. are viewed as legitimate science. Evolution is found wanting because many of its claims are about historical probabilities and are not based on controlled experiments. This view of science is probably often the result of both misguided instruction and immersion in a culture that presents science only as "THE scientific method"! . . . What is needed today is teaching that promotes a broader understanding of science and evidence.
>
> —*biology educator* Mike U. Smith *(2010)*

Starting around the time of the 1959 centennial of Darwin's *Origin of Species*, philosophers and biologists began to debate the question of whether evolutionary theory was scientific. By this the philosophers meant: does it make testable predictions? The biologists had to decide whether to answer that question (as Haldane had already done in the quotation at the beginning of this chapter), or to argue that the philosophers' criterion should not be applied to evolutionary theory (or to biology in general).

Even before 1959 there were two important experiments that confirmed predictions of the Fisher-Haldane-Wright theory under controlled conditions where both natural selection and genetic drift are expected to influence the evolution of a population: the research of Kerr and Wright (see Section 14.3) and of Dobzhansky and Pavlovsky (Section 14.16). Dobzhansky and Spassky reminded readers that the Dobzhansky-Pavlovsky experiment showed that large populations have smaller variations in genotype frequencies, as predicted, and reported another experiment confirming the prediction that "if the sizes of the foundation stocks of the populations are kept constant, the variations of the outcomes should be greater if these foundation stocks came from a source with higher genetic variability than from a genetically more uniform source." In the meantime, Lamotte (1959) had shown that the expected greater variation of smaller populations was also present in natural populations.

A different kind of prediction was made by Ford, based on Fisher's theory of balanced polymorphisms. He suggested that an association between blood groups and susceptibility to certain diseases would be found in humans. This was confirmed by Struthers and others in the 1950s (Section 14.13). Many evolutionary biologists were aware of this success, yet, like the experiments by Kerr and Wright and by Dobzhansky's group, it was rarely mentioned in the discussions with philosophers.

In 1945, philosopher Karl Popper published the first of several critiques of the Darwinian theory of evolution. His article was primarily an attack on historicism—the assumption that history is determined by inescapable laws—in the social sciences. He stated that "the recent vogue of historicism" (which he blamed for the evils of fascism and communism) " might be regarded as merely part of the vogue of evolutionism." It was therefore necessary to discuss first the use of evolutionism in the biological sciences. He asserted that there could be no such thing as a scientific law of evolution, because

> the evolution of life on earth . . . is a unique historical process. . . . We cannot hope to test a universal hypothesis or to find a natural law acceptable to science if we are confined to the observation of one unique process. Nor can the observation of one unique process help us to foresee future developments.

In 1959, philosopher Michael Scriven argued, against Popper, that evolutionary theory does indeed offer "satisfactory explanation of the past . . . even when prediction of the future is impossible." Scriven's article elicited reactions from biologists and became part of the debate about the scientific legitimacy of evolutionary theory in the 1960s.

I see two serious problems with Scriven's argument. First, he defines a category of subjects, including "a great part of biology, psychology, anthropology, history, cosmogony, engineering, economics, and quantum physics," which he calls "irregular subjects"; in these subjects, "serious errors are known to arise in the application of the available regularities to individual cases," and he implies that "prediction is excluded" in these subjects. Aside from the derogatory connotation of the word "irregular," his conception of prediction is impoverished. Quantum physics most certainly does make novel predictions that can be tested in the laboratory, and all of them have been confirmed (though not before the theory was accepted; see Chapter 8). Apparently Scriven has in mind the fact that the present position or quantum state of an individual subatomic particle is indeterminate until it is measured, hence its future position or quantum state is unpredictable without a measurement; and the *presumed* fact that cosmogony deals only with unobservable and hence unknowable events in the distant past. (He had argued in 1954 that we will never be able to determine whether the age of the universe is finite or infinite.) Moreover, strictly speaking *no* science, regular or irregular, can offer accurate prediction of the future except in the sense that it can predict more or less accurately the result of a measurement we make in the present under controlled conditions. Guesses and extrapolations about future weather, earthquakes, and so on may be useful in planning our lives, but become increasingly unreliable after a few days or weeks. Yet meteorology and seismology are respectable sciences.

Thanks to the mathematical theories of Fisher, Haldane, and Wright, and the availability of fast-breeding organisms like *Drosophila*, it was possible by the 1940s to test the predictions of evolutionary theory in precisely the same way that one could test the predictions of quantum mechanics and cosmogonical theories—though philosophers, and even some biologists, seemed unaware of this fact.

Dobzhansky had already clearly distinguished (as noted at the end of Section 14.6) between the historical approach to evolution (phylogeny) and the causal-mechanistic approach (population genetics). The second approach does use the hypothetical-deductive method; Popper and other philosophers who asserted that evolutionary theory is not falsifiable simply ignored the entire corpus of work for which Dobzhansky and his colleagues had become famous by 1959.

Second, Scriven asserts that Darwin and others have attempted "to encapsulate the principles of evolution in the form of *universal* laws and base *predictions* on them," but have failed. On the contrary, while I do not claim that biologists have established universal laws of evolution, I have shown that they have used the principles of natural selection to generate several predictions that have been tested and mostly confirmed.

A theory that governs the future behavior of a system, like Newtonian or quantum mechanics or general relativity, can also yield predictions about its present behavior. This is a crucial fact often ignored by those who say that evolutionary theory cannot make testable predictions. It is ironic that Scriven lumps Darwinism with cosmology in the category of irregular theories that can explain, but not predict. Within six years of the publication of Scriven's article, the prediction from the big bang theory (contrary to the steady state theory) that space is *now* filled with microwave radiation at a few degrees above absolute temperature was spectacularly confirmed; this confirmation played some role in the decision of proponents of steady state to switch to the big bang, although that decision was not greatly influenced by knowledge that the radiation had been predicted (Chapter 12).

A related example closer to the domain of biological evolution is the theory of the origin of elements in stars. In order to explain the synthesis from hydrogen of elements heavier than helium, Fred Hoyle predicted a previously unobserved nuclear reaction, the tri-alpha reaction (fusion of three helium nuclei to form a carbon-12 nucleus), which was subsequently confirmed in a laboratory experiment. Even though the entire process of the evolution of elements from hydrogen takes too long to follow in the laboratory, almost every individual stage can be explained theoretically and tested experimentally (Section 12.8).

Although many scientific theories do make successful predictions, that success is not necessarily a requirement for accepting those theories. It does raise the barrier for anyone who wants to replace a successful theory by an as-yet-untested one.

How Evolutionists Replied to Popper and Scriven

The biologists were too quick to concede Scriven's charge that evolutionary theory does not make testable predictions. I find it remarkable that Ernst Mayr , who had himself made a confirmed prediction from evolutionary theory (as he was pleased to remind the readers of his next book), agreed with Scriven:

> The theory of natural selection can describe and explain phenomena with considerable precision, but it cannot make reliable predictions . . . Scriven has emphasized quite correctly that one of the most important contributions to philosophy made by the evolutionary theory is that it has demonstrated the independence of explanation and prediction.

But, he continued, we are happy if "our causal explanations [also] have predictive value." Apparently what Mayr meant was that evolutionary theory can be scientific as long as it provides good explanations even if it doesn't make testable predictions; but sometimes it does make testable predictions. In 1982, he distinguished logical prediction (defined as "conformance of individual observations with a theory or scientific law") from *temporal* prediction (inference from the present to the future). The latter (what physicists refer to as prediction in advance and philosophers call novel prediction) is "much more rarely possible in the biological sciences." But three years later he argued that "in principle there is no difference between the physical and the biological sciences with respect to experiment and observation"; neither makes absolute predictions. Nevertheless, he agreed with Scriven that " the ability to predict is not a requirement for the validity of a biological theory," and stated that "the theory of natural selection can describe and explain phenomena with considerable precision, but it cannot make reliable predictions, except through such trivial and meaningless statements as, for instance, 'the fitter individuals will on the average leave more offspring.' "

Simpson also agreed with Scriven that a historical science like evolutionary biology or geology does not make predictions in the same way the physical sciences do, but argued that

> the testing of hypothetical generalizations or proposed explanations against a historical record has some of the aspects of predictive testing. . . . a conspicuous example has been the theory of orthogenesis, which . . . maintains that once an evolutionary trend begins it is inherently forced to continue . . . That plainly has consequences that would be reflected in the fossil record. As a matter of observation, the theory is inconsistent with that record. . . . A crucial historical fact or event may be deduced from a theory and search may subsequently produce evidence for or against its actual prior occurrence. That has been called "prediction" . . . what is actually predicted is not the antecedent occurrence but the subsequent discovery.

A few years later he wrote: "claims that the hypothetico-deductive method (e.g., Popper 1959) is the only one allowable in science are almost absurdly extreme, but obviously it is an allowable method. . . . Retrodictive interpretation and explanation are almost unique to the historical sciences."

The most elaborate refutation of the philosophical critics in the period before 1970 was Michael Ghiselin's book *The Triumph of the Darwinian Method*. Ghiselin accepted Popper's falsifiability criterion and insisted that Darwin's theory satisfies it:

> It seems almost unreal that, among all theories of science, the Darwinian theory of natural selection has been singled out . . . as incapable of refutation. . . . A theory is refutable, hence scientific, if it is possible to given *even one* conceivable state of affairs incompatible with its truth. Such conditions were specified by Darwin himself, who observed that the existence of an organ in one species, solely "for" the benefit of another species, would be totally destructive of his theory. That such an adaptation has never been found is a most compelling argument for natural selection.

But Ghiselin did not discuss the more recent predictions by Dobzhansky, Ford, Mayr, and Wright.

Michael Ruse pointed to the return of light-colored moths to "those areas that have introduced smokeless zones" as a fact that was "clearly expected" (and thus could have been predicted) by the synthetic theory; the case of wingless insects is also evidence for natural selection, he noted. But, with the exception of Ruse's brief remarks about wingless insects, philosophers did not discuss the confirmed predictions of natural selection mentioned in this chapter, at least in the literature of the 1960s that I have been able to examine.

In my opinion the most satisfactory response to Popper's criticism of Darwinism was the book by philosopher Stephen Toulmin, who does not mention it at all. Toulmin asserts that it is futile to seek a simple criterion such as predictive success to determine whether a theory or activity is scientific:

> The critical questions which a philosopher brings to science must be coordinated with the factual studies of history. We must prepared to scrutinize in detail a representative selection of classic theories, analysing critically the merits for which they came to be accepted, and the standards by which they established their claims to superiority.
>
> Forecasting . . . is . . . an application of science, rather than the kernel of science itself . . . a novel and successful theory may lead to no increase in our forecasting skill; while, alternatively, a successful forecasting technique may remain for centuries without any scientific basis.

Thus Darwin's theory cannot "foretell the coming-into-existence of creatures of a novel species." Yet Darwinism did forecast short-term, small-scale events:

> When Australians used myxomatosis [a viral disease] to control the rabbit population, it was forecast on the best Darwinian principles that a new strain of rabbits would become dominant, whose constitutions were more resistant to the disease . . . The correctness of this prediction has helped to confirm the merits of the Darwinian theory. And the same thing has happened in other small-scale cases involving melanism in moths, the reactions of infective micro-organisms to antibiotics, and so on.

The outcome of Popper's critique of evolutionary theory was disappointing. Popper apparently never learned about the successful predictions that evolutionists had actually made; he repeated his claim that Darwinian theory is not scientific, and even though he eventually retracted it in 1978, the damage had been done: another weapon had been handed to the creationists in their battle against evolution.

According to Robert Pennock, law professor Philip Johnson was one of the founders of the modern intelligent design movement. In his book *Darwin on Trial*, Johnson mentioned Darwin's hypothesis about wingless insects as an example of an untestable proposition. Apparently he was so ignorant of the modern literature on evolution that he was unaware that the hypothesis had actually been tested and confirmed (Section 14.7).

14.22 CONTEXT AND CONCLUSIONS

I will now analyze the most popular reasons for accepting the natural selection hypothesis—that is, the empirical evidence most often mentioned in Anglo-American books and technical reviews on evolution by biologists in the period 1941–1970. See my monograph *Choosing Selection* for quantitative tables regarding evidence mentioned.

While Kettlewell's peppered moth is the most popular overall, it did not play a very significant role in these books until after 1955. So if one wants to understand why most biologists had accepted the primacy of natural selection by 1960, one has to start with the more traditional illustrations: various kinds of protective coloration, including industrial melanism as observed before Kettlewell's research. The recent controversy about the validity of Kettlewell's work is therefore largely irrelevant to the scientific status of natural selection in the 1950s and 1960s. Similarly, the new, more rigorous research on mimicry by Jane Van Zandt Brower and others published in the late 1950s and afterward came too late to have much effect in *changing* the views of evolutionists. The same is to a large extent true for Cain and Sheppard's work on snails, and the blood group–disease connection.

Instead, we have two rather different kinds of evidence playing an important role in addition to protective coloration: on one hand, the development of insecticide-resistant pests (going back to the scale insects in the citrus groves of California in the beginning of the twentieth century), and the similar development of resistance to DDT and antibiotics in the 1940s and 1950s; and on the other hand, the experiments of Dobzhansky and others on cyclic variations of chromosome inversions in *Drosophila* and *Adalia punctata*. Both show that natural selection acts much more quickly than had previously been believed—a few years in the first case, a few weeks or months in the second.

The most striking difference is the greater importance given to protective coloration and wingless insects, and the much smaller weight for chromosome inversions, snails, sickle cell and malaria, and blood groups.

Some of Dobzhansky's early experiments were designed to test genetic drift and, contrary to his expectations, showed that the effect of drift was negligible compared to that of natural selection. As a prominent leader of the evolutionary synthesis, his conversion to the natural selection hypothesis undoubtedly influenced many other biologists. The experiments in the 1950s by Kerr and Wright and by Dobzhansky and Pavlovsky showed that the effects of genetic drift *in combination with natural selection*, predicted by the Fisher-Haldane-Wright theory, could be detected by *specially designed laboratory experiments* with small populations. But there was no credible evidence that genetic drift by itself played a significant role in evolutionary processes in nature.

The fact that old evidence going back to the period before 1930 was so frequently invoked in support of the new version of natural selection theory leads us to ask: did the new evidence really play a major role in persuading evolutionists to accept natural selection? I suggest two answers to that question.

First, it seems clear to me that the mathematical labors of Fisher, Haldane, and Wright were essential to the success of the new natural selection hypothesis and indeed to the success of the evolutionary synthesis itself. One could support this view by quoting Dobzhansky , Huxley, Simpson, and many others. But perhaps the best authority is Ernst Mayr, because he is notorious for his attempts in the 1950s to minimize the contributions of the mathematicians relative to those of the field and laboratory biologists, and his ridicule of the "beanbag genetics" of Fisher, Haldane, and (unfairly) Wright. Mayr was the senior author of a zoology textbook in which one finds, near the beginning, the flat statement:

> Fisher's (1930) demonstration that even a very small selective advantage of a new gene or gene combination would cause in due time a genetic transformation of populations was a tremendously important contribution.

At the end of our period, Mayr summed up the situation as follows:

> The modern attitude toward natural selection has two roots. One is a mathematical analysis (R. A. Fisher, J.B.S. Haldane, Sewall Wright, and others), demonstrating conclusively that even very minute selective advantages eventually lead to an accumulation in the population of the genes responsible for these advantages. The other root is the overwhelming mass of material gathered by naturalists on the effect of the environment. This evidence was given a largely Lamarckian interpretation in the days when mutations were claimed to be saltational and cataclysmic. When small mutations were discovered, and when it was realized that all variation had ultimately a mutational origin, this evidence became a powerful source of documentation for the selectionist viewpoint.

So the phenomena of protective coloration, mimicry, and development of resistance to pesticides and antibiotics *became* evidence for natural selection, at least in part because of the theoretical work of Fisher, Haldane, and Wright. Fourteen years later, in 1984, industrial melanism and the development of resistance were still considered the best empirical evidence for natural selection to present in an introductory course (see my monograph *Choosing Selection* for statistics).

This evidence has also had an impact on the ongoing controversy over the teaching of evolution and creationism in American public schools. Eugenie C. Scott, a leader of the proevolutionists in this controversy, wrote that " the evidence for the operation of natural selection is so overwhelming that both IDCs [intelligent design creationists] and YECs [young earth creationists] now accept that it is responsible for such phenomena as pesticide resistance in insects or antibiotic resistance in bacteria." When pressed to account for these phenomena, creationists like Henry Morris make an arbitrary distinction between microevolution (which undeniably does occur) and macroevolution (which, they assert, does not). They seem unaware of the research by Fisher (Section 14.2), Dobzhansky (6.6), and Rensch (cited by Simpson, Section 14.10) that make this distinction untenable.

This brings us to the second reason mentioned in Mayr's 1970 summation quoted above: H. J. Muller's discovery of radiation-induced mutations provided an independent confirmation of Fisher's geometrical argument that only very *small* mutations can contribute to evolution. Large mutations may occur, but are most likely to produce death or sterility. Those biologists who rejected Lamarckism and orthogenesis for other reasons were now forced to reject also the deVries version of neo-Darwinism, and were left with the Fisher-Haldane-Wright version of natural selection as the only plausible alternative.

Third, Dobzhansky's research on *D. pseudoobscura* forged an important link between laboratory genetics and field observations, allowing evolution to become an experimental discipline in which hypotheses about natural selection and drift could be tested.

The fourth answer, which helps to explain the success of the natural selection hypothesis, is what might be called the Lakatos effect. Imre Lakatos was one of the first philosophers of science to point out (see Section 3.2) that scientists do not simply weigh the evidence on both sides of an issue; they also compare the track records of competing theories (or what Lakatos called research programs, or a series of theories). If theory A is found to successfully predict or give a better explanation of a phenomenon previously thought to be better explained by theory B, it thereby becomes more progressive while theory B becomes more degenerating. If scientists are persuaded that all of the *recent* research has transformed evidence for B into evidence for A, they are likely to switch their allegiance to A *even if most of the evidence still favors B*, because they anticipate that A will eventually win out, and this can become a self-fulfilling prophecy.

In fact, during the 1950s and 1960s several supporters of natural selection (theory A) claimed that this was the situation. Since many characters previously thought to be nonadaptive and therefore determined by genetic drift (theory B) had been shown by more detailed research to be adaptive, while there were no examples of the opposite, *therefore* the default assumption should be that all characters are adaptive (or genetically linked to other characters that are adaptive) in the absence of definite proof that they are nonadaptive. To the extent that evolutionists accepted this claim, the natural selection program was considered progressive while the genetic drift program was considered degenerating—even though there might still be a large number of cases of apparently nonadaptive characters for which no selective basis had yet been found.

Accordingly, H. J. Muller argued that when we see phenomena that *look* orthogenetic (directed toward a goal), we should not postulate "mysterious processes" like orthogenesis just because we have not *yet* explained them by natural selection. The success of natural selection in explaining other phenomena makes it the preferred default hypothesis. Of course, the Lakatos effect can help either side; it gives an advantage to the side whose evidence is more *recent*.

The alternatives to the natural selection hypothesis included, in addition to genetic drift, Lamarckism, orthogenesis, and macromutations. Throughout the period 1930–1970, Lamarckism was almost universally dismissed (except in France, and in German paleontology) as unproven and implausible. The only advocate who was taken seriously by biologists was the psychologist William McDougall, whose experiments on the training of rats were repeated and finally discredited in 1954 after a two-decades-long study by a group led by W. E. Agar.

Orthogenesis had been popular in the 1930s, but was strongly attacked by leaders of the evolutionary synthesis as a vestige of unscientific metaphysical or vitalist thinking; Simpson persuaded some but not all paleontologists (who had often invoked orthogenesis) that their observations could be better explained by natural selection.

The view that macromutations are essential for speciation continued to attract a small minority of biologists. After 1930 the only major advocate of this view was Richard Goldschmidt, and he regarded macromutations as a supplement rather than a replacement for natural selection. Goldschmidt was too important a biologist to be ignored, so his theory had to be given the courtesy of serious consideration—followed by definitive rejection. Macromutations were not important in evolution, because (1) they are always lethal; (2) there was no solid empirical evidence that speciation worked that way; (3) there was no reason to make a qualitative distinction between micro- and macromutation because every step in between had been observed and seemed to follow the same rules as micromutations; (4) there was no need to postulate macromutations because micromutations could account for all evolutionary processes, as soon as one recognized the tremendous power of selection; and (5) research in molecular biology following the establishment of the double helix model of DNA was incompatible with the existence of macromutations that could produce new species.

As noted above, at least three predictions based on natural selection were confirmed between 1930 and 1970: Darwin's prediction that flying organisms on a windy island would evolve a wingless form; Mayr's prediction that *Drosophila* chromosome inversions would prove to be adaptive (or, if you prefer, Dobzhansky's refutation of his own expectation of the opposite result); and Ford's prediction that certain blood groups would be associated with disease. Aside from Mayr and the Ford school, evolutionists gave these successful predictions no extra credit for novelty, and did not even mention them in the debate with philosophers about whether evolutionary theory is falsifiable. Nor did they mention the confirmations of evolutionary theory's predictions about the effects of genetic drift by Kerr and Wright, by Dobzhansky and Pavlovsky, and by Lamotte.

If one wanted to make the case that evolutionists were converted to the natural selection hypothesis by following the scientific method, the best argument would have to be based on the outcome of the second prediction. Dobzhansky did, in effect, make a prediction; he did compare it with observations, and then he tested the conclusion by controlling the variables in a laboratory experiment, which indeed showed that the proposed selective agent (seasonal temperature changes) produced the predicted result. Moreover, Dobzhansky himself stated that the result of this research led him to adopt the natural selection hypothesis (which he had previously doubted), and the research was mentioned in this connection by several authors of books on evolution. Yet only a minority of authors of textbooks, popular articles, and books even mentioned this test (see my monograph *Choosing Selection* for statistics) and, as far as I know, none of them asserted that the result was persuasive *because* it was a successful *novel* prediction. Moreover, when philosophers and a few biologists criticized evolutionary theory for not being testable, no evolutionist seems to have brought forward this example in response.

I can conclude only that while biologists do make testable predictions and sometimes confirm them, they do not (despite the rhetoric of textbook writers in their introductory chapters on scientific method) really believe that predictiveness is the most important criterion for judging a new theory. On the other hand, they expect that an established theory should be able to make predictions, and that if one of those predictions is falsified, one need not necessarily abandon the theory itself but perhaps make minor adjustments to accommodate the data. In this respect they are very much like physical scientists. The difference is that at least during this period, their predictions were more qualitative than quantitative.

NOTES FOR CHAPTER 14

422 *The Revival of Natural Selection 1930–1970 This chapter is based on my monograph Choosing Selection: The Revival of Natural Selection in Anglo-American Evolutionary Biology, 1930–1970* (Philadelphia: American Philosophical Society, *Transactions*, Volume 99, part 3, 2009), vii + 183 pp., by permission of the American Philosophical Society. This monograph should be consulted for further details and references. I am greatly indebted to the pathbreaking publications of Will Provine and to his extensive critique of an earlier draft. Lindley Darden also provided a very helpful and detailed critique. I thank Francisco Ayala, John Beatty, Matt Chew, James Crow, George Garraty, Sandra Herbert, Richard Highton, Richard Lewontin, David O'Brochta, Anya Plutynski, David Rudge, Ezra Shahn, V. Betty Smocovitis, Carol Sokolski, Roger Thomas, Polly Winsor, and Nick Zimmerman for answering my questions and providing much valuable information in correspondence. Two anonymous referees for the American Philosophical Society provided extremely useful critiques. My research on this topic was supported in part by a fellowship from the John Simon Guggenheim Memorial Foundation, and by the Institute for Physical Science and Technology at the University of Maryland.

422 *"No scientific theory is worth anything"* J.B.S. Haldane, *Adventures of a Biologist* (New York: Harper, 3rd ed. 1940), pp. 7–8; "Weather," in *What I Require from Life: Writings on Science and Life from J. B. S. Haldane*, edited by Krishna Dronamraju (Oxford: Oxford University Press, 2009), pp. 54–59 (reprinted from *Daily Worker* (London), 1940), on pp. 58–59.

422 *"The fact that natural selection offers a general explanation"* A. F. Shull, G. R. Larue, and A. G. Ruthven, *Principles of Animal Biology* (New York: McGraw-Hill, 5th ed., 1941), p. 359.

422 *"What makes evolution a scientific explanation"* Eric Lander, quoted by Deborah Blum, "Putting Faith in Science," *History of Science Society Newsletter* 36, no. 3 (July 2007): 18–22, on p. 20.

Section 14.1

423 *It was the core of an evolutionary synthesis* "The evolutionary synthesis crystallized around the notion of natural selection as the main channeling agent that acted on particulate individual and genetic entities—explaining macroevolutionary patterns by microevolutionary processes . . . No originality is claimed for what has been said thus far." Richard G. Delisle, "What was really Synthesized During the Evolutionary

Synthesis? A Historiographic Proposal," *Studies in History and Philosophy of Biological and Biomedical Sciences* 42 (2011): 50–59, on p. 51.

423 *the three canonical works of modern evolutionary theory* Fisher, *The Genetical Theory of Natural Selection* (Oxford: Oxford University Press, 1930; second edition, New York: Dover, 1958); Haldane, *Causes of Evolution* (London: Longmans, 1932); Wright, "Evolution in Mendelian Populations," *Genetics* 16 (1931): 97–159.

423 *It was sometimes argued that natural selection is only a* negative *factor* T. H. Morgan, *Scientific Basis of Evolution* (New York: Norton, 1932); for other references see Brush, *Choosing Selection*, note 9; Pablo Razeto-Barry and Ramiro Frick, "Probabilistic Causation and the Explanatory Role of Natural Selection," *Studies in History and Philosophy of Biological and Biomedical Sciences* 42C (2011): 344–355.

424 *This is what biologist Stephen Jay Gould referred to as the "hardening" of the Synthesis* Gould, "The Hardening of the Modern Synthesis," in *Dimensions of Darwinism*, edited by M. Grene (Cambridge: Cambridge University Press, 1983), pp. 71–93.

424 *This study is limited in space as well as in time* In addition to the works cited in Brush, *Choosing Selection*, note 15, see Eve-Marie Engels and Thomas F. Glick, editors, *The Reception of Charles Darwin in Europe* (London and New York: Continuum, 2008); articles by Thomas Junker (Germany), Yasha Gall, and Michael B. Konashov (Russia); Edmund F. Kolchinsky (Soviet Russia); Thomas Glick (Spain); and Patrick Tort (France).

Section 14.2

424 *some important features of the evolutionary synthesis* S. S. Chetverikov, *On certain Aspects of the Evolutionary Process from the Standpoint of modern Genetics* (Placitas, NM: Genetics Heritage Press, 1997), edited and with an introduction by C. D. Mellon, translated by M. Barker from *Zhurnal Eksperimentalnoi Biologii* [new series] 2, no. 1 (1926): 3–54.

424 *the following sentence, which appears halfway through the first chapter* Fisher, *Genetical Theory*, p. 11.

425 *"bisexual reproduction will tend rapidly to produce uniformity"* Fisher, *Genetical Theory*, p.2, the words in brackets were inserted in the 1958 edition; see *The Genetical Theory of Natural Selection: A complete Variorum edition*, edited and with foreword and notes by J. H. Bennett (Oxford: Oxford University Press, 1999). Fisher evidently felt the need to emphasize that these statements, originally presented as a list of separate postulates, were logically connected.

425 *was usually blamed on the "excessive production of offspring"* Fisher, *Genetical Theory*, p. 46.

425 *"There is something like a relic of creationist philosophy"* Fisher, *Genetical Theory*, p. 47.

425 *T. H. Morgan opposed Darwinism* Bowler, "Hugo de Vries and Thomas Hunt Morgan: The Mutation Theory and the Spirit of Darwinism," *Annals of Science* 35 (1978): 55–73, on p. 55. See also G. E. Allen, "Thomas Hunt Morgan and the Problem of Natural Selection," *Journal of the History of Biology* 13 (1968): 113–139.

425 *Tennyson's famous line "Nature, red in tooth and claw"* Alfred Lord Tennyson, *In Memoriam* (London: Moxon, 1850), Section LVI, Fourth quatrain.

425 *the counterculture slogan "make love, not war"* P. A. Moody, *Introduction to Evolution* (New York: Harper, 2nd ed., 1962), p. 355; for other reference see Brush, *Choosing Selection*, note 24.

425 *an otherwise similar sexual organism would evolve much faster* Fisher, *Genetical Theory*, pp. 121–123.

426 *current explanation for the prevalence of sexual rather than asexual* Michael A. Brockhurst, "Sex, Death, and the Red Queen," *Science* 333 (2011): 166–167; L. T. Morran et al., "Running with the Red Queen: Host-Parasite Coevolution Selects for Biparental Sex," ibid. 216–218.

426 *"the whole business would be settled"* Fisher, *Genetical Theory Variorum Edition*, p. 78.

426 *Fisher's theoretical conclusion . . . was confirmed by later experiments* "The Selection of Invisible Mutations," *Proceedings of the Royal Society of London* B131 (1942): 50–64; see also Mather, "Polygenic Inheritance and Natural Selection," *Biological Reviews of the Cambridge Philosophical Society* 18 (1943): 32–64.

426 *"a single factor may be in stable equilibrium"* Fisher, *Genetical Theory*, pp. 99–100; see Fisher, "On the Dominance Ratio," *Proceedings of the Royal Society of Edinburgh* 42 (1922): 321–341 for an earlier statement of this principle.

427 *"was content that Darlington was providing in* Heredity *a ready platform"* O. S. Harman, *The Man Who Invented the Chromosome: The Life of Cyril Darlington* (Cambridge, MA: Harvard University Press, 2004), pp. 209–210.

427 *"bean bag genetics"* E. Mayr, "Where are We?" *Cold Spring Harbor Symposia on Quantitative Biology* 24 (1959): 1–14.

427 *that the variability of a species should be proportional* Fisher and E. B. Ford, "Variability of a Species," *Nature* 148 (1926): 515–516; 'The Variability of Species in the *Lepidoptera*, with Reference to Abundance and Sex," *Transactions of the Royal Entomological Society of London* 76 (1928): 367–379.

427 *Mendel's published results are too good to be true* Fisher, "Has Mendel's Work Been Rediscovered?" *Annals of Science* 1 (1936): 115–137; A. Franklin et al., *Ending the Mendel-Fisher Controversy* (Pittsburgh: University of Pittsburgh Press, 2008).

427 *more sophisticated arguments made about natural selection* Abner Shimony, "The Non-Existence of a Principle of Natural Selection," *Biology and Philosophy* 4 (1989): 255–273.

SECTION 14.3

428 *The most favorable situation for evolution is "a large population"* Wright, "Evolution," p. 158.

428 *"was probably the most influential paper he ever published"* Provine, in Sewall Wright, *Evolution: Scientific Papers*, edited by W. Provine (Chicago: University of Chicago Press, 1986), on p. 158.

428 *"The chances are good that one at least will come under the influence"* Wright, "The Roles of Mutation, Inbreeding, Crossbreeding, and Selection in Evolution," *Proceedings of the Sixth International Congress of Genetics* 1 (1932): 356–366, on page 75 of the reprint in *Evolution*, edited by G. E. Brousseau Jr., *Evolution* (Dubuque, IA: Brown, 1967).

428 *"the blood group alleles are neutral as far as is known "* Wright, "The Statistical Consequences of Mendelian Heredity in Relation to Speciation," in *The New Systematics*, edited by J. Huxley (Oxford: Clarendon Press, 1940), pp. 161–183, on p. 179.

429 *Wright's "diagrams of adaptive landscapes"* J. Hodge, "Darwinism after Mendelism: The Case of Sewall Wright's Intellectual Synthesis in his Shifting Balance Theory of Evolution (1931)," *Studies in History and Philosophy of Biological and Biomedical Sciences* 42 (2011): 30–39, on p. 31. (Hodge does not cite any examples.)

429 *whether . . . his diagrams are incoherent and meaningless* See the review by Jonathan Kaplan, "The End of the Adaptive Metaphor?" *Biology and Philosophy* 23 (2008): 625–638; also the preceding and following papers in the same issue of that journal by Massimo Pigliucci and Anya Plutynski, respectively.

429 *his analogy of Mexican jumping beans* Wright, "Physiological Genetics, Ecology of Population, and Natural Selection," in *Evolution after Darwin*, volume 2, edited by S. Tax (Chicago: University of Chicago Press, 1960), pp. 429–475.

430 *there is no reason to think there is even one position that represents* Fisher, "Average Excess and Average Effect of a Gene Substitution," *Annals of Eugenics* 11 (1941): 53–63, on p. 58.

430 *This objection was made most effectively by O. W. Richards and G. C. Robson* Richards and Robson, "The Species Problem and Evolution," *Nature* 117 (1926): 345–347, 382–384.

SECTION 14.4

430 *natural selection could account for the evolutionary change* "In 1924 I published what my colleagues generally think my most important paper . . . They contained calculations showing great intensity of natural selection in favour of dark colour in a British moth species. This was regarded as ridiculously high, but 30 years later Kettlewell found a slightly higher figure in field studies." Haldane, "An Autobiography in Brief," in *What I Require from Life: Writings on Science and Life from J. B. S. Haldane*, edited by Krishna Dronamraju (Oxford: Oxford University Press, 2009), pp. xxix–xxxv," on pp. xxxi–xxxii.

430 *Haldane's was purely prospective* Sarkar, "Evolutionary Theory in the 1920s: The Nature of the "Synthesis," *Philosophy of Science* 71 (2004): 1215–1226, on p. 1223. See also H. L. Carson, "Cytogenetics and the Neo-Darwinian Synthesis," in *The Evolutionary Synthesis*, edited by E. Mayr and W. Provine (Cambridge, MA: Harvard University Press, 1980), pp. 86–95.

431 *providing a large number of biological examples of natural selection* "In *The Causes of Evolution,* I published the first estimate of a human mutation rate. Since then, this has became a matter of international politics in connection with atom-bomb tests." Haldane, *What I Require*, p. xxxii.

431 *"I can write of natural selection with authority"* J. B. S. Haldane, *The Causes of Evolution* (London: Longmans, 1932), on p. 13.

431 *Could a morally perfect God have made the tapeworm?* Haldane, *The Causes of Evolution* (London: Longmans, 1932), on p. 159. Cf. Charles Darwin's remark: "I cannot persuade myself that a beneficent & omnipotent God would have designedly created the Ichneumonidæ with the express intention of their feeding within the living bodies of caterpillars, or that a cat should play with mice." Darwin, Letter to Asa Gray 22 May [1860] http://www.darwinproject.ac.uk/entry-2814.

SECTION 14.5

431 *attracted little attention before 1935* Ford, *Mendelism and Evolution* (London: Methuen, 1931).

431 *Muller's experiments on the production of small mutations by X-rays* Muller, "Artificial Transmutation of the Gene," *Science* 66 (1927): 84–87; J. F. Crow and S. Abrahamson, "Seventy Years ago: Mutation Becomes Experimental," *Genetics* 147 (1997): 1491–1496.

431 *"the theory of natural selection is coming back"* Shull *evolution* (New York: McGraw-Hill, 1936), pp. 142, 161, 166–183, 205, 207, 212.

431 *"capable of giving a theoretical explanation of any conceivable occurrence"* Watson et al., *Proceedings of the Royal Society of London* B121 (1936): 43–73.

431 *he invoked his own theory of macromutations* Goldschmidt, *The Material Basis of Evolution* (New Haven: Yale University Press, 1940), p. 390.

Section 14.6

432 *famous series of papers on the genetics of natural populations Dobzhansky's Genetics of Natural Populations, I–XLIII*, edited by R, C. Lewontin et al. (New York: Columbia University Press, 1995).

432 *"may represent an earlier step in the evolutionary process"* Lancefield, "A Genetic Study of Crosses of Two Races or Physiological Species of *Drosophila obscura*," *Zeitschrift für Induktive Abstammungs- und Vererbungslehre* 52 (1929): 287–317.

432 *the possibility of creating a new species in the laboratory* Karpachenko crossed broccoli with radish to make a new species, *Raphanobroussica*. See Dobzhansky, *Genetics and the Origin of Species*, 2nd ed. (New York: Columbia University Press, 1941), pp. 233, 235, 273, 325.

433 *the views of those evolutionists who still followed de Vries* See Francisco Ayala and Roberts Arp, editors, *Contemporary Debates in Philosophy of Biology* (Malden, MA: Wiley-Blackwell, 2010), Part V, "Are Microevolution and Macroevolution Governed by the same Processes," with introduction by editors (pp. 165–167), article arguing the affirmative by Michael R. Dietrich (pp. 169–179) and article arguing the negative by Douglas H. Erwin (pp. 180–193).

433 *we have examples of every conceivable intermediate step* Sewall Wright, review of *The Material Basis of Evolution* by R. Goldschmidt, in *Scientific Monthly* 53 (1941): 165–170, on p. 166.

433 *"rediscovery of the giant chromosomes"* Dobzhansky, *Genetics*, 1941 edition, p. 117. E. Heitz and H. Bauer, "Beweise für die Chromosomennatur der Kernschleifen in den Knäuelkernen von *Bibio hortulans*," *Zeitschrift für Zellforschung und Mikroskopische Anatomie* 17 (1933): 67–83; T. S. Painter, "A New Method for the Study of Chromosome Aberrations and the Plotting of Chromosome Maps in *Drosophila melanogaster*," *Genetics* 19 (1934): 175–188. According to M. J. D. White, giant chromosomes in other dipterous flies were discovered as early as 1880. Progress in genetics might have been much faster if *Drosophila* researchers had been familiar with "the work of Balbiani, Carnow, Alverdes and others on the salivary chromosomes of *Chironomus*" and had looked at salivary glands twenty years earlier. White, *Animal Cytology and Evolution* (Cambridge: Cambridge University Press, 1945), pp. vii, 3. While the rediscovery came too late to play an important role in persuading biologists to accept Morgan's chromosome theory of heredity (Chapter 13), it did come just in time to change Dobzhansky's views on the relative importance of natural selection and isolation.

433 *the rhetorical approach its author used* Ceccarelli, *Shaping Science with Rhetoric: The Cases of Dobzhansky, Schrödinger, and Wilson* (Chicago: University of Chicago Press, 2001).

433 *relations between fields in the evolutionary synthesis were not symmetrical* Darden, "Relations among Fields in the Evolutionary Synthesis," in *Integrating Scientific Disciplines*, edited by W. Bechtel (Dordrecht: Nijhoff, 1986), on pp. 121, 122.

434 *"unraveling and describing" . . . and "studies on the mechanisms"* Dobzhansky, *Genetics and the Origin of Species*, 3rd ed. (New York: Columbia University Press, 1951), p. 11.

Section 14.7

434 *"the wingless condition . . . is mainly due to the action of natural selection"* Charles R. Darwin, *On the Origin of Species by Natural Selection: or, The Preservation of Favoured Races in the Struggle for Life* (London: Murray, 1859), Chapter 5, "Effects of the Increased Use and Disuse of Parts."

435 *"We decided to study experimentally the problem posed by Darwin"* P. L'Héritier, Y. Neefs and G. Teissier, "Aptérisme des Insects et Sélection Naturelle," *Comptes Rendus hebdomadairedes Séances de l'Académie des Sciences, Paris* 204 (1937): 907–909. See also Brush, *Choosing Selection*, note 92; Claudine Petit, "Teissier, Georges," *Dictionary of Scientific Biography* 18 (1990): 901–904; Patrick Tort, "The Interminable Decline of Lamarckism in France," Chapter 17 in Engels and Glick, *Reception of Darwin in Europe*, pp. 329–353, 569–571, on pp. 342–347.

435 *Experiments by H. J. Quayle showed that there were two races* Quayle, "The Development of Resistance to Hydrocyanic Gas in Certain Scale Insects," *Hilgardia* 11 (1938): 183–225.

435 *"probably the best proof of the effectiveness of natural selection yet obtained"* Dobzhansky, *Genetics*, 1937 edition, p. 161 (citing Quayle's article as "in press").

435 *but considered this question largely academic* Dobzhansky, *Process* (New York: Columbia University Press, 1970).

436 *"melanics [darker forms] are superior"* Ford, "Problems of Heredity in the *Lepidoptera*," *Biological Reviews* 12 (1937): 461–503; "Genetic Research in the *Lepidoptera*," *Annals of Eugenics* 10 (1940): 227–252. Dobzhansky, *Genetics*, 1941 edition, p. 196.

436 *"uncritical and valueless speculation"* Sumner, "Evidence for the protective Value of changeable Coloration in Fishes," *American Naturalist* 69 (1935): 245–266. Dobzhansky, *Genetics*, 1941 edition, pp. 212–214.

Section 14.8

436 *"the most powerful force in developing the selectionist attitude"* Ford, "Some Recollections Pertaining to the Evolutionary Synthesis," in *The Evolutionary Synthesis*, edited by E. Mayr and W. Provine (Cambridge, MA: Harvard University Press, 1980), pp. 334–342, on p. 337.

436 *"One of the greatest changes in biological outlook"* "Genetics, Evolution and Human Destiny," in *Genetics in the 20th Century*, edited by L. C. Dunn (New York: Macmillan, 1951), 591–621, on p. 593.

436 *"There is still a widespread reluctance"* Huxley, *New Systematics*.

437 *we can now make "quantitative prophecies with much greater fullness"* J. Huxley, *Evolution: The modern Synthesis* (New York: Harper, 1942), p. 21.

437 *Provine argues that it is a* constriction W. B. Provine, "Progress in Evolution and Meaning in Life," in *Evolutionary Progress*, edited by M. Nitecki (Princeton, NJ: Princeton University Press, 1988), pp. 99–114.

437 *"perhaps the most important single concept of recent years"* Huxley, *Evolution*, p. 124.

437 *Huxley also recognized the Sewall Wright phenomenon of drift* Huxley, *Evolution*, p. 59.

438 *gene combinations that are truly nonadaptive are exceedingly rare* Huxley, "Genetics," p. 597.

438 *"more curious and surprising than others"* Huxley, "Evolution's Copycats," *Life* 32, no. 6 (1952): 67–76; reprinted in Huxley, *New Bottles for new Wine* (New York: Harper, 1957), pp. 137–154, on p. 141.

438 *Carl Sagan used it in his television series* Cosmos Sagan, *Cosmos* (New York: Random House, 1980); V. Betty Smocovitis, *Unifying Biology; The Evolutionary Synthesis and Evolutionary Biology* (Princeton, NJ: Princeton University Press, 1996), pp. 164–165, note 202.

438 *Huxley also asserted that evolution is* progressive Delisle, "What Was Really Synthesized During the Evolutionary Synthesis? A historiographic Proposal," *Studies in History of Biological and Biomedical Sciences* 42 (2011): 50–59, on pp. 52–53, 57.

438 *Cain argues that we should abandon the unit concept* Cain, "Rethinking the Synthesis Period in Evolutionary Studies," *Journal of the History of Biology* 42 (2009): 621–648, on pp. 621, 626.

438 *The object to process shift can also be observed* Chen, "A Different Kind of Revolutionary Change: Transformation from Object to Process," *Studies in History and Philosophy of Science* 41 (2010): 182–191.

Section 14.9

439 *formulated the synthesis from the viewpoint of taxonomy and ornithology* Mayr, *Systematics and the Origin of Species, from the Viewpoint of a Zoologist* (New York: Columbia University Press, 1942).

439 *"undoubtedly the best all-round evolutionary geneticist"* K. R. Dronamraju, *If I am to be Remembered: The Life and Work of Julian Huxley, with Selected Correspondence* (River Edge, NJ: World Scientific, 1993), p. 236.

439 *widely considered to be the father of modern evolutionary biology* L. Guterman, "Harvard's Ernst Mayr, a Pioneer in Evolutionary Biology, Dies at 100," *Chronicle of Higher Education* (February 18, 2005), A17. See also C. K. Yoon, "Ernst Mayr, Pioneer in Tracing Geography's Role in the Origin of Species, Dies at 100," *New York Times* (February 5, 2005), Section A; Erika Milam, "The Equally Wonderful Field: Ernst Mayr and Organismic Biology," *Historical Studies in the Natural Sciences* 40 (2010): 279–317.

439 *the new "mutual understanding between geneticists and systematists"* Mayr, *Systematics*, Chapter 1.

439 *"and this is exactly what we find"* Mayr, *Systematics*, p. 236.

439 *"The reduced variability of small populations"* Mayr, *Systematics*, p. 236. He later credited B. Rensch with an earlier recognition of the founder principle (Mayr, "Prologue: Some Thoughts on the History of the Evolutionary Synthesis," in Mayr and Provine, *Evolutionary Synthesis*, pp. 1–48, on p. 26).

440 *"All the properties and phenomena of macroevolution"* Mayr, *Systematics*, p. 298, see also 291.

440 *"In my 1942 book, in order to be 'modern,'"* M. Ruse, *Monad to Man: The Concept of Progress in Evolutionary Biology* (Cambridge, MA: Harvard University Press, 1996), p. 414.

440 *he made what he later called a prediction* Mayr, *Animal Species and Evolution* (Cambridge, MA: Harvard University Press, 1963), p. 207.

440 *Provine sees a turning point toward selectionism* Provine, *Sewall Wright and Evolutionary Biology* (Chicago: University of Chicago Press, 1986), p. 453.

440 *characters that have heretofore been considered as 'neutral'"* E. Mayr and E. Stresemann, "Polymorphism in the Chat Genus *Oenanthe* (Aves)," *Evolution* 4 (1950): 291–300, on p. 299.

440 *a vigorous attack on the use of the genetic drift concept* Mayr, *Animal Species*, p. 204.

440 *"Virtually every case quoted in the past as caused by genetic drift"* Mayr, *Animal Species*, p. 207. The two references in the quotation are: Mayr, "Some Evidence in Favor of a Recent Date," *Lloydia* 8 (1945): 70–83; Ford, "Polymorphism," *Biological Review* 20 (1945): 73–88.

441 *Mayr continued to maintain the validity of his own founder principle* Mayr, "Prologue," p. 26.

Section 14.10

441 *evolutionary synthesis as a theory that is compatible with . . . paleontology* "G. G. Simpson," in Mayr and Provine, *Evolutionary Synthesis*, pp. 452–463. The article is based on Simpson's answers to questions from Mayr. On the reception of the book by scientists see J. Cain, "A Matter of Perspective: Multiple Readings of George Gaylord Simpson's *Tempo and Mode in Evolution*," *Archives of Natural History* 30 (2003): 28–39.

441 *"a product of the tendency of the minds of scientists"* Simpson, *Tempo and Mode in Evolution* (New York: Columbia University Press, 1944), on p. 166.

441 *Evidence that paleontologists were abandoning orthogenesis* G. L. Jepsen, E. Mayr and G. G. Simpson, editors, *Genetics, Paleontology, and Evolution* (Princeton, NJ: Princeton University Press, 1949), on pp. 49, 58, 107–109.

441 *singlehandedly brought paleontology into the evolutionary synthesis* Simpson, *The Meaning of Evolution* (New Haven: Yale University Press, 1949); L. F. Laporte, *George Gaylord Simpson: Paleontologist and Evolutionist* (New York: Columbia University Press, 2000).

441 *was especially concerned with the role of his childhood hero Simpson* Marjorie Burgess, "Stephen Jay Gould," in *Notable Twentieth Century Scientists*, edited by Emily J. McMurray (Detroit: Gale Research, 1995), vol. 2, pp. 803–805, on p. 803.

442 *"he often constructs adaptive scenarios, in the speculative mode"* Gould, "G. G. Simpson, Paleontology, and the Modern Synthesis," in Mayr and Provine, *Evolutionary Synthesis*, pp. 153–172, on pp. 166–167.

442 *"it could not trigger any major evolutionary event"* Gould, "G. G. Simpson," p. 168. The statement he quotes from Simpson is less dogmatic: "Genetic drift is certainly not involved in all or in most origins of higher categories, even of very high categories such as classes or phyla" (Simpson, *Major Features*, p. 355). The next sentence, not quoted by Gould, is: "It is not positively known to have been involved in any instance."

442 *For Gould this was tantamount to devaluing the profession* Gould, in Mayr and Provine, *Evolutionary Synthesis*, p. 170.

442 *a resurrection of Goldschmidt's macromutations* Gould, *The Structure of Evolutionary Theory* (Cambridge, MA: Harvard University Press, 2002), pp. 451–466.

442 *Goldschmidt's systemic mutations have "never been observed"* Simpson, *The Meaning of Evolution*, revised edition (New Haven: Yale University Press, 1967), pp. 231–232.

442 *Drift "plays a role, perhaps a minor one"* G. G. Simpson and W. S. Beck, *Life: An Introduction to Biology*, 2nd ed. (New York: Harcourt, Brace and World, 1965), p. 434.

442 *Cain writes: "if PE was going to have any claim"* Joe Cain, "Ritual Patricide: Why Stephen Jay Gould Assassinated George Gaylord Simpson," in *The Paleobiological Revolution: Essays on the Growth of Modern Paleontology*, edited by David Sepkoski and Michael Ruse (Chicago: University of Chicago Press, 2009), pp. 346–363, on

p. 351. The book is "In Memory of Stephen Jay Gould, J. John Sepkoski [and] Thomas J. M. Schopf" [v.].

443 *"They tend to overstate both the novelty and the generality of their ideas"* Letter from Simpson to Bucher, July 22, 1980, in Simpson Papers, American Philosophical Society Library, series 1, folder "Bucher, John," as quoted by Cain, "Ritual Patricide," p. 356.

Section 14.11

443 *"every other significant fact about genetics needed for the synthetic theory"* G. L. Stebbins, "Botany and the Synthetic Theory of Evolution," in Mayr and Provine, *Evolutionary Synthesis*, pp. 139–152, on p. 139.

443 *it is negative only in the sense that a sculptor's creation of a statue* G. L. Stebbins Jr., *Variation and Evolution in Plants* (New York: Columbia University Press, 1950), p. 102.

443 *the issue is still being debated by philosophers* See Bence Nanay, "Natural Selection and the Limitations of Environmental Resources," *Studies in History and Philosophy of Biological and Biomedical Sciences* 41 (2010): 418–419 and works cited therein.

443 *The only substantial evidence for higher plants was that of W. B. Kemp* "Natural Selection within Plant Species as Exemplified in a Permanent Pasture," *Journal of Heredity* 28 (1937): 329–333.

444 *Stebbins had defended the Sewall Wright effect in 1944* "Evolutionary Factors and the Fossil Evidence," Letter to E. H. Colbert, June 8, 1944. *Bulletin of the Committee on Common Problems in Genetics, Paleontology and Systematics* 3, 194–1944. Reprinted in *Exploring the Borderlands: Documents of the Committee on Common Problems in Genetics, Paleontology and Systematics, 1943–1944.* Vol. 94, part 2, *Transactions of the American Philosophical Society*, edited by J. Cain, 67–70 (Philadelphia: American Philosophical Society, 2004).

444 *by 1950 he had lost his enthusiasm for it* Stebbins, *Variation*, pp. 145–147; A. C. Fabergé, "Genetics of the Scapiflora Section of Papaver, II: The Alpine Poppy," *Journal of Genetics* 45 (1943): 139–170.

Section 14.12

444 *Dobzhansky originally wanted to find a definitive empirical confirmation* Provine, *Sewall Wright and Evolutionary Biology* (Chicago: University of Chicago Press, 1986).

444 *Epling's detailed observations* C. Epling, *Contributions to the Genetics, Taxonomy, and Ecology of Drosophila pseudoobscura and its Relations, III. The Historical Background* (Washington, DC: Carnegie Institution of Washington, 1944).

444 *"each gene arrangement . . . may have different selective values"* E. Mayr, "Some Evidence in Favor of a recent Date," *Lloydia* 8 (1945): 70–83, on p. 74.

444 *some gene combinations would have a "selective advantage"* Stebbins, "Evidence for Abnormally Slow Rates of Evolution, with Particular Reference to the Higher Plants and the Genus *Drosophila*," *Lloydia* 8 (1945): 84–102.

444 *Dobzhansky himself had already come to the same conclusion Genetics*, 1951 edition; Evolutionary Oscillations in *Drosophila pseudoobscura*," in *Ecological Genetics and Evolution*, edited by R. Creed (Oxford: Blackwell, 1971), pp. 109–133.

444 *"A may be better adapted than B in summer, while B is superior to A"* Dobzhansky, *Genetics*, 1951 edition, p. 109.

445 *"The existence of previously unknown gene arrangements"* Dobzhansky, *Genetics*, 1951 edition, pp. 110–112. The original reports confirming the predictions were: Dobzhansky and A. H. Sturtevant, "Inversions in the Chromosomes of *Drosophila pseudoobscura*," *Genetics* 23 (1938): 28–64; Dobzhansky, "Discovery of a predicted Gene Arrangement in *Drosophila azteca*," *Proceedings of the National Academy of Sciences* 27 (1941): 47–50. For discussion of these experiments see David Wyss Rudge, "The Complementary Roles of Observation and Experiment: Theodore Dobzhansky's Genetics of Natural Populations, IX and XII," *History and Philosophy of the Life Sciences* 22 (2000): 167–186.

445 *"the chromosomal polymorphism had no adaptive significance"* Dobzhansky, *Genetics*, 1951 edition, p. 114.

445 *he "breaks rather abruptly with the past"* Provine, in *Dobzhansky's Genetics of Natural Populations, I–XLIII*, edited by R. C. Lewontin et al. (New York: Columbia University Press, 1981), p. 303.

445 *Dobzhansky "had concluded that the inversions were under selection"* Provine, *Sewall Wright*, pp. 389ff.

446 *This allows "great advances in rebuilding the organism"* Dobzhansky, "Genetics of Natural Populations, XIII: Recombination and Variability in Populations of Drosophila Pseudoobscura," *Genetics* 31 (1946): 269–290, on p. 288.

446 *the previous view that "adaptive evolution in nature is too slow"* Dobzhansky, "Adaptive Changes induced by Natural Selection in wild Populations of Drosophila," *Evolution* 1 (1947): 1–6, on p. 1. For a brief summary of the Russian work mentioned by Dobzhansky see Brush, *Choosing Selection*, p. 66.

Section 14.13

447 *"persons of a particular blood group are unduly susceptible to the disease"* Ford, "Polymorphism," *Biological Review* 20 (1945): 73–88, on p. 85.

447 *Ford's prediction was confirmed in the 1950s* D. Struthers, "ABO Groups of Infants and Children Dying in the West of Scotland m(1949–1951)," *British Journal of Prevention and Social Medicine* 5 (1951): 223–228.

447 *persons with blood group A are significantly more likely than those with O* I. Aird, H. H. Bentall, and J.A.F. Roberts, "A Relationship between Cancer of the Stomach and ABO Blood Groups," *British Medical Journal* 1 (1953): 799–801.

447 *these authors, together with Mehigan, credited Fisher and Ford* I. Aird, H. H. Bentall, J. A. Mehigan, and J.A.F. Roberts, "The Blood Groups in Relation to Peptic Ulceration and Carcinoma of the Colon, Rectum, Breast, and Bronchus: An Association between the ABO groups and Peptic Ulceration," *British Medical Journal* 2 (1954): 315–321.

447 *Buckwalter's group published a survey of early results* J. A. Buckwalter, E. B. Wohlwend, D. C. Colter, R. T. Tidrick, and L. A. Knowles, "Peptic Ulceration and ABO Blood Groups," *Journal of the American Medical Association* 162 (1956): 1215–1220.

447 *Ford was of course entitled to claim victory* E. B. Ford, "Evolution in Progress," in *Evolution after Darwin*, volume 1, edited by S. Tax (Chicago: University of Chicago Press, 1960), pp. 181–196, on p. 194. See also Ford, "Polymorphism in Plants, Animals, and Man," *Nature* 180 (1957): 1315–1319.

447 *"a certain balance between local isolation and cross-breeding"* S. Wright, "Fisher and Ford on 'The Sewall Wright Effect,'" *American Scientist* 39 (1951): 452–458, 479, on p. 454.

447 *Before 1950, Dobzhansky had supported Wright's view* T. Dobzhansky and M.F.A. Montague, "Natural Selection and the Mental Capacities of Mankind," *Science* 105 (1947): 587–590.

447 *Dobzhansky still wanted to explain the differences in blood groups Evolution, Genetics, and Man* (New York: Wiley, 1955), pp. 130, 143.

448 *just another "swing of the pendulum"* Dobzhansky, *Genetics of the Evolutionary Process* (New York: Columbia University Press, 1970), p. 262.

448 *The association between blood groups and disease* G. Garratty, "Blood Groups and Disease: a Historical Perspective," *Transfusion Medicine Review* 4 (2000): 291–301.

448 *theoretically the blood group polymorphism should not be stable* See articles cited in Brush, *Choosing Selection*, note 142.

Section 14.14

448 *warned that the microbes attacked by such drugs* Dubos, "Microbiology," *Annual Reviews of Biochemistry* 11 (1942): 659–678; "Trends in the Study and Control of infectious Diseases," *Proceedings of the American Philosophical Society* 88 (1944): 208–213; C. L. Moberg, "René Dubos: A Harbinger of Microbial Resistance to Antibiotics," *Perspectives in Biology and Medicine* 42 (1999): 554–580.

448 *"to establish whether the resistant bacteria always occur" The Bacterial Cell in Relation to Problems of Virulence, Immunity and Chemotherapy* (Cambridge, MA: Harvard University Press, 1945), p. 322.

448 *"one of the first clear demonstrations"* A.N.H. Creager, "Adaptation or Selection? Old Issues and New Stakes in the Postwar Debates over Bacterial Drug Resistance," *Studies in History and Philosophy of Biological and Biomedical Sciences* 38 (2007): 159–190, on p. 166.

448 *evidence that "makes it probable"* M. Demerec, "Production of Staphylococcus Strains Resistant to Various Concentrations of Penicillin," *Proceedings of the National Academy of Sciences USA* 31 (1945): 16–24, on pp. 16, 23; Creager, "Adaptation or Selection," pp. 167 ff.

448 *"penicillin-resistant gonorrhea has become more frequent"* Dobzhansky, "The Genetic Basis of Evolution," *Scientific American* 84 (1950): 32–41, on pp. 33–35.

449 *his influential book* Genetics and the Origin of Species Creager, "Adaptation or Selection?" on p. 176.

449 *bacteria could display an inherited resistance to penicillin* Lederberg and Lederberg, "Replica Plating and Indirect Selection of Bacterial Mutants," *Journal of Bacteriology* 63 (1952): 399–406; "Replica Plating and Indirect Selection of Bacterial Mutants: Isolation of Preadaptive Mutants in Bacteria By Sib Selection," *Genetics* 121 (189): 395–399. For a good textbook description of the experiment see H. Curtis, *Biology*, Special edition (New York: Worth Publishers, 1968), pp. 696–696.

449 *"was the launching pad for my own investigations"* J. Lederberg, "Introduction" in *Launching the Antibiotic Era*, edited by C. L. Moberg and Z. A. Cohn (New York: Rockefeller University Press, 1990).

449 *humans should not try to conquer nature* J. E. Cooper, "of Microbes and Men: A scientific Biography of René Jules Dubos," PhD dissertation, Rutgers University, 1998; C. L. Moberg, *René Dubos: Friend of the good Earth* (Washington, DC: American Society for Microbiology Press, 2005).

449 *"resistance can evolve whenever drugs are used"* "US Zeroes in on Pork Producers' Antibiotics Use," *New York Times* (September 15, 2010): A13, A19, on p. A13.

449 *explanations, such as "staph learned how to resist methicillin"* Harvey B. Simon, *Washington Beacon* (January 2009): 9; Robert Daum, quoted by Ranit Mishori, "A Deadly Bug Invades our Towns," *Parade* (December 7, 2008): 10, 12, on p. 10.

Section 14.15

450 *a report by Cyril Diver on his studies of two closely related species* Diver, "The Problem of Closely Related Species Living in the Same Area," in Huxley, *New Systematics*, pp. 303–328.

450 *Provine, in his biography of Wright* Provine, *Wright*, Chapter 12.

450 *They concluded that speciation should be ascribed to natural selection* A. J. Cain and P.M. Sheppard, "Selection in the Polymorphic Land Snail, *Cepaea nemorailis*," *Heredity* 4 (1950): 275–294.

450 *"he was prepared to believe that the gene frequencies"* Provine, *Wright*, pp. 441–449.

450 *Maxime Lamotte published his own results* Lamotte, "Observations sur la Sélection par les Prédateurs chez *Cepaea nemoralis*," *Journal de Conchyliologie* 90 (1950): 180–190; "Recherches sur la Structure génétique des Populations naturelles de *Cepaea nemoralis L.*," *Bulletin Biologique de la France et de la Belgique* 35 (1951): S1–S238. R. L. Milstein, "Distinguishing Drift and Selection Empirically: 'The Great Snail Debate' of the 1950s," *Journal of the History of Biology* 41 (2007): 339–367; "Concepts of Drift and Selection in 'The Great Snail Debate of the 1950s and Early 1960s," in *Descended from Darwin: Insights into American Evolutionary Studies, 1900–1970*, edited by J. Cain and M. Ruse (Philadelphia: American Philosophical Society, 2009), pp. 271–298.

450 *"so far, every supposed example of random variation"* A. J. Cain, "So-Called Non-Adaptive or Neutral Characters in Evolution," *Nature* 168 (1951): 424.

450 *"selective value should be assumed in all characters"* Carter, "Non-Adaptive Characters in Evolution," *Nature* 168 (1951): 700–701.

450 *Cain denied taking that position* A. J. Cain, "Non-Adaptive or Neutral Characters in Evolution," *Nature* 168 (1951): 1049.

450 *"Cain admits that genetic drift may be a real factor"* Carter, "Non-Adaptive or Neutral Characters in Evolution," *Nature* 168 (1951): 1049.

450 *In a 1957 book on evolution, Carter omitted genetic drift entirely* Carter, *A Hundred Years of Evolution* (London: Sidgwick and Jackson, 1957).

451 *"This injudicious act roused Philip to fury"* C. A. Clarke, "Philip MacDonald Sheppard: 27 July 1921–17 October 1976," *Biographical Memoirs of Fellows of the Royal Society of London* 23 (1977): 465–500.

451 *it was a triumph for adaptationism* S. J. Gould, *The Structure of Evolutionary Theory* (Cambridge, MA: Harvard University Press, 2002), p. 541.

451 *these fluctuations were much larger than estimated* "The Spread of a Gene in Natural Conditions in a Colony of the Moth *Panaxia dominula L*," *Heredity* 1 (1947): 143–174.

451 *complained that his theory had been misrepresented* "On the Role of Directed and Random Changes in Gene Frequency in the Genetics of Populations," *Evolution* 2 (1948): 279–294.

451 *There were further replies back and forth* Provine, *Wright*, gives a detailed account of this controversy through the 1950s; R. A. Skipper Jr. follows it through the 1990s in "The R. A. Fisher-Sewall Wright Controversy in Philosophical Focus: Theory Evaluation in Population Genetics," PhD dissertation, University of Maryland,

College Park, 2000. See also Skipper, "Revisiting the Fisher-Wright Controversy," in Cain and Ruse, *Descended*, pp. 299–322.

Section 14.16

451 *Gould extended his complaint about hardening* Gould, "The Hardening of the Modern Synthesis," in *Dimensions of Darwinism*, edited by M. Grene (Cambridge: Cambridge University Press, 1983), pp. 71–93.

451 *"Thus far no evidence has been adduced"* Dobzhansky and Pavlovsky, "An Experimental Study of Interaction Between Genetic Drift and Natural Selection," *Evolution* 11 (1957): 311–319.

452 *"The either-selection-or-drift point of view is a fallacy"* Dobzhansky and Pavlovsky, "Interaction."

452 *"quite appreciable divergence on the islands of the Scilly archipelago"* Dobzhansky and Pavlovsky, "Interaction," p. 317.

452 *This divergence was attributed entirely to selection* W. H. Dowdeswell and E. B. Ford, "The Influence of Isolation on Variability in the Butterfly Maniola jurtina L.," in *Evolution: Symposia of the Society for Experimental Biology* 7, edited by R. Brown and J. F. Danielli (New York: Academic Press, 1953), pp. 254–273; Ford, *Ecological Genetics* (London: Methuen, 1964).

452 *"we are observing the emergence of novel genetic systems"* Dobzhansky and Pavlovsky, "Interactions," p. 317.

452 *Ford strongly rejected the view* E. B Ford, *Ecological Genetics*, 4th ed. (London: Chapman and Hall, 1975), pp. 66–75.

452 *"a new swing of the pendulum"* Dobzhansky, *Evolutionary Process*, p. 262.

452 *"I began as a drifter"* Quoted by J. R. Powell, "'In the Air': Theodosius Dobzhansky's *Genetics and the Origin of Species*," *Genetics* 117 (1987): 363–366.

452 *Gould was jumping on a different pendulum* S. J. Gould, "Allometry and Size in Ontogeny and Phylogeny," *Biological Reviews* 41 (1966): 587–640, on p. 621. According to Patricia Princehouse, Julian Huxley recommended that Gould be invited to write the article; "Fallout from this review made Gould's early reputation and helped secure his professorship at Harvard and tenure at age 30." Prince-house, "Punctuated Equilibria and Speciation: What does it Mean to be a Darwinian," in *The Paleobiological Revolution: Essays on the Growth of modern Paleontology*, edited by D. Sepkoski and M. Ruse (Chicago: University of Chicago Press, 2009), pp. 149–175, on pp. 154–155.

Section 14.17

453 *"chance effects . . . will rarely matter to a whole species"* "The Theory of Evolution, Before and After Bateson," *Journal of Genetics* 56 (1958): 11–27, on p. 17.

453 *"the weakest point in Wright's argument"* "Natural Selection," in *Darwin's Biological Work*, edited by P. R. Bell (Cambridge: Cambridge University Press, 1959), pp. 101–150, on p, 141.

453 *the theory of adaptive peaks "is wholly unrealistic" Mendelism and Evolution*, 6th ed. (London: Methuen, 1957), p. 86.

453 *"large selective advantages are in fact common"* Ford, *Mendelism*, 1957 edition. See also Ford, *Ecological Genetics*, 3rd ed. (London: Chapman and Hall, 1971), pp. 36–38.

453 *natural selection explains everything while drift explains nothing* W. H. Dowdeswell and E. B. Ford, "The Influence of Isolation on Variability in the Butterfly *Maniola*

jurtina L." In *Evolution: Symposia of the Society for Experimental Biology* 7, edited by R. Brown and J. F. Danielli (New York: Academic Press, 1953), pp. 254–273. Ford, "Rapid Evolution and the Conditions Which Make It Possible," *Cold Spring Harbor Symposium on Quantitative Biology* 20 (1955): 230–238; Ford, "Evolution in Progress," in *Evolution after Darwin*, volume 1, edited by S. Tax (Chicago: University of Chicago Press, 1960), pp. 181–196.

453 *Russian evolutionists were even more selectionist* Huxley, "Evolutionary Biology and Related Subjects," *Nature* 156 (1945): 254–256.

453 *geneticists may doubt if such populations could exist* Huxley, "Genetics and major Evolutionary Change" (Review of *Tempo and Mode in Evolution* by G. G. Simpson), *Nature* 156 (1945): 3–4.

453 *"and is almost wholly effected through selection"* J. Huxley, "The Evolutionary Process," in *Evolution as a Process*, edited by J. Huxley, A. C. Hardy, and E. B. Ford (London: Allen and Unwin, 1954), pp. 1–23, on p. 3.

453 *the discovery that human blood groups are maintained in a morphism* Huxley, "Morphism and Evolution," *Heredity* 9 (1955): 1–52.

453 *drift "can have definite evolutionary consequences"* Huxley, "Introduction," in *Evolution: The Modern Synthesis*, 2nd ed. (London: Allen and Unwin, 1963); R. Olby, "Huxley's Place in Twentieth-Century Biology," in *Julian Huxley: Biologist and Statesman of Science*, edited by C. K. Waters and A. Van Helden (Houston, TX: Rice University Press, 1992), pp. 53–75, on p. 70.

453 *Mayr took notice of the proposed revival of genetic drift* E. Mayr, *Populations, Species and Evolution* (Cambridge, MA: Harvard University Press, 1970), pp. 126–128.

454 *the burden of proof lies on the shoulders of anyone who wants to invoke Processes of Organic Evolution* (Englewood Cliffs, NJ: Prentice-Hall, 1966), p. 74.

454 *"no example of differentiation between populations is known"* Stebbins, *Processes*, p. 76.

454 *"chance factors by themselves probably have little effect"* Stebbins, *Processes*, p. 77.

454 *Stebbins restated Wright's concept of adaptive peaks and valleys* Stebbins, *The Basis of Progressive Evolution* (Chapel Hill: University of North Carolina Press, 1969), pp. 124–125.

454 *the Goldschmidt macromutation theory had been definitely excluded* Stebbins, *Basis*, p. 104.

454 *"evolutionary theory now stood on a solid, quantifiable* Depew and Weber, "Innovation and Tradition in Evolutionary Theory: An Interpretive Afterword," in *Evolution at a Crossroads,* edited by Depew and Weber (Cambridge, MA: MIT Press), pp. 227–260, on p. 230.

Section 14.18

455 *Ford argued that the dark (melanic) form was more viable* Ford, "Problems of Heredity in the *Lepidoptera*," *Biological Review* 12 (1937): 461–503; "Polymorphism," *Biological Review*, 20 (1945): 73–88.

455 *a novel by Margaret Drabble* Drabble, *The Peppered Moth* (New York: Harcourt, 2001). Or recent research see Arjen E. van't Hof, Nicola Edmunds, Martina Dalíkova, František, and Ilik J. Accheri, "Industrial Melanism in British Peppered Moths Has a Singular and Recent Mutational Origin," *Science* 332 (2011): 958–960.

Section 14.19

455 *"there was universal and unanimous agreement with the conclusions"* Mayr, *The Growth of Biological Thought: Diversity, Evolution, and Inheritance* (Cambridge, MA: Harvard University Press, 1982), pp. 568–569.

456 *In a series of papers with G. G. Tiniakov* Dubinin and Tiniakov, "Seasonal Cycles and the Concentration of Inversions in Populations of *Drosophila funebris*," *American Naturalist* 79 (1945): 570–572, and other papers cited in Brush, *Choosing Selection*, note 169.

456 *"inner perfecting tendencies . . . do not belong in the realm of science"* F. B. Sumner, "Where Does Adaptation Come In?" *American Naturalist* 76 (1942): 433–444, on p. 437.

456 *the color of the mice was a result of selection by predators* F. B. Sumner, *The Life History of an American Naturalist* (Lancaster, PA: Jacques Cattell Press, 1945), pp. 239–243.

456 *"his reinterpretation of the adaptive value of the differences"* Provine, *Wright*, p. 230; see also Provine, "Francis B. Sumner and the Evolutionary Synthesis," *Studies in the History of Biology* 3 (1979): 211–240.

456 *Allee, who followed Wright's theory of evolution* W. C. Allee, *The Social Life of Animals* (New York: Norton, 1938); W. C. Allee, A. E. Emerson, O. Park, T. Park, and K. P. Schmidt, *Principles of Animal Ecology* (Philadelphia: Saunders, 1949).

456 *could find no evidence that the differences have adaptive significance* D. Lack, "The Galapagos Finches (Geospizinae): A Study in Variation," *Occasional Papers of the California Academy of Sciences* 21 (1945): 1–159, on pp. 116–117.

457 *"subspecific and specific differences in Darwin's finches are adaptive"* D. Lack, "My Life As an Amateur Ornithologist," *Ibis* 115 (1973): 421–431, on p. 429.

457 *"at least most of the seemingly non-adaptive differences in Darwin's finches"* D. Lack, preface to reprint of his *Darwin's Finches* (New York: Harper, 1961).

457 *"the immense power of natural selection"* D. Lack, *Ecological Adaptations for Breeding in Birds* (London: Methuen, 1968), on p. 310.

Section 14.20

458 *The value of textbooks is succinctly expressed* Gould, *Structure*, pp. 576–577.

459 *but forgot to remove his endorsement of de Vries* J. W. Mavor, *General Biology*, 3rd ed. (New York: Macmillan, 1947), pp. 864, 880.

459 *he discussed one example of genetic drift* Elliott, *Zoology* (New York: Appleton-Century Crofts, 3rd ed. 1963, 4th ed. 1968). The example was the religious isolate known as the Dunkers, living in Franklin County, PA. See Brush, *Choosing selection*, pp. 79–80.

459 *where he also mentions the modern synthesis* Beaver, *General Biology* (St. Louis: Mosby, 5th ed. 1958, 6th ed. 1962), p. 628.

459 *The seventh edition (1966), coauthored by George B. Noland* Beaver and Noland, *General Biology* (St. Louis: Mosby, 7th ed., 1966), p. 506.

Section 14.21

459 *"Students often hold a variety of erroneous views of science"* M. U. Smith, "Current Status of Research in Teaching and Learning Evolution: Philosophical/Epistemological Issues," *Science & Education* 19 (2010): 523–538, on p. 529.

460 *"the variations of the outcomes should be greater"* T. Dobzhansky and N. P. Spassky, "Genetic Drift and Natural Selection in Experimental Populations of Drosophila Pseudoobscura," *Proceedings of the National Academy of Sciences* 48 (1962): 148–156.

460 *the expected greater variation of smaller populations was also present* M. Lamotte, "Polymorphism of Natural Populations of *Cepaea nemoralis*," *Cold Spring Harbor Symposia on Quantitative Biology* 24 (1959): 65–84.

460 *"might be regarded as merely part of the vogue of evolutionism"* K. Popper, "The Poverty of Historicism, III," *Economica* 12 (1945): 69–89, on p. 69.

460 *there could be no such thing as a scientific law of evolution* Popper, "Poverty," p. 70.

460 *"even when prediction of the future is impossible"* Scriven, "Explanation and Prediction in Evolutionary Theory," *Science* 130 (1959): 477–482, on p. 477.

461 *we will never be able to determine whether the age of the universe is finite* Scriven, "The Age of the Universe," *British Journal for the Philosophy of Science* 5 (1954): 181–190.

461 *Darwin and others have attempted "to encapsulate the principles"* Scriven, "Explanation," p. 477.

462 *Mayr, who had himself made a confirmed prediction* Mayr, *Animal Species*, p. 207; Mayr, "Cause and Effect in Biology," *Science* 134 (1961): 1501–1506, on p. 1504.

462 *he distinguished* logical *prediction* Mayr, *Growth*, pp. 57–59.

462 *he argued that "in principle there is no difference"* Mayr, "How Biology Differs from the Physical Sciences," in Depew and Weber, *Evolution*, pp. 43–63, on p. 50.

462 *"the ability to predict is not a requirement"* Mayr, *Toward a new Philosophy of Biology: Observations of an Evolutionist* (Cambridge, MA: Harvard University Press, 1988), pp. 20, 31–32.

462 *"what is actually predicted is not the antecedent occurrence"* G. G. Simpson, *This View of Life: The World of an Evolutionist* (New York: Harcourt, Brace and World, 1964), pp. 144–145, 147.

462 *"the hypothetico-deductive method (e.g., Popper, 1959) is the only one"* Simpson, "Uniformitarianism: An Inquiry Into Principle, Theory, and Method in Geohistory and Biohistory," in *Essays in Evolution and Genetics*, edited by M. K. Hecht and W. C. Steere (New York: Appleton-Century-Crofts, 1970), pp. 85, 90.

463 *"the existence of an organ in one species, solely 'for' the benefit of another"* M. Ghiselin, *The Triumph of the Darwinian Method* (Berkeley: University of California Press, 1969), p. 63.

463 *as a fact that was "clearly expected"* Ruse, "Confirmation and Falsification of Theories of Evolution," *Scientia* 104 (1969): 329–357.

463 *Darwinism did forecast short-term, small-scale events* Toulmin, *Foresight and Understanding* (New York: Harper Torchbooks, 1961), pp. 16, 36, 25.

464 *he eventually retracted it in 1978* K. Popper, "Natural Selection and the Emergence of Mind," *Dialectica* 32 (1978): 339–355. For a general assessment of Popper's influence on our understanding of evolutionary theory see David L. Hull, "The Use and Abuse of Sir Karl Popper," *Biology and Philosophy* 14 (1999): 481–504. According to philosopher Bence Nanay, "Popper famously held that the growth of scientific knowledge and the Darwinian mechanism of trial and error elimination are analogous processes . . . until the 1960s, he used the Darwinian process as a model for understanding the growth of scientific knowledge, whereas from the 1960s on, the explanatory order was reversed: he used his new insights about the growth of scientific knowledge to say something about the real nature of Darwinian selection . . . this led him . . . to flirt with Lamarckism."

464 *one of the founders of the modern intelligent design movement* R. T. Pennock, "The Postmodern Sin of Intelligent Design Creationism," *Science & Education* 19 (2010): 757–778. "Intelligent Design Creationism is the bastard child of Christian fundamentalism and postmodernism . . . it was born through the influence of

Critical Legal Studies upon Philip Johnson, who was the godfather of the ID movement and its philosophical approach to attacking evolution."

464 *Johnson mentioned Darwin's hypothesis about wingless insects* P. E. Johnson, *Darwin on Trial*, 2nd ed. (Downers Grove, IL: Intervarsity Press, 1993).

Section 14.22

464 *new, more rigorous research on mimicry* J. V. Z. Brower, "Experimental Studies of Mimicry in Some North American Butterflies," *Evolution* 12 (1958): 32–47, 123–136, 273–285; "Experimental Studies of Mimicry, IV: the Reactions of Starlings to different Proportions of Models and Mimics," *American Naturalist* 94 (1960): 271–282.

465 *One could support this view by quoting Dobzhansky, Huxley, Simpson* See the references in Brush, *Choosing Selection*, note 207.

465 *his ridicule of the "beanbag genetics"* Mayr, "Where are We?"; see also Provine, *Wright*, pp. 480–484. On beanbag genetics see Dronamraju, *Beanbag*, which includes correspondence between Mayr and Haldane on this and other subjects. He writes: "the early Mendelians used to keep different colored beans in bags for the purpose of counting and analyzing Mendelian ratios. This method implied that genes behaved as isolated independent entities with no interaction with each other" (p. ix).

465 *"was a tremendously important contribution"* E. Mayr, E. G. Linsley, and L. Usinger, *Methods and Principles of Systematic Zoology* (New York: McGraw-Hill, 1953), p. 12.

465 *"The modern attitude toward natural selection has two roots"* Mayr, *Populations*, p. 108.

465 *industrial melanism and the development of resistance* J. A. Moore, "Science as a Way of Knowing: Evolutionary Biology," *American Zoologist* 24 (1954): 467–534, on pp. 515–516.

465 *"the evidence for the operation of natural selection is so overwhelming"* Scott, *Evolution vs. Creationism: An Introduction* (Westport, CT: Greenwood Press, 2004), p. xxiii.

465 *creationists like Henry Morris make an arbitrary distinction* Morris, *Scientific Creationism*, Public School Edition (San Diego: Creation-Life Publishers, 1974), and many articles in *Acts & Facts*, the monthly newsletter of the Institute for Creation Research. On the acceptance of microevolution by creationists see R. L. Numbers, "Ironic Heresy: How Young-Earth Creationists Came to Embrace rapid Microevolution by Means of Natural Selection," in *Darwinian Heresies*, edited by A. Lustig, R. J. Richards, and M. Ruse (Cambridge: Cambridge University Press, 2004), pp. 84–100.

466 *the natural selection program was considered progressive* See Brush, *Choosing Selection*, note 210.

466 *makes it the preferred default hypothesis* Muller, "Redintegration of the Symposium on Genetics, Populations, and Evolution," in *Genetics, Populations and Evolution*, edited by G. L. Jepsen, E. Mayr and G. G. Simpson (Princeton, NJ: Princeton University Press, 1949), pp. 421–445, on p. 440.

467 *The only advocate who was taken seriously by biologists* W. E. Agar, F. H. Drummond, O. W. Tiegs, and M. H. Gunson, "Fourth (Final) Report on a Test of McDougall's Lamarckian Experiment on the Training of Rats," *Journal of Experimental Biology* 31 (1954): 307–321.

PART FIVE

Conclusions

15

Which Works Faster: Prediction or Explanation?

> Science walks forward on two feet, namely theory and experiment. . . . Sometimes it is only one foot which is put forward first, sometimes the other, but continuous progress is only made by the use of both—by theorizing and then testing, or by finding new relations in the process of experimenting and then bringing the theoretical foot up and pushing it on beyond, and so on in unending alterations.
>
> —Robert A. Millikan, *Nobel Prize Lecture (1924)*

15.1 COMPARISON OF CASES PRESENTED IN THIS BOOK

The case studies discussed in Parts Two through Four were grouped by broad topics, and then chronologically, for ease of exposition. Now I list the results in a different order, starting with those theories accepted mostly or entirely because of confirmed predictions and ending with those accepted because of successful explanations (see Table 15.1). Obviously there is some subjective judgment involved in deciding when a theory was accepted; I have used statements in published articles and books, and (where relevant) award of prizes. The first prediction confirmed and the nature of its confirmation may be found in the section cited.

Clearly there are several factors involved in determining the time-lag numbers in the right-hand column. We can eliminate some of those factors by considering theories proposed in the same time period in the same field. For example, the light quantum hypothesis and quantum mechanics were both accepted in the mid-1920s, and several of the same physicists were involved in both cases. We can also include cases involving related theories whose reception I did not study in detail, but for which some evidence can be found. In particular, historians such as Richard Staley have concluded that Einstein's special theory of relativity was accepted by 1911 (before predictions such as the increase of mass with speed had been confirmed), on the basis of its explanation of the Michelson-Morley experiment and the aesthetic appeal of the theory itself. Conversely, simply by looking at

Table 15.1 Relative Importance of Novel Predictions in Acceptance of Theories

Theory	Proposed by	1st prediction confirmed	Theory accepted	Time lag
Light quantum	Einstein 1905	1916 (7.6)	1925	9
Relativistic QM	Dirac 1931	1932 (9.1)	*	
Nuclear forces	Yukawa 1935	1937 (9.5)	1949	12
Periodic law	Mendeleev 1869	1876 (5.5)	1882–1885	6–9
Benzene structure	Kekulé 1865, 1872	1874 (6.2)	1936–1940	62–66
MO aromaticity theory	Hückel 1931, Mulliken 1932	1978 (10.6)	1966	-12
General relativity	Einstein 1916	1919 (11.5)	1929	10
Big bang cosmology	Gamow et al. 1948	1965 (12.11)	1974	9
Chromosome	Morgan 1910–1911	1914 (13.4)	1930	16
Natural selection (Mendelian)	Fisher et al. 1930–32	1937 (14.7)	1940	3
Quantum mechanics	Heisenberg, 1925, Schrödinger 1926	1927 (8.9)	1927	1

*Never completely accepted, replaced by quantum electrodynamics (Tomonaga, Schwinger, Feynman) in the 1940s.

the lists of Nobel Prize winners we may infer that a number of predictions based on the old quantum theory were confirmed (Section 3.5), but the hypotheses leading to those predictions did not form a coherent self-consistent theory and therefore the old quantum theory was never fully accepted before it was replaced by quantum mechanics (Section 8.1).

In this chapter I summarize and compare the reception of four theories: the old quantum theory, quantum mechanics, special relativity, and general relativity. The result is a tentative answer to the question in the title of this chapter. Another way of formulating the question is this: is theory ahead of experiment (so it can predict what experiment has not yet revealed), or is experiment ahead of theory (revealing facts that theory then has to explain)? The answer immediately suggests itself (as Millikan stated in the above quotation): sometimes theory is ahead, sometimes experiment is ahead.

15.2 FROM PRINCIP TO PRINCIPE

On June 28, 1914, the Archduke Francis Ferdinand of Austria-Hungary was assassinated in Sarajevo by a Serbian nationalist, Gavrilo Princip. This event was the immediate cause of World War I. As we might say today, it was like the flapping of a butterfly's wings, which led to a four-year hurricane that devastated Europe.

It also had one indirect (one might say beneficial) effect on the fate of Albert Einstein's general theory of relativity. A German astronomical expedition led by Erwin Findlay Freundlich went to the Crimea peninsula in Russia, hoping to observe the solar eclipse scheduled for August 21, 1914. They wanted to test Einstein's prediction that starlight will be deflected by an angle of 0.87 seconds near the edge of the sun. But on August 1, 1914, Germany declared war on Russia, and the Russians therefore arrested the German astronomers as enemy aliens, preventing them from making observations. Had the astronomers done so with sufficient accuracy, they would have found that the deflection is actually 1.74 seconds—twice as much as the prediction from Einstein's theory.

Einstein later revised his general theory, and predicted on November 18, 1915 that the deflection should be 1.74 seconds. Another expedition led by astronomer Arthur S. Eddington went to the island of Principe (in the Gulf of Guinea off the west coast of central Africa) to observe a solar eclipse that was to occur on May 29, 1919. Fortunately for science, the war had ended on November 11, 1918, so such observations could be made without risk of military interference.

Eddington analyzed the observations and announced on November 6, 1919 that Einstein's (new) prediction had been confirmed. The result was enormous publicity for Einstein and his theory, starting the next day when the *Times* of London proclaimed a revolution in science started by "one of the greatest achievements in human thought."

The theory was incomprehensible to almost everyone, but involved tantalizing ideas like the fourth dimension and the curvature of space-time. Einstein himself proved to be a journalist's dream: he was handsome, gave quotable answers to questions, espoused causes like Zionism and peace, answered letters from schoolchildren, and seemed to have accomplished the extraordinary feat of bringing the Germans and the British together, at least in science, after a bitterly fought war. According to his biographer, physicist Abraham Pais, "the *New York Times Index* contains no mention of him until November 9, 1919. From that day until his death, not one single year passed without his name appearing in that paper, often in relation to science, more often in relation to other issues." Einstein acquired a more sinister side after the atomic bomb confirmed his equation $E = mc^2$.

One factor that may have contributed to Einstein's fame is the large number of books and articles by scientists written to explain relativity to the public. According to historian Peter J. Bowler, in early twentieth-century Great Britain a scientist like Eddington could write for the public without compromising his reputation among other scientists, as long as he continued to produce high-quality research. Many of those books were also published in the United States. The situation seems to have changed after World War II, at least in America, judging by the criticism and disrespect inflicted on scientists like George Gamow, Carl Sagan, and James Watson.

For whatever reasons, Einstein remained the most famous scientist in the world long after his death and was named person of the century by *Time* magazine in 1999.

Eddington's confirmation of the light-bending prediction was controversial among astronomers; he seemed to have cherry-picked the data that supported the theory, of which he was known to be an enthusiastic advocate. Replication by more objective observers, preferably ones who had no strong opinions about the validity of the theory, was needed. This was supplied by Robert Trumpler of Lick Observatory in California, who traveled to Australia to observe an eclipse in 1922. The results, analyzed by Trumpler and W. W. Campbell, announced on April 12, 1923, again confirmed Einstein's 1.74-second prediction.

Einstein had also predicted, in 1907, that the wavelength of light coming from atoms in a strong gravitational field (for example, at the surface of the sun) would be greater than light from the same atoms in a terrestrial laboratory. This is now known as the gravitational redshift. In 1907 Einstein thought the solar redshift would be too small to measure, but in a later paper (1911) he was somewhat more optimistic.

Attempts to measure the solar redshift gave conflicting results, but C. E. St. John at the Mt. Wilson Observatory in California concluded that Einstein's prediction was correct. Then in 1925 W. S. Adams, also at Mt. Wilson, announced that he had observed the gravitational redshift of the star Sirius B, which according to Eddington's theory of stellar structure has a very high density. These results, along with the explanation of the variation of Mercury's perihelion (place where it is closest to the sun) and the second confirmation of the light-bending prediction, led most astronomers to accept the general theory of relativity by 1930.

The story does not end there; new tests of general relativity, and criticisms of the old tests, continue to be reported. As I write this chapter, my colleague physicist and historian Francis Everitt has just announced his long-awaited confirmation of the geodetic and frame-dragging effects predicted by Einstein's theory. But I have to limit myself to a finite number of years.

To summarize: 15 years after Einstein proposed his general theory of relativity, the experts were satisfied that it had passed three empirical tests: light bending, advance of Mercury's perihelion, and gravitational redshift. Two of these tests were predictions in advance; the third, Mercury's perihelion motion, was an explanation of a previously known but mysterious fact.

15.3 CAN EXPLANATION BE BETTER THAN PREDICTION?

The first confirmation of Einstein's light-bending prediction in 1919 caused a sensation. Einstein quickly became the most famous scientist in the world. People who had no knowledge of his theory and made no effort to understand it proclaimed themselves to be supporters of relativity. Other physicists and astronomers who previously rejected or ignored relativity were now forced to take it seriously. But some of them argued that light bending could be explained by other causes such as refraction in a (hypothetical) extended atmosphere of the sun, without having to give up accepted theories of the nature of space, gravity, and light.

Logically, the critics were right. If theory T_1 entails (predicts) fact X, and X is observed to be true, one cannot correctly conclude that T must be true. Such a conclusion would be an example of what philosophers call the fallacy of affirming the consequent (Section 1.2). It is possible that some other theory T_2 or T_3 also entails X.

In science, a critic who proposes an alternative theory must defend it against objections. Thus an extended solar atmosphere dense enough to account for the bending of light would also cause comets to slow down as they pass the sun, but they don't. It took a few years for supporters of relativity to shoot down the proposed alternative explanations of light bending, but by 1930 the game was over.

As for the Mercury perihelion advance: astronomers had already had several decades to explain it and failed. For example, changing the exponent in the law of gravity (e.g., from 2 to 2.01) might account for Mercury's motion, but only at the exorbitant cost of sacrificing the excellent agreement of other planetary motions with Newton's theory. So, once Einstein had published his explanation, it was quickly accepted by most astronomers and physicists. (The Mercury effect was also considered by the experts to be stronger evidence than light bending because it involved a deeper part of the theory; light bending could easily be explained, and had already been explained a century earlier, by the Newtonian particle theory of light, except for a factor of two).

15.4 SPECIAL THEORY OF RELATIVITY: EXPLAINING "NOTHING"

I discussed general relativity first because it illustrates the so-called scientific method: make a hypothesis, then deduce predictions that can be tested. In fact it was the confirmation of Einstein's prediction of light bending by Eddington's 1919 eclipse observation that inspired the philosopher Karl Popper to propose falsifiability as a criterion for being scientific (Section 1.2). Popper was impressed by the contrast between relativity and theories like psychoanalysis, Marxism, and Darwinism—which could explain any given facts but could never be disproved. It was clear to him that if the eclipse test had failed to confirm Einstein's theory, the theory would have been discarded by scientists.

But now we must go back in time to 1905, and ask: what were the confirmed predictions (in advance) that led scientists to accept the special theory of relativity?

According to historian Richard Staley, who has thoroughly studied all the relevant historical evidence, "Einstein's special theory came to be widely accepted by 1911 without *any* experiment being regarded as offering uncontroversial and definitive proof of his approach."

The most persuasive experimental evidence for special relativity before 1911 was the null result of the Michelson-Morley experiment of 1887. This and earlier experiments showed that one cannot determine the absolute motion of the earth—that is, one cannot measure its motion relative to a hypothetical light-transmitting ether. Einstein himself did not cite any experimental evidence in his 1905 paper, and physicist and historian Gerald Holton has shown that (contrary to what used to be said) he did not develop his theory *in order to* explain Michelson-Morley. Einstein did, however, give this as the only empirical support for his theory in a review article published in 1907.

A theory that only explains why a certain experiment gives the result zero is not much use in science. What else can it do?

The first experiment to provide positive support for special relativity was done by physicist Alfred Heinrich Bucherer at Bonn University in Germany. His measurements of the mass of electrons at high speeds were the first to provide definitive support for Einstein's formula, at a time when experiments by Walter Kaufmann gave results closer to those derived from Max Abraham's rival theory. Because of the disagreements between these and other experiments and the difficulty of doing the measurements accurately to distinguish between the predictions of the two theories, the issue was not settled until 1914, when Kaufmann himself conceded that Einstein's theory had been confirmed.

It may seem strange that a radical new theory like relativity could have been accepted by physicists entirely on the basis of its explanation of negative results. In fact there was another reason: theoretical physicists were impressed by the generality, universality, and mathematical elegance of the theory, especially as formulated in terms of four-dimensional geometry by Hermann Minkowski. Here we have another factor influencing the acceptance of a theory: it is so beautiful that it must be true.

15.5 THE OLD QUANTUM THEORY: MANY THINGS ARE PREDICTED, BUT FEW ARE EXPLAINED

Physicist Eugene Wigner, in a famous paper published in 1960, pointed out the "unreasonable effectiveness" of mathematics in the physical sciences. One may formulate an equation to describe a familiar situation, and suddenly find that an unfamiliar—and perhaps undesirable—physical situation appears when one solves the equation.

That's what happened to Max Planck in 1900: he derived an equation for black-body radiation and found that the equation, when mated with Ludwig Boltzmann's formula for entropy, implied that radiation is composed of *particles*. Planck, as a staunch supporter of the *wave* theory of electromagnetic radiation, could not believe what the mathematics was trying to tell him. As Thomas Kuhn pointed out in 1978, Planck did not propose a physical quantum theory; he used quantization only as a convenient method of approximation.

In his 1905 paper on light, which I consider the beginning of quantum theory, Einstein discussed many phenomena. But the paper is most famous for the quantum theory of the *photoelectric effect*. The equation derived from this theory was experimentally confirmed by Robert A. Millikan. But, like Planck, Millikan refused to accept the idea that light or electromagnetic radiation in general can have a particle (atomistic) nature, in addition to its well-established wave nature.

For some physicists, the definitive proof of the quantum nature of radiation was the Compton effect. This effect was predicted theoretically and confirmed experimentally by Arthur Holly Compton. Compton assumed that X-rays act like particles when they collide with electrons. The result of the collision can then be described simply by using the laws of conservation of momentum and energy. At the same time, the X-rays can be treated as waves, and the change in their wavelength is a simple function of the angle between incident and scattered rays.

Compton's own experiment confirmed this hypothesis in 1923. Moreover, his theory led to the prediction that a recoil electron should also emerge with appropriate momentum and energy. This was observed two months later by C.T.R. Wilson.

Compton is one of the few physicists who has explicitly stated in public that one should get more credit for a confirmed prediction *in advance* than for a retrodiction or explanation of a known fact. In particular, he argued that he himself should get more credit for his discovery of the Compton effect, including the recoil electron, than Einstein deserved for his confirmed theory of the photoelectric effect. He wrote: "Since the idea of light quanta was invented primarily to explain the photoelectric effect, the fact that it does so very well is no great evidence in its favor." The quantum theory should get more credence for predicting a phenomenon "for which it had not been especially designed."

Compton's claim for extra credit has not been endorsed by either physicists or historians, perhaps because Einstein did not invent the quantum to explain the photoelectric effect, and did predict an equation for that effect that was not previously known.

I will briefly mention three other predictions of the old quantum theory, just to illustrate that theory was indeed ahead of experiment in the 1910s:

1. Einstein's prediction (1907) that specific heats of solids go to zero as T goes to zero (confirmed by Walther Nernst in 1911).
2. Niels Bohr predicted from his atomic model (1913) that electrons with energy E passing through a gas at low pressure produce no radiation until E is greater than a critical value (derived from his theory). Then, radiation is produced corresponding to the energy difference between the ground state and an excited state (confirmed by James Franck and Gustav Hertz in 1914).
3. Arnold Sommerfeld (1915–1916) generalized the Bohr model to include elliptical orbits, and predicted a relativistic correction because electrons in those orbits would sometimes have higher speeds than those in circular orbits. The corresponding change in the spectrum was confirmed by Friedrich Paschen in 1916.

Sommerfeld's prediction turned out to be an excellent example of the fallacy of affirming the consequent. From 1916 to 1925 it was considered important evidence for both special relativity and the Bohr model. But when quantum mechanics was introduced by Heisenberg and Schrödinger, along with the electron spin hypothesis of Uhlenbeck and Goudsmit, it was found that Sommerfeld's formula could be derived from the new theory without using relativity (at least not directly). Since the Bohr model was now known to be wrong—though very fruitful—the confirmation of the original Sommerfeld prediction was no longer considered evidence for relativity.

15.6 QUANTUM MECHANICS: MANY THINGS ARE EXPLAINED, BUT PREDICTIONS ARE CONFIRMED TOO LATE

By 1925 the old quantum theory was a disgraceful mess: a collection of ad hoc hypotheses, each one able to predict one kind of phenomenon, but inconsistent with the others. Thus, having started with the simple postulate that energy comes in *integer* multiples of a quantum ($nh\nu$), physicists were forced to postulate half-quanta ($[n + ½][h\nu]$) for the anomalous Zeeman effect. Worse, the Bohr model, which seemed to work so well for one-electron atoms, broke down completely as soon as one more electron was added, so that one could not even calculate accurately the ionization potential of helium.

Experiment, stimulated by the quantum hypothesis, was now ahead of theory.

In some alternative universe, Louis de Broglie's (1923, 1925) hypothesis about the wave nature of electrons might have provided a confirmed prediction inspiring the development of a new wave mechanics for subatomic particles. In our own universe the experiments of C. J. Davisson and his colleagues were both too early and too late. His early experiments with C. H. Kunsman (1921) antedated the publication of de Broglie's hypothesis, and thus deprived de Broglie of the full glory of making a prediction in advance. By the time Davisson had learned about wave mechanics and, with L. H. Germer, redesigned his diffraction experiment to make a more accurate test (1927), the game was over: quantum mechanics had already been accepted by the experts in atomic physics. The Davisson-Germer experiment did play an important role in persuading other physicists to accept the new theory. Yet as Schrödinger himself pointed out, the experiment was *not* a confirmation of his own theory, but of de Broglie's.

How could a radical new theory, first published by Werner Heisenberg in July 1925 and (in a different but essentially equivalent form) by Erwin Schrödinger in 1926, be accepted by 1927?

First, Niels Bohr gave it his public blessing in December 1925. Max Born, Pascual Jordan, Paul Dirac, and Wolfgang Pauli immediately started working on Heisenberg's theory. Arnold Sommerfeld became a strong and influential advocate for wave mechanics, using his seminar to educate several stars of the next generation including Hans Bethe, Walter Heitler, Fritz London, and Linus Pauling.

There was a veritable gold rush to extract as many results as possible from this fertile theory. The best indicator of the immediate impact of quantum mechanics on research is given in a paper by historians A. Kozhevnikov and C. Novick (1989). They cite 203 papers on quantum mechanics (mostly reporting original research) submitted for publication from July 1925 through February 1927. There were 80 authors from 14 countries.

Quantum mechanics quickly explained most of the puzzles that could not be solved by the old quantum theory, such as the mysterious half quantum numbers. The helium atom, the crucial gateway to more complicated atoms, was finally conquered by physicist Egil Hylleraas (in 1928–1929). This success was the most frequently mentioned reason for accepting quantum mechanics in monographs and review articles published in the period 1929–1932. In 1927 physicists Walter Heitler and Fritz London applied quantum mechanics to the hydrogen molecule, showing how a bond could form between two hydrogen atoms with a minimum energy at a distance close to the observed value. This would be a good start on understanding molecules in general (or quantum chemistry).

Two predictions-in-advance should be mentioned, even though they did not influence the acceptance of the theory:

1. *Ortho and para hydrogen.* Diatomic molecules like H_2 can have two forms because the spins of their two nuclei can be aligned parallel or antiparallel. This was one of the achievements for which Heisenberg received the Nobel Prize (the other was matrix mechanics), though his part in the discovery was indirect and he did not even mention it in his Nobel lecture.
2. *Stark effect intensities* (effect of electric fields on spectral lines). Laura Chalk, a graduate student working with J. Stuart Foster at McGill University, measured the intensities of the Stark components in the spectrum of hydrogen, especially those for which the values predicted by Schrödinger's equation disagreed with Stark's experimental values. Aside from a very brief announcement in 1926, Foster and Chalk did not publish their final results, confirming quantum mechanics, until 1929.

The Foster-Chalk experiment was certainly one of the first tests of a prediction of quantum mechanics (if not *the* first; see the calculation of the hydrogen spectrum by Sommerfeld and Unsold, confirmed by Kent et al. in 1927, Sections 8.7 and 8.9). Has anyone ever heard of it? Chalk seems to be completely unknown to most historians of physics and to physicists interested in publicizing the achievements of women.

The fact that quantum mechanics was accepted by experts in atomic physics before any of its predictions-in-advance had been confirmed was noted by physicist Karl K. Darrow in October 1927. It "has captivated the world of physics in a few brief months," not because of its successful predictions or its superior agreement with experience but "because it seems natural or sensible or reasonable or elegant or beautiful." Like relativity, it was so beautiful it had to be true. In the same year I. I. Rabi, an Austrian-born American physicist, received his PhD from Columbia University; decades later, looking back on those days, he said in a lecture:

> During the first period of its existence, quantum mechanics didn't predict anything that wasn't already predicted before . . . The results that came out of quantum mechanics had to a large degree been previously anticipated.

Based on this statement, in the June 2007 issue of *Physics Today* I challenged readers to "find evidence that the confirmation of *any* prediction in advance, other than electron diffraction, led *any* physicist to accept quantum mechanics before 1928." So far, no one has done so.

The lack of any confirmed predictions-in-advance did not prevent physicists from recognizing the tremendous value of quantum mechanics—with one exception. Have you ever wondered why it took more than five years for Heisenberg and Schrödinger to get the Nobel Prize? C. W. Oseen, the chair of the committee in the Swedish Academy that screened nominations for the physics prize, was primarily responsible for the delay. Before 1932, despite nominations and private communications from leading physicists, Oseen argued that quantum mechanics did not deserve the prize since it had not made any successful predictions-in-advance and therefore did not represent new knowledge. (Ironically, this was the same person who was responsible for the award of the Nobel Prize to Einstein for his equation of the photoelectric effect, since the rest of the committee refused to honor relativity.)

Oseen finally changed his mind in 1932 because of Carl D. Anderson's discovery of the positron, predicted by Paul Dirac from his relativistic version of quantum mechanics. Heisenberg received the Prize in 1932, while Dirac and Schrödinger shared the 1933 prize (Anderson had to wait until 1936).

15.7 MILLIKAN'S WALK

We may consider the quantum-relativity revolution as a single historical event composed of four parts, taking place during a limited time period (1905–1930) and involving many of the same scientists. Taking time as one variable and the two-valued parameter (Q for quantum, R for relativity) as the other, we see a rough antisymmetrical structure: P (prediction), E (explanation) for Q and E, and P for R (see Table 15.2).

Table 15.2 The Quantum-Relativity Revolution

Quantum	P →	→	→	→	* E →	*
Relativity	E →	→*	P → →	→	→	→ *
	1905	1910	1915	1920	1925	1930

Here P means a theory was proposed that made several predictions-in-advance but gave few or no explanations; E means a theory that offered several explanations but few predictions-in-advance. The * indicates the approximate date when the theory was accepted by experts. Notice that E÷ * is generally faster than P ÷ *; this is because by the time the new theory is introduced, alternative theories have already failed. P is slower because the opponents try to explain the predicted new facts by their own theories, and the new theory is accepted only after the alternatives have been refuted.

The results of this study suggest four generalizations that may be applicable to other cases in the history of science:

1. Within a single subfield there is an alternation between periods when theory is ahead, with theories being evaluated mostly by the success of their predictions, and periods when experiments are ahead, and theories are evaluated mostly by their ability to explain (retrodict) previously known facts.
2. Evaluation by prediction-testing generally takes longer to produce a consensus than evaluation by explanation. (You first have to wait for the prediction to be tested, then you have to shoot down all the alternative theories that could have predicted the same result.)

Both relativity and quantum mechanics made some important predictions that were not fully confirmed until years or decades after those theories were accepted for other reasons.

3. In either case, a theory that is considered beautiful and gives a unified account of several types of phenomena is more likely to be accepted.
4. Any statement to the effect that scientists follow a single method based on proposing hypotheses and testing predictions-in-advance is refuted by the most important revolution of twentieth-century physics.

NOTES FOR CHAPTER 15

491 *Which Works Faster: Prediction or Explanation?* Most of this chapter is based on my Pais Prize Lecture at the American Physical Society meeting in Washington, DC, February 14, 2010.

491 The biography by Abraham Pais, *'Subtle is the Lord . . . ': The Science and the Life of Albert Einstein* (Oxford: Oxford University Press, 1982), is an excellent introduction to this subject for readers not put off by equations. Helge Kragh, *Quantum*

Generations, A History of Physics in the Twentieth Century (Princeton, NJ: Princeton University Press) places the quantum-relativity revolution in a broader context.

The best source for scholarly research is the wonderful scholarly edition of Einstein's published and unpublished works: *The Collected Papers of Albert Einstein*, edited by John Stachel et al. (Princeton, NJ: Princeton University Press, 1987–). Volumes 1–12 cover writings and correspondence through 1921. Volume 11 is a cumulative index, bibliography, list of correspondence, chronology, and errata to the first 10 volumes. English translations of selected items are published in separate volumes (which do not include the editorial notes and commentaries).

491 *"Science walks forward on two feet"* Robert A. Millikan, "The Electron and the Light-Quant from the Experimental Point of View," Nobel Lecture, May 23, 1924 (on receiving the Physics Prize for 1923), in *Nobel Lectures including Presentation Speeches and Laureates' Biographies, Physics 1922–1941* (Singapore: World Scientific, 1998), pp, 54–66, on p. 54.

491 *Einstein's special theory of relativity was accepted by 1911* Staley, *Einstein's Generation: The Origins of the Relativity Revolution* (Chicago: University of Chicago Press, 2008), p. 23. See also Helge Kragh, *Quantum Generations*, p. 93.

Section 15.2

492 *This event was the immediate cause of World War I* Francis Ferdinand (1863–1914), "Austrian archduke, heir apparent (after 1889) of his great-uncle, Emperor Francis Joseph. . . Laboring to transform the dual Austro-Hungarian Monarchy into a triple monarchy including a Slavic kingdom under Croatian leadership, he won the enmity of both the Pan-Serbians and the Pan-Germans"; Gavrilo Princip (1895–1918), "Serbian political agitator, born in Bosnia. As a high school student and a member of the Serbian nationalist secret society, Union or Death (known as the Black Hand), he assassinated Archduke Francis Ferdinand and his wife at Sarajevo in 1914. His act precipitated World War I. Princip died of tuberculosis in an Austrian prison. He remains a Serbian hero." Both entries from *New Columbia Encyclopedia* (1975 edition). For many more details see Vladimir Dedijer, *The Road to Sarajevo* (New York: Simon & Schuster, 1966). John F. Burns, "In Sarajevo, Divisions that drove Assassin have yet to heal," *New York Times*, June 28, 2014, p. A12.

493 *A German astronomical expedition led by Erwin Findlay Freundlich* Freundlich (1885–1964) was born in Germany. He was serving as an assistant at the Royal Observatory in Berlin in 1911 when Einstein asked him to remeasure the motion of the planet Mercury; he confirmed earlier observations that disagreed with the advance of Mercury's perihelion as calculated from Newtonian theory. Freundlich then organized an expedition to the Crimea to observe the solar eclipse scheduled for 1914, in order to test Einstein's prediction of gravitational light bending. Thanks to Gavrilo Princip, the outbreak of World War I prevented Freundlich from making a measurement which, had it been done accurately, would have shown that Einstein's prediction was too small by a factor of one-half. Freundlich continued to support Einstein and publicize his theories in the next few years. See Eric G. Forbes, "Freundlich, Erwin Finlay," *Dictionary of Scientific Biography*, vol. V (New York: Scribner, 1972), pp. 181–184, and the Freundlich-Einstein correspondence published in *The Collected Papers of Albert Einstein*, vols. 5 (1993) and 8 (1998).

493 *Eddington went to the island of Principe* John Earman and Clark Glymour, "Relativity and Eclipses: The British Eclipse Expeditions of 1919 and Their Predecessors," *Historical Studies in the Physical Science* 11, pt. 1 (1980): 49–85.

493 *a scientist like Eddington could write for the public* Bowler, *Science for All: The Popularization of Science in early Twentieth-Century Britain* (Chicago: University of Chicago Press, 2009). See the review by Melinda Baldwin in *Science*, 326 (2009): 1347–1348.

493 *Replication by more objective observers* Jeffrey Crelinstein, Jeffrey, *Einstein's Jury: The Race to Test Relativity* (Princeton, NJ: Princeton University Press, 2006); Daniel Kennefick, "Testing Relativity from the 1919 Eclipse—A Question of Bias," *Physics Today* 62, no. 3 (March 2009): 37–42.

494 *Attempts to measure the solar redshift gave conflicting results* John Earman and Clark Glymour, Clark. "The Gravitational Red Shift as a Test of General Relativity: History and Analysis," *Studies in History and Philosophy of Science* 11 (1980): 175–214; Norriss S. Hetherington, *Science and Objectivity: Episodes in the History of Astronomy* (Ames: Iowa State University Press, 1988).

494 *Everitt has just announced* C. W. F. Everitt et al., "Gravity Probe B: Final Results of a Space Experiment to Test General Relativity," *Physical Review Letters* 106, no. 222101 (2011): 1–6. The experiment tested two predictions derived in 1960 by L. I. Schiff: "an ideal gyroscope in orbit around the Earth would undergo two relativistic precessions with respect to a distant inertial frame: 1) a geodetic drift in the orbit plane due to motion through the space-time curved by the Earth's mass; 2) a frame-dragging due to the Earth's rotation" (Everitt et al. p. 1). See Dennis Overbye, "52 Years and $750 Million Prove Einstein was Right," *New York Times* (May 5, 2011): A7. Steven K. Blau, "Gravity Probe B Concludes its 50-Year Quest," *Physics Today* 64, no. 7 (July 2011): 14–16.

494 *it had passed three empirical tests* Clifford M., Will, *Theory and Experiment in Gravitational Physics* (Cambridge: Cambridge University Press, 1981; *Was Einstein Right? Putting General Relativity to the Test* (New York: Basic Books, 1986).

SECTION 15.5

496 *the "unreasonable effectiveness"* *"The Unreasonable Effectiveness of Mathematics in the Natural Sciences," Communications on Pure and Applied Mathematics* 13 (1960): 1–14.

496 *Planck did not propose a* physical *quantum theory* Thomas S. Kuhn, *Black-Body Theory and the Quantum Discontinuity 1894–1912* (Oxford: Clarendon Press, 1978). See the sentence in Planck's 1900 paper "If the ratio . . . is not an integer," quoted in Section 7.2.

496 *He wrote: "since the idea of light quanta"* A. H. Compton, "Light Waves or Light Bullets," *Scientific American* 133 (October 1925): 246–247.

SECTION 15.6

498 *The helium atom . . . was finally conquered by physicist Egil Hylleraas* See text and notes for last 2 pages of Section 8.10.

498 *Heitler and Fritz London applied quantum mechanics to the hydrogen molecule* See text and notes for first 3 pages of Section 10.1.

498 *Two predictions-in-advance should be mentioned* See text and notes for first 3 pages of Section 8.9.

499 *"quantum mechanics didn't predict anything"* S. G. Brush, "Remembering Rabi: A Challenge . . . " *Physics Today* 60, no. 6 (June 2007): 10.

499 *Oseen argued that quantum mechanics did not deserve the prize* Friedman, Robert Marc *The Politics of Excellence: Behind the Nobel Prize in Science* (New York: Freeman,

2001), pp. 171–174. Oseen's stance of strong opposition to quantum mechanics is partly contradicted in a letter from Oseen to Richard von Mises in 1930, urging that an account of wave mechanics should be included in the second edition of the Frank-von Mises book *Die Differential- und Integralgleichungen der Mechanik und Physik* (1925). In response, Frank and von Mises asked Guido Beck to write an article on wave mechanics for their second edition. See Reinhard Siegmund-Schultze, "Philipp Frank, Richard von Mises, and the Frank-Mises," *Physics in Perspective* 9 (2007): 26–57.

Selected Bibliography

Includes works cited more than once in a chapter.

Aiton, E. J. *The Vortex Theory of Planetary Motions*. London: MacDonald, 1972. 2.3
Alpher, Ralph A., Hans Bethe, and George Gamow. "The Origin of Chemical Elements." *Physical Review* [series 2] 73 (1948): 803–804.
Alpher, Ralph A. and Robert Herman. "Evolution of the Universe." *Nature* 162 (1948): 774–775. 12.8
Alpher, Ralph A. and Robert Herman. "Remarks on the Evolution of the Expanding Universe." *Physical Review* [series 2] 75 (1949): 1089–1095. 12.8
Anderson, Carl D. "Early Work on the Positron and Muon." *American Journal of Physics* 29 (1961): 825–830. 9.1
Andrade, E. N. Da C. *The Structure of the Atom*. London: Bell, 3rd ed. 1926. 8.1
Arabatzis, Theodore. *Representing Electrons: A Biographical Approach to Theoretical Entities*. Chicago: University of Chicago Press, 2005. 2.5
AHQP (Archive for History of Quantum Physics). Collection of documents and tape-recorded interviews, held at the American Philosophical Society, Philadelphia, with copies at the Center for History of Physics, College Park, Maryland, and several other repositories. 8.7
Bacciagaluppi, Guido, and Antony Valentini. *Quantum Theory at the Crossroads: Reconsidering the 1927 Solvay Conference*. Cambridge: Cambridge University Press, 2009. 8.3
Barnes, Barry. *Interests and the Growth of Knowledge*. London: Routledge, 1977. 4.2
——, David Bloor and John Henry. *Scientific Knowledge: A Sociological Analysis*. Chicago: University of Chicago Press, 1996. 4.3
Beatty, John, Robert Brandon, Elliott Sober, and Sandra D. Mitchell, "Symposium: Are there Laws of Biology?" In *PSA 1996*, Part II, *Supplement to [the journal] Philosophy of Science* volume 64, edited by Lindley Darden (1997): S432–S479. 13.2
Ben-David, Joseph. "Sociology of Scientific Knowledge" in *The State of Sociology: Problems and Prospects*, edited by James F. Short, 40–59. Beverly Hills, CA: Sage, 1981. 4.4
Bernstein, Jeremy. "A Palette of Particles." *American Scientist* 100 (2012): 146–155. 9.1
Biagioli, Mario, editor. *The Science Studies Reader*. New York: Routledge, 1999. 4.1
Bloor, David. *Knowledge and Social Imagery*. London: Routledge & Kegan Paul, 1976. 4.2
Bohr, Niels. "On the Constitution of Atoms and Molecules." *Philosophical Magazine* [series 6] 26 (1913): 1–25, 476–502, 857–875. 2.5
——. *Collected Works*, edited by K. Stolzenburg and others. Amsterdam: North-Holland. Vols. 2–6, 1981–1985. 7.7, 8.1

Bondi, H., and T. Gold. "The Steady State Theory of the Expanding Universe." *Monthly Notices of the Royal Astronomical Society* 108 (1948): 242–270. 12.6
Boorse, H. A. and L. Motz. *World of the Atom*. New York: Basic Books, 1966. 8.5
Born, Max, editor. *The Born-Einstein Letters*. New York: Walker, 1971. 7.7, 8.7
Born, Max, Werner Heisenberg, and Pascual Jordan. "Zur Quantenmechanik II." *Zeitschrift für Physik* 35 (1926): 557–615. 8.3
Boss, Valentin, *Newton and Russia: The Early Influence, 1698–1796*. Cambridge, MA: Harvard University Press, 1972. 1.2.3
Boudrieu, Pierre. *Science of Science and Reflexivity*. Chicago: University of Chicago Press, 2004. 4.2
Bowler, Peter J. *The Non-Darwinian Revolution: Reinterpreting a Historical Myth*. Baltimore: John s Hopkins University Press, 1988. 2.4
Bridges, Calvin B. "Non-Disjunction of the Sex Chromosomes of Drosophila." *Journal of Experimental Zoology* 15 (1913): 587–606. 13.4
———. "Non-Disjunction as Proof of the Chromosome Theory of Heredity." *Genetics* 1 (1916): 1–52, 107–163. 13.4
Brown, Laurie M. and Helmut Rechenberg. *The Origin of the Concept of Nuclear Forces*. Bristol, UK: Institute of Physics Publishing, 1996. 9.4, 9.5
Brown, Richard C. *Are Science and Mathematics Socially Constructed? A Mathematician Encounters Postmodern Interpretations of Science*. Singapore: World Scientific, 2009. 4
Brush, Stephen G. *The Kind of Motion We Call Heat: A History of the Kinetic Theory of Gases in the 19th Century*. Amsterdam: North-Holland, 1976. 1.2, 5.1, 7.1
———. "Nettie M. Stevens and the Discovery of Sex Determination by Chromosomes." *Isis* 69 (1978): 163–172. 13.3
———. "The Chimerical Cat: Philosophy of Quantum Mechanics in Historical Perspective." *Social Studies of Science* 10 (1980): 393–447. 4.2
———. *Statistical Physics and the Atomic Theory of Matter, from Boyle and Newton to Landau and Onsager*. Princeton, NJ: Princeton University Press, 1984. 1.2, 7, 8.3
———. "Prediction and Theory Evaluation: The Case of Light Bending." *Science* 246 (1989): 1124–1129. 11
———. "Prediction and Theory Evaluation: Alfvén on Space Plasma Phenomena." *Eos: Transactions of the American Geophysical Union* 71 (1990): 19–33. 1.1, 1.9
———. "Alfvén's Programme in Solar System Physics." *IEEE Transactions on Plasma Science* 20 (1992): 577–589. 1.1, 1.9
———. "Prediction and Theory Evaluation: Subatomic Particles." *Rivista di Storia della Scienza*, Serie II, Vol. 1, no. 2 (1993): 47–152. 9
———. "Prediction and Theory Evaluation: Cosmic Microwaves and the Revival of the Big Bang." *Perspectives on Science* 1 (1993): 565–603. 12.8
———. *Transmuted Past: The Age of the Earth and the Evolution of the Elements from Lyell to Patterson*. New York: Cambridge University Press, 1996. 12.2, 12.7
———. "The Reception of Mendeleev's Periodic Law in America and Britain." *Isis* 87 (1996): 595–628. 5
———. "Why was Relativity Accepted?" *Physics in Perspective* 1 (1999): 184–214. 11
———. "Dynamics of Theory Change in Chemistry: Part 1: The Benzene Problem 1865–1945." *Studies in History and Philosophy of Science* 30 (1999): 21–79. 6, 10
———. "Dynamics of Theory Change in Chemistry: Part 2. Benzene and Molecular Orbitals, 1945–1980." *Studies in History and Philosophy of Science* 30 (1999): 263–302. 10
———. "Postmodernism versus Science versus Fundamentalism: An Essay Review." *Science Education* 84 (2000): 114–122. 4.4

Brush, Stephen G. "Is the Earth Too Old? The Impact of Geochronology on Cosmology, 1929–1952." In *The Age of the Earth from 4004 BC to AD 2002*, edited by C. L. E. Lewis and S. J. Knell, 157–175. London: Geological Society Special Publications, no. 190, 2011. 12

———. "How Theories became Knowledge: Morgan's Chromosome Theory of Heredity in America and Britain." *Journal of the History of Biology* 35 (2002): 471–535. 13, 13.12

———. "How Theories Became Knowledge: Why Science Textbooks Should be Saved." In *Who Wants Yesterday's Papers: Essays on the Research Value of Printed Materials in the Digital Age*, edited by Yvonne Carignan et al. Lanham, MD: Scarecrow Press, 2005. 1.1

———. "How Ideas Became Knowledge: The Light Quantum Hypothesis, 1905–1935." *Historical Studies in the Physical and Biological Sciences* 37 (2007): 205–246. 7.3, 7.8

———. *Choosing Selection: The Revival of Natural Selection in Angle-American Evolutionary Biology, 1930–1970.* Philadelphia: American Philosophical Society, *Transactions*, vol. 99, part 3, 2009. 14

———, and Lanfranco Belloni. *The History of Modern Physics: An International Bibliography*. New York: Garland, 1983. 8.1

Cain, Joe. "Ritual Patricide: Why Stephen Jay Gould Assassinated George Gaylord Simpson." In *The Paleobiological Revolution: Essays on the Growth of Modern Paleontology*, edited by David Sepkoski and Michael Ruse, 346–363. Chicago: University of Chicago Press, 2009. 14.10

———, and Michael Ruse, editors. *Descended from Darwin: Insights into American Evolutionary Studies, 1900–1970.* Philadelphia: American Philosophical Society, 2009. 14.15

Calaprice, Alice, editor. *The Ultimate Quotable Einstein*. Princeton, NJ: Princeton University Press, 2011. 11

Cassidy, D. C. "Cosmic Ray Showers, High Energy Physics, and Quantum Field Theories: Programmatic Interactions in the 1930s." *Historical Studies in the Physical Sciences* 12 (1981): 1–39. 9.5

Chalmers, Alan. *The Scientist's Atom and the Philosopher's Stone: How Science Succeeded and Philosophy Failed to gain Knowledge of Atoms.* (Boston Studies in the Philosophy of Science, 279) Dordrecht: Springer, 2009. 5

Chubin, Daryl E. and Edward J. Hackett, *Peerless Science: Peer Review and U. S. Science Policy*. Albany, NY: SUNY Press, 1990. 4.2

Collins, H. M. and Robert Evans. "The Third Wave of Science Studies: Studies of Expertise and Experience." *Social Studies of Science* 32 (2003): 581–611. 4.3

Collins, H. M. and Robert Evans, "King Canute Meets the Beach Boys: Responses to the Third Wave." *Social Studies of Science* 33 (2003): 401–417. 4.3

Compton, Arthur Holly. "A Quantum Theory of the Scattering of X-rays by light Elements," *Physical Review* [series 2] 21 (1923): 483–502. 7.7

———. *Scientific Papers of Arthur Holly Compton: X-Rays and other Studies*, edited by Robert S. Shankland. Chicago: University of Chicago Press, 1973. 7.7

Copernicus, Nicholas. *De Revolutionibus orbius coelestium*, 1543. *On the Revolutions,* edited by Jerzy Dobrzycki, translation and commentary by Edward Rosen. Baltimore: Johns Hopkins University Press, 1978. 2.2

Creager, A.N.H. "Adaptation or Selection? Old Issues and new Stakes in the present Debates over Bacterial Drug Resistance." *Studies in History and Philosophy of Biological and Biomedical Sciences* 38 (2007): 159–190. 14.14

Cutcliff, Stephen H. and Carl Mitcham, editors. *Visions of STS: Counterpoints in Science, Technology and Society Studies.* Albany, NY: State University of New York Press, 2001. 4.3

Darrigol, Olivier. *From C-Numbers to Q-Numbers: The Classical Analogy in the History of Quantum Theory*. Berkeley: University of California Press, 1992. 8.7

Daston, Lorraine, editor. *Biographies of Scientific Objects*. Chicago: University of Chicago Press, 2000. 4.4

———. "Science Studies and the History of Science." *Critical Inquiry* 35 (2009): 798–803. 4.1

Darwin, Charles Robert. *On the Origin of Species by Means of Natural Selection, or the Preservation of Favoured Races in the Struggle for Life*. London: John Murray, 1859; 6th ed. 1872 2.4

De, B. "A 1972 Prediction of Uranian Rings Based on the Alfvén Critical Velocity Effect." *Moon & Planets* 18 (1978): 339–342. 1.1

Dobrzycki, Jerzy, editor. *The Reception of Copernicus' Heliocentric Theory: Proceedings of a Symposium Organized by the Nicolas Copernicus Committee of the International Union of the History and Philosophy of Science, Toru_, 1972*. Dordrecht: Reidel Publishing Company, and Warsaw: Ossolineum, the Polish Academy of Sciences Press, 1972. 2.2

Dobzhansky, Theodosius. *Genetics and the Origin of Species*. New York: Columbia University Press, 2nd edition, 1941. 6.6

———. *Genetics and the Origin of Species*. New York: Columbia University Press, 3rd edition, 1951. 6.6

———. *Genetics of the Evolutionary Process*. New York: Columbia University Press, 1970. 14.13, 14.16

———, and Olga Pavlovsky, "An Experimental Study of Interaction Between Genetic Drift and Natural Selection." *Evolution* 11 (1957): 311–319. 14.16

Doing, Park. "Give Me a Laboratory and I will Raise a Discipline: The Past, Present, and Future Politics of Laboratories in STS." In Hackett et al., editors, *Handbook of Science and Technology Studies*, Third Edition (Cambridge, MA: MIT Press, 2008), 279–295. 4.3

Dongen, Jeroen van. "Emil Rupp, Albert Einstein, and the Canal Ray Experiments on Wave-Particle Duality." *Historical Studies in the Physical and Biological Sciences* 37, Supplement (2007): 73–119. 7.10

———. *Einstein's Unification*. Cambridge: Cambridge University Press, 2010. 11.1, 11.5

Donovan Arthur, Larry Laudan, and Rachel Laudan, eds. *Scrutinizing Science: Empirical Studies of Scientific Change*. Boston: Kluwer, 1988. Augmented paperback edition, Baltimore: Johns Hopkins University Press, 1992. 3.1

Eckert, Michael. "Propaganda in Science: Sommerfeld and the Spread of the Electron Theory of Metals." *Historical Studies in the Physical Sciences* 17, part 2 (1987): 191–233. 8.5

Einstein, Albert. "Über einen die Erzeugung und Verwandlung des Lichtes betreffenden heuristischen Gesichtspunkt." *Annalen der Physik* 17 (1905): 132–148. 7.3

———. "Zur Elektrodynamik bewegter Körper. *Annalen der Physik* 17 (1905):891–921. 11.1

———. "Ist die Trägheit eines Körpers von seinem Energieinhalt abhängig?" *Annalen der Physik* 18 (1905): 639–641. 11.1

———. "Die Grundlage der allgemeinen Relativitätstheorie." *Annalen der Physik* 49 (1916): 769–822. 11.2

———. *Über die spezielle und die allgemeine Relativitätstheorie (Gemein-verstandlich)*. Braunschweig: Druck und Verlag von Friedr. Vieweg & Sohn, 1917. 11.2

———. *Relativity: The Special and General Theory. A Popular Exposition*. Translated by Robert W. Lawson. London: Methuen, 1920. 11.2

Einstein, Albert. *The Collected Papers of Albert Einstein.* Edited by John Stachel, Martin J. Klein, A. J. Kox, Jürgen Renn, Robert Schulmann, Michel Janssen, József Illy, Christopher Lehner, Diana Kormos Buchwald, Daniel J. Kennefick, Tilman Sauer, Ze'ev Rosenkranz, Virginia Iris Holmes, Rudy Hirschmann, Osik Moses, Benjamin Aronin, Jennifer Stolper, and others. Princeton, NJ: Princeton University Press, 1987—(volume 12, correspondence January–December 1921, published in 2009). 3.3, 7.8, 7.11, 11, 11.2, 11.7, 15

Engels, Eve-Marie, and Thomas F. Glick, editors. *The Reception of Charles Darwin in Europe.* New York: Continuum, 2008. 2.4, 14.1

Farmelo, Graham. *The Strangest Man: The Hidden Life of Paul Dirac, Mystic of the Atom.* New York: Basic Books, 2009. 8.7

Fisher, R. A. *The Genetical Theory of Natural Selection.* Oxford: Oxford University Press, 1930; second edition, New York: Dover, 1958. 14.1

———. *The Genetical Theory of Natural Selection: A Complete Variorum Edition*, edited by J. H. Bennett. Oxford: Oxford University Press, 1999. 14.2

Ford, E. B. "Polymorphism." *Biological Review* 20 (1945): 73–88. 14.13

———. *Mendelism and Evolution*, 6th edition. London: Methuen, 1957. 14.17

Forman, Paul. "The Doublet Riddle and Atomic Physics *circa* 1924." *Isis* 59 (1968): 156–174. 8.1, 8.2

———. "Alfred Landé and the Anomalous Zeeman Effect." *Historical Studies in the Physical Sciences* 2 (1970): 153–261. 8.1, 8.2

———. "Weimar Culture, Causality, and Quantum Theory, 1918–1927: Adaptation by German Physicists and Mathematicians to a Hostile Intellectual Environment." *Historical Studies in the Physical Sciences* 3 (1971): 1–115. 4.2

Foster, J. Stuart, "A Quarter-Century of Research in Physics" (Presidential Address), *Transactions of the Royal Society of Canada* 43 (1949): 1–13. 8.9

Gavroglu, Costas and Ana Simões. *Neither Physics nor Chemistry: A History of Quantum Chemistry.* Cambridge, MA: MIT Press, 2012. 10, 10.5

Giere, Ronald N. *Explaining Science*. Chicago: University of Chicago Press, 1988. 1.3.3

Gilder, Louisa. *The Age of Entanglement: When Quantum Physics Was Reborn.* New York: Knopf, 2008; Vintage paperback reprint, 2009. 8.1

Glashow, S. L. and B. Bova. *Interactions: A Journey through the Mind of a Particle Physicist and the Matter of this World.* New York: Warner, 1988. 9

Glick, Thomas F., editor. *The Comparative Reception of Darwinism.* Chicago: University of Chicago Press, 1974. 2.4

Glick, Thomas F., editor. *The Comparative Reception of Relativity.* Dordrecht and Boston: Reidel, 1987. 11

Glick, Thomas F. and M. G. Henderson, "The Scientific and Popular Receptions of Darwin, Freud, and Einstein." In *The Reception of Darwinism in the Iberian World*, edited by Glick et al., 229–238. Dordrecht: Kluwer, 2001. 2.1

Goenner, Hubert, Jürgen Renn, Jim Ritter and Tilman Sauer, editors. *The Expanding Worlds of General Relativity.* Boston: Birkhäuser, 1999. 11.6

Goldberg, Stanley. *Understanding Relativity: Origin and Impact of a Scientific Revolution.* Boston: Birkhäuser, 1984. 11.4

Golinski, Jan. *Making Natural Knowledge: Constructivism and the History of Science.* New York: Cambridge University Press, 1998. 4.2

Gordin, Michael D. *A Well-Ordered Thing: Dmitri Mendeleev and the Shadow of the Periodic Table.* New York: Basic Books, 2004. 5

Gould, Stephen Jay. "G. G. Simpson, Paleontology, and the Modern Synthesis." In Mayr and Provine, *Evolutionary Synthesis* (Cambridge, MA: Harvard University Press, 1980), pp. 153–177. 14.10

———. *The Structure of Evolutionary Theory.* Cambridge, MA: Harvard University Press, 2002. 14.15

Hackett, Edward J., et al., editors. *The Handbook of Science and Technology Studies.* Third Edition, published in cooperation with the Society for Social Studies of Science. Cambridge, MA: MIT Press, 2008. 4.3

Hacking, Ian. *The Social Construction of What?* Cambridge, MA: MIT Press, 1999. 4.3

Haldane, J. B. S. *The Causes of Evolution.* London: Longmans, 1932. 14.1, 14.4,

———. *What I Require from Life: Writings on Science and Life from J. B. S. Haldane, edited by Krishna Dronamraju.* Oxford: Oxford University Press, 2009. 13.4, 14.4

Heilbron, J. L. "Quantum Historiography and the Archive for History of Quantum Physics." *History of Science* 7 (1968): 90–111. 8.1

———, Bruce Wheaton et al. *Literature on the History of Physics in the 20th Century.* Berkeley, CA: Office for History of Science and Technology, University of California, 1981. 8.1

Heisenberg, Werner. "Über die quantentheoretische Umdeutung kinematischer und mechanischer Beziehungen." *Zeitschrift für Physik* 33 (1925): 879–893. 8.3

———. "Mehrkörperproblem und Resonance in der Quantenmechanik," *Zeitschrift für Physik* 38 (1926): 411–426; 41 (1927): 239–267. 8.9

———, and Pascual Jordan. "Anwendung der Quantenmechanik auf das Problem der anomalen Zeemaneffekte." *Zeitschrift für Physik* 37 (1926): 263–277. 8.3

Hess, David J. *Science Studies: An Advanced Introduction.* New York: New York University Press, 1997. 4.3

Holton, Gerald. *Thematic Origins of Scientific Thought, Kepler to Einstein.* Cambridge, MA: Harvard University Press, revised edition 1988. 11.9

Holton, Gerald, and Stephen G. Brush. *Physics, the Human Adventure: From Copernicus to Einstein and Beyond.* New Brunswick, NJ: Rutgers University Press, 2001. 2.2, 2.3, 5

Hoyle, Fred. "A new Model for the Expanding Universe." *Monthly Notices of the Royal Astronomical Society* 108 (1948): 372–382.

Hu, Danian. *China and Albert Einstein: The Reception of the Physicist and his Theory in China, 1917–1979.* Cambridge, MA: Harvard University Press, 2005. 11.10, 11.11

Hubble, E. P. "A Relation between Distance and Radial Velocity Among the Extra-Galactic Nebulae." *Proceedings of the National Academy of Sciences of the USA* 15 (1929): 168–173. 12.1

———. "The Realm of the Nebulae." *Scientific Monthly* 39 (1939): 193–202. 12.5

Huxley, Julian, editor. *The New Systematics.* Oxford: Clarendon Press, 1940. 14.3, 14.8

Hylleraas, Egil A., "Reminiscences from Early Quantum Mechanics of Two-Electron Atoms." *Reviews of Modern Physics* 35 (1963): 421–431. 8.9, 8.10

Jackman, L. M., F. Sondheimer, Y. Amiel, D. A. Ben-Efraim, Y. Gaoni, R. Wolovsky, and A. A. Botherner-By. "The Nuclear Magnetic Resonance Spectroscopy of a Series of Annulenes and Dehydro-annulenes." *Journal of the American Chemical Society* 84 (1962): 4307–4312. 10.6

Jacob, Margaret. "Science Studies after Social Construction: The Turn toward the Comparative and the Global," in V. B. Bonnell and L. Hunt, editors, *Beyond the Cultural Turn: New Directions in the Study of Society and Culture*, 95–120. Berkeley: University of California Press, 1999.

Jammer, Max. *The Conceptual Development of Quantum Mechanics*. New York: McGraw-Hill, 1966; 2d ed., New York: Tomash/AIP Press, 1989. 7.7, 8.1

Jasanoff, Sheila et al., editors. *Handbook of Science and Technology Studies*. Thousand Oaks, CA: Sage Publications, 1995.

Jasanoff, Sheila. "Symposium. Breaking the Waves in Science Studies: Comment on H. M. Collins and Robert Evans, 'The Third Wave of Science Studies,'" *Social Studies of Science* 33 (2003): 389–400. 4.3

Johnson, Francis R. *Astronomical Thought in Renaissance England: A Study of the English Scientific Writings from 1500 to 1645*. Baltimore: Johns Hopkins University Press, 1937; reprinted, New York: Octagon Books, 1968. 2.2

Jordan, Pascual. "Early Years of Quantum Mechanics: Some Reminiscences." In *The Physicist's Conception of Nature*, edited by Jagdish Mehra, 294–299. Dordrecht: Reidel, 1973. 8.7

Kangro, Hans. *Planck's Original Papers in Quantum Physics*. London: Taylor and Francis, 1972. 7.2

Karachalios, Andreas. *Erich Hückel (1896–1980). From Physics to Quantum Chemistry. (Boston Studies in Philosophy of Science,* 283). Berlin: Springer, 2010. 10.1

Kekulé, August. "Sur la Constitution des Substances Aromatiques." *Bulletin de la Société Chimique de Paris* 3 (1865): 98–110. 6.1

———. "Note sur quelques produits de Substitution de la Benzine." *Bulletin de l'Academie Royale de Belgique* [ser. 2] 19 (1865): 551–563. 6.1

———. "Untersuchungen über aromatische Verbindungen," [Liebig's] *Annalen der Chemie und Pharmacie* 137 (1865): 129–196. 6.1

———. "Über einige Condensationsproducte des Aldehydes." [Liebig's] *Annalen der Chemie und Pharmacie* 162 (1872): 77–124, 309–320. 6.1

Kent, Norton A., Lucien B. Taylor, and Hazel Pearson. "Doublet-Separation and Fine Structure of the Balmer Lines of Hydrogen." *Physical Review* 30 (1927): 266–283. 8.7

Klein, M. J., editor. *Letters on Wave Mechanics: Schrödinger, Planck, Einstein, Lorentz*. New York: Philosophical Library, 1967. 8.7

Knorr-Cetina, Karin, and Michael Mulkay, editors. *Science Observed: Perspectives on the Social Study of Science*, 239–2566. London: SAGE, 1983. 4.3, 4.4

Koertge, Noretta, editor. *A House Built on Sand: Exposing Postmodernist Myths about Science*. New York: Oxford University Press, 1998. 4.3

Kohn, David, editor. *The Darwinian Heritage*. Princeton, NJ: Princeton University Press. 1985. 2.4

Koyré, Alexandre, *Newtonian Studies*. Cambridge, MA: Harvard University Press, 1965. 2.3

Kragh, Helge. "The Fine Structure of Hydrogen and the Gross Structure of the Physics Community, 1916–1926." *Historical Studies in the Physical Sciences* 15, part 2 (1985): 67–125. 8.1

———. *Dirac: A Scientific Biography*. Cambridge: Cambridge University Press, 1990. 8.7, 8.9

———. *Cosmology and Controversy: The Historical Development of Two Theories of the Universe*. Princeton, NJ: Princeton University Press, 1996. 12

Kragh, Helge. "Social Constructivism, the Gospel of Science, and the Teaching of Physics." *Science & Education* 7 (1998): 231–243. 4.2

———. *Quantum Generations: A History of Physics in the Twentieth Century*. Princeton, NJ: Princeton University Press, 1999. 15

———. "The early Reception of Bohr's Atomic Theory (1913–1915): A Preliminary Investigation." *Reposs: Research Publications on Science Studies*, 9. Aarhus: Department of Science Studies, University of Aarhus, 2010; available online at www.ivs.au.dk/reposs. 2.5

Kuhn, Thomas S. *The Copernican Revolution: Planetary Astronomy in the Development of Western Thought*. Cambridge, MA: Harvard University Press, 1957. 2.2

———. *The Structure of Scientific Revolutions*. Chicago: University of Chicago Press, 1962; 2nd edition, 1970. 4.2

———. *Black-Body Theory and the Quantum Discontinuity* 1894–1912. Oxford: Clarendon Press, 1978. 7.2, 15

———. *The Road since Structure*. Chicago: University of Chicago Press, 2000. 4.2

Lakatos, Imre. "Falsification and the Methodology of Scientific Research Programmes." In *Criticism and the Growth of Knowledge*, edited by I. Lakatos and A. Musgrave, 91–196. New York: Cambridge University Press, 1970. 3.2

———, and Elie Zahar. "Why did Copernicus' Research Program Supersede Ptolemy's?" In Robert S. Westman, editor, *The Copernican Achievement*. Berkeley: University of California Press, 1975, pp. 354–383. 2.2

Latour, Bruno. "One More Turn after the Social Turn." In *The Social Dimensions of Science*, edited by Ernan McMullin, 272–294. Notre Dame, IN: University of Notre Dame Press, 1992. 4.3

———. *Pandora's Hope: Essays on the Reality of Science Studies*. Cambridge, MA: Harvard University Press, 1999. 4.4

———. "Why Has Critique Run out of Steam? From Matters of Fact to Matters of Concern." *Critical Inquiry* 30 (2004): 225–248. 4.3

———. *On the Modern Cult of the Factish Gods*. Durham, NC: Duke University Press, 2010. 4.2

———, and Steve Woolgar. *Laboratory Life: The Social Construction of Scientific Facts*. Beverly Hills, CA: Sage Publications, 1979. Second edition, *Laboratory Life: The Construction of Scientific Facts*. Princeton, NJ: Princeton University Press, 1986. 4.2

Lecoq de Boisbaudran, Paul Émile. "Caractères chimiques et spectroscopiques d'un nouveau métal, le gallium, découvert dans und blende de la mine de Pierrefitte, Vallée d'Argelès Pyrénées." *Comptes Rendus Hebdomadaires des Sciences de l'Académie des Sciences, Paris* 81 (1875): 493–495. 5.5

———. "Sur un nouveau métal, le gallium." *Annales de Chimie* [series 5] 10 (1877): 100–141.

Leigh, Egbert Giles Jr. "Ronald Fisher and the Development of Evolutionary Theory. I. The Role of Selection." *Oxford Surveys in Evolutionary Biology* 3 (1986): 187–223. 3.5

Lemaître, Georges. "Un Universe homogène de Masse constant et de Rayon Croissant, rendant Compte de la Vitesse Radiale des Nebuleuses Extra-Galactique." *Annales de la Société Scientifique de Bruxelles*, Serie A, 47 (1927): 45–59. English translation (partial), "A Homogeneous Universe of Constant Mass and Increasing Radius Accounting for the Radial Velocity of Extra-Galactic Nebulae," *Monthly Notices of the Royal Astronomical Society* 91 (1931): 483–490. 12.1

———. "The Beginning of the World from the Point of View of the Quantum Theory." *Nature* 127 (1931): 706. 12.8

Lemaître, Georges. "L'Expansion de l'Espace." *Revue des Questions Scientifiques* 100 (1931): 391–410. 12.8

Magio, Koffi. "The Reception of Newton's Gravitational Theory by Huygens, Varignon, and Maupertuis: How Normal Science may be Revolutionary." *Perspectives on Science* 11 (2003): 135–169. 2.3

———. *Systematics and the Origin of Species from the Viewpoint of a Zoologist*. New York: Columbia University Press, 1942. 14.8, 14.9

Mayr, Ernst. "Where are We?" *Cold Spring Harbor Symposium on Quantitative Biology* 24 (1959): 1–14. 14.2

———. *Animal Species and Evolution*. Cambridge, MA: Harvard University Press, 1963. 14.8, 14.9

———. *Populations, Species, and Evolution*. Cambridge, MA: Harvard University Press, 1970. 14.17

———. "Prologue: Some Thoughts on the History of the Evolutionary Synthesis." In Mayr and Provine, editors, *The Evolutionary Synthesis* (Cambridge, MA: Harvard University Press, 1980), pp. 1–48. 14.9

———. *The Growth of Biological Thought: Diversity, Evolution, and Inheritance*. Cambridge, MA: Harvard University Press, 1982. 14.19

———, and Will Provine, editors. *The Evolutionary Synthesis*. Cambridge, MA: Harvard University Press, 1980. 14.4, 14.8

Mehra, Jagdish, and Helmut Rechenberg. *The Historical Development of Quantum Theory*, 6 vols. New York: Springer, 1982–2001. 8.1

Mendeleev, Dmitrii. "Das periodische Gesetzmässigkeit der chemische Elemente," *Liebigs Annalen der Chemie und der Pharmacie (Supplement)* 8 (1871): 133–229. 5.3

Meyenn, K. Von, editor. *Wolfgang Pauli Wissenschaftliche Briefwechsel mit Bohr, Einstein, Heisenberg, u. a.*, Vol. II. Berlin: Springer, 1985. 9.5

Mikulak, Maxim William. *Relativity Theory and Soviet Communist Philosophy (1922–1960)*. PhD dissertation, Columbia University, 1965. 11.10

Miller, Arthur I. *Albert Einstein's Special Theory of Relativity, Emergence (1905) and early Interpretation (1905–1911)*. Reading, MA: Addison-Wesley, 1981. 11.3, 11.6

Morgan, Thomas H., Alfred H. Sturtevant, Herman J. Muller, and Calvin B. Bridges. *The Mechanism of Mendelian Heredity*. New York: Holt, 1915. IV.I.4

Moss, Jean Dietz. *Novelties in the Heavens: Rhetoric and Science in the Copernican Controversy*. Chicago: University of Chicago Press, 1993. 2.2

Mott, N. F. "The Collision Between Two Electrons," *Proceedings of the Royal Society of London* A126 (1930): 259–267. 8.9

Navarro Brotóns, Victor. "The Reception of Copernicus in sixteenth-century Spain: The Case of Diego de Zuñiga." *Isis* 86 (1995): 52–78. 2.2

Newton, Isaac. *Philosophiae Naturalis Principia Mathematica*. London, 1st ed. 1687, 2nd ed. 1713, 3rd ed. 1726. *The Principia: Mathematical Principles of Natural Philosophy*, translated by I. Bernard Cohen and Anne Whitman. Berkeley: University of California Press, 1999. 2.3

Niaz, Mansoor, and Arelys Maza. *Nature of Science in General Chemistry Textbooks*. Dordrecht: Springer, 2010. 1.7

Numbers, Ronald L. *Darwinism Comes to America*. Cambridge, MA: Harvard University Press, 1998. 2.4

Nussbaumer, Harry, and Lynda Bieri. *Discovering the Expanding Universe*. New York: Cambridge University Press, 2009. 12

Pais, Abraham. *"Subtle is the Lord":The Science and the Life of Albert Einstein*. Oxford: Oxford University Press, 1982. 1.8

———. *Niels Bohr's Times*. Oxford: Clarendon Press, 1991. 8.1

Patterson, C. C. "The Isotopic Composition of Meteoric, Basaltic and Oceanic Leads, and the Age of the Earth." *Proceedings of the Conference on Nuclear Processes in Geologic Settings, Williams Bay, Wisconsin, September 21–23, 1953*, pp. 36–40. 12.7

Peebles, P.J.E., L. A. Page Jr., and R. B. Partridge, editors. *Finding the Big Bang*. New York: Cambridge University Press, 2009. 12

Penzias, A. A. and R. W. Wilson, "A Measurement of Excess Antenna Temperature at 4080 Mc/s," *Astrophysical Journal* 142 (1965): 419–421. 12.11

Pickering, Andy. *Constructing Quarks: A Sociological History of Particle Physics.* Chicago: University of Chicago Press, 1984. 4.2

Planck, Max. "Zur Theorie des Gesetze der Energieverteilung im Normalspectrum." *Verhandlungen der Deutschen Physikalische Gesellschaft* 2 (1900) 237–245. 13.2

———. *Die Entstehung und bisherige Entwicklung der Quantentheorie* (Nobel Prize Lecture). Leipzig: Barth, 1920. 7.2

[Planck, Max] Kangro, Hans. *Planck's Original Papers in Quantum Physics.* London: Taylor and Francis, 1972. 7.2

Platt, John R. "Strong Inference." *Science* 146 (1964): 347–353. 10.7

Popper, Karl. *Logik der Forschung: Erkenntnistheorie der modernen Naturwissenschaft.* Vienna: J. Springer, 1935, pub. 1934. English translation, *The Logic of Scientific Discovery.* London: Hutchinson, 1959. 1.2

———. "The Poverty of Historicism, III." *Economica* 12 (1945): 69–89. 14.21

———. "Autobiography of Karl Popper," in *The Philosophy of Karl Popper,* edited by P. A. Schilpp, pp. 3–181. LaSalle, IL: Open Court, 1974. 3.5, 3.7

Provine, William B. *Sewall Wright and Evolutionary Biology.* Chicago: University of Chicago Press, 1986. 13.2, 14.8

Przibram, K., editor. *Schrödinger-Planck-Einstein-Lorentz: Briefe zur Wellenmechanik,* Vienna" Springer-Verlag, 1963. 8.7 (See also Klein, Martin J.)

Punnett, R. C. "Linkage Groups and Chromosome Number in Lathyrus." *Proceedings of the Royal Society of London* B102 (1927): 236–238. 13.5

Rechenberg, Helmut, and Laurie M. Brown. "On the Origin of the Concept of Nuclear Forces (Fundamental Theories of Nuclear Forces). IV. Yukawa's heavy Quantum and the Mesotron (1935–1937)." Preprint, 1992. 9.5

Rocke, Alan J. "Kekulé's Benzene Theory and the Appraisal of Scientific Theories." In *Scrutinizing Science: Empirical Studies of Scientific Change,* edited by Arthur Donovan, Larry Laudan, and Rachel Laudan, pp. 145–161. Dordrecht: Kluwer, 1988. 6.4

———. *Image and Reality: Kekulé, Kopp, and the scientific Imagination.* Chicago: University of Chicago Press, 2010. 6

Sandage, Alan. "Edwin Hubble, 1889–1953." *Journal of the Royal Astronomical Society of Canada* 83 (1989): 351–362. 12.5

Savage, C. W., editor. *Scientific Theories* (Minnesota Studies in the Philosophy of Science, XIV) Minneapolis: University of Minnesota Press, 1990. 3.2

Scerri, Eric R. *The Periodic Table: Its Story and its Significance.* Oxford: Oxford University Press, 2006. 5

Scriven, Michael. "Explanation and Prediction in Evolutionary Theory." *Science* 130 (1959): 477–482. 14.21

Septième Conseil de Physique [Solvay Conference]. *Structure et Propriétes des Noyaux Atomiques. Rapports et Discussions du Septième Conseil de Physique tenu a Bruxelles du 22 au 29 Octobre 1933 sous les Auspices de l'Institut International de Physique Solvay.* Paris: Gauthier-Villars, 1934. 9.2

Seth, Suman. *Crafting the Quantum: Arnold Sommerfeld and the Practice of Theory, 1890–1926.* Cambridge, MA: MIT Press, 2010. 8.1

Shull, A. F. *Heredity.* New York: McGraw-Hill, 1926. 13.10

Simões, A. I. *Converging Trajectories, Diverging Traditions: Chemical Bond, Valence, Quantum Mechanics and Chemistry, 1927–1937.* PhD dissertation, University of Maryland, 1993. 10.2, 10.4

Simpson, G. G. *Tempo and Mode in Evolution*. New York: Columbia University Press, 1944. 14.10

———. *The Major Features of Evolution*. New York: Columbia University Press, 1953. 14.10

Slezak, Peter. "Sociology of Scientific Knowledge and Science Education." In *Constructivism in Science Education: A Philosophical Examination,* edited by Michael R. Matthews, 159–188. Dordrecht: Kluwer, 1998. 4.4

Smith, Robert W. *The Expanding Universe: Astronomy's "Great Debate" 1900–1931.* Cambridge: Cambridge University Press, 1982. 12

Smocovitis, V. Betty. *Unifying Biology: The Evolutionary Synthesis and Evolutionary Biology*. Princeton, NJ: Princeton University Press, 1996. 14.8,

Sokal, Alan D. "Transgressing the Boundaries: Toward a Transformative Hermeneutics of Quantum Gravity." *Social Text*, nos. 46/47 (1996): 217–252. 4.3

———. "A Physicist Experiments with Cultural Studies." *Lingua Franca* 6, no. 4 (1996): 62–64. 4.3

Solvay Conference (7th, 1933), *see* Septième Conseil de Physique.

Sondheimer, Franz, Reuven Wolovsky and Yakov Amiel. "Unsaturated Macrocyclic Compounds. XXIII. The Synthesis of the Fully Conjugated Macrocyclic Polyenes Cycloöctadecanonaene ([18] Annulene), Cyclotetracosadodecaeno ([24] Annulene), and Cyclotriacontapenta-decaene ([30] Annulene." *Journal of the American Chemical Society*, 84 (1962): 274–284.

Stachel, John. "Bohr and the Photon." In *Quantum Reality, Relativistic Causality, and Closing the Epistemic Circle*, edited by Wayne C. Myrvold and Joy Christian. [n. p.] Springer, 2009. 7.7, 7.8

Staley, Richard. *Einstein's Generation: The Origin of the Relativity Revolution.* Chicago: University of Chicago Press, 2008. 15

Stebbins, G. L. Jr. *Variation and Evolution in Plants.* New York: Columbia University Press, 1950. 14.11

———. *Processes of organic Evolution.* Englewood Cliffs, NJ: Prentice-Hall, 1966. 14.17

———. *The Basis of Progressive Evolution.* Chapel Hill, NC: University of North Carolina Press, 1969. 14.17

Stevens, Nettie M. *Studies in Spermatogenesis with especial Reference to the "Accessory Chromosome."* Washington, DC: Carnegie Institution of Washington, 1905, Publication No. 36. 13.3

Stimson, Dorothy. *The Gradual Acceptance of the Copernican Theory of the Universe.* New York: Baker & Taylor, 1917. 2.2

Sulloway, Frank J. *Born to Rebel: Birth Order, Family Dynamics, and Creative Lives.* New York: Pantheon Books, 1996. 2.2, 2.4

Sutton, Geoffrey V. *Science for a Polite Society: Gender, Culture, and the Demonstration of Enlightenment*. Boulder, CO: Westview, 1995. 2.3

Swinburne, R. G. "The Presence-and-Absence Theory." *Annals of Science* 18 (1962): 131–145. 13.8

Tatole, Victoria. "Notes on the Reception of Darwin's Theory in Romania." Chapter 25 in *The Reception of Charles Darwin in Europe*, edited by Eve-Marie Engels and Thomas F. Glick, pp. 463–479, 587–590. New York: Continuum, 2008. 2.4

Taton, R. and C. Wilson, editors. *The General History of Astronomy*, vol. 2, *Planetary Astronomy from the Renaissance to the Rise of Astrophysics*, Part B, *The Eighteenth and Nineteenth Centuries*. New York: Cambridge University Press, 1995. 2.3

Thomas, Jerry. "John Stuart Foster, McGill University and the Renascence of Nuclear Physics in Montreal, 1935–1950." *Historical Studies in the Physical Sciences* 14 (1984): 357–377. 8.9

Thrower, Norman J. W., editor. *Standing on the Shoulders of Giants*. Berkeley: University of California, Press, 1990. 2.3

Tierney, John. "Economic Optimism? Yes, I'll Take That Bet." *New York Times/Science Times*, December 28, 2010: D1, D3.

van der Waerden, B. L., editor. *Sources of Quantum Mechanics*. Amsterdam: North-Holland, 1967. 8.3

Walter, Scott. "Minkowski, Mathematicians, and the Mathematical Theory of Relativity." In *The Expanding Worlds of General Relativity*, edited by Hubert Goenner, Jürgen Renn, Jim Ritter, and Tilman Sauer, 45–86. Boston: Birkhäuser, 1999. 11.6

Weiner, Charles, editor. *History of Twentieth Century Physics*. New York: Academic Press, 1977. 8.7

Westman, Robert S. "The Melanchthon Circle, Rheticus, and the Wittenberg Interpretation of the Copernican Theory." *Isis* 66 (1975): 165–193. 2.2

———, editor, *The Copernican Achievement*. Berkeley: University of California Press, 1975. 2.2

Wright, Sewall. "Evolution in Mendelian Populations." *Genetics* 16 (1931): 97–159.14.1

Yearley, Steven. *Making Sense of Science: Understanding the Social Study of Science*. London: Sage, 2005. 4.2

Zahar, Elie. "Why Did Einstein's Programme Supersede Lorentz's?" *British Journal for the Philosophy of Science* 24 (1975): 95–123, 223–262. 3.2

Zammito, John H. *A Nice Derangement of Epistemes: Post-Positivism in the Study of Science from Quine to Latour*. Chicago: University of Chicago Press, 2004. 4

Zuckerman, H. and R. K. Merton. "Patterns of Evaluation in Science: Institutionalization, Structure and Functions of the Referee System." *Minerva* 9 (1971): 66–100. Reprinted in Merton's *The Sociology of Science* (Chicago: University of Chicago Press, 1973). 4.2

Index